Good Calories,
BAD CALORIES

Good Calories, BAD CALORIES

*Challenging the Conventional Wisdom on Diet,
Weight Control, and Disease*

GARY TAUBES

ALFRED A. KNOPF NEW YORK 2007

THIS IS A BORZOI BOOK
PUBLISHED BY ALFRED A. KNOPF

Library of Congress Cataloging-in-Publication Data

Taubes, Gary.
Good calories, bad calories : challenging the conventional wisdom
on diet, weight control, and disease / Gary Taubes.—1st ed.
p. cm.
ISBN 978-1-4000-4078-0
"Borzoi book."
Includes bibliographical references and index.
1. Low-carbohydrate diet. 2. Reducing diets. 3. Weight loss.
4. Carbohydrates, Refined—Physiological effect. 5. Nutritionally
induced diseases. I. Title.
RM237.73.T38 2007   613.2'63—dc22   2007006794

Manufactured in the United States of America
Published October 2, 2007
Second Printing Before Publication

FOR

SLOANE AND HARRY, MY FAMILY

Contents

Prologue

A BRIEF HISTORY OF BANTING

Farinaceous and vegetable foods are fattening, and saccharine matters are especially so. . . . In sugar-growing countries the negroes and cattle employed on the plantations grow remarkably stout while the cane is being gathered and the sugar extracted. During this harvest the saccharine juices are freely consumed; but when the season is over, the superabundant adipose tissue is gradually lost.

THOMAS HAWKES TANNER, *The Practice of Medicine*, 1869

WILLIAM BANTING WAS A FAT MAN. In 1862, at age sixty-six, the five-foot-five Banting, or "Mr. Banting of corpulence notoriety," as the *British Medical Journal* would later call him, weighed in at over two hundred pounds. "Although no very great size or weight," Banting wrote, "still I could not stoop to tie my shoe, so to speak, nor attend to the little offices humanity requires without considerable pain and difficulty, which only the corpulent can understand." Banting was recently retired from his job as an upscale London undertaker; he had no family history of obesity, nor did he consider himself either lazy, inactive, or given to excessive indulgence at the table. Nonetheless, corpulence had crept up on him in his thirties, as with many of us today, despite his best efforts. He took up daily rowing and gained muscular vigor, a prodigious appetite, and yet more weight. He cut back on calories, which failed to induce weight loss but` did leave him exhausted and beset by boils. He tried walking, riding horseback, and manual labor. His weight increased. He consulted the best doctors of his day. He tried purgatives and diuretics. His weight increased.

Luckily for Banting, he eventually consulted an aural surgeon named William Harvey, who had recently been to Paris, where he had heard the great physiologist Claude Bernard lecture on diabetes. The liver secretes glucose, the substance of both sugar and starch, Bernard had reported, and it was this glucose that accumulates excessively in the bloodstream of diabetics. Harvey then formulated a dietary regimen based on Bernard's revelations. It was well known, Harvey later explained, that a diet of only meat and dairy would check the secretion of sugar in the urine of a diabetic. This in turn suggested that complete abstinence from sugars and starches might do the same. "Knowing too that a saccharine and farinaceous diet is used to fatten certain animals," Harvey wrote, "and that in diabetes the whole of the fat of the body rapidly disappears, it occurred to me that excessive obesity might be allied to diabetes as to its cause, although widely diverse in its development; and that if a purely animal diet were useful in the latter disease, a combination of animal food with such vegetable diet as contained neither sugar nor starch, might serve to arrest the undue formation of fat."

Harvey prescribed the regimen to Banting, who began dieting in August 1862. He ate three meals a day of meat, fish, or game, usually five or six ounces at a meal, with an ounce or two of stale toast or cooked fruit on the side. He had his evening tea with a few more ounces of fruit or toast. He scrupulously avoided any other food that might contain either sugar or starch, in particular bread, milk, beer, sweets, and potatoes. Despite a considerable allowance of alcohol in Banting's regimen—four or five glasses of wine each day, a cordial every morning, and an evening tumbler of gin, whisky, or brandy—Banting dropped thirty-five pounds by the following May and fifty pounds by early 1864. "I have not felt better in health than now for the last twenty-six years," he wrote. "My other bodily ailments have become mere matters of history."

We know this because Banting published a sixteen-page pamphlet describing his dietary experience in 1863—*Letter on Corpulence, Addressed to the Public*—promptly launching the first popular diet craze, known farther and wider than Banting could have imagined as Bantingism. His *Letter on Corpulence* was widely translated and sold particularly well in the United States, Germany, Austria, and France, where according to the *British Medical Journal,* "the emperor of the French is trying the Banting system and is said to have already profited greatly thereby." Within a year, "Banting" had entered the English language as a verb meaning "to diet." "If he is gouty, obese, and nervous, we strongly recommend him to 'bant,' " suggested the *Pall Mall Gazette* in June 1865.

The medical community of Banting's day didn't quite know what to make of him or his diet. Correspondents to the *British Medical Journal* seemed occasionally open-minded, albeit suitably skeptical; a formal paper was presented on the efficacy and safety of Banting's diet at the 1864 meeting of the British Medical Association. Others did what members of established societies often do when confronted with a radical new concept: they attacked both the message and the messenger. The editors of *The Lancet,* which is to the *BMJ* what *Newsweek* is to *Time,* were particularly ruthless. First, they insisted that Banting's diet was old news, which it was, although

Banting never claimed otherwise. The medical literature, wrote *The Lancet*, "is tolerably complete, and supplies abundant evidence that all which Mr. Banting advises has been written over and over again." Banting responded that this might well have been so, but it was news to him and other corpulent individuals.

In fact, Banting properly acknowledged his medical adviser Harvey, and in later editions of his pamphlet he apologized for not being familiar with the three Frenchmen who probably should have gotten credit: Claude Bernard, Jean Anthelme Brillat-Savarin, and Jean-François Dancel. (Banting neglected to mention his countrymen Alfred William Moore and John Harvey, who published treatises on similar meaty, starch-free diets in 1860 and 1861 respectively.)

Brillat-Savarin had been a lawyer and gourmand who wrote what may be the single most famous book ever written about food, *The Physiology of Taste*, first published in 1825.* In it, Brillat-Savarin claimed that he could easily identify the cause of obesity after thirty years of talking with one "fat" or "particularly fat" individual after another who proclaimed the joys of bread, rice, and potatoes. He added that the effects of this intake were exacerbated when sugar was consumed as well. His recommended reducing diet, not surprisingly, was "more or less rigid abstinence from everything that is starchy or floury."

Dancel was a physician and former military surgeon who publicly presented his ideas on obesity in 1844 to the French Academy of Sciences and then published a popular treatise, *Obesity, or Excessive Corpulence, The Various Causes and the Rational Means of Cure*. Dancel's thinking was based in part on the research of the German chemist Justus von Liebig, who, at the time, was defending his belief that fat is formed in animals primarily from the ingestion of fats, starches, and sugars, and that protein is used exclusively for the restoration or creation of muscular tissue. "All food which is not flesh—all food rich in carbon and hydrogen—must have a tendency to produce fat," wrote Dancel. "Upon these principles only can any rational treatment for the cure of obesity satisfactorily rest." Dancel also noted that carnivores are never fat, whereas herbivores, living exclusively on plants, often are: "The hippopotamus, for example," wrote Dancel, "so uncouth in form from its immense amount of fat, feeds wholly upon vegetable matter—rice, millet, sugar-cane, &c."

---

* When the first American edition of *The Physiology of Taste* was published in 1865, it was entitled *The Handbook of Dining, or Corpulence and Leanness Scientifically Considered*, perhaps to capitalize on the Banting craze.

The second primary grievance that *The Lancet*'s editors had with Banting, which has been echoed by critics of such diets ever since, was that his diet could be dangerous, and particularly so for the credibility of those physicians who did not embrace his ideas. "We advise Mr. Banting, and everyone of his kind, not to meddle with medical literature again, but be content to mind his own business," *The Lancet* said.

When Bantingism showed little sign of fading from the scene, however, *The Lancet*'s editors adopted a more scientific approach. They suggested that a "fair trial" be given to Banting's diet and to the supposition that "the sugary and starchy elements of food be really the chief cause of undue corpulence."

Banting's diet plays a pivotal role in the science of obesity—and, in fact, chronic disease—for two reasons. First, if the diet worked, if it actually helped people lose weight safely and keep it off, then that is worth knowing. More important, knowing whether "the sugary and starchy elements of food" are "really the chief cause of undue corpulence" is as vital to the public health as knowing, for example, that cigarettes cause lung cancer, or that HIV causes AIDS. If we choose to quit smoking to avoid the former, or to use condoms or abstinence to avoid the latter, that is our choice. The scientific obligation is first to establish the cause of the disease beyond reasonable doubt. It is easy to insist, as public-health authorities inevitably have, that calories count and obesity must be caused by overeating or sedentary behavior, but it tells us remarkably little about the underlying process of weight regulation and obesity. "To attribute obesity to 'overeating,' " as the Harvard nutritionist Jean Mayer suggested back in 1968, "is as meaningful as to account for alcoholism by ascribing it to 'overdrinking.' "

After the publication of Banting's "Letter on Corpulence," his diet spawned a century's worth of variations. By the turn of the twentieth century, when the renowned physician Sir William Osler discussed the treatment of obesity in his textbook *The Principles and Practice of Medicine*, he listed Banting's method and versions by the German clinicians Max Joseph Oertel and Wilhelm Ebstein. Oertel, director of a Munich sanitorium, prescribed a diet that featured lean beef, veal, or mutton, and eggs; overall, his regimen was more restrictive of fats than Banting's and a little more lenient with vegetables and bread. When the 244-pound Prince Otto von Bismarck lost sixty pounds in under a year, it was with Oertel's regimen. Ebstein, a professor of medicine at the University of Göttingen and author of the 1882 monograph *Obesity and Its Treatment,* insisted that fatty foods were crucial because they increased satiety and so decreased fat accumulation. Ebstein's diet allowed no sugar, no sweets, no potatoes, limited bread, and a few green vegetables, but "of meat *every* kind may be eaten, and fat meat especially." As for Osler himself, he advised obese women to "avoid taking too much food, and particularly to reduce the starches and sugars."

The two constants over the years were the ideas that starches and sugars—i.e., carbohydrates—must be minimized to reduce weight, and that meat, fish, or fowl would constitute the bulk of the diet. When seven prominent British clinicians, led by Raymond Greene (brother of the novelist Graham Greene), published a textbook entitled *The Practice of Endocrinology** in 1951, their prescribed diet for obesity was almost identical to that recommended by Banting, and that which would be prescribed by such iconoclasts as Herman Taller and Robert Atkins in the United States ten and twenty years later.

Foods to be avoided:

1. Bread, and everything else made with flour . . .
2. Cereals, including breakfast cereals and milk puddings
3. Potatoes and all other white root vegetables
4. Foods containing much sugar
5. All sweets . . .

You can eat as much as you like of the following foods:

1. Meat, fish, birds
2. All green vegetables
3. Eggs, dried or fresh
4. Cheese
5. Fruit, if unsweetened or sweetened with saccharin, except bananas and grapes

"The great progress in dietary control of obesity," wrote Hilde Bruch, considered the foremost authority on childhood obesity, in 1957, "was the recognition that meat . . . was not fat producing; but that it was the innocent foodstuffs, such as bread and sweets, which lead to obesity."

The scientific rationale behind this supposed cause and effect was

---

* Endocrinology is the study of the glands that secrete hormones and the hormones themselves.

based on observation, experimental evidence, and maybe the collected epiphanies and anecdotes of those who had successfully managed to bant. "The overappropriation of nourishment seen in obesity is derived in part from the fat ingested with the food, but more particularly from the carbohydrates," noted James French in 1907 in his *Textbook of the Practice of Medicine.* Copious opinions were offered, but no specific hypotheses. In his 1940 monograph *Obesity and Leanness,* Hugo Rony, director of the Endocrinology Clinic at the Northwestern University Medical School in Chicago, reported that he had carefully questioned fifty of his obese patients, and forty-one professed a "more or less marked preference for starchy and sweet foods; only 1 patient claimed preference for fatty foods." Rony had one unusual patient, "an extremely obese laundress," who had no taste for sweets, but "a craving for laundry starch which she used to eat by the handful, as much as a pound a day. . . ." So maybe carbohydrates are *fattening* because that's what those with a tendency to gain weight eat to excess.

To others, carbohydrates carry some inherent quality that makes them uniquely *fattening.* Maybe they induce a continued sensation of hunger, or even a specific hunger for more carbohydrates. Maybe they induce less satiation per calorie consumed. Maybe they somehow cause the human body to preferentially store away calories as fat. "In Great Britain obesity is probably more common among poor women than among the rich," Sir Stanley Davidson and Reginald Passmore wrote in the early 1960s in their classic textbook *Human Nutrition and Dietetics,* "perhaps because foods rich in fat and protein, which satisfy appetite more readily than carbohydrates, are more expensive than the starchy foods which provide the bulk of cheap meals."

This belief in the fattening powers of carbohydrates can be found in literature as well. In Tolstoy's *Anna Karenina,* for instance, written in the mid-1870s, Anna's lover, Count Vronsky, abstains from starches and sweets in preparation for what turns out to be the climactic horse race. "On the day of the races at Krasnoe Selo," writes Tolstoy, "Vronsky had come earlier than usual to eat beefsteak in the officers' mess of the regiment. He had no need to be in strict training, as he had very quickly been brought down to the required weight of one hundred and sixty pounds, but still he had to avoid gaining weight, and he avoided starchy foods and desserts." In Giuseppe di Lampedusa's *The Leopard,* published in 1958, the protagonist, Prince Fabrizio, expresses his distaste for the plump young ladies of Palermo, while blaming their condition on, among other factors, "the dearth of proteins and the overabundance of starch in the food."

This was what Dr. Spock taught our parents and our grandparents in the first five decades, six editions, and almost 50 million copies of *Baby and Child Care,* the bible of child-rearing in the latter half of the twentieth century. "Rich desserts," Spock wrote, and "the amount of plain, starchy foods (cereals, breads, potatoes) taken is what determines, in the case of most people, how much [weight] they gain or lose." It's what my Brooklyn-born mother taught me forty-odd years ago. If we eat too much bread or too much spaghetti, we will get fat. The same, of course, is true of sweets. For over a century, this was the common wisdom. "All popular 'slimming regimes' involve a restriction in dietary carbohydrate," wrote Davidson and Passmore in *Human Nutrition and Dietetics,* offering this advice: "The intake of foods rich in carbohydrate should be drastically reduced since over-indulgence in such foods is the most common cause of obesity." "The first thing most Americans do when they decide to shed unwanted pounds is to cut out bread, pass up the potatoes and rice, and cross spaghetti dinners off the menu entirely," wrote the *New York Times* personal-health reporter, Jane Brody, in her 1985 best-selling *Good Food Book.*

But by that time there had been a sea change. Now even Brody herself was recommending a diet rich in potatoes, rice, and spaghetti for the same purpose. "We need to eat more carbohydrates," Brody declared. "Not only is eating pasta at the height of fashion. . . . It can help you lose weight." The carbohydrate had become *heart-healthy* diet food. Now it was the butter rather than the bread, the sour cream on the baked potato that put on the pounds. The bread and the potato themselves were no longer the cause of weight gain but the cure. When a committee of British authorities compiled their "Proposals for Nutritional Guidelines for Health Education in Britain" in 1983, they had to explain that "the previous nutritional advice in the UK to limit the intake of all carbohydrates as a means of weight control now runs counter to current thinking. . . ."

This was one of the more remarkable conceptual shifts in the history of public health. As clinical investigators were demonstrating the singular ability of carbohydrate-restricted diets to generate significant weight loss without hunger,* the mainstream medical establishment was insisting, as in a 1973 editorial by the American Medical Association, that the diets

---

* By 1973, there had been six major conferences or symposiums dedicated solely to research on obesity: at Harvard and at Iowa State University in the early 1950s; in Falsterbo, Sweden, in 1963, hosted by the Swedish Nutrition Foundation; at the University of San Francisco in 1967; the inaugural meeting of the British Obesity Association in London in 1968; and an international meeting in Paris in 1971. In all six, carbohydrate-restricted diets were portrayed as uniquely effective at inducing weight loss.

were dangerous fads—"bizarre concepts of nutrition and dieting [that] should not be promoted to the public as if they were established scientific principles."

Just four months *after* the AMA publicly censured the use of these diets in *The Journal of the American Medical Association,* obesity researchers from around the world gathered in Bethesda, Maryland, for the first conference on obesity ever hosted by the National Institutes of Health. The only talk on the dietary treatment of obesity was presented by Charlotte Young, a well-known dietitian and nutritionist at Cornell University who had been studying and treating obesity for twenty years. Young first discussed the work of Margaret Ohlson, chair of nutrition at Michigan State University, who had tested carbohydrate-restricted diets in the early 1950s. "The diets developed by Ohlson," reported Young, "gave excellent clinical results as measured by freedom from hunger, allaying of excessive fatigue, satisfactory weight loss, suitability for long term weight reduction and subsequent weight control." She then presented the results of her research at Cornell, testing Banting-like diets on overweight young men. As in the other reports over the last century, she noted, her subjects seemed to lose weight by restricting only sugars and starches, without feeling any particular sense of hunger. Moreover, the less carbohydrates in their diets, the greater their weight loss, even though all her subjects were eating equivalent amounts of calories and protein. "No adequate explanation could be given," Young reported, implying that further scientific research might be important to clarify this issue.

None would be forthcoming, and a century of empirical evidence would be rendered irrelevant, as the AMA's spin on Banting's low-carbohydrate diet as fad was quickly adopted as the conventional wisdom, one that has been adhered to faithfully ever since. Dietary fat had been identified as a probable cause of heart disease, and low-fat diets were now being advocated by the American Heart Association as the means of prevention. At the same time, the low-fat diet as the ideal treatment for weight loss was adopted as well, even though a low-fat diet was, by definition, high in the very carbohydrates that were once considered fattening.

This transformation is all the more remarkable because the medical authorities behind it were concerned with heart disease, not obesity. They presented no dramatic scientific data to support their beliefs, only ambiguous evidence, none of which addressed the efficacy of low-fat diets in weight loss. What they did have was the *diet-heart hypothesis,* which proposed that the excessive consumption of fat in our diets—particularly saturated fats—raises cholesterol levels and so causes atherosclerosis, heart disease, and untimely death. The proponents of this theory believed that Americans—and later the entire developed world—had become gluttons. Americans ate too much of everything—particularly fat—because we could afford to, and because we could not or would not say no. This over-nutrition was certainly the cause of obesity. Eating too many calories was the problem, and since fat contains more than twice as many calories per gram as either protein or carbohydrates, "people who cut down on fat usually lose weight," as the *Washington Post* reported in 1985.

A healthy diet, by definition, had suddenly become a low-fat diet. Beginning in the late 1980s with publication of *The Surgeon General's Report on Nutrition and Health,* an entire research industry arose to create palatable nonfat fat substitutes, while the food industry spent billions of dollars marketing the less-fat-is-good-health message. The U.S. Department of Agriculture's (USDA's) booklet on dietary guidelines, and its ubiquitous Food Guide Pyramid, recommended that fats and oils be eaten "sparingly," while we were now to eat six to eleven servings per day of the pasta, potatoes, rice, and bread once considered uniquely fattening.

The reason for this book is straightforward: despite the depth and certainty of our faith that saturated fat is the nutritional bane of our lives and that obesity is caused by overeating and sedentary behavior, there has always been copious evidence to suggest that those assumptions are incorrect, and that evidence is continuing to mount. "There is always an easy solution to every human problem," H. L. Mencken once said—"neat, plausible, and wrong." It is quite possible, despite all our faith to the contrary, that these concepts are such neat, plausible, and wrong solutions. Moreover, it's also quite possible that the low-fat, high-carbohydrate diets we've been told to eat for the past thirty years are not only making us heavier but contributing to other chronic diseases as well.

Consider, for instance, that most reliable evidence suggests that Americans *have* indeed made a conscious effort to eat less fat, and particularly less saturated fat, since the 1960s. According to the USDA, we have been eating less red meat, fewer eggs, and more poultry and fish; our average fat intake has dropped from 45 percent of total calories to less than 35 percent, and National Institutes of Health surveys have documented a coincident fall in our cholesterol levels. Between 1976 and 1996, there was a 40-percent decline in hypertension in America, and a 28-percent decline in the number of individuals with chronically high cholesterol levels. But the evidence does not suggest that these decreases have improved our health.

Heart-disease death rates have indeed dropped over those years. The risk of suffering a severe heart attack, what physicians call an *acute myocardial infarction*, may have diminished as well. But there is little evidence that the *incidence* of heart disease has declined, as would be expected if eating less fat made a difference. This was the conclusion, for instance, of a ten-year study of heart-disease mortality published in *The New England Journal of Medicine* in 1998, which suggested that the death rates are declining largely because doctors and emergency-medical-service personnel are treating the disease more successfully. American Heart Association statistics support this view: between 1979 and 2003, the number of inpatient medical procedures for heart disease increased 470 percent. In 2003 alone, more than a million Americans underwent cardiac catheterizations; more than a quarter-million had coronary-artery bypass surgery.

The percentage of Americans who smoke cigarettes has also dropped considerably over the years—from 33 percent of Americans over eighteen in 1979 to 25 percent fifteen years later. This should also have significantly reduced the incidence of heart disease. That it hasn't, strongly suggests we're doing something that counteracts the beneficial effect of giving up cigarettes. Indeed, if the last few decades were considered a test of the fat-cholesterol hypothesis of heart disease, the observation that the incidence of heart disease has not noticeably decreased could serve in any functioning scientific environment as compelling evidence that the hypothesis is wrong.

Throughout the world, on the other hand, the incidence of obesity and diabetes is increasing at an alarming rate. Obesity levels in the United States remained relatively constant from the early 1960s through 1980, between 12 and 14 percent of the population; over the next twenty-five years, coincident with the official recommendations to eat less fat and so more carbohydrates, it surged to over 30 percent. By 2004, one in three Americans was considered clinically obese. Diabetes rates have increased apace. Both conditions are associated with an increased risk of heart disease, which could explain why the incidence of heart disease is not decreasing. It is also possible that obesity, diabetes, and heart disease all share a single, underlying cause. The surge in obesity and diabetes occurred as the population was being bombarded with the message that dietary fat is dangerous and that carbohydrates are good for the heart and for weight control. This suggests the possibility, however heretical, that this official embrace of carbohydrates might have had unintended consequences.

I first heard this notion in 1998, when I interviewed William Harlan, then associate director of the Office of Disease Prevention at the National

Institutes of Health. Harlan told me that public-health experts like himself assumed that if they advised all Americans to eat less fat, with its densely packed calories, weights would go down. "What we see instead," he said, "is actually weights have gone up, the portion sizes have gone up, the amount we eat has gone up. . . . Foods lower in fat became higher in carbohydrates and people ate more."

The result has been a polarization on the subject of nutrition. Most people still believe that saturated fat, if not any and all fat, is the primary dietary evil—that butter, fat, cheese, and eggs will clog our arteries and put on weight—and have reduced their intakes. Public-health experts and many in the media insist that the obesity epidemic means the population doesn't take their advice and continues to shun physical activity while eating fatty foods to excess. But a large number of people have turned to the message of Banting and one remarkably best-selling diet book after another: *Eat Fat and Grow Slim* (1958), *Calories Don't Count* (1961), *The Doctor's Quick Weight Loss Diet* (1968), *Dr. Atkins' Diet Revolution* (1972), *The Complete Scarsdale Medical Diet* (1978), *The Zone* (1995), *Protein Power* (1996), *Sugar Busters!* (1998), and *The South Beach Diet* (2003). All advocate an alternative hypothesis: that carbohydrates are the problem, not fat, and if we eat less of them, we will weigh less and live longer. All have been summarily dismissed by the American Heart Association, the American Medical Association, and nutritional authorities as part of a misguided fad.

But is it? If 150 years of anecdotal evidence and observation suggest that carbohydrates are uniquely fattening, it would be unjustifiable scientifically to reject that hypothesis without compelling evidence to the contrary. Such evidence does not exist. My purpose here is to examine the data that do exist and to demonstrate how we have reached the conclusions we have and whether or not they are justified.

There is a more important issue here as well, and it extends far beyond the ideal weight-loss diet. Prior to the official acceptance of the low-fat-is-good-health dogma, clinical investigators, predominantly British, had proposed another hypothesis for the cause of heart disease, diabetes, colorectal and breast cancer, tooth decay, and half-dozen or so other chronic diseases, including obesity. The hypothesis was based on decades of eyewitness testimony from missionary and colonial physicians and two consistent observations: that these "diseases of civilization" were rare to nonexistent among isolated populations that lived traditional lifestyles and ate traditional diets, and that these diseases appeared in these populations only after they were exposed to Western foods—in particular, sugar, flour, white rice, and maybe beer. These are known technically as *refined* carbohydrates, which are those carbohydrate-containing foods—usually sugars and starches—that have been machine-processed to make them more easily digestible.

In the early 1970s, the hypothesis that refined carbohydrates cause heart disease and other chronic diseases competed directly with the dietary-fat hypothesis of heart disease. Carbohydrates could not cause heart disease, so the argument went, because fat seemed to cause heart disease. Moreover, any diet that contained a suitably low proportion of calories as fat would, by definition, be high in carbohydrates, and vice versa. The only caveat was that the fat hypothesis was, indeed, only a hypothesis, and the evidence to support it was ambiguous at best. By the mid-1970s, the carbohydrate theory of chronic disease had been transformed into a more politically and commercially acceptable version: it wasn't the addition of refined and starchy carbohydrates to the diet that caused chronic disease, but the absence of *fiber* or *roughage,* removed in the refining process, that was responsible. This conclusion, however, has not been supported by clinical trials, which have shown that fiber has little or no effect on the incidence of any chronic disease.

We have come to accept over the past few decades the hypotheses—and that is what they are—that dietary fat, calories, fiber, and physical activity are the critical variables in obesity and leanness in health and disease. But the fact remains that, over those same decades, medical researchers have elucidated a web of physiological mechanisms and phenomena involving the singular effect of carbohydrates on blood sugar and on insulin, and the effect of blood sugar and insulin, in turn, on cells, arteries, tissues, and other hormones, that explain the original observations and support this alternative hypothesis of chronic disease.

In this book my aim is to look critically at a straightforward question to which most of us believe we know the answer: What constitutes a healthy diet? What should we eat if we want to live a long and a healthy life? To address this question, we'll examine the evidence supporting both the prevailing wisdom and this alternative hypothesis, and we'll confront the strong possibility that much of what we've come to believe is wrong.

This scenario would not be uncommon in the history of science, although, if it happened in this case, it would be a particularly dramatic and unfortunate example. If it is true, it would be because medical researchers had a relatively easy, reliable test for blood levels of cholesterol as early as 1934, and therefore fixated on the accumulation of cholesterol in the arteries as the cause of heart disease, despite considerable evidence to the contrary. By the time they developed reliable methods for measuring what are known as blood lipids, such as triglycerides, and for measuring blood levels of insulin and a condition known as insulin resistance—indicators that may be more reliable and important—a critical mass of clinicians, politicians, and health reporters had decided that dietary fat and high cholesterol levels were the cause of heart disease, and that low-fat, high-carbohydrate diets were the solution.

In science, researchers often evoke a drunk-in-the-streetlight metaphor to describe such situations: One night a man comes upon a drunk crawling on hands and knees on the pavement under a streetlight. When the man asks the drunk what he's doing, the drunk says that he's looking for his keys. "Is this where you lost them?" asks the man. "I don't know where I lost them," says the drunk, "but this is where the light is." For the past half-century, cholesterol was where the light was.

By critically examining the research that led to the prevailing wisdom of nutrition and health, this book may appear to be one-sided, but only in that it presents a side that is not often voiced publicly. Since the 1970s, the belief that saturated fat causes heart disease and perhaps other chronic diseases has been justified by a series of expert reports—from the U.S. Department of Agriculture, the Surgeon General's Office, the National Academy of Sciences, and the Department of Health in the U.K., among others. These reports present the evidence in support of the fat-cholesterol hypothesis and mostly omit the evidence in contradiction. This makes for a very compelling case, but it is not how science is best served. It is a technique used to its greatest advantage by trial lawyers, who assume correctly that the most persuasive case to a jury is one that presents only one side of a story. The legal system, however, assures that judge and jury hear both sides by requiring the presence of competing attorneys.

In the case of the fat-cholesterol hypothesis of heart disease, there has always been considerable skepticism of the hypothesis and the data. Why this skepticism is rarely made public is a major theme of this book. In fact, skeptics have often been attacked or ignored, as if disloyal at time of war. Skepticism, however, cannot be removed from the scientific process. Science does not function without it.

An underlying assumption of this book is that the evolution of medical science has suffered enormously, although unavoidably, by the degree of specialization needed to make progress. "Each science confines itself to a fragment of the evidence and weaves its theories in terms of notions suggested by that fragment," observed the British mathematician and philosopher Alfred North Whitehead. "Such a procedure is necessary by reason of the limitations of human ability. But its dangers should always be kept in mind." Researchers and clinical investigators by necessity focus their attention on a tiny fragment of the whole, and then employ the results of other disciplines to extend the implications of their own research. This means that researchers have to take on faith the critical acumen and scientific ability of those researchers whose results they are borrowing, and, as Whitehead noted, "it will usually be the case that these loans really belong to the state of science thirty or forty years earlier."

This problem is exacerbated in the study of nutrition, obesity, and chronic disease because significant observations emerge from so many diverse disciplines. Indeed, the argument can be made that, to fully understand obesity alone, researchers should have a working familiarity with the literature in clinical treatment of obesity in humans, body-weight regulation in animals, mammalian reproduction, endocrinology, metabolism, anthropology, exercise physiology, and perhaps human psychology, not to mention having a critical understanding and familiarity with the nuances of clinical trials and observational epidemiology. Most researchers and clinicians barely have time to read the journals in their own subspecialty or sub-sub-specialty, let alone the dozens of significant journals that cover the other disciplines involved. This is a primary reason why the relevant science is plagued with misconceptions propagated about some of the most basic notions. Researchers will be suitably scientific and critical when addressing the limitations of their own experiments, and then will cite something as gospel because that's what they were taught in medical school, however many years earlier, or because they read it in *The New England Journal of Medicine*. Speculations, assumptions, and erroneous interpretations of the evidence then become truth by virtue of constant repetition. It is my belief that when *all* the evidence is taken into account, rather than just a prejudicial subset, the picture that emerges will be more revealing of the underlying reality.

One consequence of this sub-specialization of modern medicine is the belief, often cited in the lay press, that the causes of obesity and the common chronic diseases are complex and thus no simple answer can be considered seriously. Individuals involved in treating or studying these ailments will stay abreast of the latest "breakthroughs" in relevant fields—the discovery of allegedly cancer-fighting phytochemicals in fruits and vegetables, of genes that predispose us to obesity or diabetes, of molecules such as leptin and ghrelin that are involved in the signaling of energy supply and demand around the body. They will assume rightfully, perhaps, that the mechanisms of weight regulation and disease are complex, and then make the incorrect assumption that the fundamental *causes* must also be complex. They lose sight of the observations that must be explained—the prevalence of obesity and chronic disease in modern societies and the relationship between them—and they forget that Occam's razor applies to this science, just as it does to all sciences: do not invoke a complicated hypothesis to explain the observations, if a simple hypothesis will suffice. By the same token, molecular biologists have identified a multitude of genes and proteins involved in the causation and spread of cancer, and so it could be argued, as well, that cancer is much more complex than we ever imagined. But to say that lung cancer, in over 90 percent of the cases, is caused by anything other than smoking cigarettes is to willfully miss the point. In this case, if refined carbohydrates and sugars are indeed the reasons why we fatten—through their effect on insulin and insulin's effect on fat accumulation—and if our goal is to prevent or remedy the disorder, the salient question is why any deeper explanation, at the moment, is necessary.

This book is divided into three parts. Part I is entitled "The Fat-Cholesterol Hypothesis" and describes how we came to believe that heart disease is caused by the effect of dietary fat and particularly saturated fat on the cholesterol in our blood. It evaluates the evidence to support that hypothesis. Part II is entitled "The Carbohydrate Hypothesis." It describes the history of the carbohydrate hypothesis of chronic disease, beginning in the nineteenth century. It then discusses in some detail the science that has evolved since the 1960s to support this hypothesis, and how this evidence was interpreted once public-health authorities established the fat-cholesterol hypothesis as conventional wisdom. Part II ends with the suggestion, which is widely accepted, that those factors of diet and lifestyle that cause us to fatten excessively are also the primary environmental factors in the cause of all of the chronic diseases of civilization. Part III, entitled "Obesity and the Regulation of Weight," discusses the competing hypotheses of how and why we fatten. It addresses whether or not the conventional wisdom that we get fat because we consume more calories than we expend—i.e., by overeating and sedentary behavior—can explain any of the observations about obesity, whether societal or individual. It then discusses the alternative hypothesis: that obesity is caused by the quality of the calories, rather than the quantity, and specifically by the effect of refined and easily digestible carbohydrates on the hormonal regulation of fat storage and metabolism.

My background is as a journalist with scientific training in college and graduate school. Since 1984, my journalistic endeavors have focused on controversial science and the excruciating difficulties of getting the right answer in any scientific pursuit. More often than not, I have chronicled the misfortunes of researchers who have come upon the wrong answer and found reason, sooner or later, to regret it. I began writing and reporting on public-health and medical issues in the early 1990s, when I realized that the research in these critically important disciplines often failed to live up to the strict standards necessary to establish reliable knowledge. In a series of lengthy articles written for the journal *Science,* I then developed the approach to the conventional wisdom of public-health recommendations that I applied in this book.

It begins with the obvious question: what is the evidence to support the current beliefs? To answer this question, I find the point in time when the conventional wisdom was still widely considered controversial—the 1970s, for example, in the case of the dietary-fat/cholesterol hypothesis of heart disease, or the 1930s for the overeating hypothesis of obesity. It is during such periods of controversy that researchers will be most meticulous in documenting the evidence to support their positions. I then obtain the journal articles, books, or conference reports cited in support of the competing propositions to see if they were interpreted critically and without bias. And I obtain the references cited by these earlier authors, working ever backward in time, and always asking the same questions: Did the investigators ignore evidence that might have refuted their preferred hypothesis? Did they pay attention to experimental details that might have thrown their preferred interpretation into doubt? I also search for other evidence in the scientific literature that wasn't included in these discussions but might have shed light on the validity of the competing hypotheses. And, finally, I follow the evidence forward in time from the point at which a consensus was reached to the present, to see whether these competing hypotheses were confirmed or refuted by further research. This process also includes interview with clinical investigators and public-health authorities, those still active in research and those retired, who might point me to research I might have missed or provide further information and details on experimental methods and interpretation of evidence.

Throughout this process, I necessarily made judgments about the quality of the research and about the researchers themselves. I tried to do so using what I consider the fundamental requirement of good science: a relentless honesty in describing precisely what was done in any particular work, and a similar honesty in interpreting the results without distorting them to reflect preconceived opinions or personal preferences. "If science is to progress," as the Nobel Prize–winning physicist Richard Feynman wrote forty years ago, "what we need is the ability to experiment, honesty in reporting results—the results must be reported without somebody saying what they would like the results to have been—and finally—an important thing—the intelligence to interpret the results. An important point about this intelligence is that it should not be sure ahead of time what must be." This was the standard to which I held all relevant research and researchers. I hope that I, too, will be judged by the same standard.

Because this book presents an unorthodox hypothesis as worthy of serious consideration, I want to make the reader aware of several additional details. The research for this book included interviews with over 600 clinicians, investigators, and administrators. When necessary, I cite or quote these individuals to add either credibility or a personal recollection to the point under discussion. The appearance of their names in the text, however, does not imply that they agree with all or even part of the thesis set forth in this book. It implies solely that the attribution is accurate and reflects their beliefs about the relevant point in that context and no other.

Lastly, I often refer to articles and reports, for the sake of simplicity and narrative flow, as though they were authored by a single relevant individual, when that is not the case. A more complete list of authors can be found using the notes and bibliography.

Part One

THE FAT-CHOLESTEROL HYPOTHESIS

Men who have excessive faith in their theories or ideas are not only ill prepared for making discoveries; they also make very poor observations. Of necessity, they observe with a preconceived idea, and when they devise an experiment, they can see, in its results, only a confirmation of their theory. In this way they distort observation and often neglect very important facts because they do not further their aim. . . . But it happens further quite naturally that men who believe too firmly in their theories, do not believe enough in the theories of others. So the dominant idea of these despisers of their fellows is to find others' theories faulty and to try to contradict them. The difficulty, for science, is still the same.

CLAUDE BERNARD, *An Introduction to the*
Study of Experimental Medicine, 1865

Chapter One

THE EISENHOWER PARADOX

In medicine, we are often confronted with poorly observed and indefinite facts which form actual obstacles to science, in that men always bring them up, saying: it is a fact, it must be accepted.

CLAUDE BERNARD, *An Introduction to the*
Study of Experimental Medicine, 1865

PRESIDENT DWIGHT D. EISENHOWER SUFFERED his first heart attack at the age of sixty-four. It took place in Denver, Colorado, where he kept a second home. It may have started on Friday, September 23, 1955. Eisenhower had spent that morning playing golf and lunched on a hamburger with onions, which gave him what appeared to be indigestion. He was asleep by nine-thirty at night but awoke five hours later with "increasingly severe low substernal nonradiating pain," as described by Dr. Howard Snyder, his personal physician, who arrived on the scene and injected Eisenhower with two doses of morphine. When it was clear by Saturday afternoon that his condition hadn't improved, he was taken to the hospital. By midday Sunday, Dr. Paul Dudley White, the world-renowned Harvard cardiologist, had been flown in to consult.

For most Americans, Eisenhower's heart attack constituted a learning experience on coronary heart disease. At a press conference that Monday morning, Dr. White gave a lucid and authoritative description of the disease itself. Over the next six weeks, twice-daily press conferences were held on the president's condition. By the time Eisenhower's health had returned, Americans, particularly middle-aged men, had learned to attend to their cholesterol and the fat in their diets. Eisenhower had learned the same lesson, albeit with counterintuitive results.

Eisenhower was assuredly among the best-chronicled heart-attack survivors in history. We know that he had no family history of heart disease, and no obvious risk factors after he quit smoking in 1949. He exercised regularly; his weight remained close to the 172 pounds considered optimal for his height. His blood pressure was only occasionally elevated. His cholesterol was below normal: his last measurement before the attack, according to George Mann, who worked with White at Harvard, was 165 mg/dl (milligrams/deciliter), a level that heart-disease specialists today consider safe.

After his heart attack, Eisenhower dieted religiously and had his cholesterol measured ten times a year. He ate little fat and less cholesterol; his meals were cooked in either soybean oil or a newly developed polyunsaturated margarine, which appeared on the market in 1958 as a nutritional palliative for high cholesterol.

The more Eisenhower dieted, however, the greater his frustration (meticulously documented by Dr. Snyder). In November 1958, when the president's weight had floated upward to 176, he renounced his breakfast of oatmeal and skimmed milk and switched to melba toast and fruit. When his weight remained high, he renounced breakfast altogether. Snyder was mystified how a man could eat so little, exercise regularly, and not lose weight. In March 1959, Eisenhower read about a group of middle-aged New Yorkers attempting to lower their cholesterol by renouncing butter, margarine, lard, and cream and replacing them with corn oil. Eisenhower did the same. His cholesterol continued to rise. Eisenhower managed to stabilize his weight, but not happily. "He eats nothing for breakfast, nothing for lunch, and therefore is irritable during the noon hour," Snyder wrote in February 1960.

By April 1960, Snyder was lying to Eisenhower about his cholesterol. "He was fussing like the devil about cholesterol," Snyder wrote. "I told him it was 217 on yesterday's [test] (actually it was 223). He has eaten only one egg in the last four weeks; only one piece of cheese. For breakfast he has skim milk, fruit and Sanka. Lunch is practically without cholesterol, unless it would be a piece of cold meat occasionally." Eisenhower's last cholesterol test as president came January 19, 1961, his final day in office. "I told him that the cholesterol was 209," Snyder noted, "when it actually was 259," a level that physicians would come to consider dangerously high.

Eisenhower's cholesterol hit 259 just six days after University of Minnesota physiologist Ancel Keys made the cover of *Time* magazine, championing precisely the kind of supposedly heart-healthy diet on which Eisenhower had been losing his battle with cholesterol for five years. It was two weeks later that the American Heart Association—prompted by Keys's force of will—published its first official endorsement of low-fat, low-cholesterol diets as a means to prevent heart disease. Only on such a diet, Keys insisted, could we lower our cholesterol and our weight and forestall a premature death. "People should know the facts," Keys told *Time*. "Then if they want to eat themselves to death, let them."

Scientists justifiably dislike anecdotal evidence—the experience of a single individual like Eisenhower. Nonetheless, such cases can raise interesting issues. Eisenhower died of heart disease in 1969, age seventy-eight. By then, he'd had another half-dozen heart attacks or, technically speaking, myocardial infarctions. Whether his diet extended his life will never be known. It certainly didn't lower his cholesterol, and so Eisenhower's experience raises important questions.

Establishing the dangers of cholesterol in our blood and the benefits of low-fat diets has always been portrayed as a struggle between science and corporate interests. And although it's true that corporate interests have been potent forces in the public debates over the definition of a healthy diet, the essence of the diet-heart controversy has always been scientific. It took the AHA ten years to give public support to Keys's hypothesis that heart disease was caused by dietary fat, and closer to thirty years for the rest of the world to follow. There was a time lag because the evidence in support of the hypothesis was ambiguous, and the researchers in the field adamantly disagreed about how to interpret it.

From the inception of the diet-heart hypothesis in the early 1950s, those who argued that dietary fat caused heart disease accumulated the evidential equivalent of a mythology to support their belief. These myths are still passed on faithfully to the present day. Two in particular provided the foundation on which the national policy of low-fat diets was constructed. One was Paul Dudley White's declaration that a "great epidemic" of heart disease had ravaged the country since World War II. The other could be called the story of *the changing American diet*. Together they told of how a nation turned away from cereals and grains to fat and red meat and paid the price in heart disease. The facts did not support these claims, but the myths served a purpose, and so they remained unquestioned.

The heart-disease epidemic vanishes upon closer inspection. It's based on the proposition that coronary heart disease was uncommon until it emerged in the 1920s and grew to become the nation's number-one killer. The epidemic was a "drastic development—paralleled only by the arrival of bubonic plague in fourteenth-century Europe, syphilis from the New World at the end of the fifteenth century and pulmonary tuberculosis at the beginning of the nineteenth century," the Harvard nutritionist Jean Mayer noted in 1975. When deaths from coronary heart disease appeared to decline after peaking in the late 1960s, authorities said it was due, at least in part, to the preventive benefits of eating less fat and lowering cholesterol.

The disease itself is a condition in which the arteries that supply blood and oxygen to the heart—known as coronary arteries because they descend on the heart like a crown—are no longer able to do so. If they're blocked entirely, the result is a heart attack. Partial blocks will starve the heart of oxygen, a condition known as ischemia. In atherosclerosis, the coronary arteries are lined by plaques or lesions, known as atheromas, the root of which comes from a Greek word meaning "porridge"—what they vaguely look like. A heart attack is caused most often by a blood clot—a thrombosis—typically where the arteries are already narrowed by atherosclerosis.

The belief that coronary heart disease was rare before the 1920s is based on the accounts of physicians like William Osler, who wrote in 1910 that he spent a decade at Montreal General Hospital without seeing a single case. In his 1971 memoirs, Paul Dudley White remarked that, of the first hundred papers he published, only two were on coronary heart disease. "If it had been common I would certainly have been aware of it, and would have published more than two papers on the subject." But even White originally considered the disease "part and parcel of the process of growing old," which is what he wrote in his 1929 textbook *Heart Disease,* while noting that "it also cripples and kills often in the prime of life and sometimes even in youth." So the salient question is whether the increasing awareness of the disease beginning in the 1920s coincided with the budding of an epidemic or simply better technology for diagnosis.

In 1912, the Chicago physician James Herrick published a seminal paper on the diagnosis of coronary heart disease—following up on the work of two Russian clinicians in Kiev—but only after Herrick used the newly invented electrocardiogram in 1918 to augment the diagnosis was his work taken seriously. This helped launch cardiology as a medical specialty, and it blossomed in the 1920s. White and other practitioners may have mistaken the new understanding of coronary heart disease for the emergence of the disease itself. "Medical diagnosis depends, in large measure, on fashion," observed the New York heart specialist R. L. Levy in 1932. Between 1920 and 1930, Levy reported, physicians at New York's Presbyterian Hospital increased their diagnosis of coronary disease by 400 percent, whereas the hospital's pathology records indicated that the disease incidence remained constant during that period. "It was after the publication of the papers of Herrick," Levy observed, that "clinicians became more alert in recognizing the disturbances in the coronary circulation and recorded them more frequently."

Over the next thirty years, recorded cases of coronary-heart-disease fatalities increased dramatically, but this rise—the alleged epidemic—had little to do with increasing incidence of disease. By the 1950s, premature deaths from infectious diseases and nutritional deficiencies had been all but eliminated in the United States, which left more Americans living long enough to die of chronic diseases—in particular, cancer and heart disease. According to the Bureau of the Census, in 1910, out of every thousand men born in America 250 would die of cardiovascular disease, compared with 110 from degenerative diseases, including diabetes and nephritis; 102 from influenza, pneumonia, and bronchitis; 75 from tuberculosis; and 73 from infections and parasites. Cancer was eighth on the list. By 1950, infectious diseases had been subdued, largely thanks to the discovery of antibiotics: male deaths from pneumonia, influenza, and bronchitis had dropped to 33 per thousand; tuberculosis deaths accounted for only 21; infections and parasites 12. Now cancer was second on the list, accounting for 133 deaths per thousand. Cardiovascular disease accounted for 560 per thousand.

Fortune magazine drew the proper conclusion in a 1950 article: "The conquering of infectious diseases has so spectacularly lengthened the life of Western man—from an average life expectancy of only forty-eight years in 1900 to sixty-seven years today—that more people are living longer to succumb to the deeper-seated degenerative or malignant diseases, such as heart disease and cancer. . . ." Sir Maurice Cassidy made a similar point in 1946 about the rising tide of heart-disease deaths in Britain: the number of persons over sixty-five, he explained, the ones most likely to have a heart attack, more than doubled between 1900 and 1937. That heart-attack deaths would more than double with them would be expected.

Another factor militating against the reality of an "epidemic" was an increased likelihood that a death would be classified on a death certificate as coronary heart disease. Here the difficulty of correctly diagnosing cause of death is the crucial point. Most of us probably have some atherosclerotic lesions at this moment, although we may never feel symptoms. Confronted with the remains of someone who expired unexpectedly, medical examiners would likely write "(unexplained) sudden death" on the death certificate. Such a death could well have been caused by atherosclerosis, but, as Levy suggested, physicians often go with the prevailing fashions when deciding on their ultimate diagnosis.

The proper identification of cause on death certificates is determined by the International Classification of Diseases, which has gone through numerous revisions since its introduction in 1893. In 1949, the ICD added a new category for arteriosclerotic heart disease.* That made a "great

---

* Arteriosclerosis is the condition in which atheroma accumulates in arteries throughout the body. The term was often used interchangeably with "atherosclerosis."

difference," as was pointed out in a 1957 report by the American Heart Association:

> The clinical diagnosis of coronary arterial heart disease dates substantially from the first decade of this century. No one questions the remarkable increase in the *reported* number of cases of this condition. Undoubtedly the wide use of the electrocardiogram in confirming clinical diagnosis and the inclusion in 1949 of Arteriosclerotic Heart Disease in the International List of Causes of Death play a role in what is often believed to be an actual increased "prevalence" of this disease. Further, in one year, 1948 to 1949, the effect of this revision was to raise coronary disease death rates by about 20 percent for white males and about 35 percent for white females.

In 1965, the ICD added another category for coronary heart disease—ischemic heart disease (IHD). Between 1949 and 1968, the proportion of heart-disease deaths attributed to either of these two new categories rose from 22 percent to 90 percent, while the percentage of deaths attributed to the other types of heart disease dropped from 78 percent to 10 percent. The proportion of deaths classified under all "diseases of the heart" has been steadily dropping since the late 1940s, contrary to the public perception. As a World Health Organization committee said in 2001 about reports of a worldwide "epidemic" of heart disease that followed on the heels of the apparent American epidemic, "much of the apparent increase in [coronary heart disease] mortality may simply be due to improvements in the quality of certification and more accurate diagnosis. . . ."

The second event that almost assuredly contributed to the appearance of an epidemic, specifically the jump in coronary-heart-disease mortality *after* 1948, is a particularly poignant one. Cardiologists decided it was time they raised public awareness of the disease. In June 1948, the U.S. Congress passed the National Heart Act, which created the National Heart Institute and the National Heart Council. Until then, government funding for heart-disease research had been virtually nonexistent. The administrators of the new heart institute had to lobby Congress for funds, which required educating congressmen on the nature of heart disease. That, in turn, required communicating the message publicly that heart disease was the number-one killer of Americans. By 1949, the National Heart Institute was allocating $9 million to heart-disease research. By 1960, the institute's annual research budget had increased sixfold.

The message that heart disease is a killer was brought to the public forcefully by the American Heart Association. The association had been founded in 1924 as "a private organization of doctors," and it remained that way for two decades. In 1945, charitable contributions to the AHA totaled $100,000. That same year, the other fourteen principal health agencies raised $58 million. The National Foundation for Infantile Paralysis alone raised $16.5 million. Under the guidance of Rome Betts, a former fund-raiser for the American Bible Society, AHA administrators set out to compete in raising research funds.

In 1948, the AHA re-established itself as a national volunteer health agency, hired a public-relations agency, and held its first nationwide fund-raising campaign, aided by thousands of volunteers, including Ed Sullivan, Milton Berle, and Maurice Chevalier. The AHA hosted Heart Night at the Copacabana. It organized variety and fashion shows, quiz programs, auctions, and collections at movie theaters and drugstores. The second week in February was proclaimed National Heart Week. AHA volunteers lobbied the press to alert the public to the heart-disease scourge, and mailed off publicity brochures that included news releases, editorials, and entire radio scripts. Newspaper and magazine articles proclaiming heart disease the number-one killer suddenly appeared everywhere. In 1949, the campaign raised nearly $3 million for research. By January 1961, when Ancel Keys appeared on the cover of *Time* and the AHA officially alerted the nation to the dangers of dietary fat, the association had invested over $35 million in research alone, and coronary heart disease was now widely recognized as the "great epidemic of the twentieth century."

Over the years, compelling arguments dismissing a heart-disease epidemic, like the 1957 AHA report, have been published repeatedly in medical journals. They were ignored, however, not refuted. David Kritchevsky, who wrote the first textbook on cholesterol, published in 1958, called such articles "unobserved publications": "They don't fit the dogma and so they get ignored and are never cited." Thus, the rise and fall of the coronary-heart-disease epidemic is still considered a matter of unimpeachable fact by those who insist dietary fat is the culprit. The likelihood that the epidemic was a mirage is not a subject for discussion.

"The present high level of fat in the American diet did not always prevail," wrote Ancel Keys in 1953, "and this fact may not be unrelated to the indication that coronary disease is increasing in this country." This is the second myth essential to the dietary-fat hypothesis—the changing-American-diet story. In 1977, when Senator George McGovern announced publication of the first *Dietary Goals for the United States*, this is the reasoning he evoked:

"The simple fact is that our diets have changed radically within the last fifty years, with great and often very harmful effects on our health." Michael Jacobson, director of the influential Center for Science in the Public Interest, enshrined this logic in a 1978 pamphlet entitled *The Changing American Diet,* and Jane Brody of the *New York Times* employed it in her best-selling 1985 *Good Food Book.* "Within this century," Brody wrote, "the diet of the average American has undergone a radical shift away from plant-based foods such as grains, beans and peas, nuts, potatoes, and other vegetables and fruits and toward foods derived from animals—meat, fish, poultry, eggs, and dairy products." That this changing American diet went along with the appearance of a great American heart-disease epidemic underpinned the argument that meat, dairy products, and other sources of animal fats had to be minimized in a healthy diet.

The changing-American-diet story envisions the turn of the century as an idyllic era free of chronic disease, and then portrays Americans as brought low by the inexorable spread of fat and meat into the American diet. It has been repeated so often that it has taken on the semblance of indisputable truth—but this conclusion is based on remarkably insubstantial and contradictory evidence.

Keys formulated the argument initially based on Department of Agriculture statistics suggesting that Americans at the turn of the century were eating 25 percent more starches and cereals, 25 percent less fats, and 20 percent less meat than they would be in the 1950s and later. Thus, the heart-disease "epidemic" was blamed on the apparently concurrent increase in meat and fat in the American diet *and* the relative decrease in starches and cereals. In 1977, McGovern's *Dietary Goals for the United States* would set out to return starches and cereal grains to their rightful primacy in the American diet.

The USDA statistics, however, were based on guesses, not reliable evidence. These statistics, known as "food disappearance data" and published yearly, estimate how much we consume each year of any particular food, by calculating how much is produced nationwide, adding imports, deducting exports, and adjusting or estimating for waste. The resulting numbers for per-capita consumption are acknowledged to be, at best, rough estimates.

The changing-American-diet story relies on food disappearance statistics dating back to 1909, but the USDA began compiling these data only in the early 1920s. The reports remained sporadic and limited to specific food groups until 1940. Only with World War II looming did USDA researchers estimate what Americans had been eating back to 1909, on the basis of the limited data available. These are the numbers on which the changing-American-diet argument is constructed. In 1942, the USDA actually began publishing regular quarterly and annual estimates of food disappearance. Until then, the data were particularly sketchy for any foods that could be grown in a garden or eaten straight off the farm, such as animals slaughtered for local consumption rather than shipped to regional slaughterhouses. The same is true for eggs, milk, poultry, and fish. "Until World War II, the data are lousy, and you can prove anything you want to prove," says David Call, a former dean of the Cornell University College of Agriculture and Life Sciences, who made a career studying American food and nutrition programs.

Historians of American dietary habits have inevitably observed that Americans, like the British, were traditionally a nation of meat-eaters, suspicious of vegetables and expecting meat three to four times a day. One French account from 1793, according to the historian Harvey Levenstein, estimated that Americans ate eight times as much meat as bread. By one USDA estimate, the typical American was eating 178 pounds of meat annually in the 1830s, forty to sixty pounds more than was reportedly being eaten a century later. This observation had been documented at the time in *Domestic Manners of the Americans,* by Fanny Trollope (mother of the novelist Anthony), whose impoverished neighbor during two summers she passed in Cincinnati, she wrote, lived with his wife, four children, and "with plenty of beef-steaks and onions for breakfast, dinner and supper, but with very few other comforts."

According to the USDA food-disappearance estimates, by the early twentieth century we were living mostly on grains, flour, and potatoes, in an era when corn was still considered primarily food for livestock, pasta was known popularly as macaroni and "considered by the general public as a typical and peculiarly Italian food," as *The Grocer's Encyclopedia* noted in 1911, and rice was still an exotic item mostly imported from the Far East.

It may be true that meat consumption was relatively low in the first decade of the twentieth century, but this may have been a brief departure from the meat-eating that dominated the century before. The population of the United States nearly doubled between 1880 and 1910, but livestock production could not keep pace, according to a Federal Trade Commission report of 1919. The number of cattle only increased by 22 percent, pigs by 17 percent, and sheep by 6 percent. From 1910 to 1919, the population increased another 12 percent and the livestock lagged further behind. "As a result of this lower rate of increase among meat animals," wrote the Federal Trade Commission investigators, "the amount of meat consumed per capita in the United States has been declining." The USDA noted further decreases in meat consumption between 1915 and 1924—the years immediately preceding the agency's first attempts to record food disappearance data—because of food rationing and the "nationwide propaganda" during World War I to conserve meat for "military purposes."

Another possible explanation for the appearance of a low-meat diet early in the twentieth century was the publication in 1906 of Upton Sinclair's book *The Jungle*, his fictional exposé on the meatpacking industry. Sinclair graphically portrayed the Chicago abattoirs as places where rotted meat was chemically treated and repackaged as sausage, where tubercular employees occasionally slipped on the bloody floors, fell into the vats, and were "overlooked for days, till all but the bones of them had gone out to the world as Anderson's Pure Leaf Lard!" *The Jungle* caused meat sales in the United States to drop by half. "The effect was long-lasting," wrote Waverly Root and Richard de Rochemont in their 1976 history *Eating in America*. "Packers were still trying to woo their customers back as late as 1928, when they launched an 'eat-more-meat' campaign and did not do very well at it." All of this suggests that the grain-dominated American diet of 1909, if real, may have been a temporary deviation from the norm.

The changing-American-diet argument is invariably used to support the proposition that Americans should eat more grain, less fat, and particularly less saturated fat, from red meat and dairy products. But the same food-disappearance reports used to bolster this low-fat, high-carbohydrate diet also provided trends for vegetables, fruits, dairy products, *and* the various fats themselves. These numbers tell a different story and might have suggested a different definition entirely of a healthy diet, if they had been taken into account. During the decades of the heart-disease "epidemic," vegetable consumption increased dramatically, as consumption of flour and grain products decreased. Americans nearly doubled (according to these USDA data) their consumption of leafy green and yellow vegetables, tomatoes, and citrus fruit.

This change in the American diet was attributed to nutritionists' emphasizing the need for vitamins from the fruits and green vegetables that were conspicuously lacking in our diets in the nineteenth century. "The preponderance of meat and farinaceous foods on my grandfather's table over fresh vegetables and fruits would be most unwelcome to modern palates," wrote the University of Kansas professor of medicine Logan Clendening in *The Balanced Diet* in 1936. "I doubt if he ever ate an orange. I know he never ate grapefruit, or broccoli or cantaloup or asparagus. Spinach, carrots, lettuce, tomatoes, celery, endive, mushrooms, lima beans, corn, green beans and peas—were entirely unknown, or rarities. . . . The staple vegetables were potatoes, cabbage, onions, radishes and the fruits—apples, pears, peaches, plums and grapes and some of the berries—in season."

From the end of World War II, when the USDA statistics become more reliable, to the late 1960s, while coronary heart-disease mortality rates supposedly soared, per-capita consumption of whole milk dropped steadily, and the use of cream was cut by half. We ate dramatically less lard (13 pounds per person per year, compared with 7 pounds) and less butter (8.5 pounds versus 4) and more margarine (4.5 pounds versus 9 pounds), vegetable shortening (9.5 pounds versus 17 pounds), and salad and cooking oils (7 pounds versus 18 pounds). As a result, during the worst decades of the heart-disease "epidemic," vegetable-fat consumption per capita in America doubled (from 28 pounds in the years 1947–49 to 55 pounds in 1976), while the average consumption of all animal fat (including the fat in meat, eggs, and dairy products) dropped from 84 pounds to 71. And so the increase in total fat consumption, to which Ancel Keys and others attributed the "epidemic" of heart disease, paralleled not only increased consumption of vegetables and citrus fruit, but of vegetable fats, which were considered heart-healthy, and a decreased consumption of animal fats.

In the years after World War II, when the newspapers began talking up a heart-disease epidemic, the proposition that cholesterol was responsible—the "medical villain *cholesterol*," as it would be called by the Chicago cardiologist Jeremiah Stamler, one of the most outspoken proponents of the diet-heart hypothesis—was considered hypothetical at best. Cholesterol itself is a pearly-white fatty substance that can be found in all body tissues, an essential component of cell membranes and a constituent of a range of physiologic processes, including the metabolism of human sex hormones.

Cholesterol is also a primary component of atherosclerotic plaques, so it was a natural assumption that the disease might begin with the abnormal accumulation of cholesterol. Proponents of the hypothesis then envisioned the human circulatory system as a kind of plumbing system. Stamler referred to the accumulation of cholesterol in lesions on the artery walls as "biological rust" that can "spread to choke off the flow [of blood], or slow it just like rust inside a water pipe so that only a dribble comes from your faucet." This imagery is so compelling that we still talk and read about artery-clogging fats and cholesterol, as though the fat of a greasy hamburger were transported directly from stomach to artery lining.

The evidence initially cited in support of the hypothesis came almost exclusively from animal research—particularly in rabbits. In 1913, the Russian pathologist Nikolaj Anitschkow reported that he could induce atherosclerotic-type lesions in rabbits by feeding them olive oil and cholesterol. Rabbits, though, are herbivores and would never consume such high-cholesterol diets naturally. And though the rabbits did develop cholesterol-filled lesions in their arteries, they developed them in their tendons and connective tissues, too, suggesting that theirs was a kind of storage disease; they had no way to metabolize the cholesterol they were force-fed. "The condition produced in the animal was referred to, often contemptuously, as the 'cholesterol disease of rabbits,' " wrote the Harvard clinician Timothy Leary in 1935.

The rabbit research spawned countless experiments in which researchers tried to induce lesions and heart attacks in other animals. Stamler, for instance, took credit for first inducing atherosclerotic-type lesions in chickens, although whether chickens are any better than rabbits as a model of human disease is debatable. Humanlike atherosclerotic lesions could be induced in pigeons, for instance, fed on corn and corn oil, and atherosclerotic lesions were observed occurring naturally in wild sea lions and seals, in pigs, cats, dogs, sheep, cattle, horses, reptiles, and rats, and even in baboons on diets that were almost exclusively vegetarian. None of these studies did much to implicate either animal fat or cholesterol.

What kept the cholesterol hypothesis particularly viable through the prewar years was that any physician could measure cholesterol levels in human subjects. Correctly interpreting the measurements was more difficult. A host of phenomena will influence cholesterol levels, some of which will also influence our risk of heart disease: exercise, for instance, lowers total cholesterol. Weight gain appears to raise it; weight loss, to lower it. Cholesterol levels will fluctuate seasonally and change with body position. Stress will raise cholesterol. Male and female hormones will affect cholesterol levels, as will diuretics, sedatives, tranquilizers, and alcohol. For these reasons alone, our cholesterol levels can change by 20 to 30 percent over the course of weeks (as Eisenhower's did in the last summer of his presidency).

Despite myriad attempts, researchers were unable to establish that patients with atherosclerosis had significantly more cholesterol in their bloodstream than those who didn't. "Some works claim a significant elevation in blood cholesterol level for a majority of patients with atherosclerosis," the medical physicist John Gofman wrote in *Science* in 1950, "whereas others debate this finding vigorously. Certainly a tremendous number of people who suffer from the consequences of atherosclerosis show blood cholesterols in the accepted normal range."

The condition of having very high cholesterol—say, above 300 mg/dl— is known as *hypercholesterolemia*. If the cholesterol hypothesis is right, then most hypercholesterolemics should get atherosclerosis and die of heart attacks. But that doesn't seem to be the case. In the genetic disorder famil- ial hypercholesterolemia, cholesterol is over 300 mg/dl for those who inherit one copy of the defective gene, and as high as 1,500 mg/dl for those who inherit two. One out of every two men and one out of every three women with this condition are likely to have a heart attack by age sixty, an observation that is often evoked as a cornerstone of the cholesterol hypoth- esis. But certain thyroid and kidney disorders will also cause hypercholes- terolemia; autopsy examinations of individuals with these maladies have often revealed severe atherosclerosis, but these individuals rarely die of heart attacks.

Autopsy examinations had also failed to demonstrate that people with high cholesterol had arteries that were any more clogged than those with low cholesterol. In 1936, Warren Sperry, co-inventor of the measurement technique for cholesterol, and Kurt Landé, a pathologist with the New York City Medical Examiner, noted that the severity of atherosclerosis could be accurately evaluated only after death, and so they autopsied more than a hundred very recently deceased New Yorkers, all of whom had died vio- lently, measuring the cholesterol in their blood. There was no reason to believe, Sperry and Landé noted, that the cholesterol levels in these indi- viduals would have been affected by their cause of death (as might have been the case had they died of a chronic illness). And their conclusion was unambiguous: "The incidence and severity of atherosclerosis are not directly affected by the level of cholesterol in the blood serum per se."

This was a common finding by heart surgeons, too, and explains in part why heart surgeons and cardiologists were comparatively skeptical of the cholesterol hypothesis. In 1964, for instance, the famous Houston heart surgeon Michael DeBakey reported similarly negative findings from the records on seventeen hundred of his own patients. And even if high cho- lesterol was *associated* with an increased incidence of heart disease, this begged the question of why so many people, as Gofman had noted in *Sci- ence*, suffer coronary heart disease despite having low cholesterol, and why a tremendous number of people with high cholesterol never get heart dis- ease or die of it.

———

Ancel Keys deserves the lion's share of credit for convincing us that choles-terol levels predict heart disease and that dietary fat is a killer. Keys ran the Laboratory of Physiological Hygiene at the University of Minnesota and considered it his franchise, as he would tell *Time* magazine, "to find out why people get sick before they got sick." He became famous during World War II by developing the K ration for combat troops—the "K," it is said, stood for "Keys." He spent the later war years doing the seminal study of human starvation, using conscientious objectors as his subjects. He then documented the experience, along with the world's accumulated knowledge on starvation, in *The Biology of Human Starvation,* a fourteen-hundred-page tome that cemented Keys's reputation. (I'll talk more about Keys's remarkable starvation study in chapter 15.)

Keys's abilities as a scientist are arguable—he was more often wrong than right—but his force of will was indomitable. Henry Blackburn, his longtime collaborator at Minnesota, described him as "frank to the point of bluntness, and critical to the point of sharpness." David Kritchevsky, who studied cholesterol metabolism at the Wistar Institute in Philadelphia and was a competitor, described Keys as "pretty ruthless" and not a likely winner of any "Mr. Congeniality" awards. Certainly, Keys was a relentless defender of his own hypotheses; he minced few words when he disagreed with a competitor's interpretation of the evidence, which was inevitably when the evidence disagreed with his hypothesis.

When Keys launched his crusade against heart disease in the late 1940s, most physicians who believed that heart disease was caused by diet implicated dietary cholesterol as the culprit. We ate too much cholesterol-laden food—meat and eggs, mostly—and that, it was said, elevated our blood cholesterol. Keys was the first to discredit this belief publicly, which had required, in any case, ignoring a certain amount of the evidence. In 1937, two Columbia University biochemists, David Rittenberg and Rudolph Schoenheimer, demonstrated that the cholesterol we eat has very little effect on the amount of cholesterol in our blood. When Keys fed men for months at a time on diets either high or low in cholesterol, it made no difference to their cholesterol levels. As a result, Keys insisted that dietary cholesterol had little relevance to heart disease. In this case, most researchers agreed.

In 1951, Keys had an epiphany while attending a conference in Rome on nutrition and disease, which focused exclusively, as Keys later recalled, on malnutrition. There he was told by a physiologist from Naples that heart disease was not a problem in his city. Keys found this comment remark-able, so he and his wife, Margaret, a medical technician whose specialty was fast becoming cholesterol measurements, visited Naples to see for themselves. They concluded that the general population was indeed heart-disease-free—but the rich were not. Margaret took blood-cholesterol readings on several hundred workers and found that they had relatively low cholesterol. They asked "a few questions about their diet," Keys recalled, and concluded that these workers ate little meat and that this explained the low cholesterol. As for the rich, "I was taken to dine with members of the Rotary Club," Keys wrote. "The pasta was loaded with meat sauce and everyone added heaps of parmesan cheese. Roast beef was the main course. Dessert was a choice of ice cream or pastry. I persuaded a few of the diners to come for examination, and Margaret found their cholesterol levels were much higher than in the workmen." Keys found "a similar picture" when he visited Madrid. Rich people had more heart disease than poor people, and rich people ate more fat.

This convinced Keys that the crucial difference between those with heart disease and those without it was the fat in the diet. A few months later, he aired his hypothesis at a nutrition conference in Amsterdam— "fatty diet, raised serum cholesterol, atherosclerosis, myocardial infarction." Almost no one in the audience, he said, took him seriously. By 1952, Keys was arguing that Americans should reduce their fat consumption by a third, though simultaneously acknowledging that his hypothesis was based more on speculation than on data: "Direct evidence on the effect of the diet on human arteriosclerosis is very little," he wrote, "and likely to remain so for some time."

Over the next half-dozen years, Keys assembled a chain of observations that became the bedrock of his belief that fat caused heart disease. He fed high-fat and medium-fat diets to schizophrenic patients at a local mental hospital and reported that the fat content dramatically raised cholesterol. He traveled to South Africa, Sardinia, and Bologna, where Margaret measured cholesterol and they assessed the fat content of the local diet. In Japan, they measured the cholesterol levels of rural fisherman and farmers; they did the same for Japanese immigrants living in Honolulu and Los Angeles. He concluded that the cholesterol/heart-disease association was not peculiar to race or nationality, not a genetic problem, but a dietary one. They visited a remote logging camp in Finland and learned that these hardworking men were plagued by heart disease. A local clinic had six patients, including three young men, who "suffered from myocardial infarction." They shared a snack with the loggers: "slabs of cheese the size of a slice of bread on which they smeared butter," Keys wrote; "they washed it down with beer. It was an object lesson for the coronary problem."

Keys bolstered his hypothesis with a 1950 report from Sweden that heart disease deaths had virtually disappeared there during the German occupation of World War II. Similar phenomena were reported in nations that had undergone severe food-rationing during the war—Finland, Norway, Great Britain, Holland, the Soviet Union. Keys concluded that the dramatic reduction in coronary deaths was caused by decreased consumption of fat from meat, eggs, and dairy products. Skeptics observed, however, that these are among many deprivations and changes that accompany food rationing and occupation. Fewer calories are consumed, for instance, and weight is lost. Unavailability of gasoline leads to increased physical activity. Sugar and refined-flour consumption decreases. Any of these might explain the reduction in heart-disease mortality, these investigators noted.

Keys encountered similar skepticism in 1953, when he argued the same proposition, using comparisons of diet and heart-disease mortality in the United States, Canada, Australia, England and Wales, Italy, and Japan. The higher the fat intake, Keys said, the higher the heart-disease rates. Americans ate the most fat and had the highest heart-disease mortality. This was a "remarkable relationship," Keys wrote: "No other variable in the mode of life besides the fat calories in the diet is known which shows anything like such a consistent relationship to the mortality rate from coronary or degenerative heart disease."

Many researchers wouldn't buy it. Jacob Yerushalmy, who ran the biostatistics department at the University of California, Berkeley, and Herman Hilleboe, the New York State commissioner of health, co-authored a critique of Keys's hypothesis, noting that Keys had chosen only six countries for his comparison though data were available for twenty-two countries. When all twenty-two were included in the analysis, the apparent link between fat and heart disease vanished. Keys had noted *associations* between heart-disease death rates and fat intake, Yerushalmy and Hilleboe pointed out, but they were just that. Associations do not imply cause and effect or represent (as Stephen Jay Gould later put it) any "magic method for the unambiguous identification of cause."

This is an irrefutable fact of logical deduction, but confusion over the point was (and still is) a recurring theme in nutrition research. George Mann, a former director of the famous Framingham Heart Study, called this drawing of associations between disease and lifestyles "a popular but not very profitable game." When the science of epidemiology was founded in 1662 by John Graunt, a London merchant who had undertaken to interpret the city's mortality records, Mann noted, even Graunt realized the danger of confusing such associations with cause and effect. "This causal-ity being so uncertain," Graunt wrote, "I shall not force myself to make any inference from the numbers."

The problem is simply stated: we don't know what other factors might be at work. Associations can be used to fuel speculation and establish hypotheses, but nothing more. Yet, as Yerushalmy and Hilleboe noted, researchers often treat such associations "uncritically or even superfi-cially," as Keys had: "Investigators must remember that evidence which is not inherently sound cannot serve even for partial support." It "is worse than useless."

Ironically, some of the most reliable facts about the diet-heart hypothesis have been consistently ignored by public-health authorities because they complicated the message, and the least reliable findings were adopted because they didn't. Dietary cholesterol, for instance, has an insignificant effect on blood cholesterol. It *might* elevate cholesterol levels in a small percentage of highly sensitive individuals, but for most of us, it's clinically meaningless.* Nonetheless, the advice to eat less cholesterol—avoiding egg yolks, for instance—remains gospel. Telling people they should worry about cholesterol in their blood but not in their diet has been deemed too confusing.

The much more contentious issues were how the quantity and type of fat influenced cholesterol levels, and, ultimately more important, whether cholesterol is even the relevant factor in causing heart disease. Keys and his wife had measured only *total* cholesterol in the blood, and he was com-paring this with the *total* amount of fat in the diet. Through the mid-1950s, Keys insisted that all fat—both vegetable and animal—elevated choles-terol. And if all fat raised cholesterol, then one way to lower it was to eat less fat. This was the basis of our belief that a healthy diet is by definition a low-fat diet. Keys, however, had oversimplified. Since the mid-1950s, researchers have known that the total amount of dietary fat has little effect on cholesterol levels.

In 1952, however, Laurance Kinsell, director of the Institute for Meta-bolic Research at the Highland–Alameda County Hospital in Oakland, California, demonstrated that vegetable oil will decrease the amount of

* Decreasing cholesterol consumption from four hundred milligrams a day, the average Ameri-can intake in the 1990s, to the three hundred milligrams a day recommended by the National Cholesterol Education Program would be expected to reduce cholesterol levels by 1 to 2 mg/dl, or a decrease of perhaps 1 percent.

cholesterol circulating in our blood, and animal fats will raise it. That same year, J. J. Groen of the Netherlands reported that cholesterol levels were independent of the *total* amount of fat consumed: cholesterol levels in his experimental subjects were lowest on a vegetarian diet with a high fat content, he noted, and highest on an animal-fat diet that had less total fat. Keys eventually accepted that animal fats tend to raise cholesterol and vegetable fats to lower it, only after he managed to replicate Groen's finding with his schizophrenic patients in Minnesota.

Kinsell and Edward "Pete" Ahrens of Rockefeller University then demonstrated that the crucial factor in controlling cholesterol was not whether the fat was from an animal or a vegetable, but its degree of "saturation," as well as what's known as the chain length of the fats. This saturation factor is a measure of whether or not the molecules of fat—known as triglycerides—contain what can be considered a full quotient of hydrogen atoms, as they do in saturated fats, which tend to raise cholesterol, or whether one or more are absent, as is the case with unsaturated fats, which tend, in comparison, to lower it. This kind of nutritional wisdom is now taught in high school, along with the erroneous idea that all animal fats are "bad" saturated fats, and all "good" unsaturated fats are found in vegetables and maybe fish. As Ahrens suggested in 1957, this accepted wisdom was probably the greatest "handicap to clear thinking" in the understanding of the relationship between diet and heart disease. The reality is that both animal and vegetable fats and oils are composed of many different kinds of fats, each with its own chain length and degree of saturation, and each with a different effect on cholesterol. Half of the fat in beef, for instance, is unsaturated, and most of that fat is the same monounsaturated fat as in olive oil. Lard is 60 percent unsaturated; most of the fat in chicken fat is unsaturated as well.

In 1957, the American Heart Association opposed Ancel Keys on the diet-heart issue. The AHA's fifteen-page report castigated researchers— including Keys, presumably—for taking "uncompromising stands based on evidence that does not stand up under critical examination." Its conclusion was unambiguous: "There is not enough evidence available to permit a rigid stand on what the relationship is between nutrition, particularly the fat content of the diet, and atherosclerosis and coronary heart disease."

Less than four years later, the evidence hadn't changed, but now a six-man ad-hoc committee, including Keys and Jeremiah Stamler, issued a new AHA report that reflected a change of heart. Released to the press in

December 1960, the report was slightly over two pages long and had no references.* Whereas the 1957 report had concluded that the evidence was insufficient to authorize telling an entire nation to eat less fat, the new report argued the opposite—"the best scientific evidence of the time" strongly suggested that Americans would reduce their risk of heart disease by reducing the fat in their diets, and replacing saturated fats with polyunsaturated fats. This was the AHA's first official support of Keys's hypothesis, and it elevated high cholesterol to the leading heart-disease risk. Keys considered the report merely an "acceptable compromise," one with "some undue pussy-footing" because it didn't insist all Americans should eat less fat, only those at high risk of contracting heart disease (overweight middle-aged men, for instance, who smoke and have high cholesterol).

After the AHA report hit the press, *Time* quickly enshrined Keys on its cover as the face of dietary wisdom in America. As *Time* reported, Keys believed that the ideal heart-healthy diet would increase the percentage of carbohydrates from less than 50 percent of calories to almost 70 percent, and reduce fat consumption from 40 percent to 15 percent. The *Time* cover story, more than four pages long, contained only a single paragraph noting that Keys's hypothesis was "still questioned by some researchers with conflicting ideas of what causes coronary disease."

* It did include a half-page of "recent scientific references on dietary fat and atherosclerosis," many of which contradicted the conclusions of the report.

Chapter Two

THE INADEQUACY
OF LESSER EVIDENCE

Another reason for the confusion and contradictions which abound in the literature concerning the etiology of coronary artery disease is the tyranny that a concept or hypothesis once formulated appears to exert upon some investigators in this field. Now to present, to emphasize, and even to enthuse about one's own theory or hypothesis is legitimate and even beneficial, but if presentation gives way to evangelistic fervor, emphasis to special pleading, and enthusiasm to bias, then progress is stopped dead in its tracks and controversy inevitably takes over. Unfortunately it must be admitted that in the quest to determine the causes of coronary artery disease, these latter deteriorations have taken place.

MEYER FRIEDMAN, *Pathogenesis of
Coronary Artery Disease*, 1969

FROM THE 1950S ONWARD, researchers worldwide set out to test Ancel Keys's hypothesis that coronary heart disease is strongly influenced by the fats in the diet. The resulting literature very quickly grew to what one Columbia University pathologist in 1977 described as "unmanageable proportions." By that time, proponents of Keys's hypothesis had amassed a body of evidence—a "totality of data," in the words of the Chicago cardiologist Jeremiah Stamler—that to them appeared unambiguously to support the hypothesis. Actually, those data constituted only half the evidence at best, and the other half did not support the hypothesis. As a result, "two strikingly polar attitudes persist on this subject, with much talk from each and little listening between," wrote Henry Blackburn, a protégé of Keys at the University of Minnesota, in 1975.

Confusion reigned. "It must still be admitted that the diet-heart relation is an unproved hypothesis that needs much more investigation," Thomas Dawber, the Boston University physician who founded the famous Framingham Heart Study, wrote in 1978. Two years later, however, he insisted the Framingham Study had provided "overwhelming evidence" that Keys's hypothesis was correct. "Yet," he noted, "many physicians and investigators of considerable renown still doubt the validity of the fat hypothesis. . . . Some even question the relationship of blood cholesterol level to disease."

Understanding this difference of opinion is crucial to understanding why we all came to believe that dietary fat, or at least saturated fat, causes heart disease. How could a proposition that incited such contention for the first twenty years of its existence become so quickly established as dogma? If two decades' worth of research was unable to convince half the investigators involved in this controversy of the validity of the dietary-fat/cholesterol hypothesis of heart disease, why did it convince the other half that they were absolutely right?

One answer to this question is that the two sides of the controversy operated with antithetical philosophies. Those skeptical of Keys's hypothesis tended to take a rigorously scientific attitude. They believed that reliable knowledge about the causes of heart disease could be gained only by meticulous experiments and relentlessly critical assessments of the evidence. Since this was a public-health issue, and any conclusions would have a very real impact on human lives, they believed that living by this scientific philosophy was even more critical than it might be if they were engaged in a more abstract pursuit. And the issue of disease prevention entailed an unprecedented need for the highest standards of scientific rigor. Preventive medicine, as the Canadian epidemiologist David Sackett had observed, targets those of us who believe ourselves to be healthy, only to tell us how we must live in order to remain healthy. It rests on the presumption that any recommendation is based on the "highest level" of evidence that the proposed intervention will do more good than harm.

The proponents of Keys's hypothesis agreed in principle, but felt they had an obligation to provide their patients with the latest medical wisdom. Though their patients might appear healthy at the moment, they could be inducing heart disease by the way they ate, which meant they should be treated as though they already had heart disease. So these doctors prescribed the diet that they believed was most likely to prevent it. They believed that withholding their medical wisdom from patients might be causing harm. Though Keys, Stamler, and like-minded physicians respected the philosophy of their skeptical peers, they considered it a luxury to wait for "final scientific proof." Americans were dying from heart disease, so the physicians had to act, making leaps of faith in the process.

This optimistic philosophy was evident early in the controversy. In October 1961, *The Wall Street Journal* reported that the NIH and the AHA were planning a huge National Diet-Heart Study that would answer the "important question: Can changes in the diet help prevent heart attacks?"

Fifty thousand Americans would be fed cholesterol-lowering diets for as long as a decade, and their health compared with that of another fifty thousand individuals who continued to eat typical American diets. This article quoted Cleveland Clinic cardiologist Irving Page saying that the time had come to resolve the conflict: "We must do something," he said. Jeremiah Stamler said resolving the conflict would "take five to ten years of hard work." The article then added that the AHA was, nonetheless, assembling a booklet of cholesterol-lowering recipes. The food industry, noted *The Wall Street Journal*, had already put half a dozen new cholesterol-lowering polyunsaturated margarines on the market. Page was then quoted saying, "Perhaps all this yakking we've been doing is beginning to take some effect." But the yakking, of course, was premature, because the National Diet-Heart Study had yet to be done. In 1964, when the study still hadn't taken place, a director of the AHA described its purpose as the equivalent of merely "dotting the final i" on the confirmation of Keys's hypothesis.

This is among the most remarkable aspects of the controversy. Keys and other proponents of his hypothesis would often admit that the benefits of cholesterol-lowering had not been established, but they would imply that it was only a matter of time until they were. "The absence of final, positive proof of a hypothesis is not evidence that the hypothesis is wrong," Keys would say. This was undeniable—but irrelevant.

The press also played a critical role in shaping the evolution of the dietary-fat controversy by consistently siding with proponents of those who saw dietary fat as an unneccessary evil. These were the researchers who were offering specific, positive advice for the health-conscious reader—eat less fat, live longer. The more zealously stated, the better the copy. All the skeptics could say was that more research was neccessary, which wasn't particularly quotable. A positive feedback loop was created. The press's favoring of articles that implied Keys's hypothesis was right helped convince the public; their belief in turn would be used to argue that the time had come to advise cholesterol-lowering diets for everyone, thus further reinforcing the belief that this advice must be scientifically defensible.

Believing that your hypothesis *must* be correct before all the evidence is gathered encourages you to interpret the evidence selectively. This is human nature. It is also precisely what the scientific method tries to avoid. It does so by requiring that scientists not just test their hypotheses, but try to prove them false. "The method of science is the method of bold conjectures and ingenious and severe attempts to refute them," said Karl Popper, the dean of the philosophy of science. Popper also noted that an infinite number of possible wrong conjectures exist for every one that happens to be right. This is why the practice of science requires an exquisite balance between a fierce ambition to discover the truth and a ruthless skepticism toward your own work. This, too, is the ideal albeit not the reality, of research in medicine and public health.

In 1957, Keys insisted that "each new research adds detail, reduces areas of uncertainty, and, so far, provides further reason to believe" his hypothesis. This is known technically as *selection* bias or *confirmation* bias; it would be applied often in the dietary-fat controversy. The fact, for instance, that Japanese men who lived in Japan had low blood-cholesterol levels and low levels of heart disease was taken as a confirmation of Keys's hypothesis, as was the fact that Japanese men in California had higher cholesterol levels and higher rates of heart disease. That Japanese men in California who had very low cholesterol levels *still* had more heart disease than their counterparts living in Japan with similarly low cholesterol was considered largely irrelevant.

Keys, Stamler, and their supporters based their belief on the compelling nature of the hypothesis supplemented *only* by the evidence in support of it. Any research that did not support their hypothesis was said to be misinterpreted, irrelevant, or based on untrustworthy data. Studies of Navajo Indians, Irish immigrants to Boston, African nomads, Swiss Alpine farmers, and Benedictine and Trappist monks all suggested that dietary fat seemed unrelated to heart disease. These were explained away or rejected by Keys.

The Masai nomads of Kenya in 1962 had blood-cholesterol levels among the lowest ever measured, despite living exclusively on milk, blood, and occasionally meat from the cattle they herded. Their high-cholesterol diets supplied nearly three thousand calories a day of mostly saturated fat. George Mann, an early director of the Framingham Heart Study, examined the Masai and concluded that these observations refuted Keys's hypothesis. In response, Keys cited similar research on the Samburu and Rendille nomads of Kenya that he interpreted as supporting his hypothesis. Whereas the Samburu had low cholesterol—despite a typical diet of five to seven quarts of high-fat milk a day, and twenty-five to thirty-five hundred calories of fat—the Rendille had cholesterol values averaging 230 mg/dl, "fully as high as United States averages." "It has been estimated," Keys wrote, "that at the time of blood sampling the percentage of calories from fats may have been 20–25 percent of calories from fat for the

Samburu and 35–40 percent for the Rendille. Such diets, consumed at a bare subsistence level, would be consistent with the serum cholesterol values achieved." Keys, however, had no reason to assume that either the Samburu or the Rendille were living at a bare subsistence level. To explain away Mann's research on the Masai, Keys then evoked more recent research suggesting that the Masai, living in nomadic isolation for thousands of years, must have somehow evolved a unique "feedback mechanism to suppress endogenous cholesterol synthesis." This mechanism, Keys suggested, would bestow immunity on the Masai to the cholesterol-raising effects of fat.

To believe Keys's explanation, we would have to ignore Mann's further research reporting that the Masai indeed had extensive atherosclerosis, despite their low cholesterol, without suffering heart attacks or any other symptoms of coronary heart disease. And we'd have to ignore still more research reporting that when the Masai moved into nearby Nairobi and began eating traditional Western diets, their cholesterol rose considerably. By 1975, Keys had relegated the Masai, and even the Samburu and the Rendille, to the sidelines of the controversy: "The peculiarities of those primitive nomads have no relevance to diet-cholesterol-CHD [coronary heart disease] relationships in other populations," he wrote.

Once having adopted firm convictions about the dangers of dietary fat based on his own limited research among small populations around the world, Keys repeatedly preached against the temptation to adopt any firm contrary convictions based on the many other studies of small populations that seemed to repudiate his hypothesis. "The data scarcely warrant any firm conclusion," he would write about such contradictory evidence. When a 1964 article in JAMA, *The Journal of the American Medical Association,* for instance, reported that the mostly Italian population of Roseto, Pennsylvania, ate copious animal fat—eating prosciutto with an inch-thick rim of fat, and cooking with lard instead of olive oil—and yet had a "strikingly low" number of deaths from heart disease, Keys said it warranted "few conclusions and certainly cannot be accepted as evidence that calories and fats in the diet are not important."

The Framingham Heart Study was an ideal example of this kind of selective thinking at work. The study was launched in 1950 under Thomas Dawber's leadership to observe in a single community aspects of diet and lifestyle that might predispose its members to heart disease—risk factors of heart disease, as they would come to be called. The factory town of Framingham, Massachusetts, was chosen because it was what Dawber called a "reasonably typical" New England town. By 1952, fifty-one hundred Framingham residents had been recruited and subjected to comprehensive physicals, including, of course, cholesterol measurements. They were then re-examined every two years to see who got heart disease and who didn't. High blood pressure, abnormal electrocardiograms, obesity, cigarette smoking, and genes (having close family with heart disease) were identified as factors that increased the risk of heart disease. In October 1961, Dawber announced that cholesterol was another one. The risk of heart disease for those Framingham men whose cholesterol had initially been over 260 mg/dl was five times greater than it was for men whose cholesterol had been under 200. This is considered one of the seminal discoveries in heart-disease research. It was touted as compelling evidence that Keys's hypothesis was correct.

But there were caveats. As the men aged, those who succumbed to heart disease were ever more likely to have low cholesterol (as had Eisenhower) rather than high cholesterol. The cholesterol/heart-disease association was tenuous for women under fifty, and nonexistent for women older. Cholesterol has "no predictive value," the Framingham investigators noted in 1971. This means women over fifty would have no reason to avoid fatty foods, because lowering their cholesterol by doing so would not lower their risk of heart disease. None of this was deemed relevant to the question of whether Keys's hypothesis was true.

The dietary research from Framingham also failed to support Keys's hypothesis. This never became common knowledge, because it was never published in a medical journal. George Mann, who left the Framingham Study in the early 1960s, recalled that the NIH administrators who funded the work refused to allow publication. Only in the late 1960s did the NIH biostatistician Tavia Gordon come across the data and decide they were worth writing up. His analysis was documented in the twenty-fourth volume of a twenty-eight-volume report on Framingham released in 1968. Between 1957 and 1960, the Framingham investigators had interviewed and assessed the diet of a thousand local subjects. They focused on men with exceedingly high cholesterol (over 300) and exceedingly low cholesterol (under 170), because these men "promised to be unusually potent in the evaluation of dietary hypotheses." But when Gordon compared the diet records of the men who had very high cholesterol with those of the men who had very low cholesterol, they differed not at all in the amount or type of fat consumed. This injected a "cautionary note" into the proceedings, as the report noted. "There is a considerable range of serum cholesterol levels within the Framingham Study Group. Something explains this inter-individual variation, but it is not diet (as measured here)."

"As measured here" encapsulates much of the challenge of scientific investigation, as well as the loophole that allowed the dietary-fat controversy to evolve into Henry Blackburn's two strikingly polar attitudes. Perhaps the Framingham investigators failed to establish that dietary fat caused the high cholesterol levels seen in the local population because (1) some other factor was responsible or (2) the researchers could not measure either the diet or the cholesterol of the population, or both, with sufficient accuracy to establish the relationship.

As it turned out, however, the Framingham Study wasn't the only one that failed to reveal any correlation between the fat consumed and either cholesterol levels or heart disease. This was the case in virtually every study in which diet, cholesterol, and heart disease were compared *within* a single population, be it in Framingham, Puerto Rico, Honolulu, Chicago, Tecumseh, Michigan, Evans County, Georgia, or Israel. Proponents of Keys's theory insisted that the diets of these populations were too homogenous, and so *everyone* ate too much fat. The only way to show that fat was responsible, they argued, was to compare entirely different populations, those with high-fat diets and those with low-fat diets. This might have been true, but perhaps fat just wasn't the relevant factor.

Ever since Sir Francis Bacon, in the early seventeenth century, scientists and philosophers of science have cautioned against the tendency to reject evidence that conflicts with our preconceptions, and to make assumptions about what assuredly would be true if only the appropriate measurements or experiments could be performed. The ultimate danger of this kind of selective interpretation, as I suggested earlier, is that a compelling body of evidence can be accumulated to support *any* hypothesis. The method of science, though, evolved to compel scientists to treat all evidence identically, including the evidence that conflicts with preconceptions, precisely for this reason. "The human understanding," as Bacon observed, "still has this peculiar and perpetual fault of being more moved and excited by affirmatives than by negatives, whereas rightly and properly it ought to give equal weight to both."

To Keys, Stamler, Dawber, and other proponents of the dietary-fat hypothesis, the positive evidence was all that mattered. The skeptics considered the positive evidence intriguing but were concerned about the negative evidence. If Keys's hypothesis was incorrect, it was only the negative evidence that could direct investigators to the correct explanation. By the 1970s, it was as if the two sides had lived through two entirely different decades of research. They could not agree on the dietary-fat hypothesis; they could barely discuss it, as Henry Blackburn had noted, because they were seeing two dramatically different bodies of evidence.

Another revealing example of selection bias was the reanalysis of a study begun in 1957 on fifty-four hundred male employees of the Western Electric Company. The original investigators, led by the Chicago cardiologist Oglesby Paul, had given them extensive physical exams and come to what they called a "reasonable approximation of the truth" of what and how much each of these men ate. After four years, eighty-eight of the men had developed symptoms of coronary heart disease. Paul and his colleagues then compared heart disease rates among the 15 percent of the men who seemingly ate the most fatty food with the 15 percent who seemingly ate the least. "Worthy of comment," they reported, "is the fact that of the 88 coronary cases, 14 have appeared in the high-fat intake group and 16 in the low-fat group."

Two decades later, Jeremiah Stamler and his colleague Richard Shekelle from Rush–Presbyterian–St. Luke's Medical Center in Chicago revisited Western Electric to see how these men had fared. They assessed the health of the employees, or the cause of death of those who had died, and then considered the diets each subject had reportedly consumed in the late 1950s. Those who had reportedly eaten large amounts of polyunsaturated fats, according to this new analysis, had slightly lower rates of coronary heart disease, but "the amount of saturated fatty acids in the diet was not significantly associated with the risk of death from [coronary heart disease]," they reported. This alone could be considered a refutation of Keys's hypothesis.

But Stamler and Shekelle knew what result they *should* have obtained, or so they believed, and they interpreted the data in that light. Their logic is worth following. "Although most attempts to document the relation of dietary cholesterol, saturated fatty acids, and polyunsaturated fatty acids to serum cholesterol concentration in persons who are eating freely have been unsuccessful," they explained, "positive results have been obtained in investigations besides the Western Electric Study." They then listed four such studies: a new version of Keys's study on Japanese men in Japan, Hawaii, and California; a study of men living for a year at a research station in Antarctica; a study of Tarahumara Indians in the Mexican highlands; and one of infants with a history of breast-feeding. To Stamler and Shekelle, these four studies provided sufficiently compelling support for Keys's hypothesis that they could interpret their own ambiguous results in a similar vein. "If viewed in isolation," they explained, "the conclusions that can be drawn from a single epidemiologic study are limited. Within the context of the total literature, however, the present observations support the conclusion that the [fat] composition of the diet affects the level of serum cholesterol and the long-term risk of death from [coronary heart disease, CHD] in middle-aged American men."

The New England Journal of Medicine published Stamler's analysis of the Western Electric findings in January 1981, and the press reported the results uncritically. "The new report," stated the *Washington Post*, "strongly reinforces the view that a high-fat, high-cholesterol diet can clog arteries and cause heart disease." Jane Brody of the *New York Times* quoted Shekelle saying, "The message of these findings is that it is prudent to decrease the amount of saturated fats and cholesterol in your diet." The Western Electric reanalysis was then cited in a 1990 joint report by the American Heart Association and the National Heart, Lung, and Blood Institute, entitled "The Cholesterol Facts," as one of seven "epidemiologic studies showing the link between diet and CHD [that] have produced particularly impressive results" and "showing a correlation between saturated fatty acids and CHD," which is precisely what it did *not* do.*

In preventive medicine, benefits without risks are nonexistent. Any diet or lifestyle intervention can have harmful effects. Changing the composition of the fats we eat could have profound physiological effects throughout the body. Our brains, for instance, are 70 percent fat, mostly in the form of a substance known as myelin that insulates nerve cells and, for that matter, all nerve endings in the body. Fat is the primary component of all cell membranes. Changing the proportion of saturated to unsaturated fats in the diet, as proponents of Keys's hypothesis recommended, might well change the composition of the fats in the cell membranes. This could alter the permeability of cell membranes, which determines how easily they transport, among other things, blood sugar, proteins, hormones, bacteria,

---

* Another of the seven was a reanalysis of a 1964 study that had compared the health and diet of Dubliners with those of their siblings who had immigrated to Boston. The 1964 incarnation of the study concluded that the Boston Irish consumed six hundred calories a day less than their Dublin siblings and 10 percent less animal fat, but weighed more and had higher cholesterol. Heart-disease rates were similar, but the Irish brothers lived longer. This study was then reinterpreted twenty years later by Lawrence Kushi, who worked in Keys's department at the University of Minnesota. Kushi concluded that those men who reportedly ate the most saturated fat and the least polyunsaturated fat in the early 1960s had slightly higher heart-disease rates in the years that followed. Though "The Cholesterol Facts" described the reanalysis as producing "particularly impressive results," Kushi himself had been less impressed: "These results," he wrote, "tend to support the hypothesis that diet is related, albeit weakly, to the development of coronary heart disease."

viruses, and tumor-causing agents into and out of the cell. The relative saturation of these membrane fats could affect the aging of cells and the likelihood that blood cells will clot in vessels and cause heart attacks.

When we consider treating a disease with a new therapy, we always have to consider potential side effects such as these. If a drug prevents heart disease but can cause cancer, the benefits may not be worth the risk. If the drug prevents heart disease but can cause cancer in only a tiny percentage of individuals, and only causes rashes in a greater number, then the trade-off might be worth it. No drug can be approved for treatment without such consideration. Why should diet be treated differently?

The Seven Countries Study, which is considered Ancel Keys's masterpiece, is a pedagogical example of this risk-benefit problem. The study is often referred to as "landmark" or "legendary" because of its pivotal role in the diet-heart controversy. Keys launched it in 1956, with $200,000 yearly support from the Public Health Service, an enormous sum of money then for a single biomedical research project. Keys and his collaborators cobbled together incipient research programs from around the world and expanded them to include some thirteen thousand middle-aged men in sixteen mostly rural populations in Italy, Yugoslavia, Greece, Finland, the Netherlands, Japan, and the United States. Keys wanted populations that would differ dramatically in diet and heart-disease risk, which would allow him to find meaningful associations between these differences. The study was *prospective,* like Framingham, which means the men were given physical examinations when they signed on, and the state of their health was assessed periodically thereafter.

Results were first published in 1970, and then at five-year intervals, as the subjects in the study aged and succumbed to death and disease. The mortality rates for heart disease were particularly revealing. Expressed in deaths per decade, there were 9 heart-disease deaths for every ten thousand men in Crete, compared with 992 for the lumberjacks and farmers of North Karelia, Finland. In between these two extremes were Japanese villagers at 66 per ten thousand, Belgrade faculty members and Rome railroad workers at 290, and U.S. railroad workers with 570 deaths per ten thousand.

According to Keys, the Seven Countries Study taught us three lessons about diet and heart disease: first, that cholesterol levels predicted heart-disease risk; second, that the amount of saturated fat in the diet predicted cholesterol levels and heart disease (contradicting Keys's earlier insistence that total fat consumption predicted cholesterol levels and heart disease with remarkable accuracy); and, third, a new idea, that monounsaturated fats protected against heart disease. To Keys, this last lesson explained why Finnish lumberjacks and Cretan villagers could both eat diets that were 40 percent fat but have such dramatically different rates of heart disease. Twenty-two percent of the calories in the Finnish diet came from saturated fats, and only 14 percent from monounsaturated fats, whereas the villagers of Crete obtained only 8 percent from saturated fat and 29 percent from monounsaturated fats. This could also explain why heart-disease rates in Crete were even lower than in Japan, even though the Japanese ate very little fat of any kind, and so very little of the healthy monosaturated fats, as well. This hypothesis could not explain many of the other relationships in the study—why eastern Finns, for instance, had three times the heart disease of western Finns, while having almost identical lifestyles and eating, as far as fat was concerned, identical diets—but this was not considered sufficient reason to doubt it. Keys's Seven Countries Study was the genesis of the Mediterranean-diet concept that is currently in vogue, and it prompted Keys to publish a new edition of his 1959 best-seller, *Eat Well and Stay Well*, now entitled *How to Eat Well and Stay Well the Mediterranean Way*.

Despite the legendary status of the Seven Countries Study, it was fatally flawed, like its predecessor, the six-country analysis Keys published in 1953 using only national diet and death statistics to support his points. For one thing, Keys chose seven countries he knew in advance would support his hypothesis. Had Keys chosen at random, or, say, chosen France and Switzerland rather than Japan and Finland, he would likely have seen no effect from saturated fat, and there might be no such thing today as the French paradox—a nation that consumes copious saturated fat but has comparatively little heart disease.

In 1984, when Keys and his colleagues published their report on the data after fifteen years of observation, they explained that "little attention was given to longevity or total mortality" in their initial results, even though what we really want to know is whether or not we will live longer if we change our diets. "The ultimate interest being prevention," they wrote, "it seemed reasonable to suppose that measures controlling coronary risk factors would improve the outlook for longevity as well as for heart attacks, at least in the population of middle-aged men in the United States for whom [coronary heart disease] is the outstanding cause of premature death." Now, however, with "the large number of deaths accumulated over the years," they realized that coronary heart disease accounted for less than one-third of all deaths, and so this "forced attention to total mortality."

Now the story changed: High cholesterol did not predict increased mortality, despite its association with a greater rate of heart disease. Saturated fat in the diet ceased to be a factor as well. The U.S. railroad workers, for instance, had a death rate from all causes lower—and so a life expectancy longer—than the Finns, the Italians, the Yugoslavs, the Dutch, and particularly the Japanese, who ate copious carbohydrates, fruits, vegetables, and fish. Only the villagers of Crete and Corfu could still expect to live significantly longer than the U.S. railroad workers. Though this could be explained by other factors, it still implied that telling Americans to eat like the Japanese might not be the best advice. This was why Keys had begun advocating Mediterranean diets, though evidence that the Mediterranean diet was beneficial was derived only from the villagers of Crete and Corfu in Keys's study, and not from those who lived on the Mediterranean coast of Yugoslavia or in the cities of Italy.

In discussions of dietary fat and heart disease, it is often forgotten that the epidemiologic tools used to link heart disease to diet were relatively new and had never been successfully put to use previously in this kind of challenge. The science of epidemiology evolved to make sense of infectious diseases, not common chronic diseases like heart disease. Though the tools of epidemiology—comparisons of populations with and without the disease—had proved effective in establishing that a disease such as cholera is caused by the presence of micro-organisms in contaminated water, as the British physician John Snow demonstrated in 1854, it is a much more complicated endeavor to employ those same tools to elucidate the subtler causes of chronic disease. They can certainly contribute to the case against the most conspicuous determinants of noninfectious diseases—that cigarettes cause lung cancer, for example. But lung cancer was an extremely rare disease before cigarettes became widespread, and smokers are thirty times as likely to get it as nonsmokers. When it comes to establishing that someone who eats copious fat might be twice as likely to be afflicted with heart disease—a very common disorder—as someone who eats little dietary fat, the tools were of untested value.

The investigators attempting these studies were constructing the relevant scientific methodology as they went along. Most were physicians untrained to pursue scientific research. Nonetheless, they decided they could reliably establish the cause of chronic disease by accumulating diet and disease data in entire populations, and then using statistical analyses to determine cause and effect. Such an approach "*seems* to furnish information about causes," wrote the Johns Hopkins University biologist Raymond Pearl in his introductory statistics textbook in 1940, but it fails, he said, to do so.

"A common feature of epidemiological data is that they are almost certain to be biased, of doubtful quality, or incomplete (and sometimes all three)," explained the epidemiologist John Bailar in *The New England Journal of Medicine* in 1980. "Problems do not disappear even if one has flawless data, since the statistical associations in almost any nontrivial set of observations are subject to many interpretations. This ambiguity exists because of the difficulty of sorting out causes, effects, concomitant variables, and random fluctuations when the causes are multiple or diffuse, the exposure levels low, irregular, or hard to measure, and the relevant biologic mechanisms poorly understood. Even when the data are generally accepted as accurate, there is much room for individual judgment, and the considered conclusions of the investigators on these matters determine what they will label 'cause' . . ."

The only way to establish cause and effect with any reliability is to do "controlled" experiments, or controlled *trials,* as they're called in medicine. Such trials attempt to avoid all the chaotic complexities of comparing populations, towns, and ethnic groups. Instead, they try to create two identical situations—two groups of subjects, in this case—and then change only one variable to see what happens. They "control" for all the other possible variables that might affect the outcome being studied. Ideally, such trials will randomly assign subjects into an experimental group, which receives the treatment being tested—a drug, for instance, or a special diet—and a group, which receives a placebo or eats their usual meals or some standard fare.

Not even randomization, though, is sufficient to assure that the only meaningful difference between the experimental group and the control group is the treatment being studied. This is why, in drug trials, placebos are used, to avoid any distortion that might occur when comparing individuals who are taking a pill in the belief that their condition might improve with individuals who are not. Drug trials are also done double-blind, which means neither subjects nor physicians know which pills are placebos and which are not. *Double-blind, placebo-controlled* clinical trials are commonly referred to in medicine as the gold standard for research. It's not that they are better than other methods of establishing truth, but that truth, in most instances, cannot be reliably established without them.

Diet trials are particularly troublesome, because it's impossible to conduct them with placebos or a double-blind. Diets including copious meat, butter, and cream do not look or taste like diets without them. It is also impossible to make a single change in a diet. Saturated fats cannot be eliminated from the diet without decreasing calories as well. To ensure that calories remain constant, another food has to replace the saturated fats. Should polyunsaturated fats be added, or carbohydrates? A single carbohydrate or mixed carbohydrates? Green leafy vegetables or starches? Whatever the choice, the experimental diet is changed in *at least* two significant ways. If saturated-fat calories are reduced and carbohydrate calories are increased to compensate, the investigators have no way to know which of the two was responsible for any effect observed. (To state that "saturated fats raise cholesterol," as is the common usage, is meaningful only if we say that saturated fat raises cholesterol compared with the effect of some other nutrient in the diet—polyunsaturated fats, for instance.)

Nonetheless, dietary trials of diet and heart disease began appearing in the literature in the mid-1950s. Perhaps a dozen such trials appeared over the next twenty years. The methods used were often primitive. Many had no controls; many neglected to randomize subjects into experimental and control groups.

Only *two* of these trials actually studied the effect of a *low-fat* diet on heart-disease rates—not to be confused with a *cholesterol-lowering* diet, which replaces saturated with polyunsaturated fats and keeps the total fat content of the diet the same. Only these two trials *ever* tested the benefits and risks of the kind of low-fat diet that the American Heart Association has recommended we eat since 1961, and that the USDA food pyramid recommends when it says to "use fats and oils sparingly." One, published in a Hungarian medical journal in 1963, concluded that cutting fat consumption to only 1.5 ounces a day reduced heart-disease rates. The other, a British study, concluded that it did not. In the British trial, the investigators also restricted daily fat consumption to 1.5 ounces, a third of the fat in the typical British diet. Each day, the men assigned to this experimental diet, all of whom had previously had heart attacks, could eat only half an ounce of butter, three ounces of meat, one egg, and two ounces of cottage cheese, and drink two ounces of skim milk. After three years, average cholesterol levels dropped from 260 to 235, but the recurrence of heart disease in the control and experimental groups was effectively identical. "A low-fat diet has no place in the treatment of myocardial infarction," the authors concluded in 1965 in *The Lancet*.

In all the other trials, cholesterol levels were lowered by changing the fat content of the diet, rather than the total amount of fat consumed. Polyunsaturated fats replaced saturated fats, without altering the calorie content. These diet trials had a profound influence on how the diet/heart-disease controversy played out.

The first and most highly publicized was the Anti-Coronary Club Trial, launched in the late 1950s by New York City Health Department Director Norman Jolliffe. The eleven hundred middle-aged members of Jolliffe's Anti-Coronary Club were prescribed what he called the "prudent diet," which included at least one ounce of polyunsaturated vegetable oil every day. The participants could eat poultry or fish anytime, but were limited to four meals a week containing beef, lamb, or pork. This made Jolliffe's prudent diet a model for future health-conscious Americans. Corn-oil margarines, with a high ratio of polyunsaturated to saturated fat, replaced butter and hydrogenated margarines, which were high in saturated fats. In total, the prudent diet was barely 30 percent fat calories, and the proportion of polyunsaturated to saturated fat was four times greater than that of typical American diets. Overweight Anti-Coronary Club members were prescribed a sixteen-hundred-calorie diet that consisted of less than 20 percent fat. Jolliffe then recruited a control group to use as a comparison.

Jolliffe died in 1961, before the results were in. His colleagues, led by George Christakis, began reporting interim results a year later. "Diet Linked to Cut in Heart Attacks," reported the *New York Times* in May 1962. "Special Diet Cuts Heart Cases Here," the *Times* reported two years later. Christakis was so confident of the prudent-diet benefits, reported *Newsweek,* that he "urged the government to heed the club results and launch an educational and food-labeling campaign to change U.S. diet habits."

The actual data, however, were considerably less encouraging. Christakis and his colleagues reported in February 1966 that the diet protected against heart disease. Anti-Coronary Club members who remained on the prudent diet had only one-third the heart disease of the controls. The longer you stayed on the diet, the more you benefited, it was said. But in November 1966, just nine months later, the Anti-Coronary Club investigators published a second article, revealing that twenty-six members of the club had died during the trial, compared with only six of the men whose diet had not been prudent. Eight members of the club died from heart attacks, but none of the controls. This appeared "somewhat unusual," Christakis and his colleagues acknowledged. They discussed the improvements in heart-disease risk factors (cholesterol, weight, and blood pressure decreased) and the significant reduction in debilitating illness "from new coronary heart disease," but omitted further discussion of mortality.

This mortality problem was the bane of Keys's dietary-fat hypothesis, bedeviling every trial that tried to assess the effects of a low-fat diet on *death* as well as disease. In July 1969, Seymour Dayton, a professor of medicine at the University of California, Los Angeles, reported the results of the largest diet-heart trial to that date. Dayton gave half of nearly 850 veterans residing at a local Veterans Administration hospital a diet in which corn, soybean, safflower, and cottonseed oils replaced the saturated fats in butter, milk, ice cream, and cheeses. The other half, the controls, were served a placebo diet in which the fat quantity and type hadn't been changed. The first group saw their cholesterol drop 13 percent lower than the controls; only sixty-six died from heart disease during the study, compared with ninety-six of the vets on the placebo diet.*

Thirty-one of the men eating Dayton's experimental cholesterol-lowering diet, however, died of cancer, compared with only seventeen of the controls. The risk of death was effectively equal on the two diets. "Was it not possible," Dayton asked, "that a diet high in unsaturated fat . . . might have noxious effects when consumed over a period of many years? Such diets are, after all, rarities among the self-selected diets of human population groups." Because the cholesterol-lowering diet failed to increase longevity, he added, it could not provide a "final answer concerning dietary prevention of heart disease."

If these trials had demonstrated that people actually lived longer on cholesterol-lowering diets, there would have been little controversy. But almost four decades later, only one trial, the Helsinki Mental Hospital Study, seemed to demonstrate such a benefit—albeit not from a low-fat diet but from a high-polyunsaturated, low-saturated-fat diet.

The Helsinki Study was a strange and imaginative experiment. The Finnish investigators used two mental hospitals for their trial, dubbed Hospital K (Kellokoski Hospital) and Hospital N (Nikkilä Hospital). Between 1959 and 1965, the inmates at Hospital N were fed a special cholesterol-lowering diet,† and the inmates of K ate their usual fare; from 1965 to 1971, those in Hospital K ate the special diet and the Hospital N inmates ate the usual fare. The effect of this diet was measured on whoever happened to be in the hospitals during those periods; "in mental hospitals turnover is usually rather slow," the Finnish investigators noted.

---

* When Dayton and his colleagues autopsied the men who died, they found no difference in the amount of atherosclerosis between those on the two diets.

† Ordinary milk was replaced with an emulsion of soybean oil in skim milk, and butter and ordinary margarine were replaced with a margarine made of polyunsaturated fats. These changes alone supposedly increased the ratio of polyunsaturated to saturated fats sixfold.

The diet seemed to reduce heart-disease deaths by half. More important to the acceptance of Keys's hypothesis, the men in the hospitals lived a little longer on the cholesterol-lowering diet. (The women did not.)

Proponents of Keys's hypothesis will still cite the Helsinki Study as among the definitive evidence that manipulating dietary fats prevents heart disease *and* saves lives. But if the lower death rates in the Helsinki trial were considered compelling evidence that the diet worked, why weren't the higher death rates in the Anti-Coronary Club Trial considered evidence that it did not?

The Minnesota Coronary Survey was, by far, the largest diet-heart trial carried out in the United States, yet it played no role in the evolution of the dietary-fat hypothesis. Indeed, the results of the study went unpublished for sixteen years, by which time the controversy had been publicly settled. The principal investigator on the trial was Ivan Frantz, Jr., who worked in Keys's department at the University of Minnesota. Frantz retired in 1988 and published the results a year later in a journal called *Arteriosclerosis,* which is unlikely to be read by anyone outside the field of cardiology.*

The Minnesota trial began in November 1968 and included more than nine thousand men and women in six state mental hospitals and one nursing home. Half of the patients were served a typical American diet, and half a cholesterol-lowering diet that included egg substitutes, soft margarine, low-fat beef, and extra vegetables; it was low in saturated fat and dietary cholesterol and high in polyunsaturated fat. Because the patients were not confined to the various mental hospitals for the entire four and a half years of the study, the average subject ate the diet for only a little more than a year. Average cholesterol levels dropped by 15 percent. Men on the diet had a slightly lower rate of heart attacks, but the women had more. Overall, the cholesterol-lowering diet was associated with an increased rate of heart disease. Of the patients eating the diet, 269 died during the trial, compared with only 206 of those eating the normal hospital fare. When I asked Frantz in late 2003 why the study went unpublished for sixteen years, he said, "We were just disappointed in the way it came out." Proponents of Keys's hypothesis who considered the Helsinki Mental Hospital Study reason enough to propose a cholesterol-lowering diet for the entire nation, never cited the Minnesota Coronary Survey as a reason to do otherwise.

---

* The results were also presented at a conference of the American Heart Association in 1975. A small chart documenting the results, without explanation, was then published as an abstract in the journal *Circulation,* along with the other abstracts from the conference.

As I implied earlier, we can only know if a recommended intervention is a success in preventive medicine if it causes more good than harm, and that can be established only with randomized, controlled clinical trials. Moreover, it's not sufficient to establish that the proposed intervention reduces the rate of only one disease—say, heart disease. We also have to establish that it doesn't increase the incidence of other diseases, and that those prescribed the intervention stay healthier *and* live longer than those who go without it. And because the diseases in question can take years to develop, enormous numbers of people have to be included in the trials and then followed for years, or perhaps decades, before reliable conclusions can be drawn.

This point cannot be unduly emphasized. An unfortunate lesson came in the summer of 2002, when physicians learned that the hormone-replacement therapy they had been prescribing to some six million post-menopausal women—either estrogen or a combination of estrogen and progestin—seemed to be doing more harm than good. The parallels to the dietary-fat controversy are worth pondering. Since 1942, when the FDA first approved hormone replacement therapy (HRT) for the treatment of hot flashes and night sweats, reams of observational studies comparing women who took hormone replacements with women who did not (just as dietary-fat studies compared populations that ate high-fat diets with populations that did not) reported that the therapy dramatically reduced the incidence of heart attacks. It was only in the 1990s that the National Institutes of Health launched a Women's Health Initiative that included the first large-scale, double-blind, placebo-controlled trial of hormone-replacement therapy. Sixteen thousand healthy women were randomly assigned to take either hormone replacement or a placebo, and then followed for at least five years. Heart disease, breast cancer, stroke, and dementia were all more common in the women prescribed hormone replacement than in those on placebos.*

The episode was an unfortunate lesson in what the epidemiologist David Sackett memorably called the "disastrous inadequacy of lesser evidence." In an editorial published in August 2002, Sackett argued that the blame lay solely with those medical authorities who, for numerous reasons, including "a misguided attempt to do good, advocate 'preventive' maneuvers that have never been validated in rigorous randomized trials.

---

* A second randomized double-blind controlled trial—the Heart and Estrogen/Progestin Replacement Study—tested hormone replacement in twenty-three hundred women who had already had heart disease. It also found no benefit from the hormones and suggested an increased risk of heart disease, at least for the first few years of taking hormone-replacement therapy.

Not only do they abuse their positions by advocating unproven 'preventives,' they also stifle dissent."

From 1960 onward, those involved in the diet-heart controversy had intended to conduct precisely the kind of study that three decades later would reverse the common wisdom about the long-term benefits of hormone-replacement therapy. This was the enormous National Diet-Heart Study that Jeremiah Stamler in 1961 had predicted would take five or ten years of hard work to complete. In August 1962, the National Heart Institute awarded research grants to six investigators—including Stamler, Keys, and Ivan Frantz, Jr.—to explore the feasibility of inducing a hundred thousand Americans to change the fat content of their diet.* In 1968, the National Institutes of Health assembled a committee led by Pete Ahrens of Rockefeller University to review the evidence for and against the diet-heart hypothesis and recommend how to proceed. The committee published its conclusions in June 1969. Even though the American Heart Association had been recommending low-fat diets for almost a decade already, Ahrens and his colleagues reported, the salient points remained at issue. "The essential reason for conducting a study," they noted, "*is because it is not known whether dietary manipulation has any effect whatsoever on coronary heart disease.*" And so they recommended that the government proceed with the trial, even though, Ahrens recalled, the committee members came to believe that any trial large enough and sufficiently well controlled to provide a reliable conclusion "would be so expensive and so impractical that it would never get done."

Two years later, the NIH assembled a Task Force on Arteriosclerosis, and it came to similar conclusions in its four-hundred-page, two-volume report. The task force agreed that a "definitive test" of Keys's dietary-fat hypothesis "in the general population is urgently needed." But these assembled experts also did not believe such a study was practical. They worried about the "formidable" costs—perhaps $1 billion—and recommended instead that the NIH proceed with smaller, well-controlled studies that might demonstrate that it was possible to lessen the risk of coronary heart disease without necessarily relying on diet to do it.

As a result, the NIH agreed to spend only $250 million on two smaller trials that would still constitute the largest, most ambitious clinical trials

---

* Frantz's Minnesota Coronary Survey was technically a pilot project for the National Diet-Heart Study.

ever attempted. One would test the hypothesis that heart attacks could be prevented by the use of cholesterol-lowering drugs. The other would attempt to prevent heart disease with a combination of cholesterol-lowering diets, smoking-cessation programs, and drugs to reduce blood pressure. Neither of these trials would actually constitute a test of Keys's hypothesis or of the benefits of low-fat diets. Moreover, the two trials would take a decade to complete, which was longer than the public, the press, or the government was willing to wait.

Chapter Three

CREATION OF CONSENSUS

In sciences that are based on supposition and opinion . . . the object is to command assent, not to master the thing itself.

FRANCIS BACON, *Novum Organum*, 1620

B Y 1977, WHEN THE NOTION THAT dietary fat causes heart disease began its transformation from speculative hypothesis to nutritional dogma, no compelling new scientific evidence had been published. What had changed was the public attitude toward the subject. Belief in saturated fat and cholesterol as killers achieved a kind of critical mass when an anti-fat, anti-meat movement evolved independent of the science.

The roots of this movement can be found in the counterculture of the 1960s, and its moral shift away from the excessive consumption represented by fat-laden foods. The subject of famine in the third world was a constant presence in the news: in China and the Congo in 1960, then Kenya, Brazil, and West Africa—where "Villagers in Dahomey Crawl to Town to Seek Food," as a *New York Times* headline read—followed by Somalia, Nepal, South Korea, Java, and India; in 1968, Tanzania, Bechuanaland, and Biafra; then Bangladesh, Ethiopia, and much of sub-Saharan Africa in the early 1970s. Within a decade, the Stanford University biologist Paul Ehrlich predicted in his 1968 best-seller, *The Population Bomb*, "hundreds of millions of people are going to starve to death in spite of any crash programs embarked upon now."

The fundamental problem was an ever-increasing world population, but secondary blame fell to an imbalance between food production and consumption. This, in turn, implicated the eating habits in the richer nations, particularly the United States. The "enormous appetite for animal products has forced the conversion (at a very poor rate) of more and more grain, soybean and even fish meal into feed for cattle, hogs and poultry, thus decreasing the amounts of food directly available for direct consumption by the poor," explained Harvard nutritionist Jean Mayer in 1974. To improve the world situation, insisted Mayer and others, there should be "a shift in consumption in developed countries toward a 'simplified' diet containing less animal products and, in particular, less meat." By doing so, we would free up grain, the "world's most essential commodity," to feed the hungry.

This argument was made most memorably in the 1971 best-seller *Diet for a Small Planet*, written by a twenty-six-year-old vegetarian named Francis Moore Lappé. The American livestock industry required twenty million tons of soy and vegetable protein to produce two million tons of beef, according to Lappé. The eighteen million tons lost in the process were enough to provide twelve urgently needed grams of protein daily to everyone in the world. This argument transformed meat-eating into a social issue, as well as a moral one. "A shopper's decision at the meat counter in Gary, Indiana would affect food availability in Bombay, India," explained the sociologist Warren Belasco in *Appetite for Change*, his history of the era.

By the early 1970s, this argument had become intertwined with the medical issues of fat and cholesterol in the diet. "How do you get people to understand that millions of Americans have adopted diets that will make them at best fat, or at worst, dead?" as the activist Jennifer Cross wrote in *The Nation* in 1974. "That the $139 billion food industry has not only encouraged such unwise eating habits in the interest of profit but is so wasteful in many of its operations that we are inadvertently depriving hungry nations of food?" The American Heart Association had taken to recommending that Americans cut back not just on saturated fat but on meat to do so. Saturated fat may have been perceived as the problem, but saturated fat was still considered to be synonymous with animal fat, and much of the fat in the American diet came from animal foods, particularly red meat.

Ironically, by 1968, when Paul Ehrlich had declared in *The Population Bomb* that "the battle to feed all humanity" had already been lost, agricultural researchers led by Norman Borlaug had created high-yield varieties of dwarf wheat that ended the famines in India and Pakistan and averted the predicted mass starvations. In 1970, when the Nobel Foundation awarded its Peace Prize to Borlaug, it justified the decision on the grounds that, "more than any other single person," Borlaug had "helped to provide bread for a hungry world."

Other factors were also pushing the public toward a belief in the evils of dietary fat and cholesterol that the medical-research community itself still considered questionable. The American Heart Association revised its dietary recommendations every two to three years and, with each revision, made its advice to eat less fat increasingly unconditional. By 1970, this prescription applied not just to those high-risk men who had already had heart attacks or had high cholesterol or smoked, but to everyone, "including infants, children, adolescents, pregnant and lactating women, and older persons." Meanwhile, the press and the public came to view the AHA as the primary source of expert information on the issue.

The AHA had an important ally in the vegetable-oil and margarine manufacturers. As early as 1957, the year Americans first purchased more margarine than butter, Mazola corn oil was being pitched to the public with a "Listen to Your Heart" campaign; the polyunsaturated fats of corn oil would lower cholesterol and so prevent heart attacks, it was said. Corn Products Company, the makers of Mazola, and Standard Brands, producers of Fleischmann's margarine, both initiated programs to educate doctors to the benefits of polyunsaturated fats, with the implicit assumption that the physicians would pass the news on to their patients. Corn Products Company collaborated directly with the AHA on releasing a "risk handbook" for physicians, and with Pocket Books to publish the revised version, in 1966, of Jeremiah Stamler's book *Your Heart Has Nine Lives*. By then, ads for these polyunsaturated oils and margarines needed only to point out that the products were low in saturated fat and low-cholesterol, and this would serve to communicate and reinforce the heart-healthy message.

This alliance between the AHA and the makers of vegetable oils and margarines dissolved in the early 1970s, with reports suggesting that polyunsaturated fats can cause cancer in laboratory animals. This was problematic to Keys's hypothesis, because the studies that had given some indication that cholesterol-lowering was good for the heart—Seymour Dayton's VA Hospital trial and the Helsinki Mental Hospital Study—had done so precisely by replacing saturated fats in the diet with polyunsaturated fats. Public-health authorities concerned with our cholesterol dealt with the problem by advising that we simply eat less fat and less saturated fat, even though only two studies had ever tested the effect of such low-fat diets on heart disease, and they had been contradictory.

It's possible to point to a single day when the controversy was shifted irrevocably in favor of Keys's hypothesis—Friday, January 14, 1977, when Senator George McGovern announced the publication of the first *Dietary Goals for the United States*. The document was "the first comprehensive statement by any branch of the Federal Government on risk factors in the American diet," said McGovern.

This was the first time that any government institution (as opposed to private groups like the AHA) had told Americans they could improve their health by eating less fat. In so doing, *Dietary Goals* sparked a chain reaction of dietary advice from government agencies and the press that reverberates still, and the document itself became gospel. It is hard to overstate its impact. *Dietary Goals* took a grab bag of ambiguous studies and speculation, acknowledged that the claims were scientifically contentious, and then officially bestowed on one interpretation the aura of established fact. "Premature or not," as Jane Brody of the *New York Times* wrote in 1981, "the *Dietary Goals* are beginning to reshape the nutritional philosophy of America, if not yet the eating habits of most Americans."

The document was the product of McGovern's Senate Select Committee on Nutrition and Human Needs, a bipartisan nonlegislative committee that had been formed in 1968 with a mandate to wipe out malnutrition in America. Over the next five years, McGovern and his colleagues—among them, many of the most prominent politicians in the country, including Ted Kennedy, Charles Percy, Bob Dole, and Hubert Humphrey—instituted a series of landmark federal food-assistance programs. Buoyed by their success fighting malnutrition, the committee members turned to the link between diet and chronic disease.

The operative force at work, however, was the committee staff, composed of lawyers and ex-journalists. "We really were totally naïve," said the staff director Marshall Matz, "a bunch of kids, who just thought, Hell, we should say something on this subject before we go out of business."* McGovern had attended Nathan Pritikin's four-week diet-and-exercise program at Pritikin's Longevity Research Institute in Santa Barbara, California. He said that he lasted only a few days on Pritikin's very low-fat diet, but that Pritikin's philosophy, an extreme version of the AHA's, had profoundly influenced his thinking.

McGovern's staff were virtually unaware of the existence of any scientific controversy. They knew that the AHA advocated low-fat diets, and that the dairy, meat, and egg industries had been fighting back. Matz and his fellow staff members described their level of familiarity with the subject as that of interested laymen who read the newspapers. They believed that the relevant nutritional and social issues were simple and obvious. Moreover, they wanted to make a difference, none more so than Nick Mottern, who

* The investigative reporter William Broad suggested another version of this story in *Science* in June 1979. He said the *Dietary Goals* constituted a last-ditch effort to save McGovern's Select Committee, which had required renewal every two years since its inception and was now facing a reorganization that would downgrade its status to a subcommittee of the Senate Committee on Agriculture. "They were fighting for their life," Cortez Enloe, editor of *Nutrition Today,* told Broad. "Their tenure was up."

would draft the *Dietary Goals* almost single-handedly. A former labor reporter, Mottern was working as a researcher for a consumer-products newsletter in 1974 when he watched a television documentary about famine in Africa, decided to do something meaningful with his life, and was hired to fill a vacant writing job on McGovern's committee.

In July 1976, McGovern's committee listened to two days of testimony on "Diet and Killer Diseases." Mottern then spent three months research-ing the subject and two months writing. The most compelling evidence, Mottern believed, was the changing-American-diet story, and this became the underlying foundation of the committee's recommendations: we should readjust our national diet to match that of the turn of the century, at least as the Department of Agriculture had guessed it to be. The less con-troversial recommendations of the *Dietary Goals* included eating less sugar and salt, and more fruits, vegetables, and whole grains.

Fat and cholesterol would be the contentious points. Here Mottern avoided the inherent ambiguities of the evidence by relying for his exper-tise almost exclusively on a single Harvard nutritionist, Mark Hegsted, who by his own admission was an extremist on the dietary-fat issue. Heg-sted had studied the effect of fat on cholesterol levels in the early 1960s, first with animals and then, like Keys, with schizophrenic patients at a mental hospital. Hegsted had come to believe unconditionally that eating less fat would prevent heart disease, although he was aware that this con-viction was not shared by other investigators working in the field. With Hegsted as his guide, Mottern perceived the dietary-fat controversy as analogous to the specious industry-sponsored "controversy" over cigarettes and lung cancer, and he equated his *Dietary Goals* to the surgeon general's legendary 1964 report on smoking and health. To Mottern, the food indus-try was no different from the tobacco industry in its willingness to sup-press scientific truth in the interests of greater profits. He believed that those scientists who lobbied actively against dietary fat, like Hegsted, Keys, and Stamler, were heroes.

Dietary Goals was couched as a plan for the nation, but these goals obvi-ously pertained to individual diets as well. Goal number one was to raise the consumption of carbohydrates until they constituted 55–60 percent of the calories consumed. Goal number two was to decrease fat consumption from approximately 40 percent, then the national average, to 30 percent of all calories, of which no more than a third should come from saturated fats. The report acknowledged that no evidence existed to suggest that reducing the total fat content of the diet would lower blood-cholesterol levels, but it justified its recommendation on the basis that, the lower the percentage of dense fat calories in the diet, the less likely people would be to gain weight,* and because other health associations—most notably the American Heart Association—were recommending 30 percent fat in diets. To achieve this low-fat goal, according to the *Dietary Goals*, Americans would have to eat considerably less meat and dairy products.

Though the *Dietary Goals* admitted the existence of a scientific controversy, it also insisted that Americans had nothing to lose by following the advice. "The question to be asked is not why should we change our diet but why not?" explained Hegsted at a press conference to announce publication of the document. "There are [no risks] that can be identified and important benefits can be expected." But this was still a hugely debatable position among researchers. After the press conference, as Hegsted recalled, "all hell broke loose. . . . Practically nobody was in favor of the McGovern recommendations."

Having held one set of hearings before publishing the *Dietary Goals*, McGovern responded to the ensuing uproar with eight follow-up hearings. Among those testifying was Robert Levy, director of the National Heart, Lung, and Blood Institute, who said that no one knew whether lowering cholesterol would prevent heart attacks, which was why the NHLBI was spending several hundred million dollars to study the question. ("Arguments for lowering cholesterol through diet," Levy had written just a year earlier, even in those patients who were what physicians would call coronary-prone, "remain primarily circumstantial.")

Other prominent investigators, including Pete Ahrens and the University of London cardiologist Sir John McMichael, also testified that the guidelines were premature, if not irresponsible. The American Medical Association argued against the recommendations, saying in a letter to the committee that "there is a potential for harmful effects for a radical long term dietary change as would occur through adoption of the proposed national goals." These experts were sandwiched between representatives from the dairy, egg, and cattle industries, who also vigorously opposed the guidelines, for obvious reasons. This juxtaposition served to taint the legitimacy of the scientific criticisms.

The committee published a revised edition of *Dietary Goals* later that year, but with only minor revisions. Now the first recommendation was to avoid being overweight. The committee also succumbed to pressure from

---

* As *Dietary Goals* explained, "Fat supplies 9 calories per gram, whereas protein and carbohydrates, the other two energy sources, supply only 4 calories per gram. . . . Consequently, particularly for those not involved in heavy physical activity, the consumption of a diet deriving 40 percent of its calories from fat may result in a continual struggle to lose weight."

the livestock industry and changed the recommendation that Americans "decrease consumption of meat" to one that said to "decrease consumption of animal fat, and choose meats, poultry, and fish which will reduce saturated fat intake."

The revised edition also included a ten-page preface that attempted to justify the committee's dietary recommendations in light of the uproar that had followed. It included a caveat that "some witnesses have claimed that physical harm could result from the diet modifications recommended in this report. . . ." But McGovern and his colleagues considered that unlikely: "After further review, the Select Committee still finds that no physical or mental harm could result from the dietary guidelines recommended for the general public." The preface also included a list of five "important questions, which are currently being investigated." The first was a familiar one: "Does lowering the plasma cholesterol level through dietary modification prevent or delay heart disease in man?"

This question would never be answered, but it no longer seemed to matter. McGovern's *Dietary Goals* had turned the dietary-fat controversy into a political issue rather than a scientific one, and Keys and his hypothesis were the beneficiaries. Now administrators at the Department of Agriculture and the National Academy of Sciences felt it imperative to get on the record.

At the USDA, Carol Foreman was the driving force. Before her appointment in March 1977 as an assistant secretary of agriculture, Foreman had been a consumer advocate, executive director of the Consumer Federation of America. Her instructions from President Jimmy Carter at her swearing-in ceremony were to give consumers a "strong, forceful, competent" spokeswoman within the USDA. Foreman believed McGovern's *Dietary Goals* supported her conviction that "people were getting sick and dying because we ate too much," and she believed it was incumbent on the USDA to turn McGovern's recommendations into official government policy. Like Mottern and Hegsted, Foreman was undeterred by the scientific controversy. She believed that scientists had an obligation to take their best guess about the diet-disease relationship, and then the public had to decide. "Tell us what you know, and tell us it's not the final answer," she would tell scientists. "I have to eat three times a day and feed my children three times a day, and I want you to tell me what your best sense of the data is right now."

The "best sense of the data," however, depends on whom you ask. The obvious candidate in this case was the Food and Nutrition Board of the

National Academy of Sciences, which determines Recommended Dietary Allowances, the minimal amount of vitamins and minerals required in a healthy diet, and was established in 1940 to advise the government on nutrition issues. The NAS and USDA drafted a contract for the Food and Nutrition Board to evaluate the recommendations in the *Dietary Goals,* according to *Science,* but Foreman and her USDA colleagues "got wind" of a speech that Food and Nutrition Board Chairman Gilbert Leveille had made to the American Farm Bureau Federation and pulled back. "The American diet," Leveille had said, "has been referred to as . . . 'disastrous'. . . . I submit that such a conclusion is erroneous and misleading. The American diet today is, in my opinion, better than ever before and is one of the best, if not *the* best, in the world today." NAS President Philip Handler, an expert on human and animal metabolism, had also told Foreman that McGovern's *Dietary Goals* were "nonsense," and so Foreman turned instead to the NIH and the Food and Drug Administration, but the relevant administrators rejected her overtures. They considered the *Dietary Goals* a "political document rather than a scientific document," Foreman recalled; NIH Director Donald Fredrickson told her "we shouldn't touch it with a ten-foot pole; we should let the crazies on the hill say what they wanted."

Finally, it was agreed that the USDA and the Surgeon General's Office would draft official dietary guidelines. The USDA would be represented by Mark Hegsted, whom Foreman had hired to be the first head of the USDA's Human Nutrition Center and to shepherd its dietary guidelines into existence.

Hegsted and J. Michael McGinnis from the Surgeon General's Office relied almost exclusively on a report by a committee of the American Society of Clinical Nutrition that had assessed the state of the relevant science, although with the expressed charge "*not* to draw up a set of recommendations." Pete Ahrens chaired the committee, along with William Connors of the University of Oregon Health Sciences Center, and it included nine scientists covering a "full range of convictions" in the various dietary controversies. The ASCN committee concluded that saturated-fat consumption was probably related to the formation of atherosclerotic plaques, but the evidence that disease could be prevented by dietary modification was still unconvincing.* The report described the spread of opinions on these issues as "considerable." "But the clear majority supported something like the McGovern committee report," according to Hegsted. On that basis,

---

* It also affirmed the suspicion that polyunsaturated fats might be dangerous, and so further diminished the role of margarines and corn oils in dietary recommendations.

Hegsted and McGinnis produced the USDA *Dietary Guidelines for Americans*, which was released to the public in February 1980.

The *Dietary Guidelines* also acknowledged the existence of a controversy, suggesting that a single dietary recommendation might not be appropriate for an entire diverse population. But it still declared in bold letters on its cover that Americans should "Avoid Too Much Fat, Saturated Fat, and Cholesterol." (The *Dietary Guidelines* did not define what was meant by "too much.")

Three months later, Philip Handler's Food and Nutrition Board released its own version of the guidelines—*Toward Healthful Diets*. It concluded that the only reliable dietary advice that could be given to healthy Americans was to watch their weight and that everything else, dietary fat included, would take care of itself. The Food and Nutrition Board promptly got "excoriated in the press," as one board member described it. The first criticisms attacked the board for publishing recommendations that ran counter to those of the USDA, McGovern's committee, and the American Heart Association, and so were seen to be irresponsible. They were followed by suggestions that the board members, in the words of Jane Brody, who covered the story for the *New York Times*, "were all in the pocket of the industries being hurt." The board director, Alfred Harper, chairman of the University of Wisconsin nutrition department, consulted for the meat industry. The Washington University nutritionist Robert Olson, who had worked on fat and cholesterol metabolism since the 1940s, consulted for the Egg Board, which itself was a USDA creation to sponsor research, among other things, on the nutritional consequences of eating eggs. Funding for the Food and Nutrition Board came from industry donations to the National Academy of Sciences. These industry connections were first leaked to the press from the USDA, where Hegsted and Foreman suddenly found themselves vigorously defending their own report to their superiors, and from the Center for Science in the Public Interest, a consumer-advocacy group run by Michael Jacobson that was now dedicated to reducing the fat and sugar content of the American diet. (As the *Los Angeles Times* later observed, the CSPI "embraced a low-fat diet as if it was a holy writ.")

The House Agriculture Subcommittee on Domestic Marketing promptly held hearings in which Henry Waxman, chairman of the Health Subcommittee, described *Toward Healthful Diets* as "inaccurate and potentially biased" as well as "quite dangerous." Hegsted was among those who testified, saying "he failed to see how the Food and Nutrition Board had reached its conclusions."

Philip Handler testified as well, summarizing the situation memorably. When the hearings were concluded, he said, the committee members might find themselves confronted by a dilemma. They might conclude, "as some have," that there exists a "thinly linked, if questionable, chain of observations" connecting fat and cholesterol in the diet to cholesterol levels in the blood to heart disease:

> However tenuous that linkage, however disappointing the various intervention trials, it still seems prudent to propose to the American public that we not only maintain reasonable weights for our height, body structure and age, but also reduce our dietary fat intakes significantly, and keep cholesterol intake to a minimum. And, conceivably, you might conclude that it is proper for the federal government to so recommend.
>
> On the other hand, you may instead argue: What right has the federal government to propose that the American people conduct a vast nutritional experiment, with themselves as subjects, on the strength of so very little evidence that it will do them any good?
>
> Mr. Chairman, resolution of this dilemma turns on a value judgment. The dilemma so posed is not a scientific question; it is a question of ethics, morals, politics. Those who argue either position strongly are expressing their values; they are not making scientific judgments.

Though the conflict-of-interest accusations served to discredit the advice proffered in *Toward Healthful Diets*, the issue was not nearly as simple as the media made it out to be and often still do. Since the 1940s, nutritionists in academia had been encouraged to work closely with industry. In the 1960s, this collaborative relationship deteriorated, at least in public perception, into what Ralph Nader and other advocacy groups would consider an "unholy alliance." It wasn't always.

As Robert Olson explained at the time, he had received over the course of his career perhaps $10 million in grants from the USDA and NIH, and $250,000 from industry. He had also been on the American Heart Association Research Committee for two decades. But when he now disagreed with the AHA recommendations publicly, he was accused of being bought. "If people are going to say Olson's corrupted by industry, they'd have far more reason to call me a tool of government," he said. "I think university professors should be talking to people beyond the university. I believe, also, that money is contaminated by the user rather than the source. All scientists need funds."

Scientists were believed to be free of conflicts if their only source of funding was a federal agency, but all nutritionists knew that if their research failed to support the government position on a particular subject, the funding would go instead to someone whose research did. "To be a dissenter was to be unfunded because the peer-review system rewards conformity and excludes criticism," George Mann had written in *The New England Journal of Medicine* in 1977. The NIH expert panels that decide funding represent the orthodoxy and will tend to perceive research interpreted in a contrarian manner as unworthy of funding. David Kritchevsky, a member of the Food and Nutrition Board when it released *Toward Healthful Diets,* put it this way: "The U.S. government is as big a pusher as industry. If you say what the government says, then it's okay. If you say something that isn't what the government says, or that may be parallel to what industry says, that makes you suspect."

Conflict of interest is an accusation invariably wielded to discredit those viewpoints with which one disagrees. Michael Jacobson's Center for Science in the Public Interest had publicly exposed the industry connections of Fred Stare, founder and chair of the department of nutrition at Harvard, primarily because Stare had spent much of his career defending industry on food additives, sugar, and other issues. "In the three years after Stare told a Congressional hearing on the nutritional value of cereals that 'breakfast cereals are good foods,' " Jacobson had written, "the Harvard School of Public Health received about $200,000 from Kellogg, Nabisco, and their related corporate foundations." Stare defended his industry funding with an aphorism he repeated often: "The important question is not who funds us but does the funding influence the support of truth." This was reasonable, but it is always left to your critics to decide whether or not your pursuit of truth has indeed been compromised. Jeremiah Stamler and the CSPI held the same opinions on what was healthy and what was not, and Stamler consulted for CSPI, so Stamler's alliance with industry—funding from corn-oil manufacturers—was not considered unholy. (By the same token, advocacy groups such as Jacobson's CSPI are rarely if ever accused of conflicts of interest, even though their entire reason for existence is to argue *one* side of a controversy as though it were indisputable. Should that viewpoint turn out to be incorrect, it would negate any justification for the existence of the advocacy group and, with it, the paychecks of its employees.)

When I interviewed Mark Hegsted in 1999, he defended the Food and Nutrition Board, although he hadn't done so in 1979, when he was defending his own report and his own job to Congress. In 1981, when the Reagan administration closed down Hegsted's Human Nutrition Center at the USDA and found no further use for his services, Hegsted returned to Harvard, where the research he conducted until his retirement was funded by Frito-Lay. By that time, the controversy over the Food and Nutrition Board's conflicts of interest had successfully discredited *Toward Healthful Diets,* and Hegsted's *Dietary Guidelines for America* had become the official government statement on the dangers of fat and cholesterol in our diet.

Once politics, the public, and the press had decided on the benefits of low-fat diets, science was left to catch up. In the early 1970s, when NIH administrators opted to forgo a $1 billion National Diet-Heart Study that might possibly be definitive and to concentrate instead on a half-dozen studies, at a third of the cost, they believed the results of these smaller studies would be sufficiently persuasive to conclude publicly that low-fat diets would prolong lives. The results of these studies were published between 1980 and 1984.

Four of these studies tried to establish relationships between dietary fat and health within populations—in Honolulu, Puerto Rico, Chicago (Stamler and Shekelle's second Western Electric study), and Framingham, Massachusetts. None of them succeeded. In Honolulu, the researchers followed seventy-three hundred men of Japanese descent and concluded that the men who developed heart disease seemed to eat slightly *more* fat and saturated fat than those who didn't, but the men who died seemed to eat slightly *less* fat and slightly *less* saturated fat than those who didn't. This observation was made in Framingham and Puerto Rico as well. In 1981, investigators from the three studies published an article in the journal *Circulation* discussing the problem. They said it posed a dilemma for dietary advice, but not an insurmountable one. The fact that the men in Puerto Rico and Honolulu who remained free of heart disease seemed to eat more starches suggested that it might be a good idea to recommend that we all eat more starch, as McGovern's *Dietary Goals* actually had. And because the advice should never be to eat more calories, we would have to eat less fat to avoid gaining weight.

When one is reading this report, it's hard to avoid the suspicion that once the government began advocating fat reduction in the American diet it changed the way many investigators in this science perceived their obligations. Those who believed that dietary fat caused heart disease had always preferentially interpreted their data in the light of that hypothesis. Now they no longer felt obliged to test any hypothesis, let alone Keys's. Rather, they seemed to consider their obligation to be that of "reconciling [their] study findings with current programs of prevention," which meant the now official government recommendations. Moreover, these studies were expensive, and one way to justify the expense was to generate evidence that supported the official advice to avoid fat. If the evidence didn't support the recommendations, then the task was to interpret it so that it did.*

The other disconcerting aspect of these studies is that they suggested (with the notable exception of three Chicago studies reported by Jeremiah Stamler and colleagues) low cholesterol levels were associated with a higher risk of cancer. This link had originally been seen in Seymour Dayton's VA Hospital trial in Los Angeles, and Dayton and others had suggested that polyunsaturated fats used to lower cholesterol might be the culprits. This was confirmed in 1972 by Swiss Red Cross researchers. In 1974, the principal investigators of six ongoing population studies— including Keys, Stamler, William Kannel of Framingham, and the British epidemiologist Geoffrey Rose—reported in *The Lancet* that the men who had developed colon cancer in their populations had "surprisingly" low levels of cholesterol, rather than the higher levels that they had initially expected. In 1978, a team of British, Hungarian, and Czech researchers reported similar findings from a sixteen-thousand-man clinical trial of a cholesterol-lowering drug. By 1980, this link between cancer and low cholesterol was appearing in study after study. The most consistent association was between colon cancer and low cholesterol in men. In the Framingham Study those men whose total cholesterol levels were below 190 mg/dl were more than three times as likely to get colon cancer as those men with cholesterol greater than 220; they were almost twice as likely to contract any kind of cancer than those with cholesterol over 280 mg/dl. This finding was met with "surprise and chagrin," Manning Feinleib, a National Heart, Lung, and Blood Institute (NHLBI) epidemiologist, told *Science*.

This association was considered sufficiently troublesome that the NHLBI hosted three workshops between 1980 and 1982 to discuss it. In this case, however, the relevant administrators and investigators did not consider it sufficient to pay attention only to the positive evidence (that low cholesterol was associated with an increased risk of cancer even in

---

* The Honolulu Heart Program offered an extreme example of this conflict in 1985. The study revealed that high-fat diets were significantly associated with a *lower* risk of total mortality, cancer mortality, and stroke mortality. On the other hand, the percentage of calories as fat and dietary-cholesterol intake were both associated with a higher risk of heart-disease death. Thus, the authors concluded that "these data provide support for the diet-heart hypothesis," albeit with a caveat: "They also suggest that men with low fat intakes have a higher total mortality rate than men with higher fat intakes."

clinical trials) and reject the negative evidence as irrelevant or erroneous, as they had when implicating high cholesteral as a cause of heart disease. Instead, they searched the literature and found a few studies—including a Norwegian study published a decade earlier in a Scandinavian journal supplement—that reported no link between low cholesterol and cancer. As a result, the NHLBI concluded that the evidence was inconsistent, only "suggestive" that "low cholesterol may be in some way associated with cancer risk," said Robert Levy after the first workshop. After the second workshop, by which time the Framingham, Honolulu, and Puerto Rico studies had reported the same association, the NHLBI administrators still considered the results inconclusive: "The findings do not represent a public health challenge; however, they do present a scientific challenge," they wrote. Levy did tell the journal *Science* that this low-cholesterol/cancer link might make those investigators who were arguing that everyone's cholesterol should be as low as possible "a little more cautious."

After the third workshop, Levy and his NHLBI colleagues concluded that the evidence still didn't imply cause and effect. They believed that high cholesterol caused heart disease and that low cholesterol was only a sign of people who might be cancer-prone, perhaps because of a genetic predisposition. This seemed like an arbitrary distinction, and it was certainly based on assumptions more than facts. The NHLBI administrators acknowledged that further research would be necessary to clarify "the perplexing inconsistencies." Still, the evidence did "not preclude, countermand, or contradict the current public health message which recommends that those with elevated cholesterol levels seek to lower them through diets."

In the early 1970s, the National Heart, Lung, and Blood Institute had bet its heart-disease prevention budget on two enormous trials that held out hope of resolving the controversy.

The first was the Multiple Risk Factor Intervention Trial, known as MRFIT and led by Jeremiah Stamler. The goal of MRFIT was to "throw the kitchen sink" at heart disease: to convince the subjects to quit smoking, lower their cholesterol, and lower their blood pressure—the multiple-risk-factor interventions. The MRFIT investigators tested the cholesterol of 362,000 middle-aged American men and found twelve thousand (the top 3 percent) whose cholesterol was so high, more than 290 mg/ml, that they could be considered at imminent risk of having a heart attack. The MRFIT investigators believed that these men were so likely to succumb to heart disease that preventive measures would be even more likely to demonstrate a benefit. (If men with lower cholesterol were included, or if women were included, the study would require a considerably greater number of subjects or a longer follow-up to demonstrate any significant benefit.) These twelve thousand men were randomly divided between a control group—told to live, eat, and address their health problems however they desired—and a treatment group—advised to quit smoking, take medication to lower their blood pressure if necessary, and eat a low-fat, low-cholesterol diet, which meant drinking skim milk, using margarine instead of butter, eating only one or two eggs a week, and avoiding red meat, cakes, puddings, and pastries. All twelve thousand were then followed for seven years, at a cost of $115 million.

The results were announced in October 1982, and a *Wall Street Journal* headline captured the situation succinctly: "Heart Attacks: A Test Collapses." There had been slightly more deaths among those men who had been counseled to quit smoking, eat a cholesterol-lowering diet, and treat their high blood pressure than among those who had been left to their own devices.*

The second trial was the $150 million Lipid Research Clinics (LRC) Coronary Primary Prevention Trial. The trial was led by Basil Rifkind of the NHLBI and Daniel Steinberg, a specialist on cholesterol disorders at the University of California, San Diego. The LRC investigators had screened nearly half a million middle-aged men and found thirty-eight hundred who had no overt signs of heart disease but cholesterol levels sufficiently high—more than 265 mg/dl—that they could be considered imminently likely to suffer a heart attack. Half of these men (the control group) were told to eat fewer eggs and less fatty meats and drink less milk, and were given a placebo pill to take daily. The other half (the treatment group) were counseled to eat the same cholesterol-lowering diet, but they were also given a cholesterol-lowering drug called cholestyramine. Both groups had been told to diet, because the LRC investigators considered it unethical to withhold all treatment from the control group, given their high cholesterol levels and high risk of heart disease. It was an odd deci-

---

* In 1997, the MRFIT investigators also reported that the men in the treatment group subsequently had *more* lung cancer than the controls. This was despite the fact that 21 percent of the men had quit smoking in the treatment group, compared with 6 percent in the usual-care group. Because it was hard to believe that quitting smoking *increased* rates of lung cancer, the MRFIT investigators suggested the possibility that the lower cholesterol levels in the treatment group "might explain [their] higher lung cancer mortality." And, indeed, serum cholesterol showed a "marginally significant inverse association" with lung-cancer mortality. Nonetheless, the MRFIT investigators concluded that this was not a likely explanation for the results.

sion for two reasons. First, the LRC trial had been approved in the early 1970s in lieu of the National Diet-Heart Study *that was necessary to establish the safety and effectiveness of a cholesterol-lowering diet;* the LRC investigators had no proof that such a diet would benefit their subjects, rather than harm them. Second, both groups were told to diet, the trial could determine only the effectiveness of the drug—the single variable that differed between them.

In January 1984, the results of the trial were published in *The Journal of the American Medical Association*. Cholesterol levels dropped by an average of 4 percent in the control group—those men taking a placebo. The levels dropped by 13 percent in the men taking cholestyramine. In the control group, 158 men suffered nonfatal heart attacks during the study and 38 men died from heart attacks. In the treatment group, 130 men suffered nonfatal heart attacks and only 30 died from them. All in all, 71 men had died in the control group and 68 in the treatment group. In other words, cholestyramine had improved by less than .2 percent the chance that any one of the men who took it would live through the next decade. To call these results "conclusive," as the University of Chicago biostatistician Paul Meier remarked, would constitute "a substantial misuse of the term." Nonetheless, these results were taken as sufficient by Rifkind, Steinberg, and their colleagues so they could state unconditionally that Keys had been right and that lowering cholesterol would save lives.

Rifkind and his collaborators also concluded that the cholesterol-lowering benefits of a drug applied to diet as well. Although the trial included only middle-aged men with cholesterol levels higher than those of 95 percent of the population, Rifkind and his colleagues concluded that those benefits "could and should be extended to other age groups and women and . . . other more modest elevations of cholesterol levels." As Rifkind told *Time* magazine, "It is now indisputable that lowering cholesterol with diet and drugs can actually cut the risk of developing heart disease and having a heart attack."

Pete Ahrens called this extrapolation from a drug study to a diet "unwarranted, unscientific and wishful thinking." Thomas Chalmers, an expert on clinical trials who would later become president of the Mt. Sinai School of Medicine in New York, described it to *Science* as an "unconscionable exaggeration of all the data." In fact, the LRC investigators acknowledged in their *JAMA* article that their attempt to ascertain a benefit from diet alone had failed.

Rifkind later explained the exaggerated claims. For twenty years, he said, those who believed in Keys's hypothesis had argued that lowering cholesterol would prevent heart attacks. They had spent hundreds of millions of dollars trying to prove it, in the face of extreme skepticism. Now they had demonstrated that lowering cholesterol had reduced heart-disease risk and maybe even saved lives. They could never prove that cholesterol-lowering *diets* would do the same—that would be too expensive, and MRFIT, which might have implied such a conclusion, had failed—but now they had established a fundamental link in the causal chain from lower cholesterol to cardiovascular health. With that, they could take the leap of faith from cholesterol-lowering drugs to cholesterol-lowering diets. "It's an imperfect world," Rifkind said. "The data that would be definitive is ungettable, so you do your best with what is available."

With publication of the LRC results, the National Heart, Lung, and Blood Institute launched what Robert Levy called "a massive health campaign" to convince the public of the benefits of lowering cholesterol, whether by diet or drug, and the media went along. *Time* reported the LRC findings in a story headlined "Sorry, It's True. Cholesterol Really Is a Killer." The article about a *drug* trial began, "No whole milk. No butter. No fatty meats. Fewer eggs . . ." In March, *Time* ran a follow-up cover story quoting Rifkind as saying that the LRC results "strongly indicate that the more you lower cholesterol and fat in your diet, the more you reduce your risk of heart disease." Anthony Gotto, president of the American Heart Association, told *Time* that if everyone went along with a cholesterol-lowering program, "we will have [atherosclerosis] conquered" by the year 2000.

The following December, the National Institutes of Health hosted a "consensus conference" and effectively put an end to thirty years of debate. Ideally, in a consensus conference an unbiased expert panel listens to testimony and arrives at conclusions on which everyone agrees. In this case, Rifkind chaired the planning committee, of which Steinberg was a member. Steinberg was then chosen to head the expert panel that would draft the consensus. The twenty speakers did include three skeptics—Ahrens, Robert Olson, and Michael Oliver, a cardiologist with the Medical Research Council in London—who argued that the wisdom of a cholesterol-lowering diet could not be established on the strength of a drug experiment, let alone one with such borderline results. A month after the conference, the NHLBI epidemiologist Salim Yusuf described the controversy to *Science* as remaining as polarized as ever: "Many people have already made up their minds that cholesterol-lowering helps, and they don't need any evidence. Many others have decided that cholesterol-lowering is not helpful, and they don't need any evidence either."

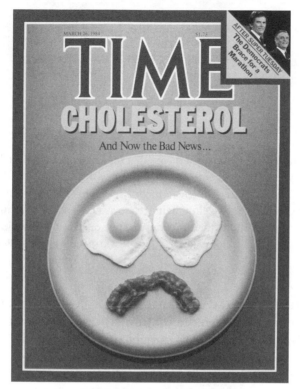

*March 1984: the results of a drug trial
are translated into the message that fatty foods
will cause heart disease. (Time* magazine © 1984
Time Inc. Reprinted by permission.)

But this was not the message of Steinberg's "consensus" panel, which was composed exclusively of lay experts and clinical investigators who "were selected to include only [those] who would, predictably, say that all levels of blood cholesterol in the United States are too high and should be lowered," as Oliver wrote in a *Lancet* editorial following the conference. "And, of course, this is exactly what was said." Indeed, the consensus conference report, written by Steinberg and his panel, revealed *no* evidence of any discord or dissent. There was "no doubt," it concluded, that low-fat diets "will afford significant protection against coronary heart disease" to every American over the age of two. The NIH Consensus Conference officially gave the appearance of unanimity where no unanimity existed. After all, if there had been a true consensus, as Steinberg himself later explained, "you wouldn't have had to have a consensus conference."

Chapter Four

THE GREATER GOOD

In reality, those who repudiate a theory that they had once proposed, or a theory that they had accepted enthusiastically and with which they had identified themselves, are very rare. The great majority of them shut their ears so as not to hear the crying facts, and shut their eyes so as not to see the glaring facts, in order to remain faithful to their theories in spite of all and everything.

MAURICE ARTHUS, *Philosophy of Scientific Investigation*, 1921

ONCE THE NATIONAL INSTITUTES OF HEALTH had declared the existence of a consensus, the controversy over dietary fat appeared to be over. A series of official government reports and guidelines that followed served to confirm it. In 1986, the NIH established the National Cholesterol Education Program (NCEP), which released its first guidelines for cholesterol reduction in October 1987. "The edict has been handed down," as the *Washington Post* reported: "total blood cholesterol should be below 200. . . . If it's above that threshold physicians must put their patients on cholesterol-lowering diets or use some of the new cholesterol-combating drugs to bring down the levels." Surgeon General C. Everett Koop's seven-hundred-page *Report on Nutrition and Health*, released in July 1988, "exhorts Americans to cut out the fat," *Time* reported. The "disproportionate consumption of food high in fats," according to the *Report on Nutrition and Health*, could be held responsible for two-thirds of the 2.1 million deaths in the United States in 1988. "The depth of the science base . . . is even more impressive than that for tobacco and health in 1964," explained Koop in the introduction, which was certainly not the case. In March 1989, the National Academy of Sciences released its version of the surgeon general's report, thirteen hundred pages long, entitled *Diet and Health: Implications for Reducing Chronic Disease Risk*. "Highest priority is given to reducing fat intake," the NAS report stated, "because the scientific evidence concerning dietary fats and other lipids and human health is strongest and the likely impact on public health the greatest."

These authoritative reports implied without foundation that yet more independent expert committees had weighed the evidence and agreed that dietary fat was a killer. But the surgeon general's report had been overseen by J. Michael McGinnis, who had been Mark Hegsted's liaison at the Surgeon General's Office when the first USDA *Dietary Guidelines* had been drafted a decade earlier. The chapter linking dietary fat to heart disease had been contracted out to the same administrators at the National Heart, Lung, and Blood Institute who had organized the NIH Consensus Conference and founded the National Cholesterol Education Program. In *Diet and Health*, the chapter assessing the hazards of fat had been drafted by three old hands in the dietary-fat controversy: Henry Blackburn, a protégé of Ancel Keys at Minnesota; Richard Shekelle, who had co-authored more than forty papers with Jeremiah Stamler; and DeWitt Goodman, who had chaired the National Cholesterol Education Program panel that had drafted the 1987 guidelines.*

In the media coverage that followed, those investigators skeptical of the underlying science seemed to have vanished from the public debate. New on the scene were public-interest groups—most notably, the Center for Science in the Public Interest and its director, Michael Jacobson—arguing that neither the NAS nor the surgeon general had gone far enough in pushing a national low-fat diet plan. Both the *Washington Post* and the *New York Times* quoted Jacobson scolding the authors of *Diet and Health* for lacking "the courage" to tell Americans straight out that a healthy lifestyle required much "greater reductions" in total fat, saturated fat, and cholesterol. In the *Post* article, Arno Motulsky, chairman of the NAS committee that compiled the report, acknowledged that one intention of *Diet and Health* was to convince Americans further of the existence of a scientific consensus on the benefits of reducing fat in the diet. "Many people may be confused by the vast amount of advice about what to eat," he said. "Some may have delayed making changes in their diets until they are more convinced that scientists have reached consensus. We hope our report will help these individuals move from inaction and complacency to action." The public face of the controversy had now shifted entirely. It was no longer about the validity of the underlying science, which was no less ambiguous than ever, but about whether Americans should be eating low-fat diets or very low-fat diets.

* The fourth author was Henry McGill, a pathologist who studied atherosclerosis in humans and in baboons, who says he had agreed unconditionally with the American Heart Association position on dietary fat since the early 1960s.

One striking fact about this evolution is that the low-fat diets now being recommended for the entire nation had only been tested twice, as I've said, once in Hungary and once in Britain, and in only a few hundred middle-aged men who had already suffered heart attacks. The results of those trials had been contradictory. The diets tested since then had been exclusively *cholesterol-lowering* diets that replaced saturated fats with unsaturated fats.

The rationale for lowering the *total* fat content of the diet to 30 percent was the tangential expectation that such a diet would help us control our weight. In 1984, the year of the NIH Consensus Conference, Robert Levy and Nancy Ernst of the NHLBI had described the state of the science this way: "There has been some indication that a low-fat diet decreases blood cholesterol levels," they wrote. "There is no conclusive proof that this lowering is independent of other concomitant changes in the diet (for example, increased dietary fiber or complex carbohydrate . . . or decreased cholesterol or saturated fatty acid level). . . . It may be said with certainty, however, that because 1 g fat provides about 9 calories—compared to about 4 calories for 1 g of protein or carbohydrate—fat is a major source of calories in the American diet. Attempts to lose weight or maintain weight must obviously focus on the content of fat in the diet." Though this was an untested conjecture (however obvious it might seem), the official healthy diet of the nation was now a low-fat diet. A new generation of diet doctors, the most influential of whom was Dean Ornish, were even prescribing 10-percent-fat diets, if not lower.

Another striking aspect of the low-fat diet recommendations is how little any individual might benefit from lowering his cholesterol.* Keys and others had argued that heart disease had to be prevented because its first symptom was often a fatal heart attack. But in twenty-four years of observation, the Framingham Heart Study had detected no relationship between cholesterol and sudden cardiac death. The likelihood of suffering a fatal first heart attack was no less for those with a cholesterol level of 180 mg/dl than for those with 250. "The lack of association between serum cholesterol level and the incidence of sudden death suggests that factors other than the atherosclerotic process may be of major importance in this manifestation of coronary artery disease," explained Thomas Dawber.

* Though women were clearly meant to adhere to the low-fat guidelines, they had not been included in any of the clinical trials. The evidence suggested that high cholesterol in women is not associated with more heart disease, as it might be in men, with the possible exception of women under fifty, in whom heart disease is exceedingly rare.

There is also little to gain from lowering cholesterol even in less cata-strophic manifestations of the disease. This was made clear in 1986, when Stamler published a reanalysis of his MRFIT data in *JAMA*. As Stamler reported it, the MRFIT investigators had continued to track the health of the 362,000 middle-aged men who had originally been screened as poten-tial candidates for MRFIT, including death certificates. Stamler reported that the cholesterol/heart-disease association applied at any level of choles-terol, and so anyone would benefit from lowering cholesterol.

Using the MRFIT data, however, it is possible to see how large or small that benefit might be (see chart, below). For every one thousand middle-aged men who had high cholesterol—between, say, 240 and 250 mg/dl—eight could expect to die of heart disease over any six-year period. For every thousand men with cholesterol between 210 and 220, roughly six could expect to die of heart disease. These numbers suggest that reducing cho-lesterol from, say, 250 to 220 would reduce the risk of dying from a heart attack in any six-year period from .8 percent (eight in a thousand) to .6 per-cent (six in a thousand). If we were to stick rigorously to a cholesterol-lowering diet for thirty years—say, from age forty to seventy, at which point

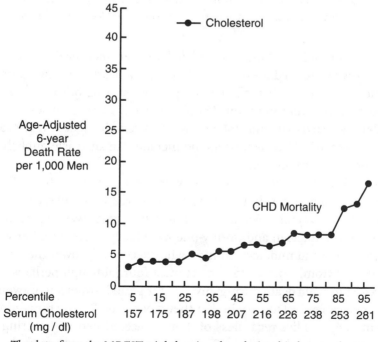

The data from the MRFIT trial showing the relationship between heart-disease mortality and cholesterol levels in the blood.

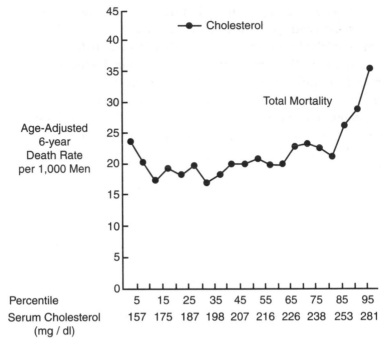

The data from the MRFIT trial showing the relationship between total mortality—i.e., death by all causes—and cholesterol levels in the blood.

high cholesterol is no longer associated with an increased risk of heart disease—we would reduce our risk of dying of a heart attack by 1 percent.

Whether we would actually live longer by lowering our cholesterol is, of course, a different question. People die from myriad causes. Though Stamler neglected to include total mortality data in his *JAMA* article, a second group of MRFIT investigators did include it in an article published in *The Lancet* just a month earlier.

Their data revealed that for every thousand men with cholesterol around 240 to 250 mg/dl, twenty to twenty-three would likely die of any cause within six years. For those whose cholesterol was approximately 220, between nineteen and twenty-one were likely to die. In other words, for every thousand middle-aged men who successfully lower their cholesterol by diet from, say, 250 to 220, at most four (although perhaps none) can expect to avoid death during any six-year period. Nineteen or twenty of these men can expect to die whether they diet or not. For the remaining 98 percent, they will live regardless of their choice. Moreover, lowering cholesterol further would not help. The death rate for men whose cholesterol is below 200 appears little different from that of men whose cholesterol falls between 200 and 250. Only for those men whose cholesterol is above 250 mg/dl does it appear that lowering cholesterol might improve the chances of living longer.

There is another way to interpret this statistical association between cholesterol, heart disease, and death. The association, as documented by Framingham, MRFIT, and other studies, only says that, the higher our cholesterol, the greater our risk of heart disease. It does not tell us whether the benefit from lowering cholesterol is shared by the *entire* population or only by a small percentage. The latter is the implicit assumption of the above analysis. But what if the benefit of lowering cholesterol is indeed shared democratically among all who do it? Perhaps we may all live longer by lowering our cholesterol. But how much longer?

Between 1987 and 1994, independent research groups from Harvard Medical School, the University of California, San Francisco, and McGill University in Montreal addressed the question of how much longer we might expect to live if no more than 30 percent of our calories came from fat, and no more than 10 percent from saturated fat, as recommended by the various government agencies. All three assumed that cholesterol levels would drop accordingly, and that this low-fat diet would have no adverse effects, which was still speculation rather than fact.

The Harvard study, led by William Taylor, concluded that men with a high risk of heart disease—such as smokers with high blood pressure— might gain one extra year of life by shunning saturated fat. Healthy non-smokers, however, might expect to gain only three days to three months. "Although there are undoubtedly persons who would choose to participate in a lifelong regimen of dietary change to achieve results of this magnitude, we suspect that some might not," the Harvard investigators noted.

The UCSF study, led by Warren Browner, was initiated and funded by the Surgeon General's Office. This study concluded that cutting fat consumption in America would *delay* forty-two thousand deaths each year, but the average life expectancy would increase by only three to four months. To be precise, a man who might otherwise die at sixty-five could expect to live an extra month if he avoided saturated fat for his entire adult life. If he lived to be ninety, he could expect an extra four months.* The McGill study, published in 1994, concluded that reducing saturated fat in the diet to 8 percent of all calories would result in an average increase in life expectancy of four days to two months.

---

* Browner's analysis also assumed that restricting dietary fat would reduce cancer deaths, which was speculative then and is even more speculative now.

Browner reported his results to the Surgeon General's Office, and only then submitted his article to *JAMA*. J. Michael McGinnis, the deputy assistant secretary for health, then wrote to *JAMA* trying to prevent publication of Browner's article, or at least to convince the editors to run an accompanying editorial that would explain why Browner's analysis should not be considered relevant to the benefits of eating less fat. "They would have liked it to come out the other way," explained Marion Nestle, who had edited the *Surgeon General's Report on Diet and Health* and had recruited Browner to do the analysis. This put Browner in the awkward position of protecting his work from his own funding agents. As he wrote McGinnis at the time, "I am sensitive to the needs of your office to put forward a consistent statement about what Americans should do, and to your dismay when a project that you have sponsored raises some questions about current policy. I am also concerned that the impacts of recommendations that apply to 240 million Americans are clearly understood. This manuscript estimates the effects of one such recommendation—altering dietary fat intake to 30 percent of calories—*based on the assumptions that underlie that recommendation*. Shooting the messenger—or creating a smoke screen—does not change those estimates." *JAMA* published Browner's article—"What If Americans Ate Less Fat?"—without an accompanying editorial.

That cholesterol-lowering provides little benefit to the individual was not unknown to the authors of these expert reports. This rationale was elucidated in *Diet and Health*, which explained that the purpose of preventive medicine in public health was to achieve the greatest good by treating entire populations rather than individuals. In this case, that meant addressing the situation of the 85 or 90 percent of the population with normal or low cholesterol. Though the actual benefit to these individuals "might be small or negligible," as *Diet and Health* explained, "because these people represent the great majority of the population, the benefit for the total population is likely to be paradoxically large."

This strategy is credited to the British epidemiologist Geoffrey Rose, a longtime veteran of the dietary-fat controversy. "The mass approach is inherently the only ultimate answer to the problem of a mass disease," Rose explained in 1981.

But, however much it may offer to the community as a whole, it offers little to each participating individual. When mass diphtheria immunization was introduced in Britain 40 years ago, even then roughly 600 children had to be immunized in order that one life would be saved—599 "wasted" immunizations for the one that was effective. . . . This is the kind of ratio that one has to accept in mass preventive medicine. A measure applied to many will actually benefit few.

When it came to dietary fat and heart disease, according to Rose's calculation, only one man in every fifty might expect to avoid a heart attack by virtue of avoiding saturated fat for his entire adult life: "Forty-nine out of fifty would eat differently every day for forty years and perhaps get nothing from it."

And thus the dilemma: "People will not be motivated to any great extent to take our advice because there is little in it for each of them, particularly in the short and medium term."* The best way around this problem, Rose explained, is to create social pressure to change. Consider young women who diet, he suggested, "not for medical reasons but because thinness is socially acceptable and obesity is not." So the task confronting public-health authorities is to create similar social pressure to induce "healthy behavior." And to do that, the benefits—or the risks of "unhealthy" behavior—have to be made to seem dramatic. "The modern British diet is killing people in their thousands from heart attacks," Rose told the BBC in 1984.

The assumption underpinning this public-health philosophy, as Rose explained in an influential 1985 *International Journal of Epidemiology* article entitled "Sick Individuals and Sick Populations," is that the entire population chronically overconsumes fat, and all of us have cholesterol levels that are unnaturally high. This is why attempts to uncover an association between fat consumption and cholesterol *within* a population like Framingham, Massachusetts, inevitably failed. Imagine, Rose suggested, if everyone smoked a pack of cigarettes every day. Any study trying to link cigarette smoking to lung cancer "would lead us to conclude that lung cancer was a genetic disease . . . since if everyone is exposed to the necessary agent, then the distribution of cases is wholly determined by individual susceptibility." The only way to escape this misconception, as with dietary fat, cholesterol, and heart disease, is to study the "differences between populations or from changes within populations over time." This "sick popu-

---

* William Taylor, the Harvard physician who had done the first of the three analyses on the questionable benefits of eating less fat, was unimpressed with this argument. "Most patients don't come into my office saying I really want to contribute to the public health statistics in this country," he said. "If they did, I'd know what to do for them."

lation" logic also explained why lowering cholesterol by 10 or 20 percent will have little effect on a single individual—just as smoking sixteen or eighteen cigarettes a day instead of twenty will do little to reduce individual lung-cancer risk—but would significantly affect the burden of heart disease across the entire population, and so should be widely recommended.

The arguments on sick populations and preventive public health are compelling, but they come with four critically important caveats.

First, Rose's logic does not differentiate between hypotheses. It would invariably be invoked to explain why studies failed to confirm Keys's fat hypothesis, and would be considered extraneous when similar studies failed to generate evidence supporting competing hypotheses. It is precisely to avoid such subjective biases that randomized controlled trials are necessary to determine which hypotheses are most likely true.

Second, as Rose observed, all public-health interventions come with potential risks, as well as benefits—unintended or unimagined side effects. Small or negligible risks to an individual will also add up and can lead to unacceptable harm to the population at large. As a result, the only acceptable measures of prevention are those that remove what Rose called "unnatural factors" and restore " 'biological normality'—that is . . . the conditions to which presumably we are genetically adapted." "Such normalizing measures," Rose explained, "may be presumed to be safe, and therefore we should be prepared to advocate them on the basis of a reasonable presumption of benefit."

This facet of Rose's argument effectively underpins all public-health recommendations that we eat low-fat or low-saturated-fat diets, despite the negligible benefits. It requires that we make assumptions about what is safe and what might cause harm, and what constitutes "biological normality" and "unnatural factors." The evidence for those assumptions will always depend as much on the observers' preconceptions and belief system as on any objective reality.

By defining "biological normality" as "the conditions to which presumably we are genetically adapted," Rose was saying that the healthiest diet is (presumably) the diet we evolved to eat. That is the diet we consumed prior to the invention of agriculture, during the two million years of the Paleolithic era—99 percent of evolutionary history—when our ancestors were hunters and gatherers. "There has been no time for significant further genetic adaptation," as the nutritionists Nevin Scrimshaw of MIT and William Dietz of the Centers for Disease Control noted in 1995. Any changes to this Paleolithic diet can be considered "unnatural factors," and so cannot be prescribed as a public-health recommendation.

The Paleolithic era, however, is ancient history, which means our conception of the typical Paleolithic diet is wide open to interpretation and bias. In the 1960s, when Keys was struggling to have his fat hypothesis accepted, Stamler's conception of the Paleolithic hunter-gatherer diet was mainly "nuts, fruits and vegetables, and small game." We only began consuming "substantial amounts of meat," he explained, and thus substantial amounts of animal fat, twenty-five thousand years ago, when we developed the skills to hunt big game. If this was the case, then we could safely recommend, as Stamler did, that we eat a low-fat diet, and particularly low in saturated fats, because animal fats in any quantity were a relatively new addition to the diet and therefore unnatural.

This interpretation, shared by Rose, was established authoritatively in 1985, the year after the NIH Consensus Conference, when *The New England Journal of Medicine* published a quantitative analysis of hunter-gatherer diets by two investigators—Boyd Eaton, a physician with an amateur interest in anthropology, and Melvin Konner, an anthropologist who had recently earned his medical degree. Eaton and Konner analyzed the diets of hunter-gatherer populations that had survived into the twentieth century and concluded that we are, indeed, genetically adapted to eat diets of 20–25 percent fat, most of which would in the past have been unsaturated. Eaton and Konner's article has since been invoked to support low-fat recommendations—in *Diet and Health*, for instance—as Rose's argument suggests it should.

But Eaton and Konner "made a mistake," as Eaton himself later said. This was only corrected in 2000, when Eaton, working now with John Speth and Loren Cordain, published a revised analysis of hunter-gatherer diets. This new analysis took into account, as Eaton and Konner's hadn't, the observation that hunter-gatherers consumed the entire carcass of an animal, not just the muscle meat, and preferentially consumed the fattest parts of the carcass—including organs, tongue, and marrow—and the fattest animals. Reversing the earlier conclusion, Eaton, Speth, and Cordain now suggested that Paleolithic diets were extremely high in protein (19–35 percent of calories), low in carbohydrates "by normal Western standards" (22–40 percent of energy), and *comparable* or *higher* in fat (28–58 percent of energy). Eaton and his new collaborators stated with certainty that those relatively modern foods that today constitute more than 60 percent of all calories in the typical American diet—cereal grains, dairy products, beverages, vegetable oils and dressings, and sugar and candy—"would have contributed virtually none of the energy in the typical hunter-gatherer diet." This latest analysis makes it seem that what Rose and the public-health authorities considered biological normality in 1985—a relatively low-fat diet—would now have to be be considered abnormal.*

The third critical caveat of Rose's logic is that it makes it effectively impossible to challenge the underlying science once it is invoked to defend a particular hypothesis, one that is said to benefit the public health. Policy and the public belief are often set early in a scientific controversy, when the subject is most newsworthy. But that's when the evidence is by definition premature and the demand for clarification most urgent. As the evidence accumulates, it may cease to support the hypothesis, but altering the conventional wisdom by then can be exceedingly difficult. (The artificial sweetener saccharine is still widely considered unhealthy, despite being absolved of any carcinogenic activity in humans over twenty years ago.) Rose's logic demonstrates why good science and public policy are often incompatible.

The fourth caveat is closely related. The philosophy of population-wide preventive medicine implies that the public health is not served by skepticism of the science or the reporting of contradictory evidence, both of which are essential to the process of science. A campaign to convince the public to embrace a public-health recommendation requires unconditional belief in the promised benefits. This was the motivation for creating the appearance of a consensus in the dietary-fat controversy and, as Arno Motulsky had told the *Washington Post*, for publishing the National Academy of Sciences *Diet and Health* report as well.

But if the underlying science is wrong—and that possibility is implied by the lack of a true consensus—then this tendency of public-health authorities to rationalize away all contradictory evidence will make it that much harder to get the science right. Once these authorities insist that a consensus exists, they no longer have motivation to pursue further research. Indeed, to fund further studies is to imply that there is still uncertainty. But the public's best interest will be served only by the kind of skeptical inquiry and attention to negative evidence that are necessary to learn the truth. "If the public's diet is going to be decided by popularity polls and with diminishing regard for the scientific evidence," remarked Pete Ahrens in 1979, "I fear that future generations will be left in ignorance of the real merits, as well as the possible faults, in any given dietary regimen aimed at prevention of [coronary heart disease]."

---

* Melvin Konner has doubts about the conclusions. "Boyd and I probably did underestimate the amount of meat in the Paleolithic diet based on our extrapolations for hunter-gatherers," he said. "I just don't think it's nearly as extreme as this paper claims."

Among the more conspicuous examples of the kind of scientific and social quagmire to which the logic of sick populations and preventive public health can lead is the proposition that dietary fat causes breast cancer. This possibility was suggested in 1976 in George McGovern's "Diet and Killer Disease" hearings, and then was cited in *Dietary Goals for the United States* as one reason why Americans should eat a low-fat diet (30-percent fat calories) as opposed to a cholesterol-lowering diet, in which the total fat content itself doesn't change. By 1982, the proposition that dietary fat causes cancer was considered so likely true that a National Academy of Sciences report entitled *Diet, Nutrition, and Cancer* not only recommended that Americans cut fat consumption to 30 percent, but noted that the evidence was sufficiently compelling that it "could be used to justify an even greater reduction." In 1984, the American Cancer Society released its first cancer-fighting, low-fat-diet prescription, and then both *The Surgeon General's Report on Nutrition and Health* and *Diet and Health* embraced the hypothesis.

The proposition had emerged originally from the same international comparisons that led to Keys's fat/heart-disease hypothesis—in particular, low rates of breast cancer and low fat consumption in Japan compared with high breast-cancer rates and high fat consumption in the United States. Moreover, when Japanese women immigrate to the United States, their breast-cancer rates quickly rise and, by the second generation, are equal to those of other American ethnic groups. As fat consumption increased in Japan from the 1950s to the early 1970s, breast-cancer rates there increased. These associations were given substance by the observation, originally made in the 1940s, that adding fat to the diet of laboratory rats promotes the growth of tumors, a phenomenon known technically as fat-induced tumorigenesis.

Considerable evidence also argued against the hypothesis. As John Higginson, founding director of the International Agency for Research on Cancer, noted in 1979, the international comparisons were as contradictory as they were confirmatory. In urban Copenhagen, breast-cancer rates were four times higher than in rural Denmark, but fat consumption was 50 percent lower. Large-population studies in Framingham; Honolulu; Evans County, Georgia; Puerto Rico; and Malmö, Sweden, had all reported low cholesterol levels associated with higher cancer rates. Since low cholesterol is allegedly the product of *low-fat* diets, it was "difficult to reconcile" this evidence, as the Framingham investigators noted in 1981, with the hypothesis that *high-fat* diets cause cancer.

The publication of the National Academy of Sciences report *Diet, Nutrition, and Cancer* in 1982 prompted the National Cancer Institute and the NAS to make funding available to test the hypothesis. A critical test would come from the Nurses Health Study, led by the Harvard epidemiologist Walter Willett, which began tracking diet, lifestyle, and disease in nearly eighty-nine thousand nurses around the country in 1982. Such a *prospective* study is no substitute for a randomized clinical control trial, but it constitutes the best that *observational* epidemiology can do. Willett and his colleagues published his first report on fat and breast cancer in January 1987 in *The New England Journal of Medicine*. Over six hundred cases of breast cancer had appeared among the eighty-nine thousand nurses over the first four years of the study. If anything, the less fat the women confessed to eating, the more likely they were to get breast cancer. In a *New York Times* article reporting the results of the study, Peter Greenwald, director of the National Cancer Institute Division of Cancer Prevention, said that the Nurses Health Study was "a good study, but not the only one," and so NCI would continue to recommend—despite what was then, by far, the best evidence available—that Americans eat less fat to prevent breast cancer. Eight months later, NCI researchers themselves published the results of a study similar to the Nurses Health Study but smaller, also suggesting that eating *more* fat and *more* saturated fat correlated with *less* breast cancer. The NCI study went virtually unnoticed, as *Science* later noted, "perhaps because no one wanted to hear the message that a promising avenue of research was turning into a blind alley, and perhaps because it swam against the 'medically politically correct' idea that fat is bad."

In 1992, Willett published the results from eight years of observation of the Nurses cohort. Fifteen hundred nurses had developed breast cancer, and, once again, those who ate less fat seemed to have more breast cancer. In 1999, the Harvard researchers published fourteen years of observations. By then almost three thousand nurses had contracted breast cancer, and the data still suggested that eating fatty foods (even those with copious saturated fat) might protect against cancer. For every 5 percent of saturated-fat calories that replaced carbohydrates in the diet, the risk of breast cancer decreased by 9 percent. This certainly argued against the hypothesis that excessive fat consumption caused breast cancer.

Despite this accumulation of contradictory evidence, Peter Greenwald and the administrators at NCI refused to let their hypothesis die. This was Rose's philosophy at work. After Willett's publication of the first Nurses Health Study results, Greenwald and his NCI colleagues had responded with an article in *JAMA* entitled "The Dietary Fat–Breast Cancer Hypothesis Is Alive." The NCI administrators argued that any study that generated evidence refuting the hypothesis could be flawed. The existence of any positive evidence, they argued, even if it came from admittedly rudimentary studies—in other words, studies that almost *assuredly* were flawed—was sufficient to keep such a critical hypothesis alive.

The only evidence that Greenwald and his collaborators considered "indisputable" was that laboratory rats fed "a high-fat, high-calorie diet have a substantially higher incidence of mammary tumors than animals fed a low-fat, calorie-restricted diet." In this they were right, but they did not rule out the possibility that it was the calories or whatever caused weight gain (what they implied by the adjective "high-calorie") and not the dietary fat itself that was to blame, which was very likely the case. Even in 1982, when the authors of *Diet, Nutrition, and Cancer* had reviewed the animal evidence for *fat*-induced tumor growth, it had been less than indisputable. Adding fat to the diets of lab rats certainly induced tumors or enhanced their growth, but the most effective fats by far at this carcinogenesis process were polyunsaturated fats—saturated fats had little effect unless "supplemented with" polyunsaturated fats. This raised questions about the applicability of these observations to Western diets, which were traditionally low in polyunsaturated fats, at least until the 1960s, when the AHA started advocating polyunsaturated fats as a tool to lower cholesterol. Adding fat to rat chow also caused the rodents to gain weight, which was among the foremost reasons why obesity researchers came to believe that dietary fat caused human obesity. But it was hard to determine in these experiments whether it was the fat or the weight gain itself that led to increased tumor growth.

This laboratory evidence that dietary fat caused breast cancer began to evaporate as soon as *Diet, Nutrition, and Cancer* was published, and researchers could get funding to study it. By 1984, David Kritchevsky, one of the authors of *Diet, Nutrition, and Cancer,* had published an article in *Cancer Research* reporting on experiments that had been explicitly designed to separate out the effects of fat and calories on cancer, at least in rats. As Kritchevsky reported, low-fat, high-calorie diets led to more tumors than high-fat, low-calorie diets, and tumor production was shut down entirely in underfed rats, regardless of how fatty their diet was. Kritchevsky later reported that if rats were given only 75 percent of their typical daily calorie requirements, they could eat five times as much fat as usual and still develop fewer tumors. Mike Pariza of the University of Wisconsin had published similar results in 1986 in the *Journal of the National Cancer Institute.* "If you restrict calories just a little bit," Pariza later said, "you completely wipe out this so-called fat enhancement of cancer." This observation has been confirmed repeatedly. Demetrius Albanes of the National Cancer Institute later described the data as "overwhelmingly striking." And he added: "Those data have very largely been ignored and strongly downplayed."

By 1997 when the World Cancer Research Fund and American Institute for Cancer Research released a seven-hundred-page report titled *Food, Nutrition and the Prevention of Cancer,* the assembled experts could find neither "convincing" nor even "probable" reason to believe that fat-rich diets increased the risk of cancer. A decade later still, Arthur Schatzkin, chief of the nutritional epidemiology branch at the National Cancer Institute, described the accumulated results from those trials designed to test the hypothesis as "largely null."

Nonetheless, the pervasive belief that eating fat causes breast cancer has persisted, partly because it once seemed undeniable. Purveyors of health advice just can't seem to let go of the notion. When the American Cancer Society released its nutrition guidelines for cancer prevention in 2002, the document still recommended that we "limit consumption of red meats, especially those high in fat," because of the same epidemiologic associations that had generated the fat-cancer hypothesis thirty years earlier. By 2006, with the next release of cancer-prevention guidelines by the American Cancer Society, the ACS was acknowledging that "there is little evidence that the total amount of fat consumed increases cancer risk." But we were still advised to eat less fat and particularly meats ("major contributors of total fat, saturated fat and cholesterol in the American diet"), because "diets high in fat tend to be high in calories and may contribute to obesity, which in turn is associated with increased risks of cancers." (Saturated fats, in particular, the ACS added, "may have an effect on increasing cancer risk," a statement that seemed to be based solely on the belief that if saturated fat causes heart disease it probably causes cancer as well.)

Belief in the hypothesis persists also because of the time lag involved in research of this nature. In 1991, the National Institutes of Health launched the $700 million Women's Health Initiative to test the hypothesis (and also the hypothesis that hormone-replacement therapy protects against heart disease and cancer). The WHI investigators enrolled forty-nine thousand women, aged fifty to seventy-nine. They randomly assigned twenty-nine thousand to eat their usual diets, and twenty thousand were prescribed a low-fat diet. The goal was to induce these women to consume only 20 percent of their calories from fat; to do this, they were told to eat more vegetables and fresh fruits, as well as whole grains, in case fiber was beneficial as well. If the diet succeeded in preventing breast cancer, or any chronic disease, the WHI investigators wouldn't know if it was because these women ate less fat or because they ate more fruits, vegetables, and grains. It's conceivable that a diet of fruits, vegetables, grains, and *more* fat, or of vegetables and fruits but *less* grains, could be even more protective. The women on the diet also consumed fewer calories—averaging 120 calories a day less than the controls over the eight years of the study.* So, similarly, if this diet appeared to prevent cancer, the WHI investigators wouldn't know whether it did so because it contained less fat (or more fruits and vegetables) or fewer calories. To induce those on the diet to stick to it for the better part of a decade, the WHI investigators provided them with an intensive nutritional and behavioral-education program. The women assigned to eat their usual diets received no such attention, which means they would be considerably less likely to change their lives in other ways that might also have an effect on breast cancer—to exercise or maintain their weight, stay away from sweets, refined flour, fast-food joints, and smoky bars. This disparity in counseling is known as an *intervention* effect, and it is precisely to avoid such an effect that drug trials must be done with placebos and double-blind.

All of these effects would be expected to bias the trial in favor of observing a beneficial effect where none exists, but the WHI trial still came up negative. In the winter of 2006, the WHI investigators reported that those women who were eating what we today consider the essence of a healthy diet—little fat, lots of fiber, considerable fruits, vegetables, and whole grains, fewer calories—had no less breast cancer than those who ate their typical American fare. (The women on the diet had no less heart disease, colon cancer, or stroke, either.) The results confirmed those of every study that had been done on diet and breast cancer since 1982. This, however, was still not generally perceived as a definitive refutation of the hypothesis. Rose's logic of preventive medicine held fast (it still does). In a press release on the findings, NHLBI Director Elizabeth Nabel stated, "The results of this study do not change established recommendations on disease prevention." In editorials that accompanied the WHI articles in *JAMA*, in virtually every press report, and even in the World Health Organization's official statement on the trial, it was said that this *particular* study may have failed to show a beneficial effect of a low-fat, high-fiber diet on breast cancer (and heart disease, stroke, colon cancer, and weight), but

---

* They did not, however, lose any weight because of this, which is paradoxical, and an issue we will discuss later.

that was not a reason to disbelieve the hypotheses. (The WHO press release was entitled "The World Health Organization Notes the Women's Health Initiative Diet Modification Trial, but Reaffirms That the Fat Content of Your Diet Does Matter.") Rather than enumerate the ways the WHI trial was biased to find a positive relationship, which was one facet of the controversy in the early 1990s over whether the trial should be funded to begin with, the WHI investigators and those like-minded now enumerated all the reasons why the study might have failed to find an effect.

At the core of all such ongoing scientific controversies is the inability to measure accurately the phenomenon at issue—the effect of dietary fat, for instance, on heart disease or cancer—either because it is negligible or nonexistent, or because the epidemiological tools available lack sufficient resolution for the task. Even clinical trials, unless done with meticulous attention to detail, double-blind, and placebo-controlled, cannot do the job. And if fat consumption has no effect whatsoever on heart disease or breast cancer, the available clinical and epidemiological tools will always be incapable of demonstrating such a fact, because it is impossible in science to prove the nonexistence of a phenomenon. So the effect of saturated fat on heart disease—or the benefit of replacing saturated fat in the diet with carbohydrates or unsaturated fats—will remain beyond the realm of science to demonstrate *unambiguously*. Investigators and public-health authorities will continue to base their conclusions on their personal assessment of the totality of the data or the consensus of opinion among their colleagues.

One challenge in this kind of controversy is to determine whether those skeptical of the established wisdom are incapable of accepting reality, closed-minded, or self-serving, or whether their skepticism is well founded. In other words, is the evidence invoked to support the established wisdom the product of sound scientific thinking and reasonably unambiguous, in which case the skeptics are wrong, or is it what Francis Bacon would have called "wishful science," based on fancies, opinions, and the exclusion of contrary evidence, in which case the skeptics are right to be so skeptical? Bacon offered one viable suggestion for differentiating. Good science, he observed, is rooted in reality, and so it grows and develops, and the evidence grows increasingly more compelling, whereas wishful sciences remain "stuck fast in their tracks," or "rather the reverse, flourishing most under their first authors before going downhill ever since."

Wishful science eventually devolves to the point where it is kept alive simply by the natural reluctance of its advocates to recognize or acknowledge error, rather than compelling evidence that it is right. "These are cases where there is no dishonesty involved," explained the Nobel Prize–winning chemist Irving Langmuir in a celebrated 1953 lecture, "but where people are tricked into false results by a lack of understanding about what human beings can do to themselves in the way of being led astray by subjective effects, wishful thinking or threshold interactions." Whereas good science would blossom over time, Langmuir noted, this "pathological science" would not. The most concise statement of this philosophy may be an unwritten rule of experimental physics credited originally to Wolfgang Panofsky, a former Manhattan Project physicist and presidential science adviser. "If you throw money at an effect and it doesn't get larger," Panofsky said, "that means it is not real."

That has certainly been the case with the dietary-fat/breast-cancer hypothesis. The relationship between dietary fat, cholesterol, and heart disease is more complicated, because the hypothesis constitutes three independent propositions: first, that lowering cholesterol prevents heart disease; second, that eating less fat or less saturated fat not only lowers cholesterol and prevents heart disease but, third, that it *prolongs* life.

Since 1984, the evidence that cholesterol-lowering drugs, particularly those known as *statins*, are beneficial—proposition number one—has certainly blossomed, particularly regarding people at high risk of heart attack. These drugs reduce serum-cholesterol levels dramatically, and they seem to prevent heart attacks, although whether they actually do so by lowering cholesterol levels or by other means as well is still an open question. ("Most drugs have multiple actions," notes the University of Washington biostatistician Richard Kronmal. Saying that statins reduce heart-disease risk by lowering cholesterol, he adds, is like "saying that aspirin reduces heart-disease risk by reducing headaches.") There is also a legitimate question as to whether they will prolong the life of anyone who is not in imminent danger of having a heart attack, but new trials consistently seem to confirm their benefits. All this may be irrelevant to the question of a healthy diet, however, because there is no compelling reason we should believe that a drug and a diet will have equivalent effects on our health, even if they both happen to lower cholesterol.

The evidence supporting the second and third propositions—that eating less fat, or less saturated fat, makes for a healthier and longer life—has remained stubbornly ambiguous. The message of the 1984 consensus conference and the ensuing expert reports was that the benefits of low-fat diets were effectively indisputable, and so pursuing further research on these questions was unnecessary. This in turn led to the ubiquitous belief in the validity of Keys's hypothesis and the unwholesome nature of saturated fat, but the reality is that since the early 1980s the evidence has become progressively less compelling.

Keys's own experience stands as an example. In the early 1950s, Keys had based his dietary-fat hypothesis of heart disease to a great extent on the congruence between the changing-American-diet story and the appearance of a heart-disease epidemic. By the early 1970s, however, he had publicly acknowledged that the heart-disease epidemic may indeed have been a mirage. There was "no basis" to make the claim, he admitted, that trends in heart-disease mortality in the United States reflect changes in the consumption of any item in the diet.

In the late 1950s, Keys supported his fat hypothesis with the disparity in fat consumption, cholesterol levels, and heart-disease mortality he found among Japanese men living in Japan, Hawaii, and Los Angeles. This association was then confirmed, more or less, in his Seven Countries Study, in which the Japanese villagers still had remarkably little fat in their diets, low cholesterol levels, and fewer heart-disease deaths over ten years than any other population with the exceptions of those of the islands of Crete and Corfu and the village of Velika Krsna in what is now Serbia. By the mid-1990s, however, the Japanese contingent of the Seven Countries Study, led by Yoshinori Koga, reported that fat intake in Japan had increased from the 6 percent of calories they had measured in the farming village of Tanushi-maru thirty-five years earlier, to 22 percent of calories. "There have been progressive increases in consumption of meats, fish and shellfish and milk," they reported. Mean cholesterol levels rose in the community from 150 mg/dl to nearly 190 mg/dl, which is only 6 percent lower than the average American values (202 mg/dl as of 2004). Yet this change went along with a "remarkable reduction" in the incidence of strokes and *no* change in the incidence of heart disease. In fact, the chance that a Japanese man of any particular age would die of heart disease had steadily diminished since 1970. "It is suggested that dietary changes in Tanushi-maru in the last thirty years have contributed to the prevention of cardiovascular disease," Koga and his colleagues concluded.

In the late 1950s, Keys had dismissed the possibility that misdiagnosis might have contributed to the extremely low heart-disease death rates in Japan they had observed initially. In 1984, Keys reversed himself, saying that the Japanese cardiologists who had worked with his Seven Countries Study "might have been misled by the local physicians who signed the death certificates and provided details."

Three years later, Keys acknowledged to the *New York Times* that he had re-evaluated his hypothesis. "I've come to think that cholesterol is not as important as we used to think it was," he said, "Let's reduce cholesterol by reasonable means, but let's not get too excited about it."

As in Japan, increases in fat consumption with coincident decreases in heart disease have occurred recently in Spain and Italy, which has prompted the observation that the French paradox—a nation that eats a high-fat diet and has little heart disease—has evolved into the French-Italian-Spanish paradox.* Through the mid-1990s, according to John Powles, an epidemiologist with the British Institute of Public Health, France and Italy both showed *declines* in death rates from stroke and heart disease that were greater than those in most European countries, while the decline in mortality in Spain lagged only slightly behind. And studies of Mediterranean immigrants to Australia suggest that the low heart-disease rates of these immigrants fall even lower in Australia, despite a considerable increase in their meat consumption.

In the late 1970s, the World Health Organization launched a research project known as MONICA, for "MONItoring CArdiovascular disease," that was similar in concept to Keys's Seven Countries Study but considerably larger. The study tracked heart disease and risk factors in thirty-eight populations in twenty-one countries—a total population of roughly six million people, which unlike previous studies included both men and women. Hugh Tunstall-Pedoe, the MONICA spokesman, has described the project as "far and away the biggest international collaborative study of cardiovascular disease ever carried out" and noted that, "whatever the results, nobody else has better data." By the late 1990s, MONICA had recorded 150,000 heart attacks and analyzed 180,000 risk-factor records. Its conclusion: heart-disease mortality was declining worldwide, but that decline was *independent* of cholesterol levels, blood pressure, or even smoking habits.

The MONICA investigators suggested reasons why their study might not have confirmed Keys's hypothesis, among them the possibility, as Tunstall-Pedoe noted, that with populations "the contribution of classical risk factors is swamped by that of other dietary, behavioral, environmental, or developmental factors." He also discussed something that may have contributed initially to the widespread belief in Keys's hypothesis: the tendency to publish or pay attention to only that evidence that confirms the

---

* This paradox could also include Switzerland. In 1979, Swiss public-health authories reported that cardiovascular mortality had undergone a "suprising decline" in Switzerland between 1951 and 1976, during a period in which the Swiss increased their consumption of animal fats by 20 percent.

existing beliefs about heart disease and risk factors. "If you do a study in your population and you show a perfect correlation between risk factors and heart disease, you rush off and publish it. If you don't, unless you have great confidence in yourself, you worry that perhaps you didn't measure something properly, or perhaps you'd better keep quiet, or perhaps there's something you haven't thought about. And by doing this, there is a risk of myths' becoming self-perpetuating." "There are people," Tunstall-Pedoe said, "who want to believe that if we find anything less than 100-percent correlation between traditional risk factors and trends in heart disease, we are somehow traitors to the cause of public health, and what we say should be suppressed, and we should be ashamed of ourselves. Whereas we are asking a perfectly reasonable question, and we came up with results. That is what science is about."

In the two decades since the NIH, the surgeon general, and the National Academy of Sciences first declared that all Americans should consume low-fat diets, the research has also failed to support the most critical aspect of this recommendation: that such diets will lead to a longer and healthier life. On the contrary, it has consistently indicated that these diets may cause more harm than good. In 1986, the year before the National Cholesterol Education Program recommended cholesterol-lowering for every American with cholesterol over 200 mg/dl, the University of Minnesota epidemiologist David Jacobs visited Japan, where he learned that Japanese physicians were advising patients to *raise* their cholesterol, because low cholesterol levels were linked to hemorrhagic stroke. At the time, Japanese men were dying from stroke almost as frequently as American men were succumbing to heart disease. Jacobs looked for this *inverse* relationship between stroke and cholesterol in the MRFIT data and found it there, too. And the relationship transcended stroke: men with very low cholesterol seemed prone to premature death; below 160 mg/dl, the lower the cholesterol, the shorter the life.

In April 1987, the Framingham investigators provided more reason to worry when they finally published an analysis of the relationship between cholesterol and all mortality. After thirty years of observation, there was a significant association between high cholesterol and premature death for men under fifty. But for those over fifty, both men and women, life expectancy showed no association with cholesterol. This suggested, in turn, that if low cholesterol *did* prevent heart disease, then it must raise the risk of dying from other causes.

This was compounded by what may have been the single most striking result in the history of the cholesterol controversy, although it passed without comment by the authorities: those Framingham residents whose cholesterol *declined* over the first fourteen years of observation were *more* likely to die prematurely than those whose cholesterol remained the same or increased. They died of cardiovascular disease more frequently as well. The Framingham investigators rejected the possibility that the drop in cholesterol itself was diet-related—the result of individuals' following AHA recommendations and eating low-fat diets. Instead, they described it as a "spontaneous fall," and insisted that it must be caused by other diseases that eventually led to death, but they offered no evidence to support that claim.

The association between low cholesterol and higher mortality prompted administrators at the National Heart, Lung, and Blood Institute once again to host a workshop and discuss it. Researchers from nineteen studies around the world met in Bethesda, Maryland, in 1990 to report their results. The data were completely consistent (see charts on following page): when investigators tracked all deaths, not just heart-disease deaths, it was clear that men with cholesterol levels above 240 mg/dl tended to die prematurely because of their increased risk of heart disease. Those whose cholesterol was below 160 mg/dl tended to die prematurely with an increased risk of cancer, respiratory and digestive diseases, and trauma. As for women, if anything, the *higher* their cholesterol, the *longer* they lived.*

The proponents of Keys's hypothesis said the results could not be meaningful. The excess deaths at low cholesterol levels *had* to be due to pre-existing conditions; chronic illness leads to low cholesterol, they concluded, not vice versa, and then the individuals die from the illnesses, which confuses the mortality issue. This was the assumption the Framingham researchers had made. At the one end of the population distribution of cholesterol, low cholesterol is the effect and disease is the cause. At the other end of the distribution, high cholesterol is the cause and disease is the effect. This, of course, is a distinction based purely on assumptions rather than actual evidence, and one consistent with the universal recommendations to lower cholesterol by diet. When NIH Administrator Basil Rifkind offered this interpretation during my interview with him in 1999,

---

* "Among women high blood cholesterol is not associated with all-cause mortality nor even with cardiovascular mortality," wrote UCSF epidemiologist Steve Hulley and his collaborators in a 1992 *Circulation* editorial about these data, entitled "Health Policy on Blood Cholesterol: Time to Change Directions." "We are coming to realize that the results of cardiovascular research in men, which represents the great majority of the effort thus far, may not apply to women."

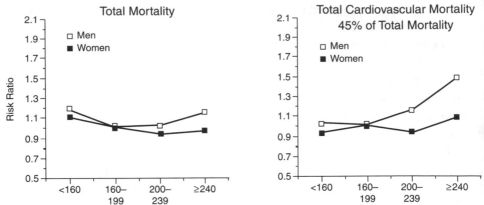

The relationship between blood cholesterol (horizontal axes) and all deaths (total mortality) or just heart disease deaths, as reported in a 1990 NIH conference.

he pointed to the report of the 1990 conference as the definitive document in support of it. But the report, which Rifkind co-authored, states unequivocally that this interpretation was *not* supported by the available evidence.

In an alternate interpretation, both ends of the cholesterol distribution are treated identically. Whether high or low, either our cholesterol levels directly increase mortality or they're a symptom of an underlying disorder that itself increases our risk of disease and death. In both cases, diet leads to disease, although whether it does so directly, via its effect on cholesterol, or through other mechanisms would still be an open question. In this interpretation, what a cholesterol-lowering diet does to cholesterol levels, and what that in turn does to arteries, may be only one component of the diet's effect on health. So lowering cholesterol by diet *might* help prevent heart disease for some individuals, but it *might* also raise susceptibility to other conditions—such as stroke and cancer—or even cause them. This is what had always worried those investigators who were skeptical of Keys's hypothesis. "Questions should be pursued about biological mechanisms that might help explain low [total cholesterol] : disease associations," noted the report from the 1990 NHLBI workshop. Nonetheless, public-health recommendations to eat low-fat diets and lower cholesterol would remain inviolate and unconditional.

In 1964, when the physicist Richard Feynman presented what would become a renowned series of lectures at Cornell University, he observed that it was a natural condition of scientists to be biased or prejudiced toward their beliefs. That bias, Feynman said, would ultimately make no difference, "because if your bias is wrong a perpetual accumulation of experiments will perpetually annoy you until they cannot be disregarded any longer." They could be disregarded, he said, only if "you are absolutely sure ahead of time" what the answer must be.

In the case of Keys's hypothesis, the annoying evidence was consistently disregarded from the beginning. Because the *totality of evidence* was defined as only those data that confirmed the hypothesis, Keys's hypothesis would always appear monolithic. Annoying observations could not force a reanalysis of the underlying assumptions, because each of those observations would be discarded immediately as being inconsistent with the totality of the evidence. This was a self-fulfilling phenomenon. It was unlikely, however, to lead to reliable knowledge about either the cause of heart disease or the routes to prevention. It did not mean the hypothesis was false, but its truth could never be established, either.

One other method can be employed to judge the validity of the hypotheses that dietary fat or saturated fat causes heart disease, and that cholesterol-lowering diets prevent it. This is a technique known as meta-analysis, viewed as a kind of last epidemiological resort in these kinds of medical and public-health controversies: if the existing studies give ambiguous results, the true size of a benefit or harm may be assessed by pooling the data from all the studies in such a way as to gain what's known as statistical power. Meta-analysis is controversial in its own right. Investigators can choose, for instance, which studies to include in their meta-analysis, either consciously or subconsciously, based on which ones are most likely to give them the desired result.

For this reason, a collaboration of seventy-seven scientists from eleven countries founded the Cochrane Collaboration in 1993. The founders, led by Iain Chalmers of Oxford University, believed that meta-analyses could be so easily biased by researchers' prejudices that they needed a standardized methodology to minimize the influence of such prejudice, and they needed a venue that would allow for the publication of impartial reviews. The Cochrane Collaboration methodology makes it effectively impossible for researchers to influence a meta-analysis by the criteria they use to include or exclude studies. Cochrane Collaboration reviews must include all studies that fit a prespecified set of criteria, and they must exclude all that don't.

In 2001, the Cochrane Collaboration published a review of "reduced or modified dietary fat for preventing cardiovascular disease." The authors combed the literature for all possibly relevant studies and identified twenty-seven that were performed with sufficient controls and rigor to be considered meaningful.* These trials encompassed some ten thousand subjects followed for an average of three years each. The review concluded that the diets, whether low-fat or cholesterol-lowering, had *no* effect on longevity and not even a "significant effect on cardiovascular events." There was only a "suggestion" of benefit from the trials lasting more than two years. In 2006, the Cochrane Collaboration published a review of multiple-risk-factor interventions—including lowering blood pressure and cholesterol—for the prevention of coronary heart disease. In this case, thirty-nine trials were identified of which ten (comprising over nine-hundred thousand patient years of observation) included sufficient data and were carried out with sufficient rigor to draw meaningful inferences. "The pooled effects suggest multiple risk factor intervention has no effect on mortality," the authors concluded. Although, once again, a "small" benefit of treatment, perhaps "a 10 percent reduction in CHD mortality," may have been missed, they added.

If we believe in Rose's philosophy of preventive medicine, this suggestion of benefit or the possibility that even a "small" benefit was missed still constitutes sufficient motivation to advocate cholesterol-lowering diets to the entire population, as indeed the authors of the first Cochrane review suggested. We could also assume that if a suggestion of a benefit can be induced after two years on such a diet, we might do considerably better after ten or twenty years, although we would still need trials to test that assumption.

We might also compare this conclusion to the original predictions of Keys's hypothesis in the mid-1950s. When Keys first suggested that eating fat caused heart disease, as we discussed, he did so partly on the basis of the experience in wartime Europe, where food shortages of a few years' duration coincided with *dramatic* decreases in the incidence of heart disease. Keys had attributed those decreases to the reduced availability of meat, eggs, and dairy products. Other investigators pointed out that the war changed many other aspects of diet and lifestyle. Mortality from infectious diseases, diabetes, tuberculosis, and cancer all dropped during the war. Still, to Keys it was the fat, particularly saturated fat, that was crucial. "A major lesson gained from World War II," he wrote in 1975, "is the proof that in a very few years the incidence of CHD [coronary heart disease]

---

* Conspicuously absent from the final analysis, because it was not a "randomized" trial and so the results could not be trusted, was the famous Helsinki Mental Hospital Study that had been cited by three generations of investigators, including *The Surgeon General's Report on Nutrition and Health* and the National Academies of Science *Diet and Health* report, as providing the most compelling evidence that cholesterol-lowering diets lowered mortality, not just heart disease.

could drop to a level of the order of one-fourth the preceding rate." If this were indeed possible, or even vaguely possible based on the collective European experience during World War II, as the evidence indeed suggested, then something considerably more profound had been happening than was reflected in only the "suggestion" of a reduction in mortality seen in the clinical trials of cholesterol-lowering diets. Other factors of diet or lifestyle that had changed during wartime must have played far more significant roles in improving the health of the populations.

Part Two

THE CARBOHYDRATE
HYPOTHESIS

The world is gradually going carbohydrate. That is because there are more people than there have ever been before (one would like to add "or ever will be again") so there must be more food. You can get about eight times as many calories from an acre of corn as you can from the flesh of pigs fed on this same corn. Because of population pressure, certain sections of the world are progressively using more of the vegetable and less of animal materials. This means that the carbohydrates, from sugar and cereals particularly, are increasing steadily in quantity. One does not need to view this with alarm, but it is not amiss to point out that this tendency is not the best road to health. Not that starches and sugars are harmful, but they are low in the essentials we must have for good health. If the proportion of carbohydrates is high then the amount of something else of greater importance is low. Nutrition is a six-way teeter-totter. Have you ever tried to balance such a device?

C. C. FURNAS AND S. M. FURNAS, *Man, Bread & Destiny:*
The Story of Man and his Food, 1937

Chapter Five

DISEASES OF CIVILIZATION

The potato took 200–250 years, in spite of organized encouragement, to become accepted in England. It took only fifty years in Ireland. Maize and cassava have come to be accepted in parts of Africa in considerably less time. . . . Tea, white bread, rice and soft drinks have entered many African dietaries in even shorter time and the extent to which they have spread and their consequences to nutrition have been rather severe.

F. T. SAI, Food and Agricultural Organization
regional nutrition officer for Africa, 1967

O N APRIL 16, 1913, ALBERT SCHWEITZER arrived at Lambaréné, a small village in the interior lowlands of West Africa, to establish a missionary hospital on the banks of the Ogowe River. Attended by his wife, Hélène, who had trained as a nurse, he began treating patients the very next morning. Schweitzer estimated that he saw almost two thousand patients in the first nine months, and then averaged thirty to forty a day and three operations a week for the better part of four decades. The chief complaints, at least in the beginning, were endemic diseases and infections: malaria, sleeping sickness, leprosy, elephantiasis, tropical dysentery, and scabies.

Forty-one years after Schweitzer's arrival, and a year and a half after he received the Nobel Peace Prize for his missionary work, Schweitzer encountered his first case of appendicitis among the African natives. Appendicitis was not the only Western disease to which the natives seemed to be resistant. "On my arrival in Gabon," he wrote, "I was astonished to encounter no cases of cancer. . . . I can not, of course, say positively that there was no cancer at all, but, like other frontier doctors, I can only say that if any cases existed they must have been quite rare." In the decades that followed, he witnessed a steady increase in cancer victims. "My observations inclined me to attribute this to the fact that the natives were living more and more after the manner of the whites."

As Schweitzer had suggested, his experience was not uncommon for the era. In 1902, Samuel Hutton, a University of Manchester–trained physician, began treating patients at a Moravian mission in the town of Nain, on the northern coast of Labrador, or about as far from the jungles of West Africa as can be imagined, in both climate and the nature of the indigenous population. As Hutton told it, his Eskimo patients fell into two categories: There were those who lived isolated from European settlements and ate a traditional Eskimo diet. "The Eskimo is a meat eater," he wrote, "the vegetable part of his diet is a meager one." Then there were those Eskimos living in Nain or near other European settlers who had taken to consuming a "settler's dietary," consisting primarily of "tea, bread, ship's biscuits, molasses, and salt fish or pork." Among the former, European diseases were uncommon or remarkably rare. "The most striking is cancer," noted Hutton on the basis of his eleven years in Labrador. "I have not seen or heard of a case of malignant growth in an Eskimo." He also observed no asthma and, like Schweitzer, no appendicitis, with the sole exception of a young Eskimo who had been "living on a 'settler' dietary." Hutton observed that the Eskimos who had adopted the settlers' diet tended to suffer more from scurvy, were "less robust," and endured "fatigue less easily, and their children are puny and feeble."

What both Schweitzer and Hutton had witnessed during their missionary years was a "nutrition transition," a term now commonly used to describe a population's Westernization in diet, lifestyle, and health status. The World Health Organization recently described the current version of the nutrition transition this way:

> Changes in the world food economy have contributed to shifting dietary patterns, for example, increased consumption of an energy-dense diet high in fat, particularly saturated fat, and low in carbohydrates. This combines with a decline in energy expenditure that is associated with a sedentary lifestyle. . . . Because of these changes in dietary and lifestyle patterns, diet-related diseases—including obesity, diabetes mellitus, cardiovascular disease, hypertension and stroke, and various forms of cancer—are increasingly significant causes of disability and premature death in both developing and newly developed countries.

This is little more than an updated version of the changing-American-diet story Ancel Keys and others had invoked to advocate low-fat diets: we eat fewer carbohydrates and ever more fat then we did in some idealized past, and we pay the price in chronic disease. Keys's reference point was the American diet circa 1909 (as portrayed by USDA estimates), or the Japanese or Mediterranean diets of the 1950s. When it was suggested to Keys that other nutrition transitions, including those witnessed by

Schweitzer and Hutton, could be edifying, he argued that not enough was known about the diets or about the health of those isolated populations for us to draw reliable conclusions. He also insisted that in many of these populations—particularly the Inuit—relatively few individuals were likely to live long enough to develop chronic disease, so little could be learned.

This argument, too, has taken on the aura of undisputed truth. This could be called the "nasty, brutish, and short" caveat, after Thomas Hobbes's pithy interpretation of the state of primitive lives. But earlier generations of physicians had the advantage of observing conditions of nutrition and health considerably further back on what anthropologists refer to as the curve of modernization. In this sense, their job was easier: noting the absence of a disease in a population, or the appearance of diseases in a previously unaffected population—the transition from healthy populations to sick populations, as Geoffrey Rose would put it—is an observation less confounded with diagnostic and cultural artifacts than are the comparisons of disease rates among populations all of which are afflicted.

Most of these historical observations came from colonial and missionary physicians like Schweitzer and Hutton, administering to populations prior to and coincidental with their first substantial exposure to Western foods. The new diet inevitably included carbohydrate foods that could be transported around the world without spoiling or being devoured by rodents on the way: sugar, molasses, white flour, and white rice. Then *diseases of civilization,* or *Western diseases,* would appear: obesity, diabetes mellitus, cardiovascular disease, hypertension and stroke, various forms of cancer, cavities, periodontal disease, appendicitis, peptic ulcers, diverticulitis, gallstones, hemorrhoids, varicose veins, and constipation. When any diseases of civilization appeared, all of them would eventually appear.

This led investigators to propose that all these diseases had a single common cause—the consumption of easily digestible, refined carbohydrates. The hypothesis was rejected in the early 1970s, when it could not be reconciled with Keys's hypothesis that fat was the problem, an attendant implication of which was that carbohydrates were part of the solution. But was this alternative carbohydrate hypothesis rejected because compelling evidence refuted it, or for reasons considerably less scientific?

The original concept of diseases of civilization dates to the mid-nineteenth century, primarily to Stanislas Tanchou, a French physician who served with Napoleon before entering private practice and studying the statistical distribution of cancer. Tanchou's analysis of death registries led him to conclude that cancer was more common in cities than in rural areas, and that the incidence of cancer was increasing throughout Europe. "Cancer, like insanity," he said, "seems to increase with the progress of civilization." He supported this hypothesis with communications from physicians working in North Africa, who reported that the disease had once been rare or nonexistent in their regions, but that the number of cancer cases was "increasing from year to year, and that this increase stands in connection with the advance of civilization."

By the early twentieth century, such reports had become the norm among physicians working throughout Africa. They would typically report a few cancers in towns where the "natives mingled with Europeans" and had copied their "dietetic and other domestic practices," but not in those areas where lifestyles and diets remained traditional. These reports, often published in the *British Medical Journal, The Lancet,* or local journals like the *East African Medical Journal,* would typically include the length of service that the author had undergone among the natives, the size of the native population served by the hospital in question, the size of the local European population, and the number of cancers diagnosed in both. F. P. Fouché, for instance, district surgeon of the Orange Free State in South Africa, reported to the *BMJ* in 1923 that he had spent six years at a hospital that served fourteen thousand natives. "I never saw a single case of gastric or duodenal ulcer, colitis, appendicitis, or cancer in any form in a native, although these diseases were frequently seen among the white or European population."

In 1908, the Smithsonian Institution's Bureau of American Ethnology published the first significant report on the health status of Native Americans. The author was the physician-turned-anthropologist Aleš Hrdlička, who served for three decades as curator of the Division of Physical Anthropology at the National Museum in Washington (now the Smithsonian's National Museum of Natural History). In a 460-page report entitled *Physiological and Medical Observations Among the Indians of Southwestern United States and Northern Mexico,* Hrdlička described his observations from six expeditions he had undertaken. "Malignant diseases," he said, "if they exist at all—that they do would be difficult to doubt—must be extremely rare." He had not encountered "unequivocal signs of a malignant growth on an Indian bone." Hrdlička also noted that he saw only three cases of "organic heart trouble" among more than two thousand Native Americans he examined, and "not one pronounced instance of advanced arterial sclerosis." Varicose veins were rare, and hemorrhoids infrequent. "No case of appendicitis, peritonitis, ulcer of the stomach, or of any grave disease of the liver was observed," he wrote.

Hrdlička considered the possibility, which Keys would raise fifty years later, that these Native Americans were unaffected by chronic disease because their life expectancy was relatively short; he rejected it because the evidence suggested that they lived as long as or longer than the local whites.

In 1910, Hrdlička's field observations on cancer were confirmed by Isaac Levin, a Columbia University pathologist, who surveyed physicians working for the Indian Affairs Bureau on reservations throughout the Midwestern and Western states. Levin's report, entitled "Cancer Among the North American Indians and Its Bearing upon the Ethnological Distribution of the Disease," discussed the observations of 107 physicians who had responded to his survey, with their names, locations, size of practice, duration of practice, and number of cancers diagnosed: Chas. M. Buchannan, for instance, practiced fifteen years among two thousand Indians with an average life expectancy of fifty-five to sixty years and saw only one case of cancer; Henry E. Goodrich, practicing for thirteen years among thirty-five hundred Indians, saw not a single case. Levin's survey covered over 115,000 Native Americans treated by agency doctors for anywhere from a few months to two decades and produced a total of twenty-nine documented cases of malignant tumors.

The two most comprehensive attempts to deal with the question of cancer in isolated populations were in *The Natural History of Cancer, with Special Reference to Its Causation and Prevention*, published in 1908 by W. Roger Williams, a fellow of the British Royal College of Surgeons, and *The Mortality from Cancer Throughout the World*, published in 1915 by the American statistician Fredrick Hoffman. In *The Natural History of Cancer*, Williams marched from continent to continent, region to region. In Fiji, for instance, in 1900, among 120,000 aborigines, Melanesians, Polynesians, and "Indian coolies," there were only two recorded deaths from malignant tumors. In Borneo, a Dr. Pagel wrote that he had been in practice for ten years and had never seen a case. Williams also documented the rising mortality from cancer that Tanchou had reported in the developed nations. In the United States, the proportional number of cancer deaths rose dramatically in the latter part of the nineteenth century: in New York, from thirty-two per thousand deaths in 1864 to sixty-seven in 1900; in Philadelphia, from thirty-one in 1861 to seventy in 1904.

Hoffman dedicated the better part of his career to making sense of these observations. He began his cancer studies as chief statistician of the

Prudential Insurance Company and continued them as part of an investigation of the Committee on Statistics of the American Society for the Control of Cancer (a predecessor of the American Cancer Society, of which Hoffman was a founder). In *The Mortality from Cancer Throughout the World* and then again in *Cancer and Diet*, his 1937, seven-hundred-plus-page update of the evidence, Hoffman concluded that cancer mortality was increasing "at a more or less alarming rate throughout the entire world," and this could only partially be explained by new diagnostic practices and the aging of the population.

Hoffman could not explain away the observations that physicians like Schweitzer and Hutton had made around the world and that both he and Williams had documented so comprehensively. In 1914, Hoffman himself had surveyed physicians working for the Bureau of Indian Affairs. "Among some 63,000 Indians of all tribes," he reported, "there occurred only 2 deaths from cancer as medically observed during the year 1914."

"There are no known reasons why cancer should not occasionally occur among any race or people, even though it be of the lowest degree of savagery or barbarism," Hoffman wrote.

> Granting the practical difficulties of determining with accuracy the causes of death among non-civilized races, it is nevertheless a safe assumption that the large number of medical missionaries and other trained medical observers, living for years among native races throughout the world, would long ago have provided a more substantial basis of fact regarding the frequency of occurrence of malignant disease among the so-called "uncivilized" races, if cancer were met with among them to anything like the degree common to practically all civilized countries. Quite to the contrary, the negative evidence is convincing that in the opinion of qualified medical observers cancer is exceptionally rare among primitive peoples.

Through the 1930s, this evidence continued to accumulate, virtually without counterargument. By the 1950s, malignancies among the Inuit were still considered sufficiently uncommon that local physicians, as in Africa earlier in the century, would publish single-case reports when they did appear. One 1952 article, written by three physicians from Queen's University in Ontario, begins with the comment "It is commonly stated that cancer does not occur in the Eskimos, and to our knowledge no case has so far been reported." In 1975, a team of Canadian physicians published an analysis of a quarter-century of cancer incidence among Inuit in the western and central Arctic. Though lung and cervical cancer had "dramatically increased" since 1949, they reported, the incidence of breast cancer was still "surprisingly low." They could not find a single case in an Inuit patient before 1966; they could find only two cases between 1967 and 1974.

These missionary and colonial physicians *did* often diagnose tumors and other diseases of civilization in local whites, and among natives who were working for European households and industries. In August 1923, for instance, A. J. Orenstein reported in the *British Medical Journal* on his experience as a superintendent of sanitation for the Rand mines in South Africa: "In a series of one hundred consecutive necropsies on native mine laborers conducted by myself in the latter part of 1922 and the first two months of 1923, two cases of carcinoma were observed—one was carcinoma of the pancreas and glands of the neck in a native male of the Shangaan race, age about 40, the other was a case of carcinoma involving practically the whole of the liver, in a native male of the same race, age about 25." The reports from these physicians were a reminder of how dramatic the course of the disease could be, and evidence against the argument that sophisticated diagnostic technology, unavailable in these outposts, was required to diagnose cancer. In 1923, George Prentice, who worked in Nyasaland, in southern central Africa, described one native patient with an inoperable breast tumor in the *British Medical Journal:* "It ran an uninterrupted course," Prentice wrote, "completely destroyed the breast, then the soft structure of the chest wall, and then ate through the ribs; when I last saw the negress in her village, I could see the heart pulsating. That was just before her death."

The absence of malignant cancer in isolated populations prompted questions about why cancer did develop elsewhere. One early hypothesis was that meat-eating was the problem, and that primitive populations were protected from cancer by eating mostly vegetarian diets. But this failed to explain why malignancies were prevalent among Hindus in India—"to whom the fleshpot is an abomination"—and rare to absent in the Inuit, Masai, and other decidedly carnivorous populations. (This hypothesis "hardly holds good in regard to the [American] Indians," as Isaac Levin wrote in 1910. "They consume a great deal of food [rich in nitrogen—i.e., meat], frequently to excess.")

By the late 1920s, the meat-eating hypothesis had given way to the notion that it was overnutrition in general, in conjunction with modern processed foods, lacking the vital elements necessary for health, that were to blame. These were those foods, as Hoffman put it, "demanding conservation or refrigeration, artificial preservation and coloring, or processing otherwise to an astonishing degree." As a result of these modern processed foods, noted Hoffman, "far-reaching changes in bodily functioning and metabolism are introduced which, extending over many years, are the causes or conditions predisposing to the development of malignant new growths, and in part at least explain the observed increase in the cancer death rate of practically all civilized and highly urbanized countries."

White flour and sugar were singled out as particularly noxious, because these had been increasing dramatically in Western diets during the latter half of the nineteenth century, coincident with the reported increase in cancer mortality. (They would also be implicated in the growing incidence of diabetes, as we'll discuss, and appendicitis.) Moreover, arguments over the nutritive value and appeal of white flour and sugar had been raging since the early nineteenth century.

Flour is made by separating the outer layers of the grain, containing the fiber—the indigestible carbohydrates—and virtually all of the vitamins and protein, from the starch, which is composed of long chains of glucose molecules. White sugar is made by removing the juice containing sucrose from the surrounding cells and husk of the cane plant or sugar beet. In both cases, the more the refining, the whiter the product, and the lower the vitamin, mineral, protein, and fiber content. The same is true for white rice, which goes through a similar refining process.

This might seem obviously disadvantageous, but white flour had its proponents. It was traditionally considered "more attractive to the eye," as Sir Stanley Davidson and Reginald Passmore observed in their textbook *Human Nutrition and Dietetics* (1963). It was preferred by bakers for its baking properties, and because it contains less fat than wholemeal flour it is less likely to go rancid and is more easily preserved. Millers preferred it because the leftover *bran* from refining rice and wheat (as with the molasses left over from refining sugar) could be sold profitably for livestock feed and industrial uses. Nutritionists also argued that white flour had better "digestibility" than whole-meal, because the presence of fiber in the latter prevented the complete digestion of any protein or carbohydrates that were attached. White flour's low protein, vitamin, and mineral content also made it "less liable than whole meal flour to infestation by beetles and the depredation of rodents," as Davidson and Passmore wrote.

It wasn't until the mid-nineteenth century that white flour became suitably inexpensive for popular consumption, with the invention of roller mills for grinding the grain. Until then, only the privileged classes ate white flour, and the poor ate wholemeal. Sugar was also a luxury until the mid-nineteenth century, when sugar-beet cultivation spread throughout the civilized world. In 1874, with the removal of tariffs on sugar importation in Britain, sugar consumption skyrocketed and led to the eventual development of the biscuit, cake, chocolate, confectionery, and soft-drink industries. By the beginning of World War I, the English were already eating more than ninety pounds of sugar per capita per year—a 500-percent increase in a single century—and Americans more than eighty pounds. Not until the mid-twentieth century did mechanical rollers begin replacing hand-pounding of rice in Asian nations, so that the poor could eat polished white rice instead of brown.

Explorers would carry enormous quantities of white flour, rice, and sugar on their travels and would trade them or give them away to the natives they met along the way.* In *The Voyage of the Beagle*, Darwin tells how the expedition's members persuaded Aborigines in Australia to hold a dancing party with "the offer of some tubs of rice and sugar." As early as 1892, the Barrow Eskimos were already described as having "acquired a fondness for many kinds of civilized food, especially bread of any kind, flour, sugar, and molasses." These foods remained primary items of trade and commerce with isolated populations well into the twentieth century.†

Until the last few decades, the nutritional debate over the excessive refining of flour and sugar had always been about whether the benefits of digestibility and the pleasing white color outweighed any potential disadvantages of removing the protein, vitamins, and minerals. In late-nineteenth-century England, the physician Thomas Allinson, head of the Bread and Food Reform League, wrote: "The true staff of life is whole meal bread." Allinson was among the first to suggest a relationship between refined carbohydrates and disease. "One great curse of this country," wrote Allinson, "is constipation of the bowel which is caused in great measure by white bread. From this constipation come piles, varicose veins, headaches, miserable feelings, dullness and other ailments. . . . As a consequence pill factories are now an almost necessary part of the state." Allinson's chain of cause and effect from white bread to constipation to chronic disease was given credibility in the late 1920s by the innovative and eccentric Scottish surgeon Sir Arbuthnot Lane in a book entitled *The Prevention of the Dis-*

---

* In *Across Australia*, Baldwin Spencer and F. J. Gillen describe embarking on an expedition through central Australia in the late 1890s with eight thousand pounds of flour (forty bags, each weighing two hundred pounds) and seven hundred pounds of sugar.

† A typical diet of one Australian Aborigine settlement, according to a joint American/Australian expedition in 1948, "consisted of white flour, rice, tea and sugar, buffalo and beef."

eases Peculiar to Civilization. The hypothesis would hold a tight grasp on a school of British medical researchers for decades to come.

The preferred explanation for how sugar, white flour, and white rice might perpetrate disease emerged from a great era of nutritional research in the early twentieth century. In 1912, the Polish-born biochemist Casimir Funk coined the term "vitamine" (the "e" was later dropped) and speculated that vitamins B_1, B_2, C, and D were necessary for human health. During the next quarter-century, researchers continued to discover new vitamins essential to health and identified a host of diseases—such as beriberi, pellagra, rickets, and scurvy—as caused by specific vitamin deficiencies. Beriberi results from a deficiency in thiamine (vitamin B_1), which is lost in the refining of polished rice and white flour. This led to the suggestion that even a disease like cancer could be a kind of deficiency disease, caused by vitamin starvation, as the journalist (and future homeopath) J. Ellis Barker called it in his book *Cancer: How It Is Caused, How It Can Be Prevented* (1924).

The Scottish nutritionist Robert McCarrison was perhaps the leading proponent of the hypothesis that the chronic illnesses of civilization could be attributed to "the extensive use of vitamin-poor white flour and to the inordinate use of vitamin-less sugar." McCarrison had founded a laboratory in India that would later become the National Institute of Nutrition and had spent nine years working in the Himalayas, "amongst isolated races far removed from the refinements of civilization," as he explained in a 1921 lecture at the University of Pittsburgh. "During the period of my association with these peoples," he wrote, "I never saw a case of asthenic dyspepsia, of gastric or duodenal ulcer, of appendicitis, of mucous colitis, or of cancer, although my operating list averaged over 400 operations a year." McCarrison attributed their good health to several factors, including a diet of "the unsophisticated foods of Nature." "I don't suppose that . . . as much sugar is imported into their country in a year as is used in a moderately sized hotel of this city in a single day," he said.

McCarrison's research included a comparative study of the diets and physiques of the disparate populations and religious groups on the Indian subcontinent. The "physique of northern races of India," McCarrison wrote, "is strikingly superior to that of the southern, eastern, and western races." Once again, he attributed the difference to the vitamins and nutrients present in the northern-Indian diet but not elsewhere. They ate well-balanced diets, with milk, butter, vegetables, fruit, and meat—and ate their wheat ground course as wholemeal flour, which "preserves all the nutrients with which Nature has endowed it." "White flour, when used as the staple article of diet," wrote McCarrison, "places its users on the same level as the rice-eaters of the south and east of India. They are faced with the same problem; they start to build up their dietaries with a staple of relatively low nutritive value." He also fed rats and mice in his laboratory on diets of these different populations and reported that the rats fared best on those diets containing "in abundance every element and complex for normal nutrition" and fared worst on those "excessively rich in carbohydrates, and deficient in suitable protein, mineral salts and vitamins."

By World War II, this rising tide of research on essential vitamins led the United States to decree that millers had to enrich white flour with vitamin B, iron, and nicotinic acid. In England, the government acted in similar fashion a decade later. The concept of "protective foods," containing the requisite protein, vitamins, and minerals for a healthy diet—fresh meat, fish, eggs, milk, fruits, and vegetables—now became the orthodox wisdom. During a century of debate, no one seems to have considered whether the properties of these refined foods—flour, sugar, and white rice—could have an impact on human health other than through the protein, fiber, vitamins, and minerals removed. Thirty years later, that would turn out to be the case, but by that time much of this original research on diseases of civilization would have been forgotten.

Chapter Six

DIABETES AND THE
CARBOHYDRATE HYPOTHESIS

The consumption of sugar is undoubtedly increasing. It is generally recognized that diabetes is increasing, and to a considerable extent, its incidence is greatest among the races and the classes of society that consume most sugar. There is a frequently discussed, still unsettled, question regarding the possible role of sugar in the etiology of diabetes. The general attitude of the medical profession is doubtful or negative as regards statements in words. . . . But the practice of the medical profession is wholly affirmative.

FREDERICK ALLEN, *Studies Concerning
Glycosuria and Diabetes,* 1913

Sugar and candies do not cause diabetes, but contribute to the burden on the pancreas and so should be used sparingly. . . . Carbohydrates are best taken in starchy forms: fruits, vegetables and cereals. The absorption is slower and the functional strain minimal.

GARFIELD DUNCAN, *Diabetes Mellitus and Obesity,* 1935

O F ALL THE DISEASES OF CIVILIZATION that may have been linked to the consumption of sugar and the refining of carbohydrates, diabetes was certainly a prime suspect. Here is a disease in which a conspicuous manifestation is the body's inability to use for fuel the carbohydrates in the circulation—known as blood sugar or, more technically, *glucose* or *serum glucose*. This glucose accumulates in the bloodstream, effectively overflows the kidneys, and spills over into the urine, causing a condition referred to as *glycosuria*. One symptom is a constant hunger, specifically for sugar and other easily digestible carbohydrates. Another is frequent urination, and the urine not only smells like sugar but tastes like it. For this reason, diabetes was often known as the sugar sickness. Hindu physicians two thousand years ago suggested it was a disease of the rich, caused by indulgence in sugar, which had only recently arrived from New Guinea, as had flour and rice.

"This ancient belief has a point in its favor," noted the American diabetologist Frederick Allen in his 1913 textbook *Studies Concerning Glycosuria and Diabetes*. "It originated before the time of organic chemistry, and there was no way for its authors to know that flour and rice are largely carbohydrate, and that carbohydrate in digestion is converted into the sugar which appears in the urine. This definite incrimination of the principal carbohydrate foods is, therefore, free from preconceived chemical ideas, and is based, if not on pure accident, on pure clinical observation."

By the end of the nineteenth century, researchers had established that the pancreas was responsible for the disease. By the 1920s, insulin was discovered and found to be essential for the utilization of carbohydrates for energy. Without insulin, diabetic patients could still mitigate the symptoms of the disease by restricting the starches and sugar in their diet. And yet diabetologists would come to reject categorically the notion that sugar and refined carbohydrates could somehow be responsible for the disease—another example of powerful authority figures winning out over science.

In the era that predated the discovery of insulin—a hormone that plays the crucial role in the carbohydrate hypothesis we will be discussing—the leading authorities on the treatment of diabetes could be divided into three groups: those firmly convinced that sugar and other carbohydrates played no causative role (among them Carl von Noorden, the pre-eminent German authority); those who thought the evidence ambiguous (including the German internist Bernhard Naunyn) and wouldn't put the blame on sugar itself but would concede, as Allen remarked, that "large quantities of sweet foods and the maltose of beer" favored the disease onset; and unequivocal believers (Raphaël Lépine of France was one), who would also note that vegetarian, beer-drinking Trappist monks frequently became diabetic, as did laborers in sugar factories.

Those diabetologists who believed that a connection existed argued that the glucose resulting from the digestion of sugar and refined carbohydrates passed with exceptional ease into the blood, and so it was easy to imagine that it might tax the body's ability to use it. Add sugar to the diet of someone whose ability to assimilate carbohydrates is already borderline or damaged in some way, and that person might pass from an apparently healthy condition to one that is pathological. In such cases, explained Allen, "in the absence of any radical difference between diabetes and non-diabetic conditions, the assumption of a possible production by sugar is logical. . . . A sufficiently excessive indulgence may presumably weaken the assimilative power of individuals in whom this power is normal or slightly reduced."

This scenario seemed to explain the fact that glycosuria will often vanish when mild diabetics fast or refrain from eating sugar and other high-carbohydrate foods. It also explained why some individuals could eat sugar, flour, and white rice for a lifetime and never get diabetes, but others, less able to assimilate glucose, would become diabetic when they consumed too many refined carbohydrates. Anything that slowed the digestion of these carbohydrates (like eating carbohydrates in unrefined forms) and so reduced the strain on the pancreas, the organ that secretes insulin in response to rising blood sugar, or anything that increased the assimilation of glucose without the need for insulin (excessive physical activity), might help prevent the disease itself. "If he is a poor laborer he may eat freely of starch," Allen wrote, "and dispose safely of the glucose arising from it, because of the slower process of digestion and assimilation of starch as compared with free sugar, and because of the greater efficiency of combustion in the muscles due to exercise. If he is well-to-do, sedentary, and fond of sweet food, he may, with no greater predisposition, become openly diabetic."

Diabetes seemed very much to be a disease of civilization, absent in isolated populations eating their traditional diets and comparatively common among the privileged classes in those nations in which the rich ate European diets: Sri Lanka (then Ceylon), Thailand, Tunisia, and the Portuguese island of Madeira, among others.* In China, diabetes was reportedly absent among the poor, but "the rich ones, who eat European food and drink sweet wine, suffer from it fairly often."

To British investigators, it was the disparate rates of diabetes among the different sects, castes, and races of India that particularly implicated sugar and starches in the disease. In 1907, when the British Medical Association held a symposium on diabetes in the tropics at its annual conference, Sir Havelock Charles, surgeon general and president of the Medical Board of India, described diabetes among "the lazy and indolent rich" of India as a "scourge." "There is not the slightest shadow of a doubt," said Charles's colleague Rai Koilas Chunder Bose of the University of Calcutta, "that with the progress of civilization, of high education, and increased wealth and prosperity of the people under the British rule, the number of diabetic cases has enormously increased." The British and Indian physicians work-

---

* In 1938, C. P. Donnison confirmed this observation in his book *Civilization and Disease*, using British Colonial Office yearly medical reports, which listed hospital inpatient diagnoses in all the British colonies. Many of the colonial physicians, wrote Donnison, reported that diabetes had never been seen in their local native populations. "Others say they have seen an odd case or two during many years experience." In those populations that had been more influenced by civilization, he continued, "a greater incidence is recorded."

ing in India agreed that the Hindus, who were vegetarians, suffered more than the Christians or the Muslims, who weren't. And it was the Bengali, who had taken on the most trappings of the European lifestyle, and whose daily sustenance, noted Charles, was "chiefly rice, flour, pulses* and sugars," who suffered the most—10 percent of "Bengali gentlemen" were reportedly diabetic. (In comparison, noted Charles, only eight cases of diabetes had been diagnosed among the seventy-six thousand British officials and soldiers working in India at the time—an incidence rate of .01 percent.)

Sugar and white flour were also obvious suspects in the etiology of diabetes, because the dramatic increase in consumption of these foodstuffs in the latter decades of the nineteenth century in the United States and Europe coincided with dramatic increases in diabetes incidence and mortality. Unlike heart disease, diabetes was a relatively straightforward diagnosis. After the introduction of a test for sugar in the urine in the 1850s, testing for diabetes became ever more common in hospitals and life-insurance exams, and as life insurance itself became popular, physicians increasingly diagnosed mild diabetes in outwardly healthy individuals, so the incidence numbers rose. As with coronary heart disease, the diagnostic definition of diabetes changed over the years, as did the relevant statistical analyses, so no conclusions can be considered definitive.

Nonetheless, the numbers were compelling. In 1892, according to William Osler in *Principles and Practice of Medicine,* only ten diabetics had been diagnosed among the thirty-five thousand patients treated at Johns Hopkins Hospital. At Massachusetts General Hospital in Boston, only 172 patients had been diagnosed as diabetic out of nearly fifty thousand admitted between 1824 and 1898; only eighteen of those were under twenty years old, and only three under ten, suggesting that childhood diabetes was an extremely rare diagnosis. Between 1900 and 1920, according to Haven Emerson, director of the Institute of Public Health at Columbia University, the death rate from diabetes, despite improved treatment of the disease, had increased by as much as 400 percent in American cities. It had increased fifteen-fold since the end of the Civil War. Emerson reported proportional increases in diabetes mortality in Great Britain and France and suggested they were due to the increased consumption of sugar, combined with an increasingly sedentary lifestyle. Moreover, diabetes rates had dropped precipitously during World War I in populations that had faced food shortages or rationing. "It is apparent," wrote Emerson in 1924, "that rises and falls in the sugar consumption are followed with fair regu-

* Such as peas, beans, and lentils.

larity within a few months by similar rises and falls in the death rates from diabetes."

The hypothesis that sugar and refined carbohydrates were responsible might have survived past the 1930s, but Elliott Joslin refused to believe it, and Joslin's name was by then "synonymous" with diabetes in the United States. Joslin may once have ranked beneath Frederick Allen in the hierarchy of American diabetologists, but Allen's reputation had been built on his starvation cure for diabetes, which was only marginally effective, and rendered unnecessary once insulin was discovered in 1921. Joslin achieved lasting fame by pioneering the use of insulin as a treatment. From the 1920s onward, Joslin's textbook *The Treatment of Diabetes Mellitus* and his *Diabetic Manual* were the bibles of diabetology.

When Emerson presented his evidence that rising sugar consumption was the best explanation for the rise in diabetes incidence, Joslin rejected it. He said that increased sugar consumption had been offset in America by decreasing apple consumption, and that the carbohydrates in apples were effectively identical to table sugar as far as diabetics were concerned. (This wasn't the case, but Joslin had little reason to believe otherwise in the 1920s.) Emerson countered with U.S. Department of Agriculture data reporting an actual increase in apple consumption in the relevant decades, but Joslin was unyielding.

Joslin found it inconceivable that sugar or any other refined starch could have a unique property that other carbohydrates did not. They all broke down to glucose after digestion, or glucose and fructose, in the case of table sugar. The insulin-releasing cells of the pancreas (known as β cells), which are dysfunctional in diabetes, respond only to the glucose. Early on in his career, Joslin, like Ancel Keys thirty years later, found the Japanese diet to be compelling evidence for the salubrious nature of carbohydrate-rich diets. "A high percentage of carbohydrate in the diet does not appear to predispose to diabetes," he wrote in 1923, since the Japanese ate such a diet and had an extremely low incidence of diabetes. He acknowledged that the rising death rate from diabetes in the United States coincided with rising sugar consumption, and that diabetes mortality and sugar consumption "must stand in some relation," but the Japanese experience argued against causality. He considered a rising incidence of obesity to be one factor in the increasing prevalence of diabetes, and decreasing physical activity, caused by the increasing mechanization of American life, to be another. A third factor, as the Japanese experience suggested, was a diet that was fat-rich and carbohydrate-poor.

Joslin effectively based his belief primarily on the work of a single researcher: Harold Himsworth of University College Hospital, London. To Joslin, Himsworth's "painstakingly accumulated" data constituted compelling evidence that a *deficiency* of carbohydrates and an *excess* of fat bring on diabetes. It was Himsworth's research and Joslin's faith in it that led a half-century of diabetologists to believe unconditionally that diabetes is not caused by the consumption of sugar and refined carbohydrates.

The two scientists effectively piggybacked on one another. In the post–World War II editions of Joslin's textbook, he cited a 1935 article by Himsworth as the support for the statement that increased fat consumption explained the rising incidence of diabetes.* Himsworth's article, in turn, rejected the hypothesis that sugar caused diabetes by citing a 1934 article by Joslin and a 1930 article by C. A. Mills of the University of Cincinnati. Joslin's 1934 article also depended almost entirely on Mills's article. Mills's article had stated "that there is no evidence in support" of the sugar-diabetes hypothesis; he had based this statement almost entirely on the observation that in Norway, Australia, and elsewhere sugar consumption rose from 1922 through the end of that decade but diabetes mortality did not. Other investigators, however, Joslin included, noted that the discovery of insulin in 1921 naturally led to a temporary leveling off of the otherwise rising tide of diabetes mortality. (On the other hand, as Mills noted, "of the thirteen countries highest in consumption of sugar, eleven are found among the thirteen highest in death rate from diabetes.")

Himsworth's achievements in clinical research were notable. He may have been the first researcher to differentiate between juvenile diabetes, caused by the inability of the pancreas to produce sufficient insulin and now known as insulin-dependent or Type 1 diabetes, and non-insulin-dependent diabetes, or Type 2, primarily a disease of adults, linked to excess weight and characterized by an insensitivity to insulin. Himsworth would later be knighted for his research contributions. Regrettably, his epidemiology was not as good as his clinical research.

Himsworth had first become convinced that diabetes was caused by fatty diets after asking his patients about their eating habits prior to their diabetes diagnosis and being told they had consumed "a smaller proportion of carbohydrate and a greater proportion of fat" than did healthy individuals.

Like Joslin, Himsworth considered all carbohydrates to be equivalent, sugar included; they could all be treated under one nutritional category when comparing diet and disease trends in populations. So Himsworth's

---

* Joslin also cited a 1936 article by Himsworth in *The Lancet,* but this latter article, if anything, tended to implicate carbohydrates as a cause of diabetes.

strongest argument was also the Japanese/American comparison. Whereas Joslin used it to exonerate sugar and high-carbohydrate diets, Himsworth used it to implicate fat and low-carbohydrate diets. Himsworth found the correlation between trends in diabetes mortality and the rising tide of fat consumption in England and Wales to be "striking" (the same word that Emerson had used to describe the correlation between trends in diabetes mortality and sugar consumption in the United States). "The progressive rise in diabetic mortality in Western countries during the last fifty years coincides with a gradual change towards higher fat and lower carbohydrate diets," Himsworth wrote. "The diabetic mortality rate is high in countries whose diets tend to be high in fat and poor in carbohydrate; and low where the opposite tendency prevails. The fall in diabetic mortality in World War I was related to a fall in fat and rise in carbohydrate intake. . . . Diabetic mortality rises with economic position and, simultaneously, dietary habits change so that a greater proportion of fat and less carbohydrate is taken." All of these observations, however, could also have been explained by variations in the consumption of sugar and white flour.

To defend his theory, Himsworth had to render irrelevant the conflicting evidence—the experience of isolated populations, for example, eating their traditional diets. "There appears to be unanimous agreement," he wrote, "that the incidence of diabetes mellitus is very low in the lower social grades of coloured races resident in their native lands, but there is evidence that when these races are transplanted to westernized countries the diabetic mortality rate rapidly rises." Himsworth's interpretation was that the original diets of these populations were fat-poor and carbohydrate-rich and became higher in fat when these people moved into urban environments. Himsworth acknowledged that the Masai ate a diet that "contained the highest proportion of fat of any recorded diet" and did not appear to suffer from diabetes, but he considered this evidence "so scanty that no opinion can be expressed."

Finally, Himsworth had to deal with the reported absence of diabetes among the Inuit. He acknowledged that his hypothesis implied that the Inuit *should* have an extremely high incidence of diabetes, which they did not. (There were three reported cases of confirmed diabetes among a population of sixteen thousand Alaskan Eskimos in 1956.) Rather than suggest that the Inuit died too young to get diabetes—Ancel Keys used the early-death rationale a quarter-century later to explain away their reported freedom from heart disease and cancer—Himsworth suggested they did not *actually* eat high-fat diets, despite all reports to the contrary. He cited two journal articles. One, he wrote, implied that the Inuit on Baffin Island ate a diet of only 48 percent fat calories, not that much higher than the average Englishman. The other, from 1930, reported that the "fisherfolk" of Labrador and northern Newfoundland subsisted on a diet of 21 percent fat calories and 70 percent carbohydrates, which meant a diet only slightly higher in fat than those eaten in Southeast Asian countries. (Himsworth did both these authors a disservice by suggesting that they believed that the Eskimo diets were carbohydrate-rich rather than fat-rich. The former article noted that the Eskimo "in his natural state eats practically only flesh," of which "in cold weather . . . one-third to one-half [by serving, not calories] may be taken as fat." The "fisherfolk" discussed in the latter article were not Eskimos, as Himsworth assumed, but those "of English and Scotch descent." Half of their daily calories came from white flour purchased at the local trading post, the author reported. Another quarter came from hard bread, rolled oats, molasses, and sugar.) "It would thus appear," Himsworth concluded, "that the most efficient way to reduce the incidence of diabetes mellitus amongst individuals predisposed to develop this disease would be to encourage the consumption of a diet rich in carbohydrate and to discourage them from satisfying their appetite with other types of food."

Once Joslin embraced Himsworth's fat hypothesis, it became the conventional wisdom among diabetologists and the mainstream medical community in the United States. In the 1946 and 1959 editions of his textbook, Joslin allotted the suggestion that sugar and refined carbohydrates play a role in diabetes less than a page and a half. In the 1971 edition, edited by Joslin's colleagues a decade after his death and renamed *Joslin's Diabetes Mellitus*, the subject had vanished entirely.

Oddly, Himsworth himself acknowledged that his own hypothesis was difficult to defend. In a 1949 lecture to the British Royal College of Physicians, Himsworth described the "paradox" of his fat hypothesis: "Though the consumption of fat has no deleterious influence on [the ability to metabolize glucose], and fat diets actually reduce the susceptibility of animals to diabetogenic agents, the incidence of human diabetes is correlated with the amount of fat consumed." Himsworth even suggested that dietary fat might not be the culprit after all, or that perhaps "other, more important, contingent variables" tracked with fat in the diet. He suggested that total calories played a role, because of the intimate association of diabetes and obesity, and because "in the individual diet, though not necessarily in national food statistics, fat and calories tend to change together." He did not mention sugar, which tends to change together with fat in *both* national food statistics and individual diets.

Despite Joslin's unconditional rejection of the hypothesis, investigators outside the United States continued to publish reports that implicated sugar specifically in the etiology of diabetes. In 1961, the Israeli diabetologist Aharon Cohen of Hadassah University reported that this was the best explanation for the pattern of diabetes seen in Jews who had immigrated to Israel from Yemen. In 1954, Cohen had spoken with Joslin, who had argued that diabetes was primarily caused by an inherited predisposition. Cohen, however, had spent the preceding years studying the dramatic differences in diabetes incidence among Native American tribes, and also treating diabetes among the refugees who had flooded into Israel with the end of World War II, and believed otherwise. As Cohen recalled the conversation, Joslin had effectively challenged him to test his belief by systematically examining the Israeli immigrant populations, and that's what Cohen did. Over the next five years, Cohen and his collaborators examined fifteen thousand Israelis living in a belt from Jerusalem to Beersheba. He concentrated on Yemenite Jews, because he had two distinct, contrasting populations to work with. One had arrived in 1949, fifty-thousand strong, flown in a legendary yearlong airlift known as Operation Magic Carpet. The other had lived in Israel since the early 1930s. Cohen was "astonished" that he found only three cases of diabetes in his examinations of five thousand Yemenites who had come in 1949. The incidence of diabetes was nearly fifty times as great in the earlier arrivals, and comparable to other populations in Israel, New York, and elsewhere. Other studies had also demonstrated, as Cohen put it, "a significantly greater prevalence" of coronary heart disease, hypertension, and high cholesterol among the Yemenites who had been in Israel for a quarter-century or more.

Cohen and his collaborators interviewed the more recent immigrants about their diets, both in Israel and in Yemen. They concluded that sugar consumption was the one noteworthy difference that might explain the increased incidence of diabetes, and perhaps the coronary heart disease, hypertension, and high cholesterol, too. "The quantity of sugar used in the Yemen had been negligible," Cohen wrote; "almost no sugar was consumed. In Israel there is a striking increase in sugar consumption, though little increase in total carbohydrates."

In New Zealand, Ian Prior, a young cardiologist who would later become the nation's most renowned epidemiologist, studied a population of five hundred Maoris living in an isolated valley of the North Island, thirty-five miles from the nearest town. Despite a physically active life— certainly by the standards of modern-day Europe or the United States—the Maoris, as Prior reported in 1964, had a remarkably high incidence of diabetes, heart disease, obesity, and gout. Sixty percent of the middle-aged women were overweight; over a third were obese. Sixteen percent had heart disease, and 11 percent had diabetes. Six percent of the men had diabetes. The staples of the Maori diet, Prior reported, were bread, flour, biscuits, breakfast cereals, sugar (over seventy pounds per person a year), and potatoes. There was also "beer, ice-cream, soft drinks, and sweets." Tea was the common beverage, "taken with large amounts of sugar by the majority."

In South Africa, George Campbell, who began his career as a general practitioner in Natal and then ran the diabetic clinic at the King Edward VIII Hospital in Durban, focused on a population of Indian immigrants living in the Natal region and on the local Zulu population. In the early 1950s, according to Campbell, his patients fell into two distinct categories of disease. The local whites suffered from diabetes, coronary thrombosis, hypertension, appendicitis, gall-bladder disease, and other diseases of civilization. The rural Zulus did not. In 1956, Campbell spent a year working at the Hospital of the University of Pennsylvania in Philadelphia and was "absolutely staggered by the difference in disease spectrum" between the black population in Philadelphia and the rural Zulus. Among the blacks of Philadelphia, he saw the same disorders that characterized his white patients in Durban.

After returning to South Africa, Campbell went to work at the King Edward VIII Hospital, which served exclusively the "non-white" population, and admitted some sixty thousand patients a year while administering to six hundred thousand outpatients. Once again, says Campbell, he was struck by the "remarkable difference in the spectrum of disease," in this instance between the urbanized Zulus, who were appearing with the same spectrum of diseases he had seen among the blacks of Philadelphia, and what he called their "country cousins" who still lived in rural areas. The Natal Indian population became the primary subject of Campbell's research when he realized that four out of every five of his diabetic patients came from that impoverished Indian community.

The ancestors of these Natal Indians had arrived in South Africa in the latter half of the nineteenth century to work as indentured laborers on the local sugar plantations. When Campbell began studying them in the late 1950s, over 70 percent lived below the poverty line, and many still worked for the sugar industry. Campbell and other researchers carried out half a dozen health surveys of this Natal Indian population. The incidence of diabetes among middle-aged men in some of the villages ran as high as 33 percent. It was nearly 60 percent among the ward patients and outpatients at the King Edward VIII Hospital. In ten years of operation, Campbell's clinic treated sixty-two hundred Indian diabetics, out of a local Indian population of only 250,000. A "veritable explosion of diabetes is taking place in these people," Campbell wrote, "in whom the incidence of the disease is now almost certainly the highest in the world." Campbell contrasted this with the numbers in India itself, where the average incidence of diabetes across the entire country was approximately 1 percent. This disparity between the incidence of diabetes in India and the incidence among the Indians of Natal ruled out a genetic predisposition to diabetes as a meaningful explanation.

For the Natal Indians, working primarily in and around sugar plantations, Campbell considered sugar the obvious suspect for their diabetes. He reported that the per-capita consumption of sugar in India was around twelve pounds yearly, compared with nearly eighty pounds for these working-class Natal Indians. The fat content of the diet in Natal was also very low, which seemed to rule out fat as the culpable nutrient. Excessive calorie consumption couldn't be to blame, according to Campbell, because some of these impoverished Natal Indians were living on as little as sixteen hundred calories a day—"a figure in many countries which would be regarded almost as a *starvation wage*"—and yet they "were enormously fat and suffered from undoubted diabetes proven by blood tests."

Campbell also found the disparities in diabetes prevalence and sugar consumption between urban and rural Zulus to be telling. The urban Zulu population, as hospital records demonstrated, was beset by diabetes. But in "thousands" of physical examinations performed on rural Zulus, Campbell wrote, "no case of diabetes has ever been discovered in any of them." Studies of a rural Zulu population in 1953 and an urban population in Durban in 1957, wrote Campbell, concluded that the former were eating six pounds of sugar a year each, compared with more than eighty pounds for the latter. The fat content of the diet in both populations was very low— less than 20 percent of the total calories—which again seemed to rule out fat as the culpable nutrient. By 1963, according to the South African Cane Growers Association, the urban Zulus were eating almost ninety pounds of sugar per person annually, while the rural Zulus were eating forty pounds each (a sixfold increase in a decade).

"In the last few years sugar intake has risen drastically in Natal," wrote Campbell, "because of very efficient advertising and because sugar has obviously reached as high an addictive status in our non-White people as in the Whites. . . . All [sugar]cane workers get a weekly ration of 1½ lb. And it is estimated that they can augment this by chewing sugar cane to the extent of ½–1 lb. *daily!*"

These sugarcane cutters, in whom, as Campbell noted, "diabetes is virtually absent," turned out to be pivotal, in that later generations of diabetologists would cite them as compelling evidence that diabetes was *not* caused by eating sugar. Campbell, however, believed it was the refining of the sugar, which allowed for its quick consumption and metabolism, that did the damage; chewing sugarcane resulted in a slow intake of sugar that he believed would be relatively benign. Moreover, cane cutters would cut and move by hand as much as seven tons of sugarcane each day, which required an extraordinary effort that suggested to Campbell—as it had to Frederick Allen a half-century before—that a physically active lifestyle might ward off the danger of excessive sugar consumption, perhaps by burning the sugar as fuel to maintain the necessary "huge output of energy" before it could do its damage. "There are few occupations in the world," Campbell wrote, "which entail such hard physical exertion as that involved in the cutting, moving, and stacking of sugar cane."

Campbell also believed that diabetes required time to manifest itself. The cane cutters had been receiving their refined-sugar ration for only a decade at most. From his medical histories of the diabetic Zulus at his clinic, Campbell found what he called a "remarkably constant period in years of exposure to town life" before rural Zulus who had moved permanently into Durban developed diabetes. "The peak 'incubation period' in 80 such diabetics," he wrote, "lay between 18 and 22 years." Thus, Campbell suggested that diabetes would appear in a population to any extent only after roughly two decades of excessive sugar consumption, just as lung cancer from cigarettes appears on average after two decades of smoking. He also suggested that, if international statistics were any indication, the kind of diabetes epidemic they were experiencing among Natal Indians— or, for that matter, most Westernized nations—required a consumption of sugar greater than seventy pounds per person each year.

Campbell appears to be the first diabetologist to propose seriously an incubation period for diabetes. Joslin's textbooks suggest he believed that if sugar consumption caused diabetes the damage could be done quickly—in a single night of "acute excess." In arguing against the sugar theory of diabetes, Joslin said that no one to his knowledge had ever developed the disease after drinking the sugar solution used in a type of diabetes test known as a glucose-tolerance test.* By the same logic, you could imagine that smoking a pack of cigarettes in an evening might cause lung cancer within the next few weeks in the rare unfortunate first-time

---

* Although, he noted in the 1946 edition of his textbook, "Dr. F. G. Brigham tells me Mrs. K. with multiple sclerosis developed diabetes after starting in to eat candy to gain weight."

smoker. That it has not been known to happen does not imply that tobacco is not a potent carcinogen.

In the early 1960s, Campbell began corresponding with a retired physician of the British Royal Navy, Surgeon Captain Thomas Latimore "Peter" Cleave. In 1966, they published *Diabetes, Coronary Thrombosis and the Saccharine Disease,* a book in which they argued that all the common chronic diseases of Western societies—including heart disease, obesity, diabetes, peptic ulcers, and appendicitis—constituted the manifestations of a single, primary disorder that could be called "refined-carbohydrate disease." Because sugar was the primary carbohydrate involved, and the starch in white flour and rice is converted into blood sugar in the body, they opted for the name *saccharine disease* ("saccharine," in this instance, meant "related to sugar" and rhymes with "wine," in their usage, not "win," as the artificial sweetener does).

After the book was published, Campbell returned to working exclusively on diabetes. Cleave tried to convince the medical establishment of the strength of evidence linking chronic diseases to the refining of carbohydrates, with little success. One biostatistician who insisted the idea should be taken seriously was Sir Richard Doll, director of the Statistical Research Unit of Britain's Medical Research Council, who wrote the introduction to *Diabetes, Coronary Thrombosis and the Saccharine Disease.* In the early 1950s, Doll had published the seminal studies linking cigarettes to lung cancer. Doll later said of Cleave's research, "His ideas deserved a lot more attention than they got."

The primary obstacle to the acceptance of Cleave's work was that he was an outsider, with no recognizable pedigree. He had spent his entire career with the British Royal Navy, retiring in 1962, after spending the last decade directing medical research at the Institute of Naval Medicine. Much of Cleave's early career was spent in British naval hospitals in Singapore, Malta, and elsewhere, which gave him firsthand experience of how chronic-disease incidence could differ between nations.

Cleave's nutritional education was furthered by the experience of his brother, Surgeon Captain H. L. Cleave, who spent the war years imprisoned by the Japanese in Hong Kong and then Tokyo. In the Hong Kong prison, peptic ulcers were a plague. The diets in these camps were predominantly white rice. Until vitamin-B supplements were distributed, beriberi was also a problem. After two years, many of the prisoners, including Cleave's brother, were transferred to a camp outside Tokyo, where the ulcers vanished. In the Tokyo POW camps, the rice was brown, lightly milled, with unmilled barley and millet added.

In the decades after the war, Cleave became an obsessive letter-writer, corresponding with hundreds of physicians around the world, requesting information on disease rates and the occurrence and appearance of specific diseases. His 1962 book on peptic ulcers contained page after page of testimony from physicians reporting the relative absence of ulcers in those populations where sugar, white flour, and white rice were hard to come by.

Cleave's intuition was to reduce the problem of nutrition and chronic disease to its most elementary form. If the primary change in traditional diets with Westernization was the addition of sugar, flour, and white rice, and this in turn occurred shortly before the appearance of chronic disease, then the most likely explanation was that those processed, refined carbohydrates were the cause of the disease. Maybe if these carbohydrates were added to any diet, no matter how replete with the essential protein, vitamins, minerals, and fatty acids, it would lead to chronic diseases of civilization. This would explain why the same diseases appeared after Westernization in cultures that lived almost exclusively on animal products—the Inuit, the Masai, and Samburu nomads, Australian Aborigines, or Native Americans of the Great Plains—as well as in primarily agrarian cultures like the Hunza in the Himalayas or the Kikuyu in Kenya.

Cleave would later be disparaged for suggesting that all chronic diseases of civilization have a single primary cause, but he insisted that it was naïve to think otherwise. Though it may seem odd, he considered dental cavities the chronic-disease equivalent of the canary in the mine. If cavities are caused primarily by eating sugar and white flour, and cavities appear first in a population no longer eating its traditional diet, followed by obesity, diabetes, and heart disease, then the assumption, until proved otherwise, should be that the other diseases were also caused by these carbohydrates.

Diabetes, obesity, coronary heart disease, gallstones and gall-bladder disease, and cavities and periodontal disease are intimately linked. As early as 1929, physicians were reporting that a fourth of their coronary-heart-disease patients also had diabetes. Diabetics, as Joslin noted, were especially prone to atherosclerosis, which became increasingly clear after the discovery of insulin. Studies in the late 1940s revealed that diabetic men were twice as likely to die of heart disease as nondiabetics; diabetic women were three times more likely. Moreover, diabetics had an exceptionally high rate of gallstones; and the obese had an exceptionally high rate of gall-bladder disease. As Joslin's textbook also observed, "The destruction of teeth and the supporting structures is very active just prior to the onset of diabetes," connecting cavities to the disease.

Cleave's desire for simplicity led him to theorize that any cluster of diseases so intimately associated must have a single underlying cause. Darwin's theory of evolution led Cleave to believe that endemic chronic disease must be caused by a relatively rapid change in our environment to which we had not yet adapted. He called this idea "The Law of Adaptation": species require "an adequate period of *time* for adaptation to take place to any unnatural (i.e., new) feature in the environment," he wrote, "so that any danger in the feature should be assessed by how long it has been there." The refining of carbohydrates represented the most dramatic change in human nutrition since the introduction of agriculture. "Whereas cooking has been going on in the human race for probably 200,000 years," Cleave said, "there is no question yet of our being adapted to the concentration of carbohydrates. . . . Such processes have been in existence little more than a century for the ordinary man and from an evolutionary point of view this counts as nothing at all."

Cleave believed the concentration of carbohydrates in the refining process did its damage in three ways.

First, it led to overconsumption, because of what he called the deception of the appetite-control apparatus by the density of the carbohydrates. He contrasted the "eating of a small quantity of sugar, say roughly a teaspoonful," with the same quantity in its original form—a single apple, for instance. "A person can take down teaspoonfuls of sugar fast enough, whether in tea or any other vehicle, but he will soon slow up on the equivalent number of apples. . . . The argument can be extended to contrasting the 5 oz. of sugar consumed, on the average, per head per day in [the United Kingdom] with up to a score of average-sized apples. . . . Who would consume that quantity daily of the natural food? Or if he did, what else would he be eating?"

Second, this would be exacerbated by the removal of protein from the original product. Cleave believed (incorrectly) that peptic ulcers were caused by the lack of protein necessary to buffer the gastric acid in the stomach.

Finally, the refining process increased the rate of digestion of carbohydrates, and so the onrush of blood sugar on the pancreas, which would explain diabetes. "Assume that what strains the pancreas is what strains any other piece of apparatus," wrote Cleave and Campbell, "not so much the total amount of work it is called upon to do, but the rate at which it is called upon to do it. In the case of eating potatoes, for example, the conversion of the starch into sugar, and the absorption of this sugar into the blood-stream, is a slower and gentler process than the violent one that follows the eating of [any] mass of concentrated sugar."

The link between refined carbohydrates and disease had been obscured over the years, Cleave and Campbell explained, by the "insufficient appreciation of the distinction" between carbohydrate foods in their natural state and the unnatural refined carbohydrates—treating sugar and white flour as equivalent to raw fruit, vegetables, and wholemeal flour. When researchers looked at trends between diet and disease, as Himsworth and Joslin had done with diabetes and Keys and a later generation of researchers would do with heart disease and even cancer, they would measure only fat, protein, and total carbohydrate consumption and fail to account for any potential effect of refined carbohydrates. Occasionally, they might include sugar consumption in their analyses, but they would rarely make a distinction between wholemeal bread and white flour, between brown rice and white. In most cases, cereal grains, tubers, vegetables, and fruits, and white sugar, flour, rice, and beer, were all included under the single category of carbohydrate. "While the consumption of all carbohydrates may not be moving appreciably with the rise or fall in the incidence of a condition," Cleave and Campbell explained, "the consumption of the refined carbohydrates may be moving decisively."

Cleave first made this point in 1956, when he published his hypothesis in an article that also contested Joslin's belief that the increased incidence of diabetes in the twentieth century was unrelated to sugar consumption. Had Joslin or Himsworth charted sugar consumption separately from that of all carbohydrates, Cleave wrote, "what was the opposite of a relationship between diabetes mortality and carbohydrate consumption would become a very close relationship."* (See chart on following page.)

Cleave had identified one of the fundamental flaws of modern nutrition and chronic-disease epidemiology. Greater affluence inevitably takes populations through a nutrition transition that represents a congruence of fundamental changes in diet. Meat consumption tends to increase, and so saturated fat increases as well. Grain consumption decreases, and so carbohydrate consumption as a whole decreases. But the carbohydrates consumed are more highly refined: white rice replaces brown, white flour

---

* This close relationship temporarily diverged at the end of World War II, when sugar rationing was relaxed. As Cleave noted, however, this coincided with the introduction of penicillin into clinical use to treat the infections that often kill adult diabetics. Diabetes management and control also improved dramatically with the development of the standard insulin syringe in 1944, and long-acting insulin two years later.

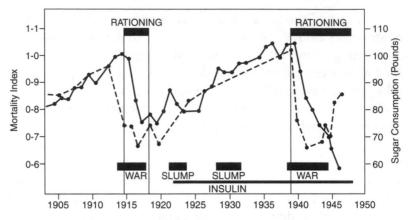

Peter Cleave's chart showing the relationship between diabetic mortality rate (with the 1938 rate equal to 1) and the amount of sugar consumed per capita in England and Wales. The dotted line is sugar consumption. The solid line is diabetes mortality.

replaces wholemeal; sugared beverages and candy spark a dramatic increase in sugar consumption. As a result, whenever investigators tested the hypothesis that chronic disease was caused by high fat intake or even high animal-fat intake or low carbohydrate intake, the refinement of the carbohydrates would *confound* the results. The changing-American-diet story led Ancel Keys and others to insist that fat caused heart disease and to advise eating low-fat, high-carbohydrate diets because, as the diagnosis of coronary heart disease increased over the century, carbohydrate consumption apparently decreased, while the total fat available for each American increased from 100 pounds per person per year to almost 130 pounds.* But the greatest single change in the American diet was in fact the spectacular increase in sugar consumption from the mid-nineteenth century onward, from less than 15 pounds a person yearly in the 1830s to 100 pounds by the 1920s and 150 pounds (including high-fructose corn syrup) by the end of the century. In effect, Americans replaced a good portion of the whole grains they ate in the nineteenth century with refined carbohydrates.

Despite the century of debate in the United Kingdom on the merits of white flour and wholemeal flour and the potential evils of sugar, it

---

* Although, as we noted earlier (p. 13), the amount of animal fat Americans ate decreased during this period, and so the increased total fat consumption was entirely due to the increased consumption of vegetable fats.

wouldn't be until the 1990s that epidemiologists began to delineate between refined and unrefined carbohydrates in their dietary analyses. Even in 1989, when the National Academy of Sciences published its seven-hundred-page *Diet and Health* report, the authors made little attempt to differentiate refined carbohydrates from unrefined, other than occasionally to note studies in which sugar intake by itself was studied.

When Keys linked the low-fat, high-carbohydrate diet of the Japanese in the late 1950s to the extremely low incidence of heart disease, he paid no attention to sugar consumption. Fat consumption in Japan was extremely low, as were heart-disease rates, and so he concluded that the lower the fat the better. But the consumption of sugars in Japan was very low, too—less than forty pounds per person per year in 1963, and still under fifty pounds in 1980—equivalent to the yearly per-capita consumption recorded in the United States or in the United Kingdom a century earlier.

The remarkable health of the islanders of Crete and Corfu in Keys's Seven Countries Study—and thus the supposedly salubrious effects of the Mediterranean diet itself—could also be explained by the lack of sugar and white flour. Despite the popularity of the Mediterranean diet today, our understanding of what exactly such a diet is—particularly in Crete and Corfu, where Keys's study had documented such remarkably low mortality rates—is based on only two dietary surveys: Keys's study itself, which analyzed the Cretan diet in 1960, and a Rockefeller Foundation study in 1947. According to the Seven Countries Study, the Cretan diet circa 1960 included a total of only sixteen pounds a year of sugar, honey, pastries, and ice cream. According to the Rockefeller study, the Cretan diet included only ten pounds a year of sugar and sweets, and the considerable bread consumed was *all* wholemeal. The reported benefits of the Mediterranean diet, therefore, could be attributed to the fish, olive oil, and vegetables consumed, as it is today, but they could also be due to the minimal quantities of sugar and the absence of white flour.

This lack of concern for any potential health-related difference between vegetables and starches, on the one hand, and refined starches and sugars, on the other, has haunted cancer research as well. Speculation that dietary fat caused breast, colon, and prostate cancer began in the 1970s, with the same international comparisons that led to the heart-disease hypothesis in the 1950s. Cancer epidemiologists simply compared carbohydrate, protein, and fat contents of diets in different countries with the mortality from various cancers. And these investigators, too, concluded that differences in cancer rates could be explained by differences in fat consumption and animal-fat consumption, particularly between Japan and the United States.

They did not serve science well by ignoring sugar consumption and the difference between refined and unrefined carbohydrates.

These preliminary studies then prompted hundreds of millions of dollars of studies that failed to confirm the initial hypothesis that fat or animal fat led to cancer. (Even in the past few years, similar studies have attributed rising cancer rates in China to the increased consumption of fat, while again paying no attention to sugar or the refinement of the carbohydrates in the diets.) In 1975, Richard Doll and Bruce Armstrong published a seminal analysis of diet and cancer, in which they noted that, the higher the sugar intake in different nations, the higher both the incidence of and mortality from cancer of the colon, rectum, breast, ovary, uterus, prostate, kidney, nervous system, and testicles.* Still, epidemiologists fixated on the fat-cancer hypothesis and made no attempt to measure the refined carbohydrates in the populations they studied. As a result, a joint 1997 report of the World Cancer Research Fund and the American Institute for Cancer Research, entitled *Food, Nutrition and the Prevention of Cancer,* said this:

> The degree to which starch is refined in diets, particularly when the intake of starch is high, may itself be an important factor in cancer risk, as may the volume of refined starches and sugars in diets. Epidemiological studies have not, however, generally distinguished between degrees of refining or processing of starches, and there are, as yet, no reliable epidemiological data specifically on the effects of refining on cancer risk.

Cleave's saccharine-disease hypothesis may be intuitively appealing, but it is effectively impossible to test without a randomized controlled trial. If Cleave was right, then epidemiologists comparing populations or individuals with and without chronic disease have to take into account not just sugar consumption but flour, and whether that flour is white or whole-grain, and whether rice is polished or unpolished, white or brown, and even how much beer is consumed compared with, say, red wine or hard liquor. They might have to distinguish between table sugar and the sugar in soft drinks and fruit juices. Just as fats are now divided into saturated, monounsaturated, and polyunsaturated (and, ideally, into the various subcategories, including stearic acid and oleic acid), carbohydrates have to be separated into subcategories as well. It would be easy, Cleave suggested, to gather together the twenty fattest people in any neighborhood and find

---

* There is even a plausible biological mechanism to explain how refined carbohydrates and sugars could cause or exacerbate cancer. See Chapter 13.

that "they wouldn't have a sweet tooth among them, and they wouldn't like sugar"—they would all be beer drinkers. "Beer is full of malt sugar and enormously fattening," he said.

It may have been these complications that led indirectly to a pared-down version of Cleave's hypothesis—one that would receive far more publicity—blaming coronary heart disease, diabetes, and other chronic diseases effectively on sugar alone. So said John Yudkin, who, unlike Cleave, was a prominent figure in the nutrition-research community. In 1953, he had founded the department of nutrition at Queen Elizabeth College in London, the first dedicated department of nutrition in Europe. In the late 1950s, Yudkin began advocating a very low-carbohydrate diet for weight loss and wrote a popular diet book, *This Slimming Business*. He believed starches and sugars brought nothing of nutritional importance to the diet except calories—sugar was the worst offender—and so they were the obvious nutrient to remove from a weight-loss diet.

Yudkin entered the heart-disease debate in 1957, after Keys published his first series of papers claiming a "remarkable relationship" between fat consumption and coronary heart disease. Yudkin was among those who had taken Keys to task for the limitations of his analysis and his overinterpretation of very limited and unreliable data. Yudkin noted that many factors correlate with heart-disease deaths (not just dietary-fat consumption) one of which happened to be sugar consumption. Yudkin paid attention only to the trends of diet and disease in developed nations, and to heart disease and obesity, rather than the whole slew of chronic diseases, and he decided that sugar itself was the fundamental problem. (Yudkin would distance himself from Cleave by refusing to use the term "refined carbohydrates," because it "gives the impression that white flour has the same ill effects as sugar," which he considered grossly misleading.) Through the 1960s, Yudkin published the results of a series of experiments implicating sugar in heart disease. He fed sugar and starch to rats, mice, chickens, rabbits, and pigs, and reported that the sugar, depending on the particular animal involved, raised some combination of cholesterol, triglycerides, and insulin levels. Triglycerides are a form of fat molecule found in the blood, and a series of researchers beginning with Pete Ahrens at Rockefeller University and Margaret Albrink of Yale had suggested that triglyceride levels were a better predictor of heart disease than was cholesterol. (Diabetics, as Joslin had noted, all too often died of atherosclerosis, and they, too, inevitably had high levels of triglycerides.) Yudkin also fed high-sugar diets to college students and reported that it raised their cholesterol and particularly their triglycerides; their insulin levels rose, and their blood cells became stickier, which he believed could explain the blood clots that seemed to precipitate heart attacks.

By the early 1970s, the medical-research community was taking Yudkin's hypothesis seriously. But now the very existence of Keys's hypothesis was the primary obstacle to the acceptance of Yudkin's. If one was right, than the other was very likely wrong. The European research community tended to be open-minded on this question. "Although there is strong evidence that dietary fats, particularly the saturated ones, play an important role in the etiology of [coronary heart disease], there is no proof that they are the only or the main culprit," wrote Roberto Masironi, an Italian cardiologist who would become president of the European Medical Association. "As regards the relationship of sugars to cardiovascular diseases, it must be borne in mind that these nutrients have common metabolic pathways with fats. Disturbances in carbohydrate metabolism may be responsible for abnormal fat metabolism and may therefore act as a causative factor in the development of atherosclerosis and of coronary disease."

In the United States, however, Keys's hypothesis ruled. Keys himself went after Yudkin in a letter that he first distributed widely to investigators in 1970, before it was published in the journal *Atherosclerosis*. Keys called Yudkin's arguments for the role of sugar in heart disease "tendentious" and his evidence "flimsy indeed"; he treated Yudkin as a figure of ridicule. What made Keys's critique so ironic, though, is that virtually every argument that he invoked to criticize Yudkin's hypothesis had been used in the past as criticism of his own. Most were equally valid for both, and spoke to the flaws in the epidemiologic evidence—the use of international cause-of-death statistics and food-consumption data or dietary-recall surveys to draw conclusions about cause and effect—rather than to the actual validity of the hypotheses. Keys's case against Yudkin eventually came to rest almost entirely on his invocation of the Seven Countries Study as support for *his* hypothesis. In fact, the Seven Countries Study had been one of the very few studies that had measured sugar consumption in its populations, and sugar indeed turned out to predict heart-disease rates as well as saturated fat did.

By the early 1970s, Keys's dietary-fat hypothesis of heart disease, despite the ambiguity of the evidence, was already being taught in textbooks and in medical schools as most likely true. After Yudkin retired in 1971, his hypothesis effectively retired with him. His university replaced him with Stewart Truswell, a South African nutritionist who was among the earliest to insist publicly that Keys's fat theory of heart disease was assuredly correct and that it was time to move on to modifying the diets of the public at large accordingly. Truswell believed it was more important for the prevention of heart disease to convince the public to eat more onions, for their reported ability to alter the "tendency to thrombosis," than to eat less sugar.

Yudkin spent his first year of retirement writing a book on his sugar theory, published in 1972 and entitled *Pure, White and Deadly* in England and *Sweet and Dangerous* in the American edition. It did not serve to move the medical-research community closer to embracing either Yudkin or his theory. By the late 1970s, to study the potentially deleterious effects of sugar in the diet, says Sheldon Reiser—who did just that at the U.S. Department of Agriculture's Carbohydrate Nutrition Laboratory in Beltsville, Maryland— and to talk about it publicly, was to endanger your reputation. "Yudkin was so discredited," says Reiser; "he was ridiculed in a way. And anybody else who said something bad about sucrose, they'd say, 'He's just like Yudkin.' "

Chapter Seven

FIBER

The thing is, it's very dangerous to have a fixed idea. A person with a fixed idea will always find some way of convincing himself in the end that he is right.
ATLE SELBERG, winner of the 1950
Fields Medal in Mathematics

T HE HYPOTHESIS THAT SUGAR AND refined carbohydrates cause chronic disease peaked as a subject of serious consideration in late April 1973, when George McGovern's Senate Select Committee on Nutrition and Human Needs held its first hearing on diet and what the committee took to calling killer diseases. The testimony would have little impact on the content of McGovern's *Dietary Goals for Americans,* in part because none of the staff members who organized the hearings would still be working for the committee three and a half years later, when the *Dietary Goals* would be drafted. Equally important, neither McGovern nor his congressional colleagues could reconcile what they were hearing from the assembled experts with what they had now come to believe about the nutritional evils of modern diets.

The committee had initially planned a series of hearings in 1972 on dietary fat, cholesterol, and heart disease, but the plans changed because McGovern ran for president. When the committee returned to the diet-and-chronic-disease issue after McGovern's defeat, the subject that seemed most urgent—thanks in part to the publication of John Yudkin's *Sweet and Dangerous*—was sugar in the diet, diabetes, and heart disease.

The hearings were a surprisingly international affair. Aharon Cohen from Jerusalem testified on diabetes and heart disease among the Yemenite Jews. George Campbell testified on his studies of diabetes in Zulus and Natal Indians in South Africa. Peter Bennett, an NIH epidemiologist, testified on the Pima Indians of Arizona, who had the highest incidence of diabetes ever recorded at the time: half of the Pima over thirty-five years old were diabetic. "The only question that I would have," Bennett said, "is whether we can implicate sugar specifically or whether the important factor is not calories in general, which in fact turns out to be really excessive amounts of carbohydrates." Walter Mertz, chairman of the USDA Human

Nutrition Institute, testified, as did his colleague Carol Berdanier, explaining that refined sugar seemed to play particular havoc with health, at least in laboratory rats. It elevated blood sugar and triglycerides, and caused subjects to become diabetic, Berdanier said, "and they die at a very early age."

When the testimony focused on sugar and diabetes, the committee members found it compelling. They occasionally solicited suggestions as to how Americans might reduce the 120-odd pounds of sugar they were eating on average in 1973, to the less than seventy pounds that Campbell said could be safely consumed without triggering an epidemic of diabetes and obesity.

With the subject of heart disease, however, controversy arrived. Cohen testified that there was no "direct relationship" linking heart disease to dietary fats, and that he had been able to induce the same blood-vessel complications seen in heart disease merely by feeding sugar to his laboratory rats. Peter Cleave testified to his belief that the problem extended to all refined carbohydrates. "I don't hold the cholesterol view for a moment," Cleave said, noting that mankind had been eating saturated fats for hundreds of thousands of years. "For a modern disease to be related to an old-fashioned food is one of the most ludicrous things I have ever heard in my life," said Cleave. "If anybody tells me that eating fat was the cause of coronary disease, I should look at them in amazement. But, when it comes to the dreadful sweet things that are served up . . . that is a very different proposition." Yudkin blamed heart disease exclusively on sugar, and he was equally adamant that neither saturated fat nor cholesterol played a role. He explained how carbohydrates and specifically sugar in the diet could induce both diabetes and heart disease, through their effect on insulin secretion and the blood fats known as triglycerides. McGovern now struggled with the difficulty of getting some consensus on these matters.

"Are you saying that you don't think a high fat intake produces the high cholesterol count?" McGovern asked Yudkin. "Or are you even saying that a person with high cholesterol count is not in great danger?"

"Well, I would like to exclude those rare people who have probably a genetic condition in which there is an extremely high cholesterol," Yudkin responded. "If we are talking about the general population, I believe both those things that you say. I believe that decreasing the fat in the diet is not the best way of combating a high blood cholesterol. . . . I believe that the high blood cholesterol in itself has nothing whatever to do with heart disease."

"That is exactly opposite what my doctor told me," said McGovern.

"If men define situations as real," the sociologist William Isaac Thomas observed in the 1920s, "they are real in their consequences." Embracing a hypothesis based on incomplete evidence or ideological beliefs is risk enough. But this also makes it extremely difficult to entertain alternative possibilities, unless we can reconcile them with what we have now convinced ourselves is indisputable.

By the early 1970s, all potential causes of heart disease, or potentially any chronic disease, had to be capable of coexisting with the belief that dietary fat was the primary cause of coronary heart disease. The notion that refined or easily digestible carbohydrates caused chronic disease could not be so reconciled.

The evidence that led Peter Cleave to propose this alternative theory—the disparity in disease rates among populations, the intimate relationship of atherosclerosis, hypertension, obesity, and diabetes, and the apparent absence of chronic disease in populations relatively free of Western influences—had to be explained in other ways if they were to be consistent with Keys's hypothesis. Fiber, the indigestible carbohydrates in vegetables, starches, and grains, now replaced refined carbohydrates and sugar in the debate about the nutritional causes of chronic diseases. The fiber hypothesis captured the public's nutritional consciousness by virtue of the messianic efforts of a single investigator, a former missionary surgeon named Denis Burkitt, who proposed that this indigestible roughage was a requisite component of a healthy diet. The notion was consistent with Keys's hypothesis, which was not the case with Cleave's or Yudkin's hypothesis, and it resonated also with the era's countercultural leanings toward diets heavy in vegetables, legumes, and cereal grains.

Burkitt's fiber hypothesis was based originally and in its entirety on Cleave's saccharine-disease hypothesis, but simply inverted the causal agent. Rather than proclaim, as Cleave did, that chronic disease was caused by the *addition* of sugar and refined carbohydrates to diets that we had evolved naturally to eat, Burkitt laid the blame on the *subtraction* of the fiber from those evolutionarily ideal diets, which in turn led to constipation and then, through a variety of mechanisms, all the chronic diseases of civilization. The fiber deficiency itself was caused either by the removal of fiber during the refining of carbohydrates or by the consumption of refined carbohydrates in lieu of the fibrous, bulky roughage we should be eating. The fiber hypothesis and the refined-carbohydrate hypothesis of chronic disease were photographic negatives of each other, and yet the fiber hypothesis caught on immediately upon appearing in the journals. The refined-carbohydrate hypothesis, which was the only one of the two that was capable of explaining the actual evidence, remained a fringe concept.

Denis Burkitt began his career as a missionary surgeon in Uganda in 1947. In the early 1960s, he earned his renown—"one of the world's best-known medical detectives," as the *Washington Post* would call him—for his studies of a fatal childhood cancer that came to be known as Burkitt's lymphoma and would be the first human cancer ever linked to a viral cause. That discovery alerted Burkitt to the lessons to be learned by tracking the geographical distribution of disease. Burkitt spent five years gathering information about the lymphoma from hundreds of African hospitals, and made a legendary ten-thousand-mile, sixty-hospital trek from Kampala to Johannesburg and back as part of his research.

In 1966, Burkitt returned to England, where he worked as a cancer epidemiologist for the Medical Research Council. There Richard Doll told him about Cleave and his saccharine-disease hypothesis. Burkitt met with Cleave and read *Diabetes, Coronary Thrombosis and the Saccharine Disease*, which he found revelatory. Cleave possessed "perceptive genius, persuasive argument and irrefutable logic," Burkitt wrote.

> What he was saying was that many of the common diseases in post-industrialized western countries are rare throughout the third world, were rare even in England or New York until about the First World War, are equally common in black and white Americans, and therefore must be due not to our skin color or our genes, but to the way we live. Now, this made an enormous amount of sense to me because I knew from my experience in Africa that he was perfectly right saying this.

On a tour of the United States, Burkitt visited hospitals and observed, as George Campbell had a decade earlier, that African-American patients in these hospitals were often obese, diabetic, or atherosclerotic, conditions virtually nonexistent among the black Ugandans Burkitt had treated.

Burkitt considered himself in the ideal position to test Cleave's hypothesis on a wider scale. He had already established a network of 150 African hospitals, mostly missionary hospitals in rural areas, that mailed him monthly reports on their cancer cases: "I was able to ask them all: 'Do you see gallstones, appendicitis, diverticular disease, coronary heart disease. . . .' " Burkitt also sent his questionnaire to mission hospitals throughout the world, and over eight hundred faithfully returned them. The results confirmed the basics of Cleave's hypothesis. Whereas Cleave had anecdotal evidence, Burkitt recalled, he now had "anecdotal multiplied by a thousand," and it was all consistent. Moreover, he had the necessary reputation to be taken seriously, whereas Cleave did not. Cleave, Campbell, and others had been "written off as cranks," Burkitt said. "Now, just because there happened to be a Burkitt's lymphoma, when Burkitt said, 'What about looking at this,' people listened to me when they hadn't listened to far better guys."

Through the early 1970s, Burkitt published a series of articles expanding on Cleave's hypothesis. "These 'western' diseases are certainly associated geographically and in many instances tend to be related to one another in individual patients," Burkitt wrote in the *Journal of the National Cancer Institute* in 1971. "My epidemiological studies in Africa and elsewhere substantiate Cleave's basic hypothesis. Changes made in carbohydrate food may of course be only one of many etiological factors, but in some instances they would appear to be the major one."

But Burkitt was beginning to revise Cleave's hypothesis. Now Burkitt's working assumption, as he explained in the *JNCI,* was that any dietary factors responsible for benign conditions such as appendicitis or diverticulitis were likely to be responsible as well for related malignant conditions—in particular, colon and rectal cancer. Burkitt's research had led him to Thomas Allinson, who in the 1880s argued that white flour caused constipation, hemorrhoids, and other ills of modern societies. It also led him to a 1920 article by the Bristol University surgeon Arthur Rendle Short, documenting a dramatic increase in the incidence of appendicitis that Rendle Short also blamed on white flour and the lack of fiber in modern diets. Burkitt believed he could draw a direct line of causation from the absence of fiber in refined carbohydrates to constipation, hemorrhoids, appendicitis, diverticulitis, polyps, and finally malignant colon and rectal cancer.

Burkitt's African correspondents had reported that appendicitis increased dramatically in urban populations—at Burkitt's Mulago Hospital in Kampala, the number of yearly appendectomies had increased twenty-fold from 1952 to 1969—whereas polyps, diverticular disease, and colorectal cancer, all common in the United States and Europe, wrote Burkitt, were still "very rare in Africa and almost unknown in rural communities." Burkitt concluded that appendicitis, just as it appeared in Western nations typically in children, appeared in Africans, both adults and children, within a few years of the adoption of Western diets.

Burkitt focused now on constipation. He theorized that removing the fiber from cereal grains would slow the "transit time" of the stool through the colon. Not only would any carcinogens in the stool therefore have more time to inflict damage on the surrounding cells, but it was conceivable that the overconsumption of refined carbohydrates would increase the bacterial flora of the stool, and that in turn could lead to carcinogens being metabolized by the bacteria out of "normal bowel constituents." Burkitt could offer no explanation for why this might cause appendicitis, but he was confident that some combination of all these factors played a role.

In the summer of 1969, Burkitt began studying stool characteristics in available subjects. "Finished bowel transit tests on family," he recorded in his diary on July 4. The following month, he visited Alec Walker, who ran the human biology department at the South African Institute of Medical Research. Walker had been studying the rising tide of chronic diseases in urban Bantus in South Africa since the late 1940s, and he was the rare investigator who shared with Burkitt an interest in human feces and constipation. Walker had done extensive studies linking the relative lack of constipation among black convicts in the local prison, as well as the lack of appendicitis in the Bantus at large, to their traditional high-fiber diets. (Walker publicly dismissed the hypothesis that sugar or refined carbohydrates caused heart disease, but he also reported that the Bantus developed chronic disease only after they moved into the city and began consuming "more white bread, sugar, soft drinks and European liquor.") Walker had also just submitted an article to the *British Medical Journal* linking the very low mortality rates from colon cancer among the Bantus to their bowel motility, a characteristic, he wrote, that was "largely lost" among Western societies. Walker's research gave Burkitt the confidence to devote his efforts to the study of stool characteristics and bowel behavior, hoping to associate in a scientific manner fiber deficiency, constipation, and the presence of chronic diseases.

It was precisely this work that led to the fiber hypothesis and its present place in our nutritional consciousness. In 1972, Burkitt and Walker published an article in *The Lancet* supporting their theory and discussing their measurements of transit time and stool characteristics in twelve hundred human subjects. In rural areas, unaffected by industrialization, they reported, "diets containing the natural amount of fiber are eaten and result in large, soft stools that traverse the intestine rapidly. By contrast, the refined low-fiber foods of the economically-developed countries produce small firm stools which pass through the gut very slowly." Thus, the relative constipation endemic in the developed world, they suggested, appeared to play a causative role in bowel-related disorders: appendicitis, diverticulitis, and both benign and malignant tumors of the colon and rectum, all of which showed the classic distribution of diseases of civilizations. "All these diseases are very closely associated epidemiologically," Burkitt and Walker explained. "These diseases are still rare in developing countries and in rural Japan, where eating habits have changed but little, but they are all seen increasingly in Japanese who live in Hawaii and California and are increasing in Japan in those who have changed to a Western diet. In no country or region is one of these diseases common and the others rare save that appendicitis, which afflicts the young, appears about a genera-tion before the other conditions."

Within two years, Burkitt had extended his hypothesis from appendici-tis, diverticulitis, and colon cancer to all chronic diseases of civilization. In the process, Cleave's refined-carbohydrate hypothesis of saccharine dis-eases was transformed into Burkitt's fiber hypothesis of Western diseases. This transformation of the causal agent of disease from the presence of carbohydrates to the absence of fiber may have been influenced by factors other than science—Burkitt's close association with Harold Himsworth in particular. Himsworth had been secretary of the Medical Research Council when Burkitt was hired, and he had been publicly effusive about Burkitt's contributions to modern medicine. It was Himsworth's research that had been responsible for convincing diabetologists that sugar and other carbo-hydrates were *not* the cause of diabetes. Indeed, Cleave and Campbell had presented their saccharine-disease theory in the context of diabetes as a refutation of Himsworth's scholarship as much as Joslin's. That Burkitt would find Cleave's general thesis compelling but the details unacceptable in light of Himsworth's own work and beliefs is quite possible. Burkitt would often tell the story of how Himsworth had convinced him of the importance of paying attention to those factors that were absent in search-ing for the causative agents of disease. "Denis," Burkitt recalled Himsworth telling him, "do you remember the story in Sherlock Holmes when Holmes said to Watson: 'The whole clue, as I see it, to this case lies in the behavior of the dog.' And Watson said: 'But, sir, the dog did nothing at all.' 'That,' said Holmes, 'is the whole point.' And it often is in medicine. . . . The clue can lie in what is not there rather than what is there." In this case, fiber was *not* there. Burkitt also seemed motivated by the simple expedi-ency of emphasizing the positive benefits of fiber rather than the negative effects of sugar and flour, which seemed like a hopeless cause. "[Sugar] is simply an integral part of the daily diet and emphatically is here to stay," Burkitt's collaborator Alec Walker said. Better to say *Don't Forget Fibre in Your Diet*, which was the title of Burkitt's 1979 diet book, than to say, Don't eat sugar, flour, and white rice, and drink less beer.

The final transformation of Cleave's refined-carbohydrate hypothesis into Burkitt's fiber hypothesis came primarily through the efforts of Burkitt's colleague Hugh Trowell, who had spent thirty years as a missionary physician in Kenya and Uganda, beginning in 1929. This had been a time, as Trowell later explained, when scores of British doctors working for the colonial service and missionary hospitals in the Kenyan highlands had the unprecedented experience of watching the native population of "three million men, women and children . . . emerge from pre-industrial life and undergo rapid westernization." After Trowell retired to England in 1959, he published *Non-infective Disease in Africa*, which was the first rigorous attempt to draw together the entire body of medical literature on the spectrum of diseases afflicting the native population of Africa.* The Western diseases—a list almost identical to Cleave's—were conspicuous by their absence.

Trowell's experiences in East Africa had left him with the characteristic awareness of the diseases-of-civilization phenomenon. When he arrived in Kenya in 1929, he said, he had noticed that the Kenyans were all as thin as "ancient Egyptians," yet when he dined with the native tribes, they always left food at the end of the meal and fed it to their domestic animals, which suggested that their relative emaciation was not caused by food shortages or insufficient calories. During World War II, according to Trowell, a team of British nutritionists was dispatched to East Africa to figure out how to induce the Africans in the British Army to put on weight, since they would not or could not do it. "Hundreds of x-rays," Trowell recalled, "were taken of African intestines in an effort to solve the mystery that lay in the fact that everyone knew how to fatten a chicken for the pot, but no one knew how to make Africans . . . put on flesh and fat for battle. It remained a mystery." Nonetheless, by the 1950s, fat Africans were a common sight, and in 1956 Trowell himself reported the first clinical diagnosis of coronary heart disease in a native East African—a Banting-esque high-court judge (five foot two and 208 pounds) who had lived in England and had been eating a Western diet for twenty years. In 1970, Trowell returned to East Africa and described what he saw as "an amazing spectacle: the towns were full of obese Africans and there was a large diabetic clinic in every city. The twin diseases had been born about the same time and are now growing together."

Burkitt and Trowell had been friends since the late 1940s, when Burkitt

---

* John Higginson, director of the World Health Organization's International Agency on Cancer Research, later described *Non-infective Diseases in Africa* as a "brilliant review" that had been "regrettably ignored."

first arrived in Uganda. In 1970 the two began working together on Burkitt's fiber hypothesis and a textbook on diseases of civilization, which Burkitt and Trowell now called "Western diseases."* To explain how obesity could be induced by the fiber deficiency of modern refined-carbohydrate foods, Trowell reasoned that the causal factor was an increased ratio of energy to nondigestible fiber in the Western diet. Ninety-three percent of the nutrients in a typical Western diet were available for use as energy, Trowell calculated, compared with only 88 or 89 percent of those in a typical primitive diet containing copious vegetables, fruits, and wholemeal bread. The lower figure, Trowell wrote, is "the figure that is the natural, inherited evolutionary figure." Over the course of a few decades, he said, we would unknowingly eat 4 percent more calories than would be evolutionary appropriate and therefore gain weight. (Later investigators would build on this idea by adding that fibrous foods were bulky, and thus more filling, and they also took longer to chew and digest, which supposedly led to an inevitable decrease in calories consumed, at least per unit of time.) As for heart disease, Trowell accommodated Keys's logic: if the relevant epidemiology suggested that a low-fat, high-carbohydrate diet protected against heart disease, then carbohydrates obviously protect against heart disease, with the critical caveat that those carbohydrates must contain "their full complement of dietary fiber." Those "partially depleted" of fiber provide only "partial protection," Trowell said; those fully depleted, sugar and white flour, offer no protection.

More attention would have been paid to Cleave's hypothesis, Trowell explained, had Cleave accepted the validity of Keys's research and "not dismissed completely the role of saturated animal fats" in heart disease. (Burkitt later said as much, too.) Trowell didn't make the same mistake. He accepted that diets rich in fat, especially saturated fat, raise cholesterol levels in the blood and so raise heart-disease risk, but then noted that the epidemiological evidence also implicated a low consumption of starchy high-fiber foods. So both fat and the absence of fiber could be blamed. (As Cleave and Yudkin had pointed out, exactly the same evidence can be used to implicate sugar and refined carbohydrates.)

Burkitt and Trowell called their fiber hypothesis a "major modification" of Cleave's ideas, but they never actually addressed the reasons why Cleave had identified refined carbohydrates as the problem to begin with: How to

---

* "The title Western diseases is preferred to that of the diseases of civilization," they explained, "for it proved obnoxious to teach African and Asian medical students that their communities had a low incidence of these diseases because they were uncivilized."

explain the absence of these chronic diseases in cultures whose traditional diets contained predominantly fat and protein and little or no plant foods and thus little or no fiber—the Masai and the Samburu, the Native Americans of the Great Plains, the Inuit? And why did chronic diseases begin appearing in these populations only with the availability of Western diets, if they weren't eating copious fiber prior to this nutrition transition? Trowell did suggest, as Keys had, that the experience of these populations might be irrelevant to the rest of the world. "Special ethnic groups like the Eskimos," he wrote, "adapted many millennia ago to special diets, which in other groups, not adapted to these diets, might induce disease." Trowell spent three decades in Kenya and Uganda administering to the Masai and other nomadic tribes, Burkitt had spent two decades there, and yet that was the extent of the discussion.

Unlike the reaction to Cleave's hypothesis, which garnered little attention even after Cleave testified to McGovern's Select Committee, the media pounced on the fiber hypothesis almost immediately. After Trowell published a pair of articles on fiber and heart disease in *The American Journal of Clinical Nutrition* in 1972, Robert Rodale, a nationally syndicated columnist, wrote a series of articles on the research, touting fiber as the answer to heart disease and obesity. Rodale was president of Rodale Press and the Rodale Institute, both dedicated to furthering the cause of organic foods and chemical- and pesticide-free "regenerative" agriculture. Rodale saw Burkitt and Trowell's fiber hypothesis as validation of the wisdom of organic foods and the agrarian lifestyle. "The natural fiber in whole processed foods may be instrumental in keeping cholesterol levels low and preventing the onset of heart disease," he wrote.

Burkitt and Alec Walker followed up Trowell's articles with an August 1974 review in *The Journal of the American Medical Association* discussing the causal chain from fiber to constipation and "changes in gastrointestinal behavior" to the entire spectrum of Western diseases. The *Washington Post* wrote up the *JAMA* article on the day of its release, calling fiber "the tonic for our time." That December, *Reader's Digest* published an article on Burkitt and the fiber hypothesis; a year later, the magazine claimed that sales of fiber-rich products had more than doubled since the article. The breakfast-cereal industry, led by Kellogg and General Foods, immediately started pushing bran and fiber as inherent heart-healthy aspects of their products. In 1975, Burkitt and Trowell published a book, *Refined Carbohydrate Foods and Disease*.

Burkitt then spent the next decade lecturing on the dangers of fiber-poor diets. He would condemn modern diets equally for their "catastrophic drop in starch," for their high fat content—"We eat three times more fat than communities with a minimum prevalence of [Western] diseases," he would say; "We must reduce our fat!"—and for their lack of fiber, which he considered "the biggest nutritional catastrophe in [the United Kingdom] in the past 100 years."

Not everyone bought into it. For public-health authorities and health reporters, dietary fat and/or cholesterol continued to be the prime suspects in chronic disease, and dietary fat had already been linked through international comparisons to colon cancer, as well as breast cancer. Burkitt recalled memorable disputes with researchers in the United States who blamed colon cancer on dietary fat, but he insisted that the absence of fiber was responsible. Eventually, they compromised. His opponents, said Burkitt, conceded "that the fact that fat happened to be causative . . . did not preclude the possibility that fiber might be protective." Harvard nutritionist Jean Mayer also discounted the significance of fiber, after Burkitt, Walker, and Trowell's early papers sparked the "furor over fiber" in the United States. But then Mayer, too, saw the wisdom of compromise. The ideal diet, he noted, would minimize the risk of both heart disease and cancer. It would be low in fat, or at least low in saturated fat, and so would be low in meat and dairy products. And it would be high in fiber. "A good diet," Mayer wrote, "high in fruits and vegetables and with a reasonable amount of undermilled cereals—will give all you need of useful fiber." The assumption that it would lead to long life and good health, however, was based more on faith and intuition than on science.

Over the last quarter-century, Burkitt's fiber hypothesis has become yet another example of Francis Bacon's dictum of "wishful science"—there has been a steady accumulation of evidence refuting the notion that a fiber-deficient diet causes colon cancer, polyps, or diverticulitis, let alone any other disease of civilization. The pattern is precisely what would be expected of a hypothesis that simply isn't true: the larger and more rigorous the trials set up to test it, the more consistently negative the evidence. Between 1994 and 2000, two observational studies—of forty-seven thousand male health professionals and the eighty-nine thousand women of the Nurses Health Study, both run out of the Harvard School of Public Health—and a half-dozen randomized control trials concluded that fiber consumption is unrelated to the risk of colon cancer, as is, apparently, the consumption of fruits and vegetables. The results of the forty-nine-thousand-women Dietary Modification Trial of the Women's Health Initiative, published in 2006, confirmed that increasing the fiber in the diet (by eating more whole grains, fruits, and vegetables) had no beneficial effect on colon cancer, nor did it prevent heart disease or breast cancer or induce weight loss.

"Burkitt's hypothesis got accepted pretty well worldwide, quite quickly, but it has gradually been disproved," said Richard Doll, who had endorsed the hypothesis enthusiastically in the mid-1970s. "It still holds up in relation to constipation, but as far as a major factor in the common diseases of the developed world, no, fiber is not the answer. That's pretty clear."

As we have seen with other hypotheses, the belief that dietary fiber is an intrinsic part of any healthy diet has been kept alive by factors that have little to do with science: in particular, by Geoffrey Rose's philosophy of preventive medicine—that if a medical hypothesis has a chance of being true and thus saving lives, it should be treated as if it is—and by the need to give the public some positive advice about how they might prevent or reduce the risk of cancer. This was immediately evident in a *New England Journal of Medicine* editorial that accompanied back-to-back April 2000 reports on two major trials—one on fourteen hundred subjects of the Phoenix [Arizona] Colon Cancer Prevention Physicians' Network, and one $30 million trial from the National Cancer Institute—both of which confirmed that fiber had no effect on colon cancer. The editorial was written by Tim Byers, a professor of preventive medicine at the University of Colorado, who said that the two trials had been short-term and focused only on the early stages of cancer. For this reason, they should *not* be interpreted as "evidence that a high-fiber cereal supplement or a low-fat high-fiber diet is not effective in protecting against the later stages of development of colorectal cancer." Byers was wrong, in that the results certainly *were* evidence that a high-fiber diet would not protect against the later stages of colorectal cancer; they simply weren't *sufficient* evidence for us to accept the conclusion wholeheartedly as true.

Burkitt's hypothesis lived on, and it would continue to live, as the fat/breast-cancer hypothesis continued to live on, in part because the original data that led to it remained unexplained: "Observational studies around the world," wrote Byers, "continue to find that the risk of colorectal cancer is lower among populations with high intakes of fruits and vegetables and that the risk changes on adoption of a different diet, but we still do not understand why." It would always be possible to suggest, as Byers had, that the trials could have been done differently—for longer or shorter duration, on younger subjects or older subjects, with more, less, or maybe a different kind of dietary fiber—and that the results would have been more promising. The American Cancer Society and the National Cancer Institute continued to suggest that high-fiber diets, high in fruits and vegetables, might reduce the risk of colon cancer, on the basis that some evidence existed to support the hypothesis and so a prudent diet would still include these ingredients.

The media would also contribute to keeping the fiber hypothesis alive, having first played a significant role in transforming Burkitt's hypothesis into dogma without benefit of any meaningful long-term clinical trials. "Scientists have known for years that a diet rich in vegetables, fruits and fiber, and low in fat, can greatly reduce—or eliminate—the chances of developing colon cancer," as a 1998 *Washington Post* article put it—four years *after* the Harvard analysis of forty-seven thousand male health professionals suggested it was not true.

Although the *New York Times* ran articles on the negative results from the Nurses Health Study (by Sheryl Gay Stolberg) and the Phoenix and NCI fiber trials (by Gina Kolata), neither was written by the two reporters who had followed the subject for decades and traditionally wrote about diet and health for the paper: Jane Brody, who wrote the *Times* personal-health column, and Marian Burros, who had begun endorsing the benefits of fiber as a *Washington Post* reporter in the 1970s and had joined the *Times* in 1981. Rather, Burros and Brody chose to respond to the negative news about Burkitt's hypothesis by continuing to defend it with the fallback position that it still might be true in other ways. "If preventing colon cancer was the only reason to eat fiber," wrote Jane Brody after the publication of the Phoenix and NCI studies, "I would say you could safely abandon bran muffins, whole-grain cereals, beans and peas and fiber-rich fruits and vegetables and return to a pristine diet of pasty white bread. But dietary fiber . . . has myriads of health benefits." After Stolberg's 1999 report on the Nurses Health Study, the *Times* published an article by Brody entitled "Keep the Fiber Bandwagon Rolling, for Heart and Health," pointing out that fiber was certainly good for constipation and that earlier results from the Nurses Health Study had suggested that women who ate "a starchy diet that was low in fiber and drank a lot of soft drinks developed diabetes at a rate two and a half times greater than women who ate less of these foods." This, according to Brody, constituted the motivation to keep fiber in a healthy diet.

Five days after Kolata's article on the negative results from the Phoenix and NCI trials, the *Times* published an explanatory article by Kolata— "Health Advice: A Matter of Cause, Effect and Confusion"—in which she discussed why the public had come to be misled on the benefits of fiber. She suggested that one reason was the loose use of language: "Scientists and the public alike use words like 'prevents' and 'protects against' and 'lowers the risk of' when they are discussing evidence that is suggestive, and hypothesis-generating, as well as when they are discussing evidence that is as firm as science can make it." Burkitt's fiber hypothesis, she said, had been based on hypothesis-generating data—international comparisons, in particular—and had then been refuted by the best studies science could do. "Yet even in the aftermath of the high-fiber diet studies," Kolata noted, "researchers were speaking confidently about other measures people could take to 'prevent' colon cancer, like exercising and staying thin. And they were saying that there were reasons to keep eating fiber because it could 'reduce the risk' of heart disease. When asked about the evidence for these statements, the researchers confessed that it was, of course, the lower level hypothesis-generating kind."

The very next day, the *Times* ran an article by Burros entitled "Plenty of Reasons to Say, 'Please Pass the Fiber,' " in which she suggested, based on what Kolata would have called "hypothesis-generating data," that eating fiber "significantly" lowers the risk of heart attack in women, and that "fiber is also useful in preventing the development of diabetes," "helps control obesity," and "may also be useful in reducing hypertension." Less than a month later, Brody followed with an article entitled "Vindication for the Maligned Fiber Diet," noting that, although fiber had "been knocked around a bit lately, after three disappointing studies failed to find that a high-fiber diet helped to prevent colon cancer," a recent study published in the *New England Journal of Medicine* of *thirteen* subjects followed for six weeks suggested it helped them to better control their diabetes and so should be eaten on that basis. "Since diabetes greatly increases a person's risk of developing heart disease and other disorders caused by fat-clogged arteries," Brody wrote, "the results of this study are highly significant to the 14 million Americans with Type 2 diabetes." By 2004, Brody was advocating high-fiber diets solely for their alleged ability, untested, to induce long-term weight loss and weight maintenance. In effect, fiber had now detached itself from its original hypothesis and existed in a realm always a step beyond what had been tested. Cleave's hypothesis that refined carbohydrates and sugars were the problem, the single best explanation for the original data, had been forgotten entirely.

Chapter Eight

THE SCIENCE OF THE
CARBOHYDRATE HYPOTHESIS

> Forming hypotheses is one of the most precious faculties of the human mind and is necessary for the development of science. Sometimes, however, hypotheses grow like weeds and lead to confusion instead of clarification. Then one has to clear the field, so that the operational concepts can grow and function. Concepts should relate as directly as possible to observation and measurements, and be distorted as little as possible by explanatory elements.
>
> MAX KLEIBER, *The Fire of Life:*
> *An Introduction to Animal Energetics,* 1961

AFTER THE UNITED STATES EXPLORATION EXPEDITION under Captain Charles Wilkes visited the Polynesian atolls of Tokelau in January 1841, the expedition's scientists reported finding no evidence of cultivation on the atolls, and confessed their surprise that the islanders could thrive on a diet composed primarily of coconuts and fish. Tokelau came under the administration of New Zealand in the mid-1920s, but the atolls remained isolated, visited only by occasional trading ships from Samoa, three hundred miles to the north. As a result, Tokelau lingered on the fringes of Western influence. The staples of the diet remained coconuts, fish, and a starchy melon known as breadfruit (introduced in the late nineteenth century) well into the 1970s. More than 70 percent of the calories in the Tokelau diet came from coconut; more than 50 percent came from fat, and 90 percent of that was saturated.

By the mid-1960s, the population of Tokelau had grown to almost two thousand and the New Zealand government, concerned about the threat of overpopulation, initiated a voluntary migration program during which more than half the Tokelauans moved to the mainland. From 1968 to 1982, a team of New Zealand anthropologists, physicians, and epidemiologists led by Ian Prior took the opportunity to study the health and diet of the emigrants as they resettled, as well as those who remained behind on the islands as their diets were progressively Westernized. This Tokelau Island Migration Study (TIMS) was a remarkably complete survey of the health and diet of all men, women, and children of Tokelauan ancestry. It was also quite likely the most comprehensive migration study ever carried out in the history of nutrition-and-chronic-disease research.

On Tokelau, the primary changes during the course of the study came in the mid-1970s, with the establishment of a cash economy and trading posts on the atolls. The year-round availability of imported foods led to a decrease in coconut consumption to roughly half of all calories. This was offset by a sevenfold increase in sugar consumption* and a nearly sixfold increase in flour consumed—from twelve pounds per person annually to seventy pounds. The islanders also began eating canned meats and frozen foods, which they stored in freezers donated by the United Nations; by 1980, six pounds of mutton per capita, three pounds of chicken backs, and five pounds of tinned corned beef had been consumed. (In comparison, 270 pounds of fish were caught per islander in 1981.) By then, the trading ships were also delivering annually some eighteen pounds per person of crackers, biscuits, and Twisties, a cheese-flavored corn snack. Smoking increased dramatically, as did alcohol consumption.

Through the 1960s, the only noteworthy health problems on the islands had been skin diseases, asthma, and infectious diseases such as chicken pox, measles, and leprosy. (Modern medical services and a trained physician had been available in Tokelau since 1917.) In the decades that followed, diabetes, hypertension, heart disease, gout, and cancer appeared. This coincided with a *decrease* in cholesterol levels, consistent with the decrease in saturated-fat consumption. Average weights increased by twenty to thirty pounds in men and women. A similar, albeit smaller, trend was seen in Tokelauan children. The only conspicuous departure from these trends was in 1979, when the chartered passenger-and-cargo ship *Cenpac Rounder* ran aground and the islanders went five months without a food or fuel delivery. "There was no sugar, flour, tobacco and starch foods," reported the *New Zealand Herald,* "and the atoll hospitals reported a shortage of business during the enforced isolation. It was reported that the Tokelauans had been very healthy during that time and had returned to the pre-European diet of coconut and fish. Many people lost weight and felt very much better including some of the diabetics."

As for the migrants to New Zealand, the move brought "immediate and extensive changes" in diet: bread and potatoes replaced breadfruit, meat replaced fish, and coconuts virtually vanished from the diet. Fat and saturated-fat consumption dropped, to be replaced once again by carbohydrates, "the difference being due to the big increase in sucrose consump-

---

* According to records from the local trading ships, this increase was nearly tenfold between 1961 and 1980: from seven pounds per person per year to sixty-nine pounds.

tion." This coincided with an almost immediate increase in weight and blood pressure, and a decrease in cholesterol levels—all more pronounced than the increases witnessed on Tokelau. Hypertension was twice as common among the migrants as among the Tokelauans who remained on the islands. The migrants also had an "exceptionally high incidence" of "diabetes, gout, and osteoarthritis, as well as hypertension." Electrocardiographic evidence suggested that the "migrants were at higher risk for coronary heart disease than were non-migrants."

A number of factors combined to make this higher disease incidence among the migrants difficult to explain. For one thing, the Tokelauans who emigrated smoked fewer cigarettes than those who remained on the atolls, so tobacco was unlikely to explain this pattern of disease. The migrants tended to be younger, too, which should have led to the appearance of less chronic disease on the mainland. And though the weights of the Tokelauan migrants were "substantially higher" than those of the atoll-dwellers and, "in fact, obesity became a problem for some," the migrant lifestyle was definitively the more rigorous of the two. The men worked in the forest service and casting shops of the railway; the women worked in electrical-assembly plants or clothing factories, or they cleaned offices during the evening hours, and they walked "some distance to and from the shops with their purchases." Finally, the original Tokelauan diet had been remarkably high in fat and saturated fat, but the migrants consumed considerably less of both. If Keys's hypothesis was correct, the migrants should have manifested less evidence of heart disease, not more.

In fact, the migrant experience had led to an increased incidence over the entire spectrum of chronic diseases. Prior and his colleagues acknowledged that their data made this difficult to explain in any simple manner. They suggested "that a different set of relevant variables might account for observed differences in incidence." Excess weight, whatever the cause, could explain at least part of the increased incidence of hypertension, diabetes, coronary heart disease, and gout among the migrants. They appeared to get more salt in their diets than the islanders did, so that might also explain the increased incidence of hypertension, as might the stress of assimilating to a new culture. The red meat consumed on the mainland might have contributed to the increased incidence of gout as well. The greater incidence of asthma could be explained by the presence of allergens in New Zealand that were absent in Tokelau.

As in the Tokelau study, the dominant approach over the past fifty years toward understanding the chronic diseases of civilization has been to assume that they are only coincidentally related, that each disease has its unique causal factors associated with the Western diet and lifestyle, although dietary fat, saturated fat, serum cholesterol, and excess weight invariably remain prime suspects.

The less common approach to this synchronicity of diseases has been to assume, as Peter Cleave did, that related diseases have related or common causes; that they are manifestations of a single underlying disorder. Cleave called it the saccharine disease because he believed sugar and other refined carbohydrates were responsible. By this philosophy, if diabetes, coronary heart disease, obesity, gout, and hypertension appear simultaneously in populations, as they did in the Tokelauan experience, and are frequently found together in the same patients, then they are very likely to be manifestations of a single underlying pathology. If nothing else, Cleave argued, this common-cause hypothesis was the simplest possible explanation for the evidence, and thus the one that should be presumed true until compelling evidence refuted it. This was Occam's razor, and it should be the guiding principle of all scientific endeavors.

In the early 1950s, clinical investigators began to characterize the physiological mechanisms that would underlie Cleave's saccharine-disease hypothesis of chronic disease, and that could explain the appearance of diseases of civilization going back over a century—the basis, in effect, of this carbohydrate hypothesis. The research evolved in multiple threads that resulted in some of the most fundamental discoveries in heart-disease and diabetes research. Only in the late 1980s did they begin to come together, when the Stanford diabetologist Gerald Reaven proposed the name Syndrome X to describe the metabolic abnormalities common to obesity, diabetes, and heart disease, all, at the very least, exacerbated by the consumption of sugar, flour, and other easily digestible carbohydrates. Syndrome X included elevated levels of the blood fats known as triglycerides; low levels of HDL cholesterol, now known as the good cholesterol; it included hypertension, and three phenomena that are considered precursors of adult-onset diabetes—chronically high levels of insulin (hyperinsulinemia), a condition known as *insulin resistance* (a relative insensitivity of cells to insulin), and the related condition of glucose intolerance (an inability to metabolize glucose properly). Over the years, other abnormalities have been added to this list: the presence of predominantly small, dense LDL particles, and high levels of a protein called fibrinogen that increases the likelihood of blood-clot formation. Elevated uric-acid concentrations in the blood, a precursor of gout, have been linked to Syndrome X, as has a state of chronic inflammation, marked by a high concentration in the blood of a protein known as C-reactive protein.

In the last decade, Syndrome X has taken on a variety of names as authorities, institutions, and associations have slowly come to accept its validity. It is often referred to as *insulin resistance syndrome.* The National Heart, Lung, and Blood Institute belatedly recognized the existence of Syndrome X in 2001, calling it *metabolic syndrome.* It has even been referred to as *insulin resistance/metabolic syndrome X,* or *MSX,* by those investigators attempting to cover all bases.* By any name, this metabolic syndrome is as much a disorder of carbohydrate metabolism as is adult-onset diabetes, and is certainly a consequence of the carbohydrate content of the diet, particularly, as Cleave would have predicted, such refined, easily digestible carbohydrates as sugar and white flour.

It wasn't until the late 1990s that the evolving science of metabolic syndrome began to have any significant influence outside the field of diabetes, at which point the media finally began to take notice.† The potential implications of metabolic syndrome for heart disease and other chronic diseases have only just begun to be appreciated by the research community. As a result, a hypothesis that emerged from research in the 1950s as an alternative explanation for the high rates of heart disease in Western nations has been accepted by medical researchers and public-health authorities a half-century later as a minor modification to Keys's dietary-fat/cholesterol hypothesis, even though this alternative hypothesis implies that Keys's hypothesis is wrong. The bulk of the science is no longer controversial, but its potential significance has been minimized by the assumption that saturated fat is still the primary evil in modern diets.

The Tokelau experience stands as an example. The current accepted explanation for the pattern of disease among the Tokelauans is that the increased sugar and flour in their diets caused metabolic syndrome, and in turn heart disease and diabetes, at least according to Scott Grundy, who is a nutritionist and specialist in the metabolism of blood lipids at the University of Texas Southwestern Medical Center and the primary author of the 2003 cholesterol guidelines published by the National Cholesterol Education Program (NCEP). This does not mean, however, that Grundy

---

* Although Reaven deserves much of the credit for identifying the syndrome and compelling the diabetes and heart-disease research communities to take notice, I will refer to it as metabolic syndrome, because that is now the preferred public-health terminology, rather than Syndrome X, except when discussing Reaven's work in particular.

† The first time the *Washington Post* mentioned metabolic syndrome or Reaven's research was in 1999, in an article about popular weight-loss diets. The second time was in 2001, in an article that actually discussed metabolic syndrome as a risk factor for heart disease. By that time, the paper had published a couple of thousand articles that at least touched on the issue of cholesterol and heart disease.

believes that Cleave's saccharine-disease hypothesis of chronic disease is correct, or that Keys was incorrect. Rather, as he explained it, in the United States the situation was less straightforward than in Tokelau. "What you're faced with," Grundy said, "is a historical change in people's habits. Going back to the 1940s, '50s, and '60s, people ate huge amounts of butter and cheese and eggs, and they had very high LDL levels [the "bad cholesterol"] and they had severe heart disease early in life, because of such high choles- terol levels. What's happened since then is, there has been a change in population behavior, and they don't consume such high quantities of satu- rated fat and cholesterol anymore, and so LDL has come down a great deal as our diets have changed. But now . . . we have got obesity, and most of the problem is due to higher carbohydrate consumption or higher total calories. And so we're switching more to metabolic syndrome."

Grundy's explanation is a modern version of the changing-American- diet story, in this case invoked as a rationale to explain how metabolic syn- drome could be the primary cause of heart disease today, while Keys's hypothesis could still be correct, but no longer particularly relevant to our twenty-first-century health problems. Grundy's explanation allows both Keys and Cleave to be right—by suggesting that their hypotheses addressed two different but relevant nutrition transitions—and therefore does not require that we question the credibility of our public-health authorities. His explanation might be valid, but it relies on a number of disputable assumptions and a selective interpretation of the evidence. It could also be true that we faced very much the same problem fifty years ago that we do today, and that a continuing accumulation of evidence exonerates the fats in the diet and incriminates refined, easily digestible carbohydrates and starches instead. The implications are profound.

The appropriate response to any remarkable proposition in science is extreme skepticism, and the carbohydrate hypothesis of chronic disease offers no exception. But looking at the hypothesis in the context of a con- cept called *homeostasis*, which is of fundamental importance for under- standing the nature of living organisms, gives us great insight. Much of the progress in physiology in the mid-twentieth century could be described as the transferral of this "concept of the nature of the wholeness," as the Nobel Laureate chemist Hans Krebs suggested in 1971, "from the realm of philosophy and theory of knowledge to that of biochemical and physiolog- ical experimentation." Though physiologists were aware of this paradig- matic shift, clinical investigators studying chronic disease have paid little attention, which means that the greater implications of the fundamental idea of homeostasis have been slighted.

In the mid-nineteenth century, the legendary French physiologist

Claude Bernard observed that the fundamental feature of all living organisms is the interdependence of the parts of the body to the whole. Living beings are a "harmonious ensemble," he said, and so all physiological systems have to work together to assure survival. The prerequisite for this survival is that we maintain the stability of our internal environment, the *milieu intérieur,* as Bernard famously phrased it—including a body temperature between 97.3°F and 99.1°F and a blood-sugar level between 70 mg/dl and 170 to 180 mg/dl—regardless of external influences. "All the vital mechanisms, however varied they may be," Bernard wrote, "have only one object, that of preserving constant the conditions of life in the internal environment." (As the British biologist J.B.S. Haldane noted a half-century later, "No more pregnant sentence was ever framed by a physiologist.") And this stability of the *milieu intérieur* is accomplished, Bernard said, by a continual adjustment of all the components of this living ensemble "with such a degree of perfection that external variations are instantly compensated and equilibrated."

In 1926, Bernard's concept was reinvented as homeostasis by the Harvard physiologist Walter Cannon, who coined the term to describe what he called more colloquially "the wisdom of the body." "Somehow the unstable stuff of which we are composed," Cannon wrote, "had learned the trick of maintaining stability." Although "homeostasis" technically means "standing the same," both Cannon and Bernard envisioned a concept more akin to what systems engineers call a dynamic equilibrium: biological systems change with time, and change in response to the forces acting on them, but always work to return to the same equilibrium point—the roughly 98.6°F of body temperature, for instance. The human body is perceived as a fantastically complex web of these interdependent homeostatic systems, maintaining such things as body temperature, blood pressure, mineral and electric-charge concentration (pH) in the blood, heartbeat, and respiration, all sufficiently stable so that we can sail through the moment-to-moment vicissitudes of the outside world. Anything that serves to disturb this harmonic ensemble will evoke instantaneous compensatory responses throughout that work to return us to dynamic equilibrium.

All homeostatic systems, as Bernard observed, must be amazingly interdependent to keep the body functioning properly. Maintaining a constant body temperature, for example, is critical because biochemical reactions are temperature-sensitive—they will proceed faster in hotter temperatures and slower in colder ones. But not all biochemical reactions are equally sensitive, so their *rates of reaction* will not change equally with changes in temperature. A biological system like ours that runs ideally at

98.6°F can spin out of control when this temperature changes and all the myriad biochemical reactions on which it depends now proceed at different rates. Our body temperature is the product of the heat released from the chemical reactions that constitute our metabolism. It is balanced in turn by the cooling of our skin in contact with the outside air. On cold days, we will metabolically compensate to generate more heat, and so more of the calories we consume go to warming our bodies than they would on hot days. Thus, the ambient temperature immediately affects, among other things, the regulation of blood-sugar and of carbohydrate and fat metabolism. Anything that increases body heat (like exercise or a hot summer day) will be balanced by a reduction of heat generated by the cells, and so there is a decrease in fuel use by the cells. It will also be balanced by dehydration, increased sweating, and the dilation of blood vessels near the surface of the skin. These, in turn, will affect blood pressure, so another set of homeostatic mechanisms must work, among other things, to maintain a stable concentration of salts, electric charge, and water volume. As the volume of water in and around the cells decreases in response to the water lost from sweating or dehydration, our bodies respond by limiting the amount of water the kidneys excrete as urine and inducing thirst, so we drink water and replenish what we've lost. And so it goes. Any change in any one homeostatic variable results in compensatory changes in all of them.

This whole-body homeostasis is orchestrated by a single, evolutionarily ancient region of the brain known as the hypothalamus, which sits at the base of the brain. It accomplishes this orchestral task through modulation of the nervous system—specifically, the autonomic nervous system, which controls involuntary functions—and the endocrine system, which is the system of hormones. The hormones control reproduction, regulate growth and development, maintain the internal environment—i.e., homeostasis— and regulate energy production, utilization, and storage. All four functions are interdependent, and the last one is fundamental to the success of the other three. For this reason, all hormones have some effect, directly or indirectly, on fuel utilization and what's known technically as *fuel partitioning*, how fuel is used by the body in the short term and stored for the long term. Growth hormone, for example, will stimulate the mobilization of fat from fat cells to use as energy for cell repair and tissue growth.

All other hormones, however, are secondary to the role of insulin in energy production, utilization, and storage. Historically, physicians have viewed insulin as though it has a single primary function: to remove and store away sugar from the blood after a meal. This is the most conspicuous function impaired in diabetes. But the roles of insulin are many and diverse. It is the primary regulator of fat, carbohydrate, and protein metabolism; it regulates the synthesis of a molecule called glycogen, the form in which glucose is stored in muscle tissue and the liver; it stimulates the synthesis and storage of fats in fat depots and in the liver, and it inhibits the release of that fat. Insulin also stimulates the synthesis of proteins and of molecules involved in the function, repair, and growth of cells, and even of RNA and DNA molecules, as well.

Insulin, in short, is the one hormone that serves to coordinate and regulate everything having to do with the storage and use of nutrients and thus the maintenance of homeostasis and, in a word, life. It's all these aspects of homeostatic regulatory systems—in particular, carbohydrate and fat metabolism, and kidney and liver functions—that are malfunctioning in the cluster of metabolic abnormalities associated with metabolic syndrome and with the chronic diseases of civilization. As metabolic syndrome implies, and as John Yudkin observed in 1986, both heart disease and diabetes are associated with a host of metabolic and hormonal abnormalities that go far beyond elevations in cholesterol levels and so, presumably, any possible effect of saturated fat in the diet.

This suggests another way to look at Peter Cleave's saccharine-disease hypothesis, or what I'll call, for simplicity, the carbohydrate hypothesis of chronic disease. As Cleave pointed out, species need time to adapt fully to changes in their environment—whether shifts in climate, the appearance of new predators, or changes in food supply. The same is true of the internal environment of the human body—Bernard's *milieu intérieur*. By far the most dramatic change to this internal environment over the past two million years is due to the introduction of diets high in sugar and refined and other easily digestible carbohydrates. Blood-sugar levels rise dramatically after these meals; insulin levels rise in response and become chronically elevated—hyperinsulinemia—and tissues become resistant to insulin. And because half of every molecule of table sugar (technically known as *sucrose*) is a molecule of the sugar known as *fructose*, which is found naturally only in small concentrations in fruits and some root vegetables, the human body has also been confronted with having to adjust to radically large amounts of fructose. In this sense, all of the abnormalities of metabolic syndrome and the accompanying chronic diseases of civilization can be viewed as the dysregulation of homeostasis caused by the repercussions throughout the body of the blood-sugar, insulin, and fructose-induced changes in regulatory systems. (As the geneticist James Neel wrote in 1998 about adult-onset diabetes, "The changing dietary patterns of Western civilization had compromised a complex homeostatic mechanism.")

It's possible that obesity, diabetes, heart disease, hypertension, and the other associated diseases of civilization all have independent causes, as the conventional wisdom suggests, but that they serve as risk factors for each other, because once we get one of these diseases we become more susceptible to the others. It's also possible that refined carbohydrates and sugar, in particular, create such profound disturbances in blood sugar and insulin that they lead to disturbances in mechanisms of homeostatic regulation and growth throughout the entire body.

Any assumptions about regulatory mechanisms and disease, as Claude Bernard explained, have to be understood in the context of the entire harmonic ensemble. "We really must learn, then, that if we break up a living organism by isolating its different parts, it is only for the sake of ease in experimental analysis, and by no means in order to conceive them separately," Bernard wrote. "Indeed when we wish to ascribe to a physiological quality its value and true significance, we must always refer it to this whole, and draw our final conclusion only in relation to its effects in the whole." When Hans Krebs paraphrased this lesson a century later, he said that if we neglect "the wholeness of the organism—we may be led, even if we experimented skillfully, to very false ideas and very erroneous deductions."

Perhaps the simplest example of this kind of erroneous deduction is the common assumption that the cause of high blood pressure and hypertension is excess salt consumption.

Hypertension is defined technically as a systolic blood pressure higher than 140 and a diastolic blood pressure higher than 90. It has been known since the 1920s, when physicians first started measuring blood pressure regularly in their patients, that hypertension is a major risk factor for both heart disease and stroke. It's also a risk factor for obesity and diabetes, and the other way around—if we're diabetic and/or obese, we're more likely to have hypertension. If we're hypertensive, we're more likely to become diabetic and/or obese. For those who become diabetic, hypertension is said to account for up to 85 percent of the considerably increased risk of heart disease. Studies have also demonstrated that insulin levels are abnormally elevated in hypertensives, and so hypertension, with or without obesity and/or diabetes, is now commonly referred to as an "insulin-resistant state." (This is the implication of including hypertension among the cluster of abnormalities that constitute metabolic syndrome.) Hypertension is so common in the obese, and obesity so common among hypertensives, that textbooks will often speculate that it's overweight that causes hypertension to begin with. So, the higher the blood pressure, the higher the cholesterol and triglyceride levels, the greater the body weight, and the greater the risk of diabetes and heart disease.

Despite the intimate association of these diseases, public-health authorities for the past thirty years have insisted that salt is the dietary cause of hypertension and the increase in blood pressure that accompanies aging. Textbooks recommend salt reduction as the best way for diabetics to reduce or prevent hypertension, along with losing weight and exercising. This salt-hypertension hypothesis is nearly a century old. It is based on what medical investigators call biological plausibility—it makes sense and so seems obvious. When we consume salt—i.e., sodium chloride—our bodies maintain the concentration of sodium in our blood by retaining more water along with it. The kidneys should then respond to the excess by excreting salt into the urine, thus relieving both excess salt and water simultaneously. Still, in most individuals, a salt binge will result in a slight increase in blood pressure from the swelling of this water retention, and so it has always been easy to imagine that this rise could become chronic over time with continued consumption of a salt-rich diet.

That's the hypothesis. But in fact it has always been remarkably difficult to generate any reasonably unambiguous evidence that it's correct. In 1967, Jeremiah Stamler described the evidence in support of the salt-hypertension connection as "inconclusive and contradictory." He still called it "inconsistent and contradictory" sixteen years later, when he described his failure in an NIH-funded trial to confirm the hypothesis that salt consumption raises blood pressure in school-age children. The NIH has funded subsequent studies, but little progress has been made. The message conveyed to the public, nonetheless, is that salt is a nutritional evil—"the deadly white powder," as Michael Jacobson of the Center for Science in the Public Interest called it in 1978. Systematic reviews of the evidence, whether published by those who believe that salt is responsible for hypertension or by those who don't, have inevitably concluded that significant reductions in salt consumption—cutting our average salt intake in half, for instance, which is difficult to accomplish in the real world—will drop blood pressure by perhaps 4 to 5 mm Hg in hypertensives and 2 mm Hg in the rest of us. If we have hypertension, however, even if just stage 1, which is the less severe form of the condition, it means our systolic blood pressure is already elevated at least 20 mm Hg over what's considered healthy. If we have stage 2 hypertension, our blood pressure is elevated by at least 40 mm Hg over healthy levels. So cutting our salt intake in half and decreasing our systolic blood pressure by 4 to 5 mm Hg makes little difference.

Our belief in the dangers of salt in the diet is once again based on Geoffrey Rose's philosophy of preventive medicine. Public-health authorities have continued to recommend that we all eat less salt because they believe that any benefit to the individual, no matter how clinically insignificant, will have a significant impact on the public health. But this evades the scientific question that still has to be answered: if excessive salt consumption does not cause hypertension, as these clinical trials suggest it does not, then what does? Moreover, embracing a suspect public-health pronouncement serves to inhibit rigorous scientific research.

Let's recall that hypertension is a disease of civilization, an observation that dates back to the late 1920s. Just as physicians in Europe and the United States took to measuring blood pressure in their patients with the availability of an instrument that could do so easily and reliably (the sphygmomanometer), missionary and colonial physicians throughout the world took to measuring blood pressure in native populations. Within a decade, noted the British physician Cyril Donnison in 1938 in *Civilization and Disease*, hypertension was already among the best-documented examples of a disease that seemed specific to Western societies and the more affluent social classes elsewhere. The average blood pressure in isolated populations eating traditional diets was inevitably low, but not dissimilar to the average blood pressure of Europeans and Americans who had not yet reached middle age. Hypertension was never seen in these populations, and blood pressure, if anything, dropped lower with age, which is the opposite of what happens in developed nations. In 1929, Donnison reported that he had measured the blood pressure in a thousand Kenyan nomads and found it similar to that of Europeans for those men under forty, but not so after that: "It tends to come down in the African," Donnison wrote, "whereas in the white races it continues its tendency to rise until the eighth decade." The Kenyan nomads in their sixties had an average systolic blood pressure forty points lower than that of European men of the same age. Over the next forty years, these observations would be confirmed in isolated populations throughout the world.

With exposure to Western lifestyles and diets, however, blood pressure among these native populations began to rise with age, as it does in Europe and America, and the average blood pressure and the incidence of hypertension increased as well. In Kenya and Uganda, British physicians considered hypertension to be nonexistent among their African patients in the late 1930s. By the 1950s, more than 10 percent of native Africans checking into hospitals for any reason were diagnosed with clinical hypertension. That number had risen to over 30 percent by the mid-1960s. By the 1970s, hypertension was considered as frequent in the native African populations as it was in Europe or America. In some urban populations, hypertension rates as high as 60 percent were reported.

Until the salt hypothesis began receiving serious attention in the 1960s, the investigators paid little attention to nutritional explanations for the rise in blood pressure that accompanied Western diets and lifestyles. Instead, they debated whether it was the stress and tension of what they considered civilized life that led blood pressure to rise, as Donnison believed. Once the salt hypothesis raised the possibility that diet was responsible, investigators began to perceive the presence or absence of hypertension in isolated populations purely as a test of the salt hypothesis. Since hypertension only appeared in these populations when they gained access to Western diets, which frequently included salt-rich processed foods, the investigators saw their studies as confirming the salt hypothesis. By the 1990s, the absence of hypertension in isolated populations eating their traditional diets was still the most compelling evidence in support of the hypothesis.

Of course, the same societies that ate little or no salt ate little or no sugar and white flour, so the evidence supported *both* hypotheses, although the investigators were interested in only one. The notion that the refined-carbohydrate hypothesis could explain many of the other chronic changes in health among these populations was rarely discussed. In two cases—Gerald Shaper's studies of nomadic tribes in Kenya and Uganda, and Ian Prior's studies of South Pacific Islanders—the investigators first implicated refined carbohydrates as a possible cause of the emergence of hypertension in their populations, because sugar and flour constituted the conspicuous additions to the diet with Western influence. Then they embraced salt as the culprit, after they became aware that investigators in the U.S. believed salt to be the problem. In the early 1970s, when the Harvard hypertension specialist Lot Page and his colleagues set out to study "the antecedents of cardiovascular disease" in the Solomon Islands, they, too, considered their research to be solely a test of the salt hypothesis, so salt was the only aspect of the Solomon Islanders' diet that they assessed. In what came to be considered a seminal study in the field, they concluded naturally enough that suspicion of the cause of high blood pressure among the islanders "falls most heavily on salt intake."

The laboratory evidence that carbohydrate-rich diets can cause the body to retain water and so raise blood pressure, just as salt consumption is supposed to do, dates back well over a century. It has been attributed first to the German chemist Carl von Voit in 1860. In 1919, Francis Benedict, director of the Nutrition Laboratory of the Carnegie Institute of Washington, described it this way: "With diets predominantly carbohydrate there is a strong tendency for the body to retain water, while with diets predominantly fat there is a distinct tendency for the body to lose water." The context of Benedict's discussion was the weight loss that occurs in the first few weeks of any calorie- or carbohydrate-restricted diet, and particularly the latter. As Benedict pointed out, this weight loss is to a large extent water, not fat, which has to be factored into any discussion of the apparent benefits of a reducing scheme. In the late 1950s, a new generation of investigators rediscovered the phenomenon, and it was then used to rationalize the popularity of carbohydrate-restricted diets as due not to the ease of losing fat, but entirely to the water lost in the first few weeks of the diet.

The "remarkable sodium and water retaining effect of concentrated carbohydrate food," as the University of Wisconsin endocrinologist Edward Gordon called it, was then explained physiologically in the mid-1960s by Walter Bloom, who was studying fasting as an obesity treatment at Atlanta's Piedmont Hospital, where he was director of research. As Bloom reported in the *Archives of Internal Medicine* and *The American Journal of Clinical Nutrition,* the water lost on carbohydrate-restricted diets is caused by a reversal of the sodium retention that takes place routinely when we eat carbohydrates. Eating carbohydrates prompts the kidneys to hold on to salt, rather than excrete it. The body then retains extra water to keep the sodium concentration of the blood constant. So, rather than having water retention caused by taking in more sodium, which is what theoretically happens when we eat more salt, carbohydrates cause us to retain water by inhibiting the excretion of the sodium that is already there. Removing carbohydrates from the diet works, in effect, just like the antihypertensive drugs known as diuretics, which cause the kidneys to excrete sodium, and water along with it.

This water loss leads to a considerable drop in blood pressure, so much so that it led critics of these diets, such as Philip White, author of a nutrition column in the *The Journal of the American Medical Association,* to worry publicly about the "low blood pressure resulting from . . . losses of . . . fluid, sodium, and other minerals." Discussions of the treatment of obesity with very low-carbohydrate diets would address the need to retain some carbohydrates in the diet to maintain "fluid balance" and "avoid large shifts in weight due to changes in water balance." By the early 1970s, researchers had demonstrated that the water-retaining effect of carbohydrates was due to the insulin secreted, which in turn induced the kidneys to reabsorb sodium rather than excrete it, and that insulin levels were indeed higher, on average, in hypertensives than in normal individuals.

Finally, by the mid-1990s, diabetes textbooks, such as *Joslin's Diabetes Mellitus,* contemplated the likelihood that chronically elevated levels of insulin were "the major pathogenetic defect initiating the hypertensive process" in patients with Type 2 diabetes. But such speculations rarely extended to the potential implications for the nondiabetic public.

There are several possible explanations for why this phenomenon rarely entered into the discussions of hypertension and heart disease. Those investigators concerned with the dangers of hypertension might simply have considered the obesity literature or even the diabetes literature of little significance to their research, other than the obvious observation that obese and diabetic patients tend to be hypertensive and vice versa. Another possibility is that by the 1960s hypertension and high cholesterol were two of the three major risk factors associated with premature coronary heart disease (the third was smoking), so it was difficult to imagine that eating carbohydrates might be beneficial for one risk factor, cholesterol, while being detrimental for another, blood pressure.

Though this carbohydrate-induced water retention and the hypertensive effect of insulin were occasionally discussed in nutrition and dietetics textbooks—*Modern Nutrition in Health and Disease,* for example, which was published in 1951 and was in its fifth edition by the 1970s—they would appear solely in the technical context of water and electrolyte balance (sodium is an electrolyte), whereas the discussion of hypertension prevention would focus exclusively on the salt hypothesis. When they were discussed in obesity conferences after the 1960s, the implications were restricted to a very narrow range, usually as evidence against any metabolic advantage of carbohydrate-restricted diets. ("One claim which is often made for the low-carbohydrate diet is that 3,000 [calories]/day or more can be eaten and the patient will still lose weight if the carbohydrate intake is restricted," explained George Bray at the Second International Conference on Obesity in 1977. "There are no convincing studies to support this claim. On the contrary . . . it is now well-established that a low-carbohydrate diet is followed by the excretion of water and that carbohydrate ingestion leads to retention of both salt and water.") Since lower weight is associated with lower insulin levels, overweight hypertensives were advised to lose weight to reduce their blood pressure, but then low-calorie diets—usually low-fat and thus high in carbohydrates—would be recommended as the means to do it. On very rare occasions, "carbohydrate overeating" would be acknowledged as a nutritional factor involved in the genesis of hypertension, at least in obese patients, and then both carbohydrate restriction and salt restriction would be recommended as treatment. Those investigators, too, had come to assume that the salt hypothesis must be true.

Since the late 1970s, investigators have demonstrated the existence of other hormonal mechanisms by which insulin raises blood pressure—in particular, by stimulating the nervous system and the same flight-or-fight response incited by adrenaline. This was first reported by Lewis Landsberg, an endocrinologist who was then at Harvard Medical School and would later become dean of the Northwestern University School of Medicine. Landsberg showed that, by stimulating the activity of the nervous system, insulin increases heart rate and constricts blood vessels, thereby raising blood pressure. The higher the insulin level, the greater the stimulation of the nervous system, Landsberg noted. If insulin levels remained high, so Landsberg's research suggested, then the sympathetic nervous system would be constantly working to raise blood pressure. The heart-disease research community has paid attention to Landsberg's work, but has considered it relevant *only* for the obese. Because obesity is associated with higher insulin levels, and because it's now believed that obesity *causes* higher insulin levels (whereas obesity itself is allegedly caused by the consumption of excess calories of all types), any possible link to carbohydrate consumption or "carbohydrate overfeeding" is overlooked. Even Landsberg has concentrated almost exclusively on the obesity-insulin-hypertension connection and ignored the idea that the increase in insulin levels due to excessive carbohydrate consumption, or due to the consumption of refined and easily digestible carbohydrates, might have a similar effect.

One question that will be addressed in the coming chapters is why medical investigators and public-health authorities, like Landsberg, will accept the effects of insulin on chronic diseases as real and potentially of great significance, and yet inevitably interpret their evidence in ways that say nothing about the unique ability of refined and easily digestible carbohydrates to chronically elevate insulin levels. This is the dilemma that haunts the past fifty years of nutrition research, and it is critical to the evolution of the science of metabolic syndrome. As we will discuss, the observation of diseases of civilization was hardly the only evidence implicating sugar and refined carbohydrates in these diseases. The laboratory research inevitably did, too. Yet the straightforward interpretation of the evidence—from carbohydrates to the chronic elevation of insulin to disease—was consistently downplayed or ignored in light of the overwhelming belief that Keys's dietary-fat hypothesis had been proved correct, which was not the case.

The coming chapters will discuss the history of the science of metabolic syndrome both in the context of how the research was interpreted at the time, in a universe dominated by Keys's hypothesis, and then how it arguably should have been interpreted if the research community had approached this science without bias and preconceptions. The next five chapters describe the science that was pushed aside as investigators and public-health authorities tried to convince first themselves and then the rest of us that dietary fat was the root of all nutritional evils. These chapters divide the science of metabolic syndrome and the carbohydrate hypothesis into five threads, to simplify the telling (although by doing so, they admittedly oversimplify).

The first (Chapter 9) covers the research that directly challenged the fundamental premise of Keys's hypothesis that cholesterol itself is the critical component in heart disease, and instead implicated triglycerides and the kinds of molecules known as lipoproteins that carry cholesterol through the blood, both of which are effectively regulated by the carbohydrate content of the diet rather than saturated fat. The chapter then explains how this research, despite its refutation of the fat-cholesterol hypothesis, has been assimilated into it nonetheless.

The second thread (Chapter 10) follows the evolution of the science of insulin resistance and hyperinsulinemia, the condition of having chronically elevated insulin levels, and how that emerged out of attempts to understand the intimate relationship of obesity, heart disease, and diabetes and led to the understanding of metabolic syndrome and the entire cluster of metabolic and hormonal abnormalities that it entails.

The third (Chapter 11) discusses the implications of metabolic syndrome in relation to diabetes and the entire spectrum of diabetic complications.

The fourth (Chapter 12) discusses table sugar and high-fructose corn syrup, in particular, and the research suggesting that they have negative health effects that are unique among refined carbohydrate foods.

The last section of this history (Chapter 13) discusses how metabolic syndrome, and particularly high blood sugar, hyperinsulinemia, and insulin resistance, have physiological repercussions that can conceivably explain the appearance of even Alzheimer's disease and cancer.

Throughout these five chapters, the science will be more technical than has typically been the case in popular discussions of what we should eat and what we shouldn't. I believe it is impossible, though, to make the argument that nutritionists for a half century oversimplified the science to the point of generating false ideas and erroneous deductions, without discussing the science at the level of complexity that it deserves.

Chapter Nine

TRIGLYCERIDES AND THE COMPLICATIONS OF CHOLESTEROL

Oversimplification has been the characteristic weakness of scientists of every generation.

ELMER MCCOLLUM, *A History of Nutrition*, 1957

THE DANGER OF SIMPLIFYING A MEDICAL ISSUE for public consumption is that we may come to believe that our simplification is an appropriate representation of the biological reality. We may forget that the science is not adequately described, or ambiguous, even if the public-health policy seems to be set in stone. In the case of diet and heart disease, Ancel Keys's hypothesis that cholesterol is the agent of atherosclerosis was considered the simplest possible hypothesis, because cholesterol is found in atherosclerotic plaques and because cholesterol was relatively easy to measure. But as the measurement technology became increasingly more sophisticated, every one of the complications that arose has implicated carbohydrates rather than fat as the dietary agent of heart disease.

In 1950, the University of California medical physicist John Gofman wrote an article in *Science* that would be credited, albeit belatedly, with launching the modern era of cholesterol research. Gofman pointed out that cholesterol is only one of several fatlike substances that circulate through the blood and are known collectively as lipids or blood lipids. These include free fatty acids and triglycerides,* the molecular forms in which fat is found circulating in the bloodstream. These could also be players in the heart-disease process, Gofman noted, and the fact that there was no easy way to measure their concentrations in the circulation didn't change that. Both cholesterol and triglycerides are shuttled through the circulation in particles called lipoproteins, and these could also be players. The amount of cholesterol and triglycerides varies in each type of lipoprotein. So, when physicians measure total cholesterol levels, they have no way of knowing how the cholesterol itself is apportioned in individual

---

* A triglyceride molecule is composed of three fatty acids—hence, the "tri"—linked together by a glycerol molecule.

lipoproteins. It is possible, Gofman noted, that in heart disease the problem may be caused not by cholesterol but by a defect in one of these lipoproteins, or an abnormal concentration of the lipoproteins themselves.

Eventually, researchers came to identify these different classes of lipoproteins by their density. Of those that appeared to play obvious roles in heart disease, three in particular stood out even in the early 1950s. Two of these are familiar today: the low-density lipoproteins, known as LDL, the bad cholesterol, and the high-density variety, known as HDL, the good cholesterol. (This is an oversimplification, as I will explain shortly.) The third class is known as VLDL, which stands for "very low-density lipoproteins," and these play a critical role in heart disease. Most of the triglycerides in the blood are carried in VLDL; much of the cholesterol is found in LDL. That LDL and HDL are the two species of lipoproteins that physicians now measure when we get a checkup is a result of the oversimplification of the science, not the physiological importance of the particles themselves.

In 1950, the only instrument capable of measuring the density of lipoproteins was an ultracentrifuge, and the only ultracentrifuge available for this work in America was being used by Gofman at the University of California, Berkeley. Gofman was both a physician and a physical chemist by training. During World War II, he worked for the Manhattan Project, and developed a process to separate plutonium that would later be used to produce H-bombs. After the war, Gofman set out to use the Berkeley ultracentrifuge to study how cholesterol and fat are transported through the blood and how this might be affected by diet and perhaps cause atherosclerosis and heart disease.

This was the research Gofman first reported in *Science* in 1950. He described how his ultracentrifuge "fractionated" lipoproteins into different classes depending on their density, and he noted that one particular class of lipoproteins, which would later be identified as LDL,* is more numerous in patients with atherosclerosis than in healthy subjects, in men than in women, in older individuals than in younger, and particularly conspicuous in diabetics, all of which suggested a possible role in heart disease. What these low-density lipoproteins did not do, Gofman reported, was to reflect consistently the amount of cholesterol in the blood, even though they carry cholesterol within them. Sometimes total cholesterol

---

* To be precise, Gofman's *Science* paper identified IDL—i.e., intermediate-density lipoproteins—as the class associated with heart disease. He would later decide that LDL was more important than IDL. For the sake of simplicity, I've used LDL throughout.

levels would be low in his subjects, he noted, and yet the concentration of these low-density lipoproteins would be abnormally high. Sometimes total cholesterol would be high while the cholesterol contained in the low-density lipoproteins was low. "At a particular cholesterol level one person may show 25 percent of the total serum cholesterol in the form of [low-density lipoproteins], whereas another person may show essentially none in this form," Gofman wrote.

After *Science* published Gofman's article, and after aggressive lobbying on Gofman's part, the National Advisory Heart Council agreed to fund a test of his hypothesis that lipoproteins are the important factor in heart disease and that cholesterol itself is not. The test would be carried out by four research groups—led by Gofman at Berkeley, Irving Page at the Cleveland Clinic, Fred Stare and Paul Dudley White at Harvard, and Max Lauffer of the University of Pittsburgh—that collectively identified five thousand men who were free of heart disease. When heart disease eventually appeared, they would determine whether total cholesterol or Gofman's lipoproteins was the more accurate predictor.

While the three Eastern laboratories took three years to learn how to use an ultracentifruge for fractionating lipoproteins, Gofman proceeded with his own research, refined his understanding of how these lipoproteins predicted heart disease, and he then insisted that the analysis techniques be updated accordingly. The other investigators, however, were having considerable trouble duplicating Gofman's *original* analysis, and so they refused to accept any further modifications.

In 1956, the four groups published a report in the American Heart Association journal *Circulation,* with a minority opinion written by Gofman and his Berkeley colleagues and a majority opinion authored by everyone else. As the majority saw it, based on the state of Gofman's research in 1952, cholesterol was indeed a questionable predictor of heart disease risk, but the measurements of lipoproteins added little predictive power. "The lipoprotein measurements are so complex," the majority report declared, "that it cannot be reasonably expected that they could be done reliably in hospital laboratories." Gofman's minority opinion, based on the state of his research in 1955, was that LDL and VLDL, the very low-density lipoproteins, were good predictors of heart disease, but that the single best predictor of risk was an *atherogenic index,* which took into account these two lipoprotein classes measured individually and added them together. The greater the atherogenic index, the greater the risk of atherosclerosis and heart disease.

Gofman would later be vindicated, but the majority opinion prevailed at the time: studying lipoproteins held no value in the clinical management of heart disease. Gofman and his Berkeley collaborators continued the research alone through 1963, when Gofman left to establish a biomedical-research division at the Lawrence Livermore National Laboratory and spent the rest of his career working on the health effects of radiation.

Lost entirely in the contretemps were the dietary implications of Gofman's research. "While it is true that, for certain individuals, the *amount* of dietary fat is an important factor," Gofman explained, "it turns out that there are other more significant factors that need to be considered. Human metabolism is so regulated that factors other than the actual dietary intake of one of these constituents may determine the amount of that constituent that will circulate in the bloodstream. Indeed, important observations have been made which indicate that certain substances in the diet that are not fatty at all may still have the effect of increasing the concentration of the fat-bearing lipoprotein substances in the blood."

Though Gofman's studies had demonstrated that the amount of LDL in the blood can indeed be elevated by the consumption of saturated fats, it was *carbohydrates*, he reported, that elevated VLDL—containing some cholesterol and most of the triglycerides in the blood—and only by restricting carbohydrates could VLDL be lowered.

This fact was absolutely critical to the dietary prevention of heart disease, Gofman said. If a physician put a patient with high cholesterol on a low-fat diet, that might lower the patient's LDL, but it would raise VLDL. If LDL was abnormally elevated, then this low-fat diet might help, but what Gofman called the "carbohydrate factor" in these low-fat diets might raise VLDL so much that the diet would do more harm than good. Indeed, in Gofman's experience, when LDL decreased, VLDL tended to rise disproportionately. And if VLDL was abnormally elevated to begin with, then prescribing a low-fat, high-carbohydrate diet would certainly *increase* the patient's risk of heart disease.

This was why Gofman described the measurement of total cholesterol as a "false and highly dangerous guide" to the effect of diet on heart disease. Total-cholesterol measurements tell us nothing about the status of VLDL and LDL. Prescribing low-fat diets indiscriminately to anyone whose cholesterol appears to be elevated, or bombarding us with "generalizations such as 'we all eat too much fat,' or 'we all eat too much animal fat,' " would increase heart-disease risk for a large proportion of the population. "Neglect of [the carbohydrate] factor can lead to rather serious consequences," wrote Gofman in 1958, "first, in the failure to correct the diet in some individuals who are very sensitive to the carbohydrate action; and second, by allowing certain individuals sensitive to the carbohydrate action to take too much carbohydrate as a replacement for some of their animal fats."

By 1955, Pete Ahrens at Rockefeller University had come to this same conclusion, although Ahrens was specifically studying triglycerides, rather than the VLDL particles that carry the triglycerides. Ahrens was considered by many investigators to be the single best scientist in the field of lipid metabolism. He had observed how the triglycerides of some patients shoot up on low-fat diets and fall on high-fat diets. This led Ahrens to describe a phenomenon that he called *carbohydrate-induced lipemia* (an excessive concentration of fat in the blood). When he gave lectures, Ahrens would show photos of two test tubes of blood serum obtained from the same patient— one when the patient was eating a high-carbohydrate diet and one on a high-fat diet. One test tube would be milky white, indicating the lipemia. The other would be absolutely clear. The surprising thing, Ahrens would explain, was "that the lipemic plasma was obtained during the high-carbohydrate period, and the clear plasma during the high-fat regimen." (Joslin had reported the same phenomenon in diabetics thirty years earlier. "The percent of fat" in the blood, he wrote, "rises with the severity of the disease . . . and is especially related to the quantity of carbohydrate, which is being oxidized, rather than with the fat administered.")

Over the course of a decade, Ahrens had seen only two patients whose blood serum became cloudy with triglycerides after eating high-fat meals. He had thirteen in whom carbohydrates caused the lipemia. Six of those thirteen had such high triglycerides that they had originally been referred to Ahrens from physicians who had misdiagnosed them as having a genetic form of high cholesterol. Since the VLDL particles that transport triglycerides, as Gofman had noted, also carry cholesterol and so contribute to the total cholesterol in the circulation, an elevated triglyceride level can elevate total cholesterol along with it. Ahrens believed that the fat-induced lipemia was a rare genetic disorder but the carbohydrate-induced lipemia was probably "an exaggerated form of the normal biochemical process which occurs in all people on high-carbohydrate diets." In both cases, the fat in the blood would clear up when the subjects went on a low-calorie diet. To Ahrens, this explained why the carbohydrate-induced increase in triglycerides was absent in Asian populations living primarily on rice. As long as they were eating relatively low-calorie diets compared with their level of physical activity, which was inevitably the case in such impoverished populations, the combination would counter-act the triglyceride-raising effect of the carbohydrates.

The critical question was whether prolonged exposure to an abnormally high triglyceride level increased the risk of atherosclerosis. If carbohydrate-induced lipemia was as common as Ahrens believed, "especially in the areas of the world distinguished by caloric abundance and obesity," then it was important to know. If so, then having patients with high triglycerides eat less fat would only make the condition worse. By 1957, Ahrens was also warning about the dangers of oversimplifying the diet-heart science: maybe fat and cholesterol caused heart disease, or maybe it was the carbo-hydrates and triglycerides. "We know of no solid evidence on this point," wrote Ahrens, "and until the question is further explored we question the wisdom of prescribing low-fat diets for the general population."

The evidence that Ahrens was looking for came first from Margaret Albrink, who was then a young physician working with John Peters, chief of the metabolic division in the Department of Medicine at Yale Univer-sity. Once again, the available technology drove the research. Peters was renowned in the medical community for his measurements of the chemi-cal constituents of body fluids. For this purpose he had a device called an analytical centrifuge, a less sophisticated version of Gofman's ultracen-trifuge, which could quantify the triglyceride concentration of the blood. Peters's lab also analyzed blood samples for New Haven Hospital (now Yale–New Haven Hospital), so Peters suggested to Albrink that they use the analytical centrifuge to measure the triglycerides in those blood sam-ples and test the hypothesis that high triglycerides are associated with an increased risk of heart disease. Peters was a "contrarian," Albrink says; he didn't believe the cholesterol hypothesis. Nor did Evelyn Man, Peters's longtime collaborator. Albrink also worked with Wister Meigs, a Yale pro-fessor of preventive medicine who also served as company physician for the nearby American Steel and Wire Company. Meigs had been recording cholesterol levels in the plant employees, along with their family history of heart disease, diabetes, and other ailments. By 1960, Albrink, Man, and Meigs (Peters died in 1955) were comparing triglyceride and cholesterol levels of heart-disease patients from New Haven Hospital with the levels among the healthy employees of American Steel and Wire. Elevated triglyceride levels, they concluded, were far more common in coronary-heart-disease patients than high cholesterol: only 5 percent of healthy young men had elevated triglycerides, compared with 38 percent of healthy middle-aged men and 82 percent of coronary patients.

In May 1961, just a few months after the American Heart Association publicly embraced Keys's hypothesis, both Ahrens and Albrink presented their research at a meeting of the Association of American Physicians in Atlantic City, New Jersey. Both reported that elevated triglycerides were associated with an increased risk of heart disease, and that low-fat, high-carbohydrate diets raised triglycerides. The *New York Times* covered Ahrens's talk—"Rockefeller Institute Report Challenges Belief that Fat Is Major Factor"—in a story buried deep in the paper. Ahrens's data suggested that "dietary carbohydrate, not fat, is the thing to watch in guarding against [atherosclerosis and heart disease]," the *Times* reported, and this "came as something of a surprise to many of the scientists and physicians attending the meeting." Albrink's talk did not make the newspaper, but she later told a similar story about her presentation. "It just about brought the house down," she recalled. "People were so angry; they said they didn't believe it." This remained the case for much of the next decade. Albrink continued to work out the connection between carbohydrates, triglycerides, and heart disease and would present her results at conferences, where she would inevitably be attacked by proponents of Keys's hypothesis.

By the early 1970s, Albrink's interpretation of the evidence had been confirmed independently, first by Peter Kuo of the University of Pennsylvania, then by Lars Carlson of the Karolinska Institute in Stockholm, and by the future Nobel laureate Joseph Goldstein and his colleagues from the University of Washington. All three reported that high triglycerides were considerably more common in heart-disease victims than was high cholesterol. In 1967, Kuo reported in *The Journal of the American Medical Association* that he had studied 286 atherosclerosis patients, of whom 246 had been referred to him by physicians who thought their patients had the genetic form of high cholesterol. This turned out to be the case for fewer than 10 percent. The other 90 percent had carbohydrate-induced lipemia, and, for most of these patients, their sensitivity to carbohydrates had elevated both their triglyceride levels and their cholesterol. When Kuo put his patients on a sugar-free diet, he reported, with only five to six hundred calories of starches a day, both their triglyceride levels *and* their cholesterol lowered. Two months later, *JAMA* published an editorial in response to Kuo's article, suggesting that the "almost embarrassingly high number of researchers [who had] boarded the 'cholesterol bandwagon' " had done a disservice to the field. "This fervent embrace of cholesterol to the exclusion of other biochemical alterations resulted in a narrow scope of study," the editorial said. "Fortunately, other fruitful approaches have been made possible in the past few years by identification of the fundamental role of such factors as triglycerides and carbohydrate metabolism in atherogenesis."

By then, however, the science had already become secondary to more practical issues. Despite *JAMA*'s optimism that a new era was dawning, it was no longer a question of whether it was cholesterol or triglycerides that caused atherosclerosis and heart disease, whether saturated fat or carbohydrates were to blame, but which of the two hypotheses dominated the research. Here Keys's hypothesis had precedence. A generation of clinical investigators—the "cholesterol bandwagon"—had gathered an enormous amount of data, however ambiguous, on cholesterol levels and heart disease; only Albrink, Kuo, and a handful of other researchers had studied triglycerides. Only Gofman had studied the VLDL particles that transport triglycerides through the circulation.

Moreover, measuring triglycerides was still much more difficult than measuring cholesterol, and so only the rare laboratory had the facilities to do it. The National Institutes of Health, which was effectively the only source of funding for this research in the United States, had already committed its resources to three enormous studies—the Framingham Heart Study, Keys's Seven Countries Study, and the pilot programs of the National Diet-Heart Study. These studies would measure only cholesterol and so test only Keys's hypothesis. No consideration was given to any alternative hypothesis. By 1961, Keys and his collaborators in the Seven Countries Study had measured cholesterol in over ten thousand men. By 1963, they had completed the exams on another eighteen hundred men. Even had it been technically possible to include triglycerides in the measurements, or to return to the original locales and retest for triglycerides, the cost would have been astronomical. The result, as we've seen, was considered a resounding victory for Keys's fat-cholesterol hypothesis.

The research that would finally lead to a large-scale test of the carbohydrate/triglyceride/heart-disease hypothesis emerged from the National Institutes of Health in early 1967. This was a collaboration between Donald Fredrickson and Robert Levy, who would become directors of the National Institutes of Health and the National Heart, Lung, and Blood Institute respectively, and Robert Lees, then of Rockefeller University. It was published in a fifty-page, five-part series in *The New England Journal of Medicine*. First Fredrickson, Levy, and Lees proposed a simplified classification of lipoproteins (perhaps an oversimplification, they acknowledged), which divided the lipoproteins in the bloodstream into four categories: LDL, which typically carried most of the cholesterol; VLDL, which carried most of the triglycerides; the high-density lipoproteins, HDL; and chylomicrons, which carry dietary fat from the intestine to the fat tissue. Then they proposed a classification scheme for disorders of lipoprotein metabolism, each delineated by a roman numeral, that included both those of abnormally high amounts of LDL cholesterol, which they suggested might be ameliorated by low-fat diets, as well as those characterized by abnormally high triglycerides carried in VLDL, which would be ameliorated by low-carbohydrate diets.

Four of the five lipoprotein disorders described in this series were characterized by abnormally elevated levels of triglycerides in the very low-density lipoproteins. For this reason, Fredrickson, Levy, and Lees also warned against the dangers of advocating low-fat diets for all patients, because these diets increased carbohydrate consumption and so would elevate triglycerides and VLDL even further. By far the most common of the five lipoprotein disorders was the one designated Type IV, characterized by elevated VLDL triglycerides—"sometimes considered synonymous with 'carbohydrate-induced hyperlipemia,' " they wrote—and it *had* to be treated with a low-carbohydrate diet. "Patients with this syndrome," Lees later wrote, "form a sizable fraction of the population suffering from coronary heart disease."*

Because Fredrickson, Levy, and Lees had also described an innovative and inexpensive technique for measuring the triglycerides and cholesterol carried in these different lipoproteins, the NIH provided the necessary funding for five studies—in Framingham, Puerto Rico, Honolulu, Albany, and San Francisco—to measure LDL cholesterol and VLDL triglycerides in these populations and determine their significance as risk factors for heart disease. This research would take almost a decade to complete, and would constitute the first time that NIH-funded research projects would measure anything other than total cholesterol in large populations.

The new research would also mark the first time that HDL was measured in large populations, and this would further confuse the diet/heart-disease relationship. The hypothesis that HDL particles or the cholesterol in HDL protects against heart disease had first been proposed in 1951 by David Barr and Howard Eder of New York Hospital–Cornell Medical Center. It had been confirmed in a handful of small studies through the 1950s,

---

* One notable case was Theodore Cooper, who was assistant secretary for health in 1976, when he testified about "diet and killer diseases" to the Senate Select Committee on Nutrition and Human Needs. Cooper said that his personal dietary concern was with carbohydrates rather than fats. "If I have a problem, it is a tendency to gain weight," Cooper explained. "I am classified Type IV. As a Type IV, my lipid levels are much more subject to elevation if I consume large amounts of carbohydrates or alcohol."

and by Gofman in the last paper he published on lipoproteins and heart disease, as had the observation that when HDL was low triglycerides tended to be high, and vice versa, which suggested some underlying mechanism linking the two. Nonetheless, heart-disease researchers had paid little attention to HDL, as the NIH biostatistician Tavia Gordon later explained, because the idea of a "negative relation" between cholesterol and heart disease—high HDL cholesterol implies a low risk of heart disease—"simply ran against the grain." "It was easy to believe that too much cholesterol in the blood could 'overload' the system and hence increase the risk of disease," Gordon wrote, "but how could 'too much' of one part of the total cholesterol reduce the risk of disease? To admit that fact challenged the whole way of thinking about the problem." Now HDL, too, would be measured in these populations.*

The results from the five studies were released in 1977 and divided into two publications, although Gordon had done the analyses for both. One reported on a comparison of nine hundred heart-disease cases with healthy controls from all five of the populations. The other addressed the *prospective* evidence from Framingham alone—measuring triglyceride, lipoprotein, and cholesterol levels in twenty-eight hundred subjects and then waiting four years to see how well these levels predicted the appearance of heart disease. The findings were consistent. Both analyses confirmed Gofman's argument that total cholesterol said little about the risk of heart disease, and that the measurement of the triglycerides and cholesterol in the different lipoproteins was considerably more revealing. In men and women fifty and older, Gordon and his collaborators wrote in the Framingham paper, "total cholesterol per se is not a risk factor for coronary heart disease at all." LDL cholesterol was a "marginal" risk factor, they reported. Triglycerides predicted heart disease in men and women in the analysis of cases from all five studies, but only in women in the Framingham analysis.

HDL was the "striking" revelation. Both analyses confirmed that the higher the HDL cholesterol the lower the triglycerides and the risk of heart disease. The *inverse* relationship between HDL and heart disease held true for every age group from forty-year-olds to octogenarians, in both men and women, and in every ethnic group from Framingham, Massachusetts, to Honolulu. "Of all the lipoproteins and lipids measured HDL had the

* This was not because the NIH had any interest in testing the HDL/heart-disease relationship, according to Gordon, but only because Fredrickson, Levy, and Lees's new measurement technique required that the amount of cholesterol in HDL be known so that the amount in LDL could be calculated.

largest impact on risk," Gordon and his colleagues wrote. For those fifty and over, which is the age at which heart disease ceases to be a rare condition, HDL was the *only* reliable predictor of risk.

The finding that high HDL cholesterol was associated with a low risk of heart disease did not mean that raising HDL would lower risk, as Gordon and his colleagues noted, but it certainly suggested the possibility. Only a few studies had ever looked at the relationship of diet and lifestyle to HDL, and the results had suggested, not surprisingly, that anything that raised triglycerides would lower HDL, and vice versa. The "fragmentary information on what maneuvers will lead to an increase in HDL cholesterol levels," Gordon and his collaborators wrote, "suggests that physical activity, weight loss and a *low carbohydrate* intake may be beneficial" (my italics).

This is where the story now takes some peculiar turns. One immediate effect of the revelation about HDL, paradoxically, was to direct attention away from triglycerides, and with them the conspicuous link, until then, to the carbohydrate hypothesis. Gordon and his colleagues had demonstrated that when both HDL and triglycerides were incorporated into the risk equations of heart disease, or when obesity and the prediabetic condition of glucose intolerance were included in the equations along with triglycerides, the apparent effect of triglycerides diminished considerably. This result wasn't surprising, considering that low HDL, high triglycerides, obesity, and glucose intolerance all seemed to be related, but that wasn't the point. The relevant question for physicians was whether high triglycerides by themselves caused heart disease. If so, then patients should be advised to lower their triglycerides, however that might be accomplished, just as they were being told already to lower cholesterol. These risk-factor equations (known as *multivariate equations*) suggested that triglycerides were not particularly important when these other factors were taken into account, and this was how they would be perceived for another decade. Not until the late 1980s would the intimate association of low HDL, high triglycerides, obesity, and diabetes be considered significant—in the context of Gerald Reaven's Syndrome X hypothesis—but by then the heart-disease researchers would be committed to the recommendations of a national low-fat, high-carbohydrate diet.

Heart-disease researchers would also avoid the most obvious implication of the two analyses—that raising HDL offers considerably more promise to prevent heart disease than lowering either LDL or total cholesterol—on the basis that this hadn't been tested in clinical trials. Here the immediate obstacle, once again, was the institutional investment in

Keys's hypothesis. The National Institutes of Health had committed its heart-disease research budget to two ongoing studies, MRFIT and the Lipid Research Clinics Trial, which together would cost over $250 million. These studies were dedicated solely to the proposition that lowering *total* cholesterol would prevent heart disease. There was little money or interest in testing an alternative approach. Gordon later recalled that, when he presented the HDL evidence to the team of investigators overseeing MRFIT, "it was greeted with a silence that was very, how should I say it, *expressive*. One of them spoke up indicating he suspected this was a bunch of shit. They didn't know how to deal with it."

Indeed, the timing of the HDL revelations could not have been less convenient. The results were first revealed to the public in an American Heart Association seminar in New York on January 17, 1977. This was just three days after George McGovern had announced the publication of the *Dietary Goals for the United States,* advocating low-fat, high-carbohydrate diets for all Americans, based exclusively on Keys's hypothesis that coronary heart disease was caused by the effect of saturated fat on total cholesterol. If the *New York Times* account of the proceedings is accurate, the AHA and the assembled investigators went out of their way to ensure that the new evidence would not cast doubt on Keys's hypothesis or the new dietary goals. Rather than challenge the theory that excess cholesterol can cause heart disease, the *Times* reported, "the findings re-emphasize the importance of a fatty diet in precipitating life-threatening hardening of the arteries in most Americans," *which is precisely what they did not do.* According to the *Times,* saturated fat was now indicted not just for increasing LDL cholesterol, which it does, but for elevating VLDL triglycerides and lowering HDL, which it does not, and certainly not compared with the carbohydrates that McGovern's *Dietary Goals* were recommending all Americans eat instead.

In a more rational world, which means a research establishment not already committed to Keys's hypothesis and not wholly reliant on funding from the institutions that had embraced the theory, the results would have immediately prompted small clinical trials of the hypothesis that raising HDL prevented heart disease, just like those small trials that had begun in the 1950s to test Keys's hypothesis. If those confirmed the hypothesis, then longer, larger trials would be needed to establish whether the short-term benefits translated to a longer, healthier life. But the NIH administrators decided that HDL studies would have to wait. Once the Lipid Research Clinics Trial results were published in 1984, they were presented to the world as proof that lowering cholesterol by eating less fat and more carbohydrates was the dietary answer to heart disease. There was simply no room now in the dogma for a hypothesis that suggested that raising HDL (and lowering triglycerides) by eating more fat and less carbohydrates might be the correct approach. No clinical trials of the HDL hypothesis would begin in the U.S. until 1991, when the Veterans Administration funded a twenty-center drug trial. The results, published in 1999, supported the hypothesis that heart disease could be prevented by raising HDL. The drug used in the study, gemfibrozil, also lowered triglyceride levels and VLDL, suggesting that a diet that did the same by restricting carbohydrates might have a similarly beneficial effect. As of 2006, no such dietary trials had been funded.

Through the 1980s and 1990s, as our belief in the low-fat heart-healthy diet solidified, the official reports on nutrition and health would inevitably discuss the apparent benefits of raising HDL—the "good cholesterol"—and would then observe correctly that no studies existed to demonstrate this would prevent heart disease and lengthen life. By 2000, well over $1 billion had been spent on trials of cholesterol-lowering, and a tiny fraction of that amount on testing the benefits of raising HDL. Thus, any discussions about the relative significance of raising HDL versus lowering total cholesterol would always be filtered through this enormous imbalance in the research efforts. Lowering LDL cholesterol would always have the *appearance* of being more important.

It was the revelations that emerged from the two HDL publications in 1977 that led to the conventional wisdom about LDL, triglycerides, and HDL that we live with today. The National Heart, Lung, and Blood Institute and the American Heart Association responded to the new research by focusing on two pragmatic concerns: first, to keep the science sufficiently simple that it could be translated into equally simple guidelines for patient care, and, second, to reconcile these new observations with Keys's hypothesis and the $250 million worth of studies that were putting it to the test. If total cholesterol was not a risk factor for heart disease above the age of fifty, as Gordon's Framingham analysis noted, then that seemed to refute Keys's hypothesis. One immediate goal, therefore, was to make sure that those aspects of the hypothesis that had seemed reasonably certain were not discarded prematurely on the basis of findings that might also someday turn out to be erroneous.

Since both of the new analyses had concluded that LDL cholesterol was associated with a slightly increased risk of heart disease, and since up to

70 percent of the total cholesterol in the circulation may be found in LDL, the American Heart Association and the proponents of Keys's hypothesis now shifted the focus of scientific discussions from the benefits of lowering total cholesterol to the benefits of lowering LDL cholesterol. "Whatever the underlying disorder," noted the Framingham investigators in 1979, "much of what has been learned in the past about the ill effects of a high serum total cholesterol can be attributed to the associated elevated levels of LD lipoprotein. . . ."

Making LDL the "bad cholesterol" oversimplified the science considerably, but it managed to salvage two decades' worth of research, and to justify why physicians had bothered to measure total cholesterol in their patients. One consequence of this effort was an upgrading of the adjectives used to describe the predictive ability of LDL. In 1977, Gordon and his collaborators had described LDL cholesterol as a "marginal risk factor" for heart disease. Within two years, the same authors were using the identical data to describe LDL as a *"powerful* predictor of risk in subjects younger than the age of 50" and as showing "a *significant* contribution . . . to coronary heart disease in persons older than the age of 50 and practically up into the eighties." This practice has continued unabated.*

Another shift in emphasis was to incorporate HDL and some combination of triglycerides, LDL, and total cholesterol into the calculation of a "lipid profile" of heart-disease risk, a process that was initiated with the very first articles by Gordon and his collaborators. These lipid profiles allowed for the continued use of LDL or total cholesterol in the calculation of heart-disease risk, even though they added little or no predictive power to the use of HDL alone.

Ironically, these lipid profiles also provided the rationale for physicians to keep measuring total cholesterol in their patients, even though it had now been confirmed, as Gofman had noted a quarter-century earlier, that it was a dangerously unreliable predictor of risk. The reason is that LDL cholesterol itself happened to be particularly difficult to measure.† It was not the kind of measurement that physicians could easily order up for

---

* In 2003, for instance, the National Cholesterol Education Program described the shift in emphasis from total cholesterol to LDL cholesterol this way: "Many earlier studies measured only serum total cholesterol, although most of total cholesterol is contained in LDL. Thus, the *robust* relationship between total cholesterol and [coronary heart disease] found in epidemiological studies strongly implies that an elevated LDL is a *powerful* risk factor [my italics]."

† In the technique described by Fredrickson, Levy, and Lees, LDL cholesterol is not measured directly but calculated from the measurements of triglycerides, HDL cholesterol, and total cholesterol.

their patients. And since it didn't seem to matter in these lipid profiles whether it was total or LDL cholesterol that was included along with HDL—either way, HDL was the dominant predictor of risk—then, "from a practical point of view," as Gordon and his colleagues noted, "total cholesterol can substitute for LDL cholesterol" in calculating risk. Total cholesterol could be measured easily in the clinic, so physicians would continue to measure it. The evidence had dictated a complete turnabout in the science, and then pragmatic considerations had turned it about again, until the clinical management of patients and the public perception were back exactly where they had started.

The revelations about HDL had equally little influence on the institution of a national low-fat, high-carbohydrate diet. Whether or not triglycerides were an independent risk factor, once the protective nature of HDL was confirmed, then Gofman's argument of 1950 was also reaffirmed: there were at least two potential diet-related ways of preventing heart disease, and any treatment that improved the situation with one risk factor had to avoid exacerbating the situation with the other. In the 1960s, Gofman, Ahrens, Albrink, and Fredrickson, Levy, and Lees had all discussed the dangers of replacing the fat in the diet with carbohydrates because this would elevate triglycerides. Now the dangers of lowering HDL became the issue. "In the search for an optimal therapy for avoiding or correcting atherosclerosis," as the Framingham investigators noted in 1979, "the ideal lipid response would appear to be the one that raises HD lipoprotein as it lowers LD lipoprotein. Therapeutic maneuvers that affect only one of these lipoprotein particle systems in a favorable way, while adversely affecting the other, may be less promising. . . ."

Diets that lowered cholesterol by replacing saturated fat with polyunsaturated fats would have accomplished such a balancing act, but there was legitimate concern that polyunsaturated fats were carcinogenic, and so the AHA had simply recommended fat reduction in general. This meant replacing the fat calories with carbohydrates. But the "good cholesterol" in HDL would be diminished by eating more carbohydrates. By the 1980s, discussions of heart-disease prevention typically avoided this dilemma by neglecting to mention the effect of carbohydrates on HDL.* Instead,

---

* Those that did mention the effect of carbohydrates on HDL cholesterol rejected the relevance to heart disease, on the basis, as the American Heart Association explained, "that epidemiological studies have demonstrated an inverse relation between carbohydrate consumption and risk for CHD."

people were told to raise their HDL through exercise and weight loss, and then prescribed, as the American Heart Association did, low-fat, high-carbohydrate diets as the means to lose that weight.

In 1985, Scott Grundy and his colleague Fred Mattson provided what appeared to be the ideal compromise—a dietary means both to lower LDL cholesterol and to raise HDL cholesterol without consuming more carbo-hydrates or saturated fats. This was monounsaturated fats, such as the oleic acid found in olive oil, and it served to keep the focus on the fat in the diet, rather than the carbohydrates. In the 1950s, Keys had assumed that monounsaturated fats were neutral, because they had no effect on total cholesterol. But this apparent neutrality, as Grundy reported, was due to the ability of these fats simultaneously to raise HDL cholesterol *and* lower LDL cholesterol. Saturated fats raise both HDL and LDL cholesterol. Car-bohydrates lower LDL cholesterol but also lower HDL. Grundy and Matt-son's discovery of the double-barreled effect of monounsaturated fats, and particularly oleic acid, reignited the popular interest in the Mediterranean diet as the ideal heart-healthy diet, though it seemed to be heart-healthy only in some Mediterranean regions and not in others, and such diets, as even Grundy conceded, had never been tested. When they finally were tested in two clinical trials in the 1990s—the Lyon Diet Heart Trial and an Italian study known as GISSI-Prevenzione—both supported the con-tention that the diet prevented heart attacks, but neither provided evidence that it did so by either raising HDL or lowering LDL, which was how it was now alleged to work.

The observation that monounsaturated fats both lower LDL cholesterol and raise HDL also came with an ironic twist: the principal fat in red meat, eggs, and bacon is not saturated fat, but the very same monounsaturated fat as in olive oil. The implications are almost impossible to believe after three decades of public-health recommendations suggesting that any red meat consumed should at least be lean, with any excess fat removed.

Consider a porterhouse steak with a quarter-inch layer of fat. After broil-ing, this steak will reduce to almost equal parts fat and protein.* Fifty-one percent of the fat is monounsaturated, of which 90 percent is oleic acid. Saturated fat constitutes 45 percent of the total fat, but a third of that is stearic acid, which will increase HDL cholesterol while having no effect on LDL. (Stearic acid is metabolized in the body to oleic acid, according to

---

* The nutritional constituents of such a piece of relatively fatty meat can be found in the Nutrient Database for Standard Reference at the USDA Web site, along with those of thousands of other foods.

Grundy's research.) The remaining 4 percent of the fat is polyunsaturated, which lowers LDL cholesterol but has no meaningful effect on HDL. In sum, perhaps as much as 70 percent of the fat content of a porterhouse steak will improve the relative levels of LDL and HDL cholesterol, compared with what they would be if carbohydrates such as bread, potatoes, or pasta were consumed. The remaining 30 percent will raise LDL cholesterol but will also raise HDL cholesterol and will have an insignificant effect, if any, on the ratio of total cholesterol to HDL. All of this suggests that eating a porterhouse steak in lieu of bread or potatoes would actually reduce heart-disease risk, although virtually no nutritional authority will say so publicly. The same is true for lard and bacon.

"Everything should be made as simple as possible," Albert Einstein once supposedly said, "but no simpler." Our understanding of the nutritional causes of heart disease started with Keys's original oversimplification that heart disease is caused by the effect of all dietary fat on total serum cholesterol. Total cholesterol gave way to HDL and LDL cholesterol and even triglycerides. All fat gave way to animal and vegetable fat, which gave way to saturated, monounsaturated, and polyunsaturated fat, and then polyunsaturated fats branched into omega-three and omega-six polyunsaturated fats. By the mid-1980s, these new levels of complexity had still not deterred the AHA and NIH from promoting carbohydrates as effectively the antidote to heart disease, and either all fats or just saturated fats as the dietary cause.

What would now become apparent was that LDL cholesterol is little more than an arbitrary concept that oversimplifies its own complex diversity. The fact that LDL and *LDL cholesterol* are not synonymous complicates the science. Just as Gofman had reported in 1950 that cholesterol itself was divided up among different lipoproteins, and those lipoproteins had different atherogenic properties and responded differently to diet, a lipid metabolism specialist named Ronald Krauss, using Gofman's ultracentrifuge, began reporting in 1980 that low-density lipoproteins were in turn composed of different, distinct subclasses, each containing differing amounts of cholesterol, and each, once again, with different atherogenic properties and different behavior in response to the carbohydrates and fats in our diet. Although Krauss has long been considered one of the most thoughtful researchers in nutrition and heart disease—the American Heart Association has treated him as such—it's worth noting in advance that his dietary research has been almost universally ignored, precisely because of its ultimate implications for what constitutes a healthy diet and what does not.

LDL cholesterol is only a "marginal risk factor," Tavia Gordon and his colleagues had observed in 1977. In other words, little difference can be observed between the average LDL cholesterol of those with and without heart disease. Only by comparing the LDL-cholesterol and heart-disease rates between nations (with all the attendant complications of such comparisons) can conspicuous differences be seen. In the analysis from Framingham, San Francisco, Albany, Honolulu, and Puerto Rico published by Gordon and his collaborators, the average LDL cholesterol of heart-disease sufferers was only a few percentage points higher than the average of those who remained healthy. "If you look in the literature and just look at the average coronary patients," Krauss says, "their LDL-cholesterol levels are often barely discernibly elevated compared to patients who do not have coronary disease."

In the late 1940s, Gofman and his collaborators began asking why the same level of LDL cholesterol will cause heart disease in some people but not in others. Krauss and his collaborators began asking this question again, thirty years later.

Krauss himself is an idiosyncratic figure in this world. He has produced a dozen years of research suggesting that high-carbohydrate diets, for the great proportion of the population, are the nutritional cause of heart disease, and yet he has also chaired the nutrition committee of the American Heart Association and was the primary author of the 1996 and 2000 AHA nutrition guidelines. In the process, he eased the AHA away from its thirty-year-old position that the maximum fat content of a heart-healthy diet should be 30 percent of calories. Or, as Krauss remarked, he managed to put the "30-percent-fat recommendation in small print." Krauss trained as a physician in the late 1960s and then worked with Fredrickson and Levy at the NIH, where he discovered a protein known as hepatic lipase that regulates how the liver metabolizes lipoproteins. He then moved to Berkeley to practice internal medicine, and it was there, in 1976, that he began working with Gofman's ultracentrifuge and with Alex Nichols and Frank Lindgren, both of whom had collaborated with Gofman in the 1950s.

When Krauss began his research at Berkeley, he had what he calls "this conventional notion, which many people still have, that LDL is just one thing, a single entity." But that turned out not to be the case. Using data from the ultracentrifuge dating back to the early 1960s, Krauss discovered that LDL actually comes in distinct subspecies, all characterized by still finer gradations in density and size. "It was blazingly obvious. Unignorable," says Krauss.* Eventually, Krauss identified seven discrete subclasses of LDL. He also noted that the smallest and densest of the low-density lipoproteins had two significant properties: it had a strong negative correlation with HDL, and it was the subspecies that was elevated in patients with heart disease.

In the early 1980s, Krauss published three papers on what he calls the "remarkable heterogeneity of LDL," all of which, he says, were met with indifference mixed with occasional hostility. Acceptance of Krauss's research was also constrained by the fact that Gofman's ultracentrifuge had been necessary to differentiate these LDL subclasses, which meant that this, too, was not the kind of measurement that could be ordered up easily by physicians. In his later publications, Krauss described a simpler, inexpensive measurement technique, but the research was still perceived as an esoteric endeavor.

To understand the implications of this association between small, dense LDL and heart disease, it helps to picture the configuration of the low-density lipoprotein itself. Imagine it as a balloon. It has a single protein— known as apo B, for short—that serves as the structural foundation of the balloon and holds it together. It has an outer membrane that is composed of cholesterol and fats of yet another type, called phospholipids. And then, inside the balloon, inflating it, are triglycerides and more cholesterol. The size of the LDL balloon itself can vary, depending on the amount of triglycerides and cholesterol it contains. Thus, as Krauss reported, some people have mostly large, fluffy LDL, with a lot of cholesterol and triglycerides inflating the balloon, and some people have mostly smaller, denser LDL particles, with less cholesterol and triglycerides.

In the 1970s, investigators had developed yet another way to quantify the concentration of these circulating lipoproteins, in this case by counting only the *number* of apo B proteins that provide the structural foundation to the LDL balloon. Because there's only one protein per LDL particle, and because VLDL is also composed of identical apo B proteins, this technique measured the *number* of LDL and VLDL particles in a blood sample, rather than the cholesterol or triglycerides they contained. As it turned out, the number of apo B proteins, and so the total number of LDL and VLDL particles combined, is also abnormally elevated in heart-disease patients.

---

* To be precise, Krauss says, he *rediscovered* this heterogeneity of LDL: Waldo Fisher of the University of Florida, and Verne Schumaker of the University of California, Los Angeles, had discovered it independently a decade earlier, but had not pursued it further.

This was first reported in 1980 by Peter Kwiterovich, a lipid-metabolism specialist from Johns Hopkins, together with Allan Sniderman, a cardiologist from McGill University. Kwiterovich and Sniderman then collaborated with Krauss on the last of his three papers on the heterogeneity of LDL. In 1983, they reported that the disproportionate elevation in the apo B protein in heart-disease patients was due to a disproportionate elevation in the amount of the smallest and densest of the low-density lipoproteins.

This explained what Krauss had set out to understand: why two people can have identical LDL-cholesterol levels and yet one develops atherosclerosis and coronary heart disease and the other doesn't—why LDL cholesterol is only a marginal risk factor for heart disease. If we have low LDL cholesterol, but it's packaged almost exclusively in small, dense LDL particles—the smaller balloons—that translates to a higher risk of heart disease. If we have high LDL cholesterol, but it's packaged in a smaller number of large, fluffy LDL particles—the larger balloons—then our heart-disease risk is significantly lower. Small, dense LDL, simply because it is small and dense, appears to be more *atherogenic,* more likely to cause atherosclerosis. Small, dense LDL can squeeze more easily through damaged areas of the artery wall to form incipient atherosclerotic plaques. Sniderman describes small, dense LDL as the equivalent of "little bits of sand" that get in everywhere and stick more avidly. The relative dearth of cholesterol in these particles may also cause structural changes in the protein that make it easier for it to adhere to the artery wall to begin with. And because small, dense LDL apparently remains in the bloodstream longer than larger and fluffier LDL, it has more time and greater opportunities to do its damage. Finally, it's possible that LDL has to be *oxidized*—the biological equivalent, literally, of rusting—before it can play a role in atherosclerosis, and the existing evidence suggests that small, dense LDL oxidizes more easily than the larger, fluffier variety.

Through the 1980s, Krauss continued to refine this understanding of how LDL subspecies affect heart disease. He discovered that the appearance of LDL in the population falls into two distinct patterns or traits, which he called pattern A and pattern B. Pattern A is dominated by large, fluffy LDL and implies a low risk of heart disease; pattern B is the dangerous one, with predominantly small, dense LDL. Pattern B is invariably accompanied by high triglycerides *and* low HDL. Pattern A is not. In 1988, Krauss and his collaborators reported in *JAMA* that heart-disease patients were three times more likely to have pattern B than pattern A. Krauss called pattern B the *atherogenic profile.* Diabetics have the identical pattern.

The effect of diet on this atherogenic profile now became the pivotal issue. In the 1960s and most of the 1970s, the dietary goal was to lower total cholesterol. After the 1977 revelations about HDL, the best diet became the one that lowered LDL cholesterol and maybe raised HDL in the process. But if Krauss and his collaborators were right, a diet that lowers total cholesterol or LDL cholesterol can conceivably do so in a way that actually *increases* the proportion of small, dense LDL in the blood turning the healthy pattern A trait into the atherogenic pattern B. If we focus on LDL cholesterol alone, such a diet might appear to prevent heart disease. But if the size, density, and number of the LDL subspecies are indeed the important variables, the diet could in fact increase heart-disease risk.

Though pattern A and B traits appear to be strongly influenced by genetics, diet and other lifestyle factors play a critical role. In the late 1980s, Krauss began a series of clinical trials to explore the association between diet and the dangerous small, dense LDL. The results of his seven trials have been consistent: the *lower* the fat in the diet and the *higher* the carbohydrates, the smaller and denser the LDL and the more likely the atherogenic pattern B appears; that is, the more carbohydrates and the less fat, the greater the risk of heart disease.

On a diet that Krauss calls the "average American diet," with 35 percent of the calories from fat, one in three men will have the atherogenic pattern B profile. On a diet of 46 percent fat, this proportion *drops*: only one man in every five manifests the atherogenic profile. On a diet of only 10 percent fat, of the kind advocated by diet doctors Nathan Pritikin and Dean Ornish, two out of every three men will have small, dense LDL and, as a result, a predicted threefold higher risk of heart disease. The same pattern holds true in women and in children, but the percentages with small, dense LDL are lower. Krauss and his colleagues even tested the effect of types of fat on these lipoproteins, and reported that, the more saturated fat in the diet, the larger and fluffier the LDL—a beneficial effect.*

Though the concept of small, dense LDL as a risk factor for heart disease has been accepted into the orthodox wisdom, as has Krauss's atherogenic profile (although now renamed *atherogenic dyslipidemia*), his dietary research has had no perceptible influence on discussions of the dietary prevention of heart disease. The implications are so provocative that many investigators simply ignore them. Even those clinical investigators who firmly believe that small, dense LDL is indeed the atherogenic form of LDL

---

* This suggests that saturated fat elevates LDL-cholesterol levels in part by increasing the *amount* of cholesterol in the LDL, and so making larger and fluffier LDL to begin with, rather than by increasing the number of LDL particles or by increasing the number of small, dense LDL particles.

often refuse to comment on the dietary implications. "Well, I would rather not get into that," said the University of Washington epidemiologist Melissa Austin, who studies triglycerides and heart disease and has collaborated with Krauss on studies of the small, dense LDL.

Goran Walldius, a cardiologist at the Karolinska Institute in Stockholm, had the same response. Walldius is the principal investigator of an enormous Swedish study to ascertain heart-disease risk factors. The 175,000 subjects include *every* patient who received a health checkup in the Stockholm area in 1985. Blood samples were taken at the time, and Walldius and his colleagues have been following the subjects ever since, to see which measures of cholesterol, triglycerides, or lipoproteins are most closely associated with heart disease. Far and away, the best predictor of risk, as Walldius reported in 2001, was the concentration of apo B proteins, reflecting the dominance of small, dense LDL particles. Half of the patients who died of heart attacks, he reported, had normal LDL-cholesterol levels but high apo B numbers. Apo B is a much better predictor of heart disease than LDL cholesterol, Walldius said, because LDL cholesterol "doesn't tell you anything about the quality of the LDL." But when asked in an interview to comment on Krauss's research and the subject of dietary interventions that might increase the size of LDL particles, Walldius said, "I'll have to pass on that one."

The notion that carbohydrates determine the ultimate *atherogenicity* of lipoproteins is surprisingly easy to explain by the current understanding of fat-and-cholesterol transport. This model also accounts neatly for the observed relationship between heart disease, triglycerides, and cholesterol, and so constitutes another level of the physiological mechanisms underlying the carbohydrate hypothesis. The details are relatively straightforward, but, not surprisingly, they represent a radical shift from the mechanisms envisioned by Keys and others, in which coronary artery disease is caused by the simple process of saturated fat raising total-cholesterol or LDL-cholesterol levels. This is another way in which the subspecialization of medical researchers works against progress. For most epidemiologists, cardiologists, internists, nutritionists, and dieticians, their knowledge of lipoprotein metabolism dates to their medical or graduate-school training. Short of reading the latest biochemistry textbooks or the specialized journals devoted to this research, they have few available avenues (and little reason, as they see it) for keeping up-to-date, and so the current understanding of these metabolic processes escapes them. The details of lipoprotein metabolism circa 2007 remain a mystery to the great proportion of clinicians and investigators involved in the prevention of heart disease.

One key fact to remember in this discussion is that LDL and *LDL cholesterol* are not one and the same. The LDL carries cholesterol, but the amount of cholesterol in each LDL particle will vary. Increasing the LDL cholesterol is not the same as increasing the number of LDL particles.

There are two ways to increase the amount of cholesterol in LDL. One is to increase the amount of cholesterol secreted to begin with; the other is to decrease the rate of disposal of cholesterol once it's been created (which is apparently what happens when we eat saturated fat). Either method will eventually result in elevated LDL cholesterol. Joseph Goldstein and Michael Brown worked out the details of the clearance-and-disposal mechanism in the 1970s, and this work won them the Nobel Prize.

As for secretion, the key point is that most low-density lipoproteins, LDL, begin their lives as very low-density lipoproteins, VLDL. (This was one implication of the observation that both LDL and VLDL are composed of the same apo B protein, and it was established beyond reasonable doubt in the 1970s.) This is why VLDL is now commonly referred to as a *precursor* of LDL, and LDL as a *remnant* of VLDL. If the liver synthesizes more cholesterol, we end up with more total cholesterol and so more LDL cholesterol, although apparently not more LDL particles. If the liver synthesizes and secretes more VLDL, we will also end up with more LDL cholesterol but we have more LDL particles as well, and they'll be smaller and denser.

This process is easier to understand if we picture what's actually happening in the liver. After we eat a carbohydrate-rich meal, the bloodstream is flooded with glucose, and the liver takes some of this glucose and transforms it into fat—i.e., triglycerides—for temporary storage. These triglycerides are no more than droplets of oil. In the liver, the oil droplets are fused to the apo B protein and to the cholesterol that forms the outer membrane of the balloon. The triglycerides constitute the cargo that the lipo-proteins drop off at tissues throughout the body. The combination of cholesterol and apo B is the delivery vehicle. The resulting lipoprotein has a very low density, and so is a VLDL particle, because the triglycerides are lighter than either the cholesterol or the apo B. (In the same way, the more air in the hold of a ship, the less dense the ship and the higher it floats in the water.) For this reason, the larger the initial oil droplet, the more triglycerides packaged in the lipoprotein, the lower its density.

The liver then secretes this triglyceride-rich VLDL into the blood, and the VLDL sets about delivering its cargo of triglycerides around the body.

Throughout this process, known poetically as the *delipidation cascade,* the lipoprotein gets progressively smaller and denser until it ends its life as a low-density lipoprotein—LDL. One result is that *any* factor that enhances the synthesis of VLDL will subsequently increase the number of LDL particles as well. As long as sufficient triglycerides remain in the lipoprotein to be deposited in tissues, this evolution to progressively smaller and denser LDL continues. It's this journey from VLDL to LDL that explains why most men who have high LDL cholesterol will also have elevated VLDL triglycerides. "It's the overproduction of VLDL and apo B that is the most common cause of high LDL in our society," says Ernst Schaefer, director of the lipid-metabolism laboratory at the Jean Mayer USDA Human Nutrition Research Center on Aging at Tufts University. None of this, so far, is controversial; the details are described in recent editions of biochemistry textbooks.

How this process is regulated is less well established. In Krauss's model, based on his own research and that of the Scottish lipid-metabolism researcher Chris Packard and others, the rate at which triglycerides accumulate in the liver controls the size of the oil droplet loaded onto the lipoprotein, and which of two pathways the lipoprotein then follows. If triglycerides are hard to come by, as would be the case with diets low in either calories *or* carbohydrates, then the oil droplets packaged with apo B and cholesterol will be small ones. The ensuing lipoproteins secreted by the liver will be of a subspecies known as intermediate-density lipoproteins—which are less dense than LDL but denser than VLDL— and these will end their lives as relatively large, fluffy LDL. The resulting risk of heart disease will be relatively low, because the liver had few triglycerides to dispose of initially.

If the liver has to dispose of copious triglycerides, then the oil droplets are large, and the resulting lipoproteins put into the circulation will be triglyceride-rich and very low-density. These then progressively give up their triglycerides, eventually ending up, after a particularly extended life in the circulation, as the atherogenic small, dense LDL. This triglyceride-rich scenario would take place whenever carbohydrates are consumed in abundance. "I am now convinced it is the carbohydrate inducing this atherogenic [profile] in a reasonable percentage of the population," says Krauss. ". . . we see a quite striking benefit of carbohydrate restriction."

This model also explains, as Pete Ahrens suggested in 1961, why high-carbohydrate diets appear innocuous in populations that are chronically undernourished. This was inevitably the case with those Southeast Asian populations extolled by Keys and others for their low total-cholesterol levels and apparent absence of heart disease. Such populations lived on carbohydrate-rich diets out of economic necessity rather than choice. Their diets were predominantly unrefined carbohydrates because that's what they cultivated and it was all they could afford. As Ahrens had noted, the great proportion of individuals in such populations barely eked out enough calories to survive. This was true not only of Japan in the years after World War II, but of Greece and other areas of the Mediterranean as well. If these populations indeed had low cholesterol and suffered little from heart disease, a relative lack of calories and a near-complete absence of refined carbohydrates would have been responsible, not the low intake of saturated fat. In developed nations—the United States, for example— where calories are plentiful, it would be the carbohydrates pushing our metabolisms toward the production of atherogenic lipoproteins. Here, too, the saturated fat in the diet is of little significance.

Chapter Ten

THE ROLE OF INSULIN

The suppression of inconvenient evidence is an old trick in our profession. The subterfuge may be due to love of a beautiful hypothesis, but often enough it is due to a subconscious desire to simplify a confusing subject. It is not many years ago that the senior physician of a famous hospital was distinctly heard to remark, sotto voce, "medicine is getting so confusing nowadays, what with insulin and things." It is a sentiment with which almost everybody who qualified more than a quarter of a century ago is likely to sympathize. . . . But ignoring difficulties is a poor way of solving them.

RAYMOND GREENE, in a letter to
The Lancet, 1953

S cientific progress is driven as much by the questions posed as by the tools available to answer them. In the 1950s, when Ancel Keys settled on dietary fat and cholesterol as causes of heart disease, he did so because he sought to understand the disparity in disease rates among nations and what he believed was a growing epidemic of coronary heart disease in the United States. Those investigators whose research would eventually evolve into the science of metabolic syndrome—the physiological abnormalities common to obesity, diabetes, and heart disease—had different questions in mind. Why are the obese exceptionally likely to become diabetic and vice versa? Why is atherosclerosis so common with both diabetes and obesity? Are these coincidental associations, or do obesity, heart disease, and diabetes share a common cause?

In the decade after World War II, Jean Vague, a professor of medicine at the University of Marseille in France, extended these associations to what he called "android obesity," where the excess fat sits predominantly around the waist. ("Beer bellies" are the archetypal example.) Vague reported that android obesity was associated with atherosclerosis, gout, kidney stones, and adult-onset diabetes. He speculated that some type of hormonal overactivity led to overeating, and that, in turn, to an increased secretion of insulin to store away the excess calories in fat tissue. This excessive secretion of insulin might then, over the years, cause what he called *pancreatic depletion* and thus diabetes. A similar hormonal overactivity, Vague suggested, might cause atherosclerosis, either directly or by inducing the secretion of "lipoprotein molecules," as John Gofman was proposing, which would then cling to the artery walls and begin the accumulation of fats and cholesterol that is characteristic of atherosclerotic plaques.

Gofman also sought out common mechanisms to explain the association between obesity and heart disease. Because weight gain was associated with both higher blood pressure and increased triglyceride-rich VLDL, he suggested, that alone could explain why the obese had an increased risk of heart disease. But Gofman did not speculate whether weight gain elevated blood pressure and triglycerides or whether the same mechanism increased our weight and raised our blood pressure and triglycerides.

It was Margaret Albrink who extended Gofman's observations to diabetes and set the stage for the science that would eventually evolve into our current understanding of metabolic syndrome. In 1931, Albrink's advisers at Yale, John Peters and Evelyn Man, had set out to test the speculation voiced by Elliot Joslin, among others, that the atherosclerosis that plagues diabetics is caused by the fat and cholesterol in their carbohydrate-restricted diets. Man and Peters measured cholesterol in seventy-nine diabetics treated at Yale and reported in 1935 that the high-fat diets then prescribed for diabetics did *not* increase cholesterol: only nine of the seventy-nine had abnormally high cholesterol—the ones who "were extremely ill and profoundly emaciated." Man and Peters continued collecting blood samples from diabetic patients for another quarter-century. In 1962, Albrink reported that the average triglycerides in these samples had increased by 40 percent over the years, and this was accompanied by a dramatic increase in the proportion of diabetics with atherosclerotic complications—from 10 percent in the early 1930s to 56 percent by the late 1950s. This coincided with a doubling of the proportion of carbohydrates in the prescribed diabetic diet and a reduction in fat calories from 60 percent to 40 percent, in accord with the increasing suspicion that fatty diets caused heart disease. (Joslin made a similar observation in 1959.) Albrink also confirmed Gofman's observation that weight gain was accompanied by high triglyceride levels: adding ten pounds in middle age was associated with a 50 percent increase in triglycerides. Almost invariably, the greater the body fat, the higher the triglycerides in the circulation.

To Albrink, these associations implied that heart-disease research should not be guided by Keys's model but, rather, by attempts to understand what she called the "abnormal metabolic patterns" common to obesity, diabetes, and heart disease. High triglycerides characterized these abnormalities, Albrink said. She proposed that these patterns were caused or exacerbated in susceptible individuals by diets high in either calories or carbohydrates or just "purified carbohydrates." But she offered no biological mechanism to explain it.

The potential explanation arrived in the form of two insulin-related conditions, insulin resistance and chronically elevated levels of insulin in the circulation, hyperinsulinemia—a vitally important focus of our inquiry.

Through the first half of the twentieth century, little was understood of insulin beyond its role in diabetes, because no method existed to measure its concentration in the bloodstream with any accuracy. Insulin is a very small protein, technically known as a peptide, and it circulates in the blood in concentrations that are infinitesimal compared with those of cholesterol and lipoproteins. As a result, the measurement of insulin in human blood relied on a variety of arcane tests that depended on the ability of insulin to prompt the absorption of glucose by laboratory rats or even by fat or muscle tissue in a test tube. This situation changed in 1960 with the discovery by Rosalyn Yalow and Solomon Berson of a method capable of reliably measuring the concentration of insulin and other peptide hormones in human blood. In 1977, when Yalow was awarded the Nobel Prize for the discovery (Berson had died in 1972), the Nobel Foundation described Yalow and Berson's measurement technology as bringing about "a revolution in biological and medical research."

The impact on diabetes research had been immediate. Yalow and Berson showed that those who had developed diabetes as adults had levels of circulating insulin significantly higher than those of healthy individuals— a surprising finding. It had long been assumed that *lack* of insulin was the root of all diabetes. As Yalow and Berson among others also reported, the obese, too, had chronically elevated insulin levels.

By 1965, Yalow and Berson had suggested why these adult-onset diabetics could appear to be lacking insulin—manifesting the symptoms of diabetes, high blood sugar, and sugar in their urine—while simultaneously having excessive insulin in their circulation: their tissues did not respond properly to the insulin they secreted. They were *insulin-resistant*, defined by Yalow and Berson as "a state (of a cell, tissue, system or body) in which greater-than-normal amounts of insulin are required to elicit a quantitatively normal response." Because of their resistance to insulin, adult-onset diabetics had to secrete more of the hormone to maintain their blood sugar within healthy levels, and this would become increasingly difficult to achieve the longer they remained insulin-resistant.*

---

* What used to be known as juvenile-onset diabetes, which is characterized by an insulin deficit, is referred to as Type 1 or insulin-dependent diabetes mellitus, IDDM. The less severe form, which

A critical aspect of this insulin resistance, Yalow and Berson noted, is that some tissues might become resistant to insulin while others continued to respond normally, and this would determine how the damage done by the insulin resistance would manifest itself in different individuals. So "it is desirable," they wrote, "wherever possible, to distinguish generalized resistance of all tissues from resistance of only individual tissues."

From the mid-1960s onward, our understanding of the role of insulin resistance in both heart disease and diabetes was driven by the work of Stanford University diabetologist Gerald Reaven. Reaven began his investigations by measuring triglycerides and glucose tolerance in heart-attack survivors. A glucose-tolerance test is a common test given by physicians to determine if a patient is either diabetic or on the way to becoming so. The patient drinks a solution of glucose and water, and then, two hours later, the physician measures his or her blood sugar. If the blood sugar is higher than what's considered normal, it means the patient has been unable to metabolize the glucose properly—hence, *glucose intolerance*—and so either lacks sufficient insulin to deal with the glucose, or is resistant to the insulin that is secreted. In 1963, Reaven reported that heart-attack survivors invariably had both high triglycerides and glucose intolerance, and this suggested that the two conditions had a common cause. Reaven considered insulin resistance to be the obvious suspect.

Working with John Farquhar, who had studied with Pete Ahrens at Rockefeller, Reaven developed a two-part hypothesis.

The first part explained why most, if not virtually all individuals with high triglycerides had what Ahrens had called *carbohydrate-induced lipemia*. In other words, their triglyceride levels increased with carbohydrate-rich diets and decreased when fat replaced the carbohydrates. The crucial factor, Reaven explained, is that, the more carbohydrates consumed, the more insulin is needed to transport the glucose from the carbohydrates into cells where it can be used as fuel. This insulin, however, also prompts the liver to synthesize and secrete triglycerides for storage in the fat tissue. If someone who is already insulin-resistant consumes a carbohydrate-rich diet, according to Reaven's hypothesis, the person will have to secrete even more insulin to deal with the glucose, prompting in turn even greater synthesis

---

is characterized by insulin resistance rather than a lack of insulin, used to be called adult-onset diabetes. It is now called Type 2 or non-insulin-dependent diabetes mellitus or NIDDM. This is the terminology that I'll now use as well.

and secretion of triglycerides by the liver, and so even higher triglyceride levels in the blood.

This, in turn, implied part two of the hypothesis: if eating a carbohydrate-rich diet in the presence of insulin resistance will abnormally elevate triglyceride levels, then it's hard to avoid the implication that eating a carbohydrate-rich diet increases the risk of heart disease. Insulin resistance and carbohydrates will also exacerbate Type 2 diabetes, according to Reaven's hypothesis, and this would explain, as well, why these diabetics inevitably have high triglycerides. By 1967, Reaven and Farquhar had reported that triglyceride levels, insulin resistance, and insulin levels moved up and down in concert even in healthy individuals: the more insulin secreted in response to carbohydrates, the greater the apparent insulin resistance and the higher the triglycerides.

Reaven and Farquhar spent the next twenty years working to establish the validity of the hypothesis. Much of the progress came with the development, once again, of new measuring techniques: in this case, tests that allowed investigators to measure insulin resistance directly. In 1970, Reaven and Farquhar published the details of the first such insulin-resistance test, which was then followed by a half-dozen more. The best of these—the "gold standard"—was developed at the NIH in the late 1960s and then refined over the next decade by a young endocrinologist named Ralph DeFronzo. It wasn't until 1979, after DeFronzo joined the faculty at Yale Medical School and began measuring insulin resistance in human patients, that he published the details. It would take another decade for Reaven, Farqhuar, and DeFronzo, along with Eleuterio Ferrannini of the University of Pisa, among others, to convince diabetologists that resistance to insulin was the fundamental defect in Type 2 diabetes.

In 1987, the American Diabetes Association honored DeFronzo with its award for outstanding scientific achievement. A year later, Reaven received the ADA's Banting Medal for Scientific Achievement.* Reaven then gave the prestigious Banting Lecture at the ADA's annual conference and took the opportunity to extend the implications of his research. For the first time, he laid out the hypothesis of what he called Syndrome X (metabolic syndrome) and the cluster of disorders—including insulin resistance, hyperinsulinemia, high triglycerides, low HDL cholesterol, and high blood pressure—that accompanies Type 2 diabetes and obesity and plays a critical role in the genesis of heart disease even in nondiabetics. "Although this concept may seem outlandish at first blush," Reaven said,

---

* Named after Frederick Banting, the co-discoverer of insulin, a distant relative of William Banting, of corpulence notoriety.

"the notion is consistent with available experimental data." As Reaven described it, the condition of being resistant to insulin leads to both heart disease and diabetes. But not everyone with insulin resistance becomes diabetic; some continue to secrete sufficient insulin to overcome their insulin resistance, though this hyperinsulinemia causes havoc on its own, including elevating triglyceride levels, and also further exacerbating the insulin resistance—a vicious cycle.

Reaven supported his hypothesis with the results of observational studies that had already linked hyperinsulinemia, insulin resistance, and Type 2 diabetes to high triglycerides, heart disease, obesity, stroke, and hypertension. Three large-scale Framingham-like prospective studies of healthy nondiabetic populations—in Paris, Helsinki, and in Busselton, Australia—had also reported that, the higher the insulin levels, the greater the risk of heart disease.

As DeFronzo later remarked, the conclusion that hyperinsulinemia and insulin resistance were related to "a whole host of metabolic disorders" was an obvious one, but it required that clinical investigators measure insulin resistance in human patients, which would always be the obstacle in the science of metabolic syndrome. Measuring insulin resistance requires multiple tests of blood sugar while insulin levels are held constant and precise amounts of glucose are consumed or infused into the bloodstream. This is not the kind of test that physicians can do in a checkup, at least not without going far beyond the usual practice of sending a blood sample out to a laboratory for a battery of tests. As a result, when the National Cholesterol Education Program officially acknowledged the existence of Reaven's Syndrome X in 2002 (renaming it metabolic syndrome), neither insulin resistance nor hyperinsulinemia was included among the diagnostic criteria, despite being the fundamental defects in the syndrome itself.

Reaven's 1988 Banting Lecture is credited as the turning point in the effort to convince diabetologists of the critical importance of insulin resistance and hyperinsulinemia, but those investigators concerned with the genesis of heart disease paid little attention, considering anything having to do with insulin to be relevant only to diabetes. This was a natural consequence of the specialization of scientific research. Through the mid-1980s, Reaven's research had focused on diabetes and insulin, and so his publications appeared almost exclusively in journals of diabetes, endocrinology, and metabolism. Not until 1996 did Reaven publish an article on Syndrome X in the American Heart Association journal *Circulation*, the primary journal for research in heart disease. Meanwhile, his work had no influence on public-health policy or the public's dietary consciousness. Neither the 1988 *Surgeon General's Report on Nutrition and Health* nor the National Academy of Sciences's 1989 *Diet and Health* mentioned insulin resistance or hyperinsulinemia in any context other than Reaven's cautions that high-carbohydrate diets might not be ideal for Type 2 diabetics. Both reports ardently recommended low-fat, high-carbohydrate diets for the prevention of heart disease.

Even the diabetes community found it easier to accept Reaven's science than its dietary implications. Reaven's observations and data "speak for themselves," as Robert Silverman of the NIH suggested at a 1986 consensus conference on diabetes prevention and treatment. But they placed nutritionists in an awkward position. "High protein levels can be bad for the kidneys," said Silverman. "High fat is bad for your heart. Now Reaven is saying not to eat high carbohydrates. We have to eat something." "Sometimes we wish it would go away," Silverman added, "because nobody knows how to deal with it."

This is what psychologists call *cognitive dissonance,* or the tension that results from trying to hold two incompatible beliefs simultaneously. When the philosopher of science Thomas Kuhn discussed cognitive dissonance in scientific research—"the awareness of an anomaly in the fit between theory and nature"—he suggested that scientists will typically do what they have invariably done in the past in such cases: "They will devise numerous articulations and *ad hoc* modifications of their theory in order to eliminate any apparent conflict." And that's exactly what happened with metabolic syndrome and its dietary implications. The syndrome itself was accepted as real and important; the idea that it was caused or exacerbated by the excessive consumption of carbohydrates simply vanished.

Among the few clinical investigators working on heart disease who paid attention to Reaven's research in the late 1980s was Ron Krauss. In 1993, Krauss and Reaven together reported that small, dense LDL was another of the metabolic abnormalities commonly found in Reaven's Syndrome X. Small, dense LDL, they noted, was associated with insulin resistance, hyperinsulinemia, high blood sugar, hypertension, and low HDL as well. They also reported that the two best predictors of the presence of insulin resistance and the dominance of small, dense LDL are triglycerides and HDL cholesterol—the higher the triglycerides and the lower the HDL, the more likely it is that both insulin resistance and small, dense LDL are present. This offers yet another reason to believe the carbohydrate hypothesis of heart disease, since metabolic syndrome is now considered perhaps the dominant heart-disease risk factor—a "coequal partner to cigarette smoking as contributors to premature [coronary heart disease]," as the National Cholesterol Education Program describes it—and both triglycerides and HDL cholesterol are influenced by carbohydrate consumption far more than by any fat.

Nonetheless, when small, dense LDL and metabolic syndrome officially entered the orthodox wisdom as risk factors for heart disease in 2002, the cognitive dissonance was clearly present. First the National Cholesterol Education Program published its revised guidelines for cholesterol testing and treatment. This was followed in 2004 by two conference reports: one describing the conclusions of a joint NIH-AHA meeting on scientific issues related to metabolic syndrome, and the other, in which the American Diabetes Association joined in as well, describing joint treatment guidelines. Scott Grundy of the University of Texas was the primary author of all three documents. When I interviewed Grundy in May 2004, he acknowledged that metabolic syndrome was the cause of most heart disease in America, and that this syndrome is probably caused by the excessive consumption of refined carbohydrates. Yet his three reports—representing the official NIH, AHA, and ADA positions—all remained firmly wedded to the fat-cholesterol dogma. They acknowledge metabolic syndrome as an *emerging risk factor* for heart disease, but identify LDL cholesterol as "the primary driving force for coronary atherogenesis." Thus, heart disease in America, as the National Cholesterol Education Program report put it, was still officially caused by "mass elevations of serum LDL cholesterol result[ing] from the habitual diet in the United States, particularly diets high in saturated fats and cholesterol."

There was no mention that carbohydrates might be responsible for causing or exacerbating either metabolic syndrome or the combination of low HDL, high triglycerides, and small, dense LDL, which is described as occurring "commonly in persons with premature [coronary heart disease].* In the now established version of the alternative hypothesis—that metabolic syndrome leads to heart disease—the carbohydrates that had always been considered the causative agent had been officially rendered harmless. They had been removed from the equation of nutrition and chronic disease, despite the decades of research and observations suggesting the critical causal role they played.

---

* The reports do acknowledge, as the AHA-NIH-ADA conference report put it, that "very high-carbohydrate diets may accentuate *atherogenic dyslipidemia*"—i.e., small, dense LDL, high triglycerides, and low HDL—but then it recommends a high-carbohydrate, low-saturated-fat diet as the treatment.

Chapter Eleven

THE SIGNIFICANCE OF DIABETES

Does carbohydrate cause arteriosclerosis? Certainly it does if taken in such excess as to produce obesity, but except in this manner no one would attribute any such function to it. . . . Is a persistent [high blood sugar] a cause of arteriosclerosis in diabetes? It very likely is a cause because it is an abnormal condition and any abnormal state would tend to wear out the machine.

ELLIOTT JOSLIN, "Arteriosclerosis and Diabetes," 1927

DESPITE NEARLY A CENTURY'S WORTH OF therapeutic innovations, the likelihood of a diabetic's contracting coronary artery disease is no less today than it was in 1921, when insulin was first discovered. Type 2 diabetics can still expect to die five to ten years prematurely, with much of this difference due to atherosclerosis and what *Joslin's Diabetes Mellitus* has called an "extraordinarily high incidence" of coronary disease.

Diabetes specialists have historically perceived this plague of atherosclerosis among their patients as though it has little relevance to the atherosclerosis and heart disease that affect the rest of us. Textbooks would note the importance of identifying and controlling the "numerous and as yet ill-defined factors generally involved in the pathogenesis of atherosclerosis," as the 1971 edition of *Joslin's Diabetes Mellitus* did, but the implication was that the requisite revelations would emerge, as they had in the past, from heart-disease researchers, as though the flow of knowledge about heart disease could proceed only from heart-disease research to diabetology and never the other way around.

The extreme example of this thinking has been the assumption that saturated fat is the nutritional agent of heart disease in diabetics, just as it supposedly is in everyone else. "The frequent cardiovascular complications seen in past years among persons with diabetes," the 1988 *Surgeon General's Report on Nutrition and Health* says, are caused by the "traditional restriction of carbohydrate intake in persons with diabetes" and thus an increased intake of fat, "usually, saturated." This was the logic that led the American Diabetes Association, from the early 1970s, to recommend that diabetics eat *more* carbohydrates rather than less, despite a complete absence of clinical trials that might demonstrate that the benefits of doing so outweigh the risks, and the decades of clinical experience establishing carbohydrate restriction as an effective method of controlling blood sugar. If atherosclerosis was accelerated in diabetics, the thinking went, it was accelerated because they ate more saturated fat than nondiabetics. Diabetologists believed they could safely prescribe a carbohydrate-rich diet to their patients, because a diet that is low in fat will be high in carbohydrates.

But the research on metabolic syndrome suggests an entirely different scenario. If the risk of heart disease is elevated in metabolic syndrome and elevated still further with diabetes, then maybe the flow of knowledge about heart disease should proceed from diabetics, who suffer the most extreme manifestation of the disease, to the rest of us, and not the other way around. Maybe diabetics have such extreme atherosclerosis because there is something about the diabetic condition that causes the disease. Perhaps the metabolic abnormalities of the diabetic condition are the essential cause of atherosclerosis and coronary heart disease in everyone, only diabetics suffer to a greater extent.

Another way to look at this is to consider that metabolic syndrome and Type 2 diabetes lie on a continuum or a curve of physical degeneration. This curve is marked by ever-worsening disturbances of carbohydrate and fat metabolism—high insulin, insulin resistance, high blood sugar, high triglycerides, low HDL, and small, dense LDL. Atherosclerosis is one manifestation of this physical degeneration. In diabetes, the metabolic abnormalities are exacerbated—diabetics are further down the curve of physical degeneration—and the atherosclerotic process is accelerated. But we all live on the same curve. The mechanisms that cause atherosclerosis are the same in all of us; only the extent of damage differs.

Consider Keys's cholesterol hypothesis as an example of this logic. One reason we came to believe that high cholesterol is a cause of heart disease is that severe atherosclerosis is a common symptom of genetic disorders of cholesterol metabolism. If having a cholesterol level of 1,000 mg/dl—as these individuals often do—makes atherosclerosis seemingly inevitable, the logic goes, and if higher cholesterol seems to associate with higher risk of heart disease among the rest of us, then cholesterol is a cause of heart disease, and elevating cholesterol by *any* amount will increase risk. The higher the cholesterol, the greater the risk. If eating saturated fat elevates cholesterol, then that in turn causes heart disease. And this is supposedly true of diabetics as well. Keys oversimplified the science and was wrong about the true relationship of cholesterol and heart disease, but the logic itself is otherwise sound.

The same logic holds for blood pressure and heart disease. The higher the blood pressure, the greater the risk of heart disease. If salt supposedly raises blood pressure, even if only by a few percentage points, then salt is a nutritional cause of heart disease. This, too, is held to be true for diabetics. Thus, the atherogenic American diet, as now officially defined, the diet that clogs arteries and causes heart disease, is a diet high in saturated fat *and* salt.

Now let's apply the same reasoning to metabolic syndrome and diabetes. Diabetics suffer more virulent atherosclerosis and die of heart disease more frequently than those with metabolic syndrome, and much more frequently than healthy individuals who manifest neither condition. Some aspect of the diabetic condition must be the cause—most likely, either high blood sugar, hyperinsulinemia, or insulin resistance, all three of which will tend to be worse in diabetics than in those with metabolic syndrome. Indeed, the existence of metabolic syndrome tells us that these same abnormalities exist in nondiabetics, although to a lesser extent, and though individuals with metabolic syndrome suffer an increased risk of heart disease, they do so to a lesser extent than diabetics. And because dietary carbohydrates and particularly refined carbohydrates elevate blood sugar and insulin and, presumably, induce insulin resistance, the implication is that eating these carbohydrates increases heart-disease risk not only in diabetics but in healthy individuals. By this reasoning, the atherogenic American diet is a carbohydrate-rich diet. Hence, cognitive dissonance.

The logic of this argument has to be taken one step further, however, even if the cognitive dissonance is elevated with it. Both diabetes and metabolic syndrome are associated with an elevated incidence of virtually every chronic disease, not just heart disease. Moreover, the diabetic condition is associated with a host of chronic blood-vessel-related problems known as *vascular* complications: stroke, a stroke-related dementia called vascular dementia, kidney disease, blindness, nerve damage in the extremities, and atheromatous disease in the legs that often leads to amputation. One obvious possibility is that the same metabolic and hormonal abnormalities that characterize the diabetic condition—in particular, elevated blood sugar, hyperinsulinemia, and insulin resistance—may also cause these complications and the associated chronic diseases. And otherwise healthy individuals, therefore, *would be expected to increase their risk of all these conditions by the consumption of refined and easily digestible carbohydrates,* which inflict their damage first through their effects on blood sugar and insulin, and then, indirectly, through triglycerides, lipoproteins, fat accumulation, and assuredly other factors as well.

This is a fundamental tenet of the carbohydrate hypothesis: If the risk of contracting any chronic disease or condition increases with metabolic syndrome and Type 2 diabetes, then it's a reasonable hypothesis that insulin and/or blood sugar plays a role in the disease process. And if insulin and blood sugar *do* play a pathological role, then it's a reasonable hypothesis that the same conditions can be caused or exacerbated in healthy individuals by the consumption of refined and easily digestible carbohydrates and sugars.

Among the immediate examples that follow from this logic is the particularly disconcerting possibility that insulin itself causes or exacerbates atherosclerosis. Since insulin resistance and hyperinsulinemia characterize Type 2 diabetes, it's certainly possible that chronically elevated levels of insulin are the cause of the persistently high incidence of atherosclerosis in diabetics, quite aside from any other effects insulin might have on triglycerides, lipoproteins, or blood pressure. And if this is the case, then the excessive secretion of insulin—induced by the consumption of refined carbohydrates and sugars—might be responsible for causing or exacerbating atherosclerosis in those of us who are not diabetic.

This is another of those conceptions, like the ability of insulin to regulate blood pressure, that have been mostly neglected for decades, despite the profound implications if it's true. The specter of this atherogenic effect of insulin is noted briefly, for example, in the fourteenth edition (2005) of *Joslin's Diabetes Mellitus*. The Harvard diabetologist Edward Feener and Victor Dzau, president of the Duke University Health System, write that "the effects of insulin on [cardiovascular disease] in diabetes and insulin resistance are related to both systematic metabolic abnormalities and the *direct effects of insulin action on the vasculature* [blood vessels; my italics]." The second mention, by two Harvard cardiologists, acknowledges the association between insulin resistance, hyperinsulinemia, and heart disease and suggests that if insulin resistance is not the problem, then "another possibility" is that insulin itself "has direct cardiovascular effects." Nothing more is said.

The first evidence of the potential atherogenicity of insulin emerged from precisely the kind of experiments in rabbits that initially gave credibility to the cholesterol hypothesis a century ago. Rabbits fed high-cholesterol diets develop plaques throughout their arteries, but *diabetic* rabbits (Type 1) will not suffer this atherosclerotic fate no matter how cholesterol-rich their diet. Infuse insulin along with the cholesterol-laden diet, however, and plaques and lesions will promptly blossom everywhere. This phenomenon was first reported in 1949 in rabbits, and then, a few years later, in chickens, by Jeremiah Stamler and his mentor Louis Katz, and later in dogs, too. Hence, insulin itself may be "one factor in the pathogenesis of the frequent, premature, severe atherosclerosis of diabetic patients," as Stamler and his colleagues suggested.

In the late 1960s, Robert Stout of Queen's University in Belfast published a series of studies reporting that insulin enhances the transport of cholesterol and fats into the cells of the arterial wall and stimulates the synthesis of cholesterol and fat in the arterial lining. Since a primary role of insulin is to facilitate the storage of fats in the fat tissue, Stout reasoned, it was not surprising that it would have the same effect on the lining of blood vessels. In 1969, Stout and the British diabetologist John Vallance-Owen pre-empted Reaven's Syndrome X hypothesis by suggesting that the "ingestion of large quantities of refined carbohydrate" leads first to hyperinsulinemia and insulin resistance, and then to atherosclerosis and heart disease. In certain individuals, they suggested, the insulin secretion after eating these carbohydrates would be "disproportionately large." "The carbohydrate is disposed of in three sites—adipose [fat] tissue, liver and arterial wall," Stout wrote. "Obesity is produced. In the liver, triglyceride and cholesterol are synthesized and find their way into the circulation. Lipid synthesis is also stimulated in the arterial wall and is augmented by deposition of [triglycerides and cholesterol] . . . which in a few decades would reach significant proportions." In 1975, Stout and the University of Washington pathologist Russell Ross reported that insulin also stimulates the proliferation of the smooth muscle cells that line the interior of arteries, a necessary step in the thickening of artery walls characteristic of both atherosclerosis and hypertension.

This insulin-atherogenesis hypothesis is the simplest possible explanation for the intimate association of diabetes and atherosclerosis: the excessive secretion of insulin accelerates atherosclerosis and perhaps other vascular complications. It also implies, as Stout suggested, that any dietary factor—refined carbohydrates in particular—that increases insulin secretion will increase risk of heart disease. This did not, however, become the preferred explanation. Even Reaven chose to ignore it.* But Reaven's hypothesis proposed that heart disease was caused primarily by insulin resistance through its influence on triglycerides. He considered hyperinsulinemia to be a secondary phenomenon. Stout considered hyperinsulinemia the primary cause of atherosclerosis.

---

* Ralph DeFronzo, on the other hand, believes that sufficient studies have confirmed Stout's observations and that insulin itself should thus be considered an "atherogenic hormone."

Most diabetologists have believed that diabetic complications are caused by the toxic effects of high blood sugar.* The means by which high blood sugar induces damage in cells, arteries, and tissues are indeed profound, and the consequences, as the carbohydrate hypothesis implies, extend far beyond diabetes itself. This line of research is pursued by only a few laboratories. As a result, its ultimate implications and validity remain to be ascertained. But it should be considered as yet another potential mechanism by which the consumption of refined carbohydrates could cause or exacerbate the entire spectrum of the chronic diseases of civilization.

In particular, raising blood sugar will increase the production of what are known technically as *reactive oxygen species* and *advanced glycation end-products,* both of which are potentially toxic. The former are generated primarily by the burning of glucose (blood sugar) for fuel in the cells, in a process that attaches electrons to oxygen atoms, transforming the oxygen from a relatively inert molecule into one that is avid to react chemically with other molecules. This is not an ideal situation biologically. One form of reactive oxygen species is those known commonly as *free radicals,* and all of them together are known as *oxidants,* because what they do is *oxidize* other molecules (the same chemical reaction that causes iron to rust, and equally deleterious). The object of oxidation slowly deteriorates. Biologists refer to this deterioration as *oxidative stress. Antioxidants* neutralize reactive oxygen species, which is why antioxidants have become a popular buzzword in nutrition discussions.

The potential of advanced glycation end-products (AGEs) for damage is equally worrisome. Their formation can take years, but the process (glycation) begins simply, with the attachment of a sugar—glucose, for instance—to a protein without the benefit of an enzyme to orchestrate the reaction. That absence is critical. The role of enzymes in living organisms is to control chemical reactions to ensure that they "conform to a tightly regulated metabolic program," as the Harvard biochemist Frank Bunn explains. When enzymes affix sugars to proteins, they do so at particular sites on the proteins, for very particular reasons. Without an enzyme overseeing the process, the sugar sticks to the protein haphazardly and sets the stage for yet more unintended and unregulated chemical reactions.

The term *glycation* refers only to this initial step, a sugar molecule

---

* This hypothesis cannot, however, explain why atherosclerosis among diabetics has remained relatively impervious to the otherwise beneficial effects of insulin therapy to control blood sugar.

attaching to a protein, and this part of the process is reversible—if blood-sugar levels are low enough, the sugar and protein will disengage, and no damage will be done. If blood sugar is elevated, however, then the process of forming an advanced glycation end-product will move forward. The protein and its accompanying glycated sugars will undergo a series of reactions and rearrangements until the process culminates in the convoluted form of an advanced glycation end-product. These AGEs will then bind easily to other AGEs and to still more proteins through a process known as *cross-linking*—the sugars hooked to one protein will bridge to another protein and lock them together. Now proteins that should ideally have nothing to do with each other will be inexorably joined.

In the mid-1970s, Rockefeller University biochemist Anthony Cerami and Frank Bunn independently recognized that AGEs and glycation play a major role in diabetes.* Both Cerami and Bunn were initially motivated by the observation that diabetics have high levels of an unusual form of hemoglobin—the oxygen-carrying protein of red blood cells—known as hemoglobin A1c, a glycated hemoglobin. The higher the blood sugar, the more hemoglobin molecules undergo glycation, and so the more hemoglobin A1c can be found in the circulation. Cerami's laboratory then developed an assay to measure hemoglobin A1c, speculating correctly that it might be an accurate reflection of the diabetic state. Diabetics have two to three times as much hemoglobin A1c in their blood as nondiabetics, a ratio that apparently holds true for nearly all glycated proteins in the body. (The best determination of whether diabetics are successfully controlling their blood sugar comes from measuring hemoglobin A1c, because it reflects the average blood sugar over a month or more.)

Since 1980, AGEs have been linked directly to both diabetic complications and aging itself (hence the acronym). AGEs accumulate in the lens, cornea, and retina of the eye, where they appear to cause the browning and opacity of the lens characteristic of senile cataracts. AGEs accumulate in the membranes of the kidney, in nerve endings, and in the lining of arteries, all tissues typically damaged in diabetic complications. Because AGE accumulation appears to be a naturally occurring process, although it is exacerbated and accelerated by high blood sugar, we have evolved sophisticated defense mechanisms to recognize, capture, and dispose of AGEs. But AGEs still manage to accumulate in tissues with the passing years, and especially so in diabetics, in whom AGE accumulation correlates with the severity of complications.

---

* Those in Cerami's laboratory at Rockefeller University and the researchers who trained with him get credit for much of the AGE work that followed.

One protein that seems particularly susceptible to glycation and cross-linking is collagen, which is a fundamental component of bones, cartilage, tendons, and skin. The collagen version of an AGE accumulates in the skin with age and, again, does so excessively in diabetics. This is why the skin of young diabetics will appear prematurely old, and why, as the Case Western University pathologist Robert Kohn first suggested, diabetes can be thought of as a form of accelerated aging, a notion that is slowly gaining acceptance. It's the accumulation and cross-linking of this collagen version of AGEs that causes the loss of elasticity in the skin with age, as well as in joints, arteries, and the heart and lungs.

The process can be compared to the toughening of leather. Both the meat and hide of an old animal are tougher and stiffer than those of a young animal, because of the AGE-related cross-linking that occurs inevitably with age. As Cerami explains, the aorta, the main artery running out of the heart, is an example of this stiffening effect of accumulated and cross-linked AGEs. "If you remove the aorta from someone who died young," says Cerami, "you can blow it up like a balloon. It just expands. Let the air out, it goes back down. If you do that to the aorta from an old person, it's like trying to inflate a pipe. It can't be expanded. If you keep adding more pressure, it will just burst. That is part of the problem with diabetes, and aging in general. You end up with stiff tissue: stiffness of hearts, lungs, lenses, joints. . . . That's all caused by sugars reacting with proteins."

AGEs and the glycation process also appear to play at least one critical role *directly* in heart disease, by causing the oxidation of LDL particles and so causing the LDL and its accompanying cholesterol to become trapped in the artery wall, which is an early step in the atherosclerotic process. Oxidized LDL also appears to be resistant to removal from the circulation by the normal mechanisms, which would also serve to increase the LDL levels in the blood. As it turns out, LDL is particularly susceptible to oxidation by reactive oxygen species and to glycation.* In this case, both the protein portion and the lipid portion (the cholesterol and the fats) of the lipoprotein are susceptible. These oxidized LDL particles appear to be "markedly elevated" in both diabetics and in nondiabetics with atherosclerosis, and are particularly likely to be found in the atherosclerotic lesions themselves.

That glycation and AGEs are critical factors in diabetic complications and in heart disease has recently been demonstrated by experiments with compounds known as *anti-AGE compounds* or *AGE breakers*. These will reverse arterial stiffness, at least in laboratory animals, and, as one recent

---

* There's also evidence that HDL molecules can become glycated, inhibiting their function and "rendering the HDL more pro-atherogenic."

report put it, ameliorate "the adverse cardiovascular and [kidney-related] changes associated with aging, diabetes and hypertension." Whether these or similar compounds will work in humans remains to be seen.

When biochemists discuss oxidative stress, glycation, and the formation of advanced glycation end-products, they often compare what's happening to a fire simmering away in our circulation. The longer the fire burns and the hotter the flame, the more damage is done. Blood sugar is the fuel. "Current evidence points to glucose not only as the body's main short-term energy source," as the American Diabetes Association recently put it, "but also as the long-term fuel of diabetes complications."

But there is no reason to believe that glucose-induced damage is limited only to diabetics, or to those with metabolic syndrome, in whom blood sugar is also chronically elevated. Glycation and oxidation accompany every fundamental process of cellular metabolism. They proceed continuously in all of us. Anything that raises blood sugar—in particular, the consumption of refined and easily digestible carbohydrates—will increase the generation of oxidants and free radicals; it will increase the rate of oxidative stress and glycation, and the formation and accumulation of advanced glycation end-products. This means that anything that raises blood sugar, by the logic of the carbohydrate hypothesis, will lead to more atherosclerosis and heart disease, more vascular disorders, and an accelerated pace of physical degeneration, even in those of us who never become diabetic.

Chapter Twelve

SUGAR

M. Delacroix, a writer as charming as he is prolific, complained once to me at Versailles about the price of sugar, which at that time cost more than five francs a pound. "Ah," he said in a wistful, tender voice, "if it can ever again be bought for thirty cents, I'll never more touch water unless it's sweetened!" His wish was granted. . . .

JEAN ANTHELME BRILLAT-SAVARIN,
The Physiology of Taste, 1825

W HEN BIOCHEMISTS TALK ABOUT "SUGAR," they're referring to a whole host of very simple carbohydrate molecules, all of which are characterized, among other things, by their sweet taste and ability to dissolve in water. Their chemical names all end in "-ose"—glucose, fructose, and lactose, among others. When physicians talk about blood sugar, they're typically talking about glucose, although other sugars can be found in the bloodstream at very much lower concentrations. Then there's the common usage of "sugar," meaning the sweet, powdered variety that we put in our coffee or tea. This is sucrose, which in turn is constituted of equal parts glucose and fructose. In the discussion to come, when we refer to "sugar" we'll always be talking about sucrose. When we use the term "blood sugar," we'll be talking about glucose.

When nutritionists in the 1960s discussed the pros and cons of sugar and starches, their concern was whether *simple* carbohydrates were somehow more deleterious than *complex* carbohydrates of starches. Chemically, simple carbohydrates, as in sugar and highly refined flour, are molecules of one or two sugars bound together, whereas the complex carbohydrates of starches are chains of sugars that can be tens of thousands of sugars long. Complex carbohydrates break down to simple sugars during the process of digestion, but they take a while to do so, and if the carbohydrate is bound up with fiber—i.e., indigestible carbohydrates—the digestion takes even longer. Since the early 1980s, both simple and complex carbohydrates have played a role in determining the glycemic index, which is a measure of how quickly carbohydrates are digested and absorbed into the circulation and so converted into blood sugar. This concept of a glycemic index has had profound consequences on the official and public perception of the risks of starches and sugar in the diet. But it has done so by ignoring the effect of fructose—in sugar and high-fructose corn syrup—on anything other than its ability in the short term to elevate blood sugar and elicit an insulin response.

In the mid-1970s, Gerald Reaven initiated the study of glycemic index to test what he called the "traditionally held tenet" that simple carbohydrates are easier to digest than more complex carbohydrates "and that they therefore produce a greater and faster rise" in blood sugar and insulin after a meal. Reaven's experiments confirmed this proposition, but he was less interested in blood sugar than in insulin, and so left this research behind. It was taken up a few years later by David Jenkins and his student Thomas Wolever, both of whom were then at Oxford University. Over the course of a year, Wolever and Jenkins tested sixty-two foods and recorded the blood-sugar response in the two hours after consumption. Different individuals responded differently, and the variation from day to day was "tremendous," as Wolever says, but the response to a specific food was still reasonably consistent. They also tested a solution of glucose alone to provide a benchmark, which they assigned a numerical value of 100. Thus the glycemic index became a comparison of the blood-sugar response induced by a particular carbohydrate food to the response resulting from drinking a solution of glucose alone. The higher the glycemic index, the faster the digestion of the carbohydrates and the greater the resulting blood sugar and insulin. White bread, they reported, had a glycemic index of 69; white rice, 72; corn flakes, 80; apples, 39; ice cream, 36. The presence of fat and protein in a food decreased the blood-sugar response, and so decreased the glycemic index.

One important implication of Jenkins and Wolever's glycemic-index research is that it provided support for Cleave's speculations on the saccharine disease. The more refined the carbohydrates, the greater the blood-sugar and insulin response. Anything that increases the speed of digestion of carbohydrates—polishing rice, for instance, refining wheat, mashing potatoes, and particularly drinking simple carbohydrates in any liquid form, whether a soda or a fruit juice—will increase the glycemic response. Thus, the addition of refined carbohydrates to traditional diets of fibrous vegetables or meat and milk, or even fish and coconuts, could be expected to elevate blood-sugar and insulin levels in the population. And this would conceivably explain the appearance of both atherosclerosis and diabetes as diseases of civilization, through the physiological abnormalities of metabolic syndrome—glucose intolerance, hyperinsulinemia, insulin resistance, high triglycerides, low HDL, and small, dense LDL.

Jenkins and Wolever's research, first published in 1981, led to a surprisingly vitriolic debate among diabetologists on the value of the glycemic index as a guide to controlling blood sugar. Reaven argued that the concept was worthless if not dangerous: saturated fat, he argued, has no glycemic index, and so adding saturated fat to sugar and other carbohydrates will lower their glycemic index and make the combination appear benign when that might not quite be the case. "Ice cream has a great glycemic index, because of the fat," Reaven observed. "Do you want people to eat ice cream?" Reaven also disparaged the glycemic index for putting the clinical focus on blood sugar, whereas he considered insulin and insulin resistance the primary areas of concern. The best way for diabetics to approach their disease, Reaven insisted, was to restrict all carbohydrates.

Paradoxically, the glycemic index appears to have had its most significant influence not on the clinical management of diabetes but on the public perception of sugar itself. The key point is that the glycemic index of sucrose is *lower* than that of flour and starches—white bread and potatoes, for instance—and fructose is the reason why. The carbohydrates in starches are broken down upon digestion, first to maltose and then to glucose, which moves directly from the small intestine into the bloodstream. This leads immediately to an elevation of blood sugar, and so a high glycemic index. Table sugar, on the other hand—i.e., sucrose—is composed of both glucose and fructose. To be precise, a sucrose molecule is composed of a single glucose molecule bonded to a single fructose molecule. This bond is broken upon digestion. The glucose moves into the bloodstream and raises blood sugar, just as if it came from a starch, but the fructose can be metabolized only in the liver, and so most of the fructose consumed is channeled from the small intestine directly to the liver. As a result, fructose has little immediate effect on blood-sugar levels, and so only the glucose half of sugar is reflected in the glycemic index.

That sugar is half fructose is what fundamentally differentiates it from starches and even the whitest, most refined flour. If John Yudkin was right that sugar is the primary nutritional evil in the diet, it would be the fructose that endows it with that singular distinction. With an eye toward primitive diets transformed by civilization, and the change in Western diets over the past few hundred years, it can be said that the single most profound change, even more than the refinement of carbohydrates, is the dramatic increase in fructose consumption that comes with either the addition of fructose to a diet lacking carbohydrates, or the replacement of a large part of the glucose from starches by the fructose in sugar.

Because fructose barely registers in the glycemic index, it appeared to be the ideal sweetener for diabetics; sucrose itself, with the possible exception of its effect on cavities, appeared no more harmful to nondiabetics, and perhaps even less so, than starches such as potatoes that were being advocated as healthy substitutes for fat in the diet. In 1983, the University of Minnesota diabetologist John Bantle reported in *The New England Journal of Medicine* that fructose could be considered the healthiest carbohydrate. "We see no reason for diabetics to be denied foods containing sucrose," Bantle wrote. This became the official government position. The American Diabetes Association still suggests that diabetics need not restrict "sucrose or sucrose-containing foods" and can even substitute them, if desired, "for other carbohydrates in the meal plan."

In 1986, the FDA exonerated sugar of any nutritional crimes on the basis that "no conclusive evidence demonstrates a hazard." The two-hundred-page report constituted a review of hundreds of articles on the health aspects of sugar, many of which reported that sugar had a range of *potentially* adverse metabolic effects related to a higher risk of heart disease and diabetes. The FDA interpreted the evidence as inconclusive. Health reporters, the sugar industry, and public-health authorities therefore perceived the FDA report as absolving sugar of having any deleterious effects on our health.

The identical message was passed along in the 1988 *Surgeon General's Report on Nutrition and Health* and the 1989 National Academy of Sciences *Diet and Health* report. Here, too, the inconclusive studies and ambiguous evidence were considered insufficient to indict sugar as a dietary evil—innocent until proven guilty. These two reports also reviewed the dietary fat/heart-disease connection, which also constituted a collection of inconclusive studies and ambiguous evidence. Here, though, dietary fat was assumed guilty until proved innocent. And so the existence of ambiguous evidence was considered sufficient reason to condemn fat in the diet, particularly saturated fat, while the existence of ambiguous evidence was simultaneously considered reason enough to exonerate sugar.

This institutional absolution of sugar might have been relatively innocuous had it not been coincident with the introduction of a type of sugar refined from corn, rather than sugarcane or beets, known as high-fructose corn syrup, or HFCS, and specifically with what is technically known as HFCS-55, a sweetener that is 55 percent fructose and 45 percent glucose and was created to be indistinguishable from sucrose by taste when used in soft drinks. HFCS-55 entered the market in 1978. By 1985, half of the sugars consumed each year in the U.S. came from corn sweeteners, and two-thirds of that was high-fructose corn syrup. More important, the average consumption of sugars in total had started climbing steadily upward.

This rise in sugar consumption is one of the more perplexing dietary trends in the last century. Though Americans' taste in starch apparently ebbed and flowed through the twentieth century, the average yearly consumption of caloric sweeteners—a category that includes table sugar, corn sweeteners, honey, and edible syrups—remained relatively constant from the 1920s, at 110–120 pounds per capita. It began to inch upward in the early 1960s, coincident with the first introduction of fructose-enhanced corn syrups. With the introduction of HFCS-55, it increased significantly. According to USDA statistics, between 1975 and 1979 Americans consumed an annual average of 124 pounds of sugars per person. By 2000, that number had jumped to almost 150 pounds. Corn sweeteners, and particularly high-fructose corn syrup, constituted virtually every ounce of the increase. And this increase came on the heels of a period in the mid-1970s when sugar consumption per capita was decreasing, as sugar was being portrayed in the popular press as a fattening and addictive dietary nuisance.

The simplest explanation for the increase in caloric-sweetener consumption is that consumers simply failed to equate high-fructose corn syrup with the sugar that we'd been eating almost exclusively until then. Although HFCS-55 is effectively identical to sucrose upon digestion, the industry treated it, and the public perceived it, as a healthy additive, whereas sucrose carried the taint of decades of controversy. Because fructose is the predominant sugar in fruit—an apple, for instance, is roughly 6 percent fructose, 4 percent sucrose, and 1 percent glucose by weight—it is often referred to as "fruit sugar" and appears somehow healthier simply by virtue of that association. And, of course, fructose was perceived as healthy because it does not elevate blood sugar and has a low glycemic index.

As a consequence, high-fructose corn syrup could be used as the primary sweetener, and often the primary source of calories, in products that had the outward appearance of being healthy or *natural*, or were advertised as such, without revealing the products to be little more than sugar, water, and chemical flavoring. This included sports drinks such as Gatorade, the fruit juices and teas such as Snapple that appeared nationwide beginning in the late 1980s, and low-fat yogurts, which also exploded in popularity with the condemnation of fat in the diet.

By defining carbohydrate foods as good or bad on the basis of their glycemic index, diabetologists and public-health authorities effectively misdiagnosed the impact of fructose on human health. The key is the influence of glucose or fructose not on blood sugar but on the liver. Glucose goes directly into the bloodstream and is taken up by tissues and organs to use as energy; only 30–40 percent passes through the liver. Fructose passes directly to the liver, where it is metabolized almost exclusively. As a result, fructose "constitutes a metabolic load targeted on the liver," the Israeli diabetologist Eleazar Shafrir says, and the liver responds by converting it into triglycerides—fat—and then shipping it out on lipoproteins for storage. The more fructose in the diet, the higher the subsequent triglyceride levels in the blood.*

The research on this *fructose-induced lipogenesis,* as it is technically known, was carried out primarily by Peter Mayes, a biochemist at King's College Medical School in London; by Shafrir at Hebrew University–Hadassah Medical School in Jerusalem; and by Sheldon Reiser and his colleagues at the USDA Carbohydrate Nutrition Laboratory in Maryland. They began in the late 1960s and worked on it through the early 1980s. "In the 1980s," says Judith Hallfrisch, who worked with Reiser at the USDA, "people didn't even believe that elevated triglycerides were a risk factor for cardiovascular disease. So they didn't care that much about the increase in triglycerides. Everything was cholesterol." (Although sugar also seemed to raise cholesterol levels, particularly LDL, as would be expected for any nutrient that increased triglyceride synthesis in the liver. In 1992, John Bantle reported that LDL cholesterol in diabetic patients was elevated more than 10 percent on a high-fructose diet after a month, which is comparable to what can be achieved by saturated fats.)

As Peter Mayes has explained it, our bodies will gradually adapt to long-term consumption of high-fructose diets, and so the "pattern of fructose metabolism" will change over time. This is why, the more fructose in the diet and the longer the period of consumption, the greater the secretion of triglycerides by the liver. Moreover, fructose apparently blocks both the metabolism of glucose in the liver and the synthesis of glucose into glycogen, the form in which the liver stores glucose locally for later use. As a result, the pancreas secretes more insulin to overcome this glucose traffic-jam at the liver, and this in turn induces the muscles to compensate by becoming more insulin resistant. The research on this fructose-induced insulin resistance was done on laboratory animals, but it confirmed what Reiser at the USDA had observed in humans and published in 1981: given sufficient time, high-fructose diets can induce high insulin levels, high

---

* For this reason, fructose is referred to as the most *lipogenic* carbohydrate. Credit for this observation dates to 1916, to Harold Higgins of the Nutrition Laboratory of the Carnegie Institution.

blood sugar, and insulin resistance, even though in the short term fructose has little effect on either blood sugar or insulin and so a very low glycemic index. It has also been known since the 1960s that fructose elevates blood pressure more than an equivalent amount of glucose does, a phenomenon called *fructose-induced hypertension.*

Because sucrose and high-fructose corn syrup (HFCS-55) are both effectively half glucose and half fructose, they offer the worst of both sugars. The fructose will stimulate the liver to produce triglycerides, while the glucose will stimulate insulin secretion. And the glucose-induced insulin response in turn will prompt the liver to secrete even more triglycerides than it would from the fructose alone, while the insulin will also elevate blood pressure apart from the effect of fructose. "This is really the harmful effect of sucrose," says Mayes, "over and above fructose alone."

The effect of fructose on the formation of advanced glycation end-products—AGEs, the haphazard glomming together of proteins in cells and tissues—is worrisome as well. Most of the research on AGE accumulation in humans has focused on the influence of glucose, because it is the dominant sugar in the blood. Glucose, however, is the *least* reactive of all sugars, the one least likely to attach itself without an enzyme to a nearby protein, which is the first step in the formation of AGEs. As it turns out, however, fructose is significantly more reactive in the bloodstream than glucose, and perhaps ten times more effective than glucose at inducing the cross-linking of proteins that leads to the cellular junk of advanced glycation end-products. Fructose also leads to the formation of AGEs and cross-linked proteins that seem more resistant to the body's disposal mechanisms than those created by glucose. It also increases markedly the oxidation of LDL particles, which appears to be a necessary step in atherosclerosis.

This research on the health effects of fructose began to coalesce in the mid-1980s, just as nutritionists were disseminating the notion that fructose was particularly harmless because of its low glycemic index. And this official opinion has proven hard to sway.

Take, for example, the British Committee on Medical Aspects of Food Policy (known commonly as COMA), which in 1989 released a report entitled *Dietary Sugars and Human Disease,* authored by a dozen of the nation's leading nutritionists, physiologists, and biochemists and chaired by Harry Keen, who is among the most renowned British diabetologists. The COMA report discussed the evidence, including the research of Reiser, Reaven, and others, and then concluded that the health effects of sugar were insignificant. The report did so, however, with a series of contradictory assumptions. First, Keen and his colleagues concluded that the implications of fructose-induced insulin resistance and elevated triglycerides are limited to a "relatively small group of people with metabolic disorders [that] includes people with diabetes and those with certain rare inherited disorders." And so, with the exception of this small percentage of the population, they noted, yearly sugar consumption at 1986 levels, estimated at roughly a hundred pounds per capita in the United Kingdom, "carries no special metabolic risks." On the other hand, they then explained, sugar consumption does carry risk for those "members of the population consuming more than about 200 g per day," which is 160 pounds per year, or only slightly more (.4 ounce per day) than what the *average* American was eating in the year 2000 (not the top 10–20 percent, but the *average*). They next suggested that those individuals with high triglycerides, a proportion that remains unspecified but might constitute the great majority of all individuals with coronary-artery disease, should restrict their consumption of added sugars to twenty to forty pounds per year, or equivalent to the amount consumed in the U.K. in the early years of the Victorian era.

All of this was then summed up in the single statement—echoing the sentiments of the FDA Task Force, the National Academy of Sciences *Diet and Health* report, and *The Surgeon General's Report on Nutrition and Health,* which preceded it—that dietary sugar consumption could not be held responsible for causing disease: "The panel concluded that current consumption of sugars, particularly sucrose, played no direct causal role in the development of cardiovascular . . . disease, of essential hypertension, or of diabetes mellitus. . . ."

Four years later, *The American Journal of Clinical Nutrition* dedicated an entire issue to the deleterious effects of dietary fructose. A common refrain throughout the issue was the need for research that would establish at what level of sugar consumption the effects discussed—the elevation of blood pressure and triglycerides, increased insulin resistance, and even accelerated formation of advanced glycation end-products—would lead to disease. "Further studies are clearly needed to determine the metabolic alteration that may take place during chronic fructose or sucrose feeding," as the Swiss physiologists Luc Tappy and Eric Jéquier wrote.

In 2002, the Institute of Medicine of the National Academies of Science released its two-volume report on *Dietary Reference Intakes* (subtitled *Energy, Carbohydrate, Fiber, Fat, Fatty Acids, Cholesterol, Protein, and Amino Acids*), and spent twenty pages discussing the possible adverse effects of sucrose and high-fructose corn syrup. It then concluded that there was still "insufficient evidence" to set up an upper limit for sugar consumption in the healthy diet. Nor did the IOM perceive any reason to pursue further research on fructose or sucrose or high-fructose corn syrup and so, perhaps, discover sufficient evidence. In early 2007, the National Institutes of Health was funding at most half a dozen research projects that addressed, even peripherally, the health effects of dietary fructose, meaning sugar and high-fructose corn syrup in the diet.

Over the years, what little research has been done on fructose metabolism has been carried out primarily by biochemists, who have had little motivation, other than perhaps personal health, to pay attention to the nutrition literature—again, an effect of specialization. Their own articles, moreover, are published in biochemistry journals, and have little influence on the nutrition and public-health communities. For this reason, observations on the potential dangers of fructose have managed to remain dissociated from discussions of sugar itself and the role of sucrose and high-fructose corn syrup in modern diets. After the ridicule that John Yudkin received for the work that culminated in his anti-sugar polemic, *Pure, White and Deadly*—and after the FDA decided that "no conclusive evidence demonstrates a hazard" from sucrose—few researchers have appeared willing even to contemplate the possibility that sugar consumption could have harmful consequences beyond perhaps causing cavities and contributing to obesity.

Chapter Thirteen

DEMENTIA, CANCER, AND AGING

The bottom line is pretty irrefutable: What is good for the heart is good for
the brain.

RUDOLPH TANZI AND ANN PARSON,
Decoding Darkness: The Search for the Genetic Causes
of Alzheimer's Disease, 2000

WHEN IT COMES TO THE CAUSE of chronic disease, as we discussed
earlier, the carbohydrate hypothesis rests upon two simple propositions. First, if our likelihood of contracting a particular disease increases
once we already have Type 2 diabetes or metabolic syndrome, then it's a
reasonable assumption that high blood sugar and/or insulin is involved in
the disease process. Second, if blood sugar and insulin are involved, then
we have to accept the possibility that refined and easily digestible carbohydrates are as well.

This applies to Alzheimer's disease and cancer, too, since both diabetes
and metabolic syndrome are associated with an increased incidence of
these two illnesses. In both cases, critical steps in the disease process have
been linked unambiguously to insulin and blood sugar, and the relevant
research is now beginning to influence the mainstream thinking in these
fields.

Though the characteristic dementia and brain lesions of Alzheimer's
were first described a century ago, the disease only recently captured the
attention of the research community. In 1975, when the NIH was supporting hundreds of research projects on atherosclerosis and cholesterol
metabolism, it was funding fewer than a dozen on Alzheimer's and what
was then called senile dementia. This number rose gradually through the
end of the 1970s. Between 1982 and 1985, the number of Alzheimer's-
related research projects funded by the NIH quintupled.

It took another decade for researchers to begin reporting that heart
disease and Alzheimer's seem to share risk factors: hypertension, atherosclerosis, and smoking are all associated with an increased risk of Alzheimer's, as is the inheritance of a particular variant of a gene called

apolipoprotein E4 (apo E4) that also increases the risk of cardiovascular disease.* This in turn led to the notion that what's good for the heart is good for the brain, but that, of course, depends on our understanding of what exactly is good for the heart. Because Alzheimer's researchers, like diabetologists, assume that Keys's fat-cholesterol hypothesis is supported by compelling evidence, they will often suggest that cholesterol and saturated fat play a role in Alzheimer's as well. But if coronary heart disease is mostly a product of the physiological abnormalities of metabolic syndrome, as the evidence suggests, then this implicates insulin, blood sugar, and refined carbohydrates instead, a conclusion supported by several lines of research that began to converge in the last decade.

A handful of studies have suggested that Alzheimer's is another disease of civilization, with a pattern of distribution similar, if not identical, to heart disease, diabetes, and obesity. Japanese Americans, for instance, develop a pattern of dementia—the ratio of Alzheimer's dementia to the stroke-related condition known as vascular dementia—that is typically American; when Japanese immigrate to the United States, their likelihood of developing Alzheimer's disease increases considerably, while their risk of developing vascular dementia decreases. The incidence of Alzheimer's dementia in African Americans, according to research published in *JAMA* in 2001, is twice that of rural Africans, and they are three times as likely to suffer vascular dementia, again suggesting that dietary or lifestyle factors play a role in both dementias.

Studies in large populations—6,000 elderly subjects in Rotterdam, 1,500 in Minnesota, 1,300 in Manhattan, 800 Catholic nuns, priests, and brothers in the American Midwest, and 2,500 Japanese Americans in Honolulu—have suggested that Type 2 diabetics have roughly twice as much risk of contracting Alzheimer's disease as nondiabetics. Diabetics on insulin therapy, according to the Rotterdam study, had a fourfold increase in risk. Hyperinsulinemia and metabolic syndrome are also associated with an increased risk of Alzheimer's disease. And so one interpretation of these results, as the Rotterdam investigators noted in 1999, is

---

* Individuals with a single copy of this apo E4 gene are nearly three times as likely to have both heart disease and Alzheimer's than those with none. Apo E4 is a cousin of apo B, the protein component of LDL and VLDL, and it is also found in the lipoproteins that transport triglycerides and cholesterol. Because heart-disease researchers have focused on cholesterol as the cause of heart disease, Alzheimer's researchers tend also to refer to apo E4 as involved in cholesterol transport as though that were all it did, thus "point[ing] to a link between cholesterol and Alzheimer's." But this took the overly simplistic 1960s view of heart disease and used it to misdirect the Alzheimer's research.

that "direct or indirect effects of insulin could contribute to the risk of dementia."

One complicating factor in this research is that the underlying cause of dementia is exceedingly difficult to diagnose, even on autopsy. For this reason, it's possible that the research linking diabetes to a higher incidence of Alzheimer's does so because it confuses the consequences of a known complication of diabetes—vascular dementia—with an apparently increased incidence of Alzheimer's dementia. These are the two most common causes of dementia, but the actual diagnoses are not clear-cut.

Alzheimer's dementia is typically perceived as a slow, insidious process that can be identified on autopsy by the presence of neurofibrillary tangles, which are twisted protein fibers located within neurons, and amyloid plaques, which accumulate outside the neurons. Vascular dementia, a recognized complication of diabetes, is perceived as a more abrupt cognitive decline that is caused by small strokes in the blood vessels of the brain. Vascular dementia is usually diagnosed because the dementia appeared shortly after a stroke, or because an autopsy revealed the characteristic stroke-related signs of vascular damage. That vascular dementia is a complication of diabetes means that diabetics are far more likely to be diagnosed someday with vascular dementia than nondiabetics.

In cases of dementia, however, the determination of the actual cause is likely to be arbitrary. Most of us, if we live long enough, will accumulate both vascular damage and Alzheimer's plaques and tangles in our brains, even if we don't manifest any perceptible symptoms of dementia. (Similarly, most of us will have plaques in our arteries even if we don't manifest clinical signs of heart disease.) Vascular dementia and Alzheimer's dementia appear to coexist frequently, a condition known as *mixed dementia*. When dementia is present, the diagnosis of its ultimate cause is a matter of clinical judgment. This gray zone of mixed dementia was examined in a seminal study of nearly seven hundred elderly members of the Sisters of Notre Dame congregation, led by the University of Kentucky epidemiologist David Snowdon. The results suggest that, the less vascular damage we have in our brains, the more easily we can tolerate the lesions of Alzheimer's without exhibiting signs of dementia. It's the extent and location of the vascular damage in the brain, according to Snowdon, that appears to be the determining factor.

The implication is that the accumulation of damage to neurons and blood vessels is one unavoidable process of aging. There is a point when the slow accumulation of Alzheimer's lesions and vascular damage passes some threshold and manifests itself as dementia, and diabetics are always likely to reach that threshold sooner than nondiabetics, if only because they accumulate vascular damage more rapidly, even if the diabetes bestows on them no special predisposition to develop Alzheimer's plaques and tangles. So whatever dietary factors or lifestyle factors lead to Type 2 diabetes will always increase the likelihood of manifesting dementia.

Two other lines of evidence linking insulin and high blood sugar to Alzheimer's disease are directly related to the amyloid-plaque buildup that is now thought to result in the degeneration and death of neurons in the Alzheimer's-affected brain. The primary component of these plaques is a protein known as *beta-amyloid*—or just amyloid, for short—and this protein is what's left after a larger protein, a *precursor* protein, is cleaved in two. The amyloid precursor protein exists naturally in brain neurons, according to the Harvard neurologist Rudolph Tanzi, and the act of cutting it down in size to the amyloid protein appears to be a normal cellular process. A healthy brain, however, clears away amyloid efficiently after the cleavage occurs; this does not happen in Alzheimer's. The question is, why not?

One phenomenon now implicated in the process of amyloid-plaque accumulation is the accumulation of AGEs, the conglomerations of haphazardly linked proteins and sugars that are found to excess in the organs and tissues of diabetics. Because neurons ideally last a lifetime, they seem to be prime candidates for the slow accumulation of AGEs and the toxic damage they inflict. The proteins that make up the plaques and tangles of Alzheimer's are particularly long-lived themselves and so particularly susceptible. And AGEs can indeed be found buried in both the plaques and tangles of Alzheimer's and even in immature plaques, suggesting that they are involved from the very beginning of the process.

Investigators studying AGEs have proposed that Alzheimer's starts with glycation—the haphazard binding of reactive blood sugars to these brain proteins. Because the sugars stick randomly to the fine filaments of the proteins, this in turn causes the proteins to stick to themselves and to other proteins. This impairs their function and, at least occasionally, leaves them impervious to the usual disposal mechanisms, causing them to accumulate in the spaces between neurons. There they cross-link with other nearby proteins, and eventually become advanced glycation endproducts. All of this would then be exacerbated by the fact that the glycation process itself generates more and more toxic reactive oxygen species (free radicals), which in turn causes even more damage to the neurons. In theory, this is what causes the amyloid plaques and leads to the degeneration of neurons, the cell loss, and the dementia of Alzheimer's. The theory is controversial, but the identification of AGEs in the plaques and tangles of Alzheimer's is not.

The involvement of insulin in Alzheimer's can be considered the sim-

plest possible explanation for the slow, relentless development of Alzheimer's plaques in the aging brain. Insulin (in a test tube) will monopolize
the attention of the *insulin-degrading enzyme* (IDE), which normally
degrades and clears *both* amyloid proteins and insulin from around the
neurons. The more insulin available in the brain, by this scenario, the less
IDE is available to clean up amyloid, which then accumulates excessively
and clumps into plaques. In animal experiments, the less IDE available,
the greater the concentration of amyloid in the brain. Mice that lack the
gene to produce IDE develop versions of both Alzheimer's disease and
Type 2 diabetes.*

Much of the relevant research in humans on insulin and Alzheimer's
has been done by Suzanne Craft, a neuropsychiatrist at the University of
Washington. In 1996, Craft and her colleagues reported that boosting
insulin levels, at least in the short term, seems to enhance memory and
mental prowess, even in Alzheimer's patients. This linked insulin to
the biochemical regulation of memory in the brain, but it said nothing
about the long-term, chronic effects of hyperinsulinemia. In 2003, Craft
reported that when insulin was infused into the veins of elderly volunteers, the amount of amyloid in their cerebral spinal fluid increased proportionately. This implied that the level of amyloid protein in their brain
had increased as well. The older the patient, the greater the increase in
amyloid protein. As Craft sees it, if insulin levels are chronically elevated
(hyperinsulinemia), then brain neurons will be excessively stimulated to
produce amyloid proteins, and IDE will be preoccupied with removing the
insulin, so that less will be available to clean up the amyloid. "We're not
saying this is *the* mechanism for all of Alzheimer's disease," Craft says.
But "it may have a role in a significant number of people."

This evidence linking insulin, amyloid, and Alzheimer's has now
evolved to the point where it has "attendant therapeutic implications," as
the Harvard neurologists Dennis Selkoe and Rudolph Tanzi wrote in a
2004 article. "Compounds that subtly increase IDE activity," they suggested, "could chronically decrease [amyloid] levels in the human brain."
This implies that anything that decreases insulin levels over the long term
(and so increases the amount of IDE available to clean up amyloid)—

* Harvard neurologist Dennis Selkoe and others have been working to track down a gene that
seems to predispose individuals to age-related Alzheimer's, rather than the inherited early-onset
form. By February 2007, they had not found it, but they had *localized* it, in the lingo, to a chunk of
a single chromosome that was known to include the gene for insulin-degrading enzyme. This
made IDE the obvious candidate and suggested that anyone who inherited a particularly unlucky
variant of the IDE gene would have an increased likelihood of getting Alzheimer's.

including such dietary approaches as eating less carbohydrates—will achieve the same effect. This isn't to say that eating carbohydrate foods to excess is a cause of Alzheimer's, only that mechanisms have now been identified to make the hypothesis plausible.

To discuss cancer, we need to first return to the subject of cancer in isolated populations eating traditional diets. The modern incarnation of these observations begins with John Higginson, who was the founding director of the World Health Organization's International Agency for Research on Cancer (IARC), a position he would hold for two decades. In the 1950s, Higginson studied cancer incidence in native African populations and compared them with incidence in the United States and Denmark, the two nations for which equivalent data existed. With a few exceptions, Higginson reported, cancer in African natives was remarkably uncommon. This led Higginson to conclude that *most* human cancers were caused by environmental factors, and that diet and lifestyle factors were the primary suspects. "It would seem, therefore, that the majority of human cancer is potentially preventable," as the World Health Organization concluded in 1964, a view that evolved into the new orthodoxy.

Cancer epidemiologists then tried to establish what proportion of cancers these might be. Higginson suggested 70 to 80 percent of all cancers could be prevented; others said as many as 90 percent. In 1981, the Oxford epidemiologists Richard Doll and Richard Peto published the seminal work on this subject: a 120-page analysis in the *Journal of the National Cancer Institute* that reviewed the existing evidence on changes in cancer incidence over time, changes upon migration from one region of the world to another, and differences in cancer rates between communities and nations. (Colon cancer, for example, was ten times more common in rural Connecticut than in Nigeria; breast cancer was diagnosed eight times more often in British Columbia than in the non-Jewish population of Israel.) Based on this evidence, Doll and Peto concluded that *at least* 75 to 80 percent of cancers in the United States might be avoidable with appropriate changes in diet and lifestyle.

In the quarter-century since Doll and Peto published their analysis, it has been cited in nearly two thousand journal articles, and yet the fundamental implications have been largely lost. The two most important conclusions in their analysis were that man-made chemicals—in pollution, food additives, and occupational exposure—play a minimal role in human cancers, and that diet played the largest role—causing 35 percent of all cancers, though the uncertainties were considered so vast that the number could be as low as 10 percent or as high as 70 percent.

Higginson had repeatedly remarked on these two points during his tenure as director of IARC. In early reports, Higginson and the World Health Organization had referred to "extrinsic factors" and "environmental factors" as the cause of most cancers, by which they meant lifestyle and diet. The public and the environmental movement had perceived this to mean almost exclusively "man-made chemicals"—the "carcinogenic soup," as it was known in the 1960s and 1970s. "It appears that only a very small part of the total cancer burden can be directly related to industrialization," Higginson wrote. The release of industrial chemicals into the environment could not explain, for example, why the nonindustrial city of Geneva had more cancer than Birmingham, "in the polluted central valleys of England," or why prostate cancer was ten times more frequent in Sweden than in Japan.*

Nonetheless, this focus on carcinogenic chemicals as the primary cancer-causing agents in the environment also carried over to nutrition-related cancer research in the laboratory. It was assumed that whatever components of diet were responsible for cancer worked the same way that chemicals did: by inducing mutations and genetic damage in cells. When cancer researchers from around the world met in September 1976 at the Cold Spring Harbor Laboratory to discuss the origins of human cancer, the talks focused on those chemicals shown to be carcinogens in animals, and the possibility that they might be found in infinitesimal or greater amounts in human diets, drinking water, or pharmaceuticals.

By the mid-1970s, when cancer epidemiologists began to convince politicians and the public that many cancers were caused by what Peto and Doll had called the "gross aspects of diet," rather than "ingestion of traces of powerful carcinogens or precarcinogens," the focus was almost exclusively on fat, fiber, and red meat, or smoked- or salt-cured meat, as well as the possibly protective nature of vitamins, vegetables, and fruits. The low incidence of cancer in vegetarians and Seventh-day Adventists was often cited as evidence that meat is carcinogenic and that green vegetables and fruit are protective. (Although the incidence of colon cancer, for instance,

---

* Higginson held the environmental movement responsible for what he considered a willful misinterpretation of the epidemiologic observations: "If they could possibly make people believe that cancer was going to result from pollution, this would enable them to facilitate the clean-up of water, of the air, or whatever it is," he told *Science* in 1979. He was all for cleaning up the environment, he added, but "to make cancer the whipping boy for every environmental evil may prevent effective action when it does matter."

among Seventh-day Adventists was no lower than among Mormons, described by Doll and his colleague Bruce Armstrong as "among the biggest beefeaters in the United States.") For the next twenty years, conferences, textbooks, and expert reports on nutrition and cancer continued to focus exclusively on these factors, although now aided by the advances in molecular biology.

By the end of the 1990s, clinical trials and large-scale prospective studies had demonstrated that the dietary fat and fiber hypotheses of cancer were almost assuredly wrong, and similar investigations had repeatedly failed to confirm that red meat played any role.* Meanwhile, cancer researchers had failed to identify any diet-related carcinogens or mutagens that could account for any of the major cancers. But cancer epidemiologists made little attempt to derive alternative explanations for those 10 to 70 percent of diet-induced cancers, other than to suggest that overnutrition, physical inactivity, and obesity perhaps played a role.

Throughout these decades, refined carbohydrates and sugars received little or no attention in discussions of cancer causation. Peter Cleave had suggested in *The Saccharine Disease* that the refining of carbohydrates might be involved in colon cancer. John Yudkin had noted that the five nations with the highest breast-cancer mortality in women in the late 1970s (in descending order: the United Kingdom, the Netherlands, Ireland, Denmark, and Canada) had the highest sugar consumption (in descending order: the United Kingdom, the Netherlands, Ireland, Canada, and Denmark), and those with the lowest mortality rates (Japan, Yugoslavia, Portugal, Spain, and Italy) had the lowest sugar consumption (Japan, Portugal, Spain, Yugoslavia, and Italy). But in 1989, when the National Academy of Sciences published its 750-page report on *Diet and Health*, the authors spent only a single page evaluating the proposition that carbohydrates might cause cancer. "There is little epidemiologic evidence to support a role for carbohydrates per se in the etiology of cancer," they noted. They did add two caveats. One was that "no definitive conclusion is justified . . . because carbohydrates have often been reported in epidemiologic studies only as a component of total energy and not analyzed separately." The other was that Richard Doll and Bruce Armstrong had found sugar intake in international comparisons to be "positively correlated with both the incidence of and mortality from" colon, rectal, breast,

* Those clinical trials that tested the dietary-fat-and-fiber hypotheses of cancer, as we discussed earlier, replaced red meat in the experimental diets with fruits, vegetables, and whole grains. When these trials failed to confirm that fat causes breast cancer, or that fiber prevents colon cancer, they also failed to confirm the hypothesis that red-meat consumption plays a role in either.

ovarian, prostate, kidney, nervous-system, and testicular cancer, and that "other investigators have produced similar findings."

The patterns of cancer incidence, for many cancers, are similar to those of heart disease, diabetes, and obesity, which alone suggests an association between these diseases that is more than coincidental. This was the basis of Cleave's speculation, of Dennis Burkitt's, and of those cancer epidemiologists who argued that dietary fat caused breast cancer. But if dietary fat, red meat, man-made chemicals, or even the absence of fiber cannot explain the "strikingly similar" patterns of disease distribution, as the Harvard epidemiologist Edward Giovannucci remarked about colon cancer and Type 2 diabetes in 2001, then something else most likely does.

Those cancers apparently caused by diet or lifestyle and not related to tobacco use are either cancers of the gastrointestinal tract, including colon and rectal cancer, or cancers of what are technically known as *endocrine-dependent organs*—breast, uterus, ovaries, and prostate—the functions of which are regulated by hormones. This connection between these diet- and life-style-related cancers and hormones has been reinforced by the number of hormone-dependent factors linked to cancers of the breast and the endometrium (the lining of the uterus). All suggest that estrogen plays an important role. All these cancers, with the possible exception of pancreatic and prostate cancer, appear to increase in incidence with weight gain. These associations together imply both a metabolic and a hormonal connection between diet and cancer. This in turn led breast-cancer researchers to focus their attention on the likely possibility that obesity increases the incidence of breast cancer by increasing estrogen production.

The most direct evidence linking overweight or overnutrition to cancer comes from animal experiments. These date back to the eve of World War I, when Peyton Rous, who would later win a Nobel Prize, demonstrated that tumors grow remarkably slowly in semi-starved animals. This line of research lapsed until 1935, when the Cornell University nutritionist Clive McCay reported that feeding rats just barely enough to avoid starvation ultimately extended their lifespan by as much as 50 percent. Seven years later, Albert Tannenbaum, a Chicago pathologist, launched a cottage research industry after demonstrating that underfeeding mice on very low-calorie diets, as McCay had, resulted in a dramatic inhibition of "many types of tumors of divergent tissue origin." In one experiment, twenty-six of fifty well-fed mice developed mammary tumors by a hundred weeks of age— the typical lifespan of lab mice—compared with none of fifty that were allowed only minimal calories. Tannenbaum's semi-starved animals not only lived longer, but were more active, he reported, and had fewer "pathologic changes in the heart, kidneys, liver, and other organs."*

To explain this inhibitory effect, Tannenbaum considered an idea that had originated in the 1920s with Otto Warburg, a German biochemist and later Nobel Prize winner. Warburg had demonstrated that tumor cells quickly develop the ability to survive without oxygen and to generate energy by a process of fermentation rather than respiration. Fermentation is considerably less efficient, and so tumors will burn perhaps thirty times as much blood sugar as normal cells. Incipient tumors in these calorie-restricted lab animals, it was thought, cannot obtain the huge amounts of blood sugar they need to fuel mitosis—division of the nucleus—and continue proliferating.

Insulin was not considered a primary suspect until just recently, but the evidence has existed for a while. The earliest such link between a dysfunction in carbohydrate metabolism and cancer dates to 1885, when a German clinician reported that sixty-two of seventy cancer patients were glucose-intolerant. One common observation by clinical investigators over the years was that women with adult-onset (Type 2) diabetes or glucose intolerance had a higher-than-average incidence of breast cancer. By the mid-1960s, researchers were reporting that insulin acts as a promoter of growth and proliferation in both healthy and malignant tissues. Howard Temin, who later won a Nobel Prize for his cancer research, reported that cells turned malignant by a chicken virus would cease to proliferate in the laboratory unless insulin was added to the serum in which they were growing. This growth-factor effect of insulin was also demonstrated in adrenal and liver-cell cancers. Insulin "intensely stimulated cell proliferation in certain tumors," noted one 1967 report. In 1976, Kent Osborne and his colleagues at the National Cancer Institute reported that one line of particularly aggressive breast-cancer cells were "exquisitely sensitive to insulin."

By the late 1970s, researchers had also reported that malignant breast tumors had more receptors for insulin than did healthy tissue. The more insulin receptors on the surface of a cell, the more sensitive it will be to the insulin in its environment. Having a greater number of insulin receptors than healthy cells, as one report noted, might confer "a selective growth advantage to tumor cells."

---

* Tannenbaum actually compared his chronically underfed mice with control mice fed the identical diet but supplemented with cornstarch. The inhibition of cancer, as Tannenbaum noted, could have been due to "carbohydrate-restriction" rather than restriction of all calories.

"Selective growth advantage" speaks directly to the process of Darwinian evolution that is considered the controlling force in tumor development. We can think of human cells as existing in a microscopic ecosystem, living in harmony with their environment, and balanced, as are all species, between the opportunities for growth and proliferation and the processes that lead to aging and death. In such an environment, the billions of cells that eventually constitute a tumor will be the descendants of a single cell that has accumulated a series of genetic mutations, each adding to its proclivity to proliferate unfettered by any of the normal inhibitions to growth. The process in which a healthy cell eventually results in malignancy is a gradual evolution driven by a series of mutations in the DNA of the genes, each bestowing on the cell either the inclination to multiply or a breakdown in the control and repair mechanisms that have evolved to counter precisely such potentially deleterious mutations. The descendants of such a mutant cell would inherit this fitness advantage over other cells in the tissue, and so, within a few years, a single such mutant cell will leave millions of descendants. As one of those descendants in turn gains, purely by chance, yet another advantageous error or mutation, its descendants will now come to dominate.

Each new mutation-bearing cell constitutes a new species, in effect, that is better suited to prevail in its local cellular environment. Eventually, with this continued accumulation of what to the body as a whole is simply bad luck, a single cell will come to possess precisely that set of mutant genes that drive it and allow it to grow and proliferate without limit. Because each single *hit* of genetic damage alone is not sufficient to produce a cancer cell, the accumulation of just the right half-dozen hits (actually, the *wrong* half-dozen hits) takes years or decades, which is why virtually all cancers become more common as we age.

Cancer researchers now believe that these cancer-causing mutations occur as errors in the replication of DNA during the process of cell division and multiplication. Each one of us is likely to experience some ten thousand trillion cell divisions over the course of our lives, constituting an "enormous opportunity for disaster," in the words of the MIT molecular biologist Robert Weinberg, author of the textbook *The Biology of Cancer*. This suggests that cancer-causing mutations are another unavoidable side effect of aging, which is why our cells have also evolved to be exceedingly resistant to genetic damage. They have sophisticated mechanisms to search out defects in newly replicated DNA and repair them, and other mechanisms that actually prompt a cell to commit suicide—programmed cell death, in the technical terminology—if the repair mechanisms are incapable of fixing the damage that occurred during replication. Alas, with time, these programs, too, can be disabled by the proper mutations.

Within this Darwinian environment, insulin provides fuel and growth signals to incipient cancer cells. Its more lethal effects, however, might come through the actions of insulin-like growth factor (IGF). Growth hormone itself is secreted by the pituitary gland and works throughout the body; IGF is secreted both by the liver and by tissues and cells throughout the body, and it then works locally, where concentrations are highest. Most tissues require at least two growth factors to grow at an optimal rate, and IGF is almost invariably one of the two, and perhaps the primary regulator.

Insulin-like growth factor is sufficiently similar in structure to insulin that it can actually mimic its effects. IGF can stimulate muscle cells to take up blood sugar, just as insulin does, though not as well. Researchers now believe that IGF serves as the necessary intermediary between the growth hormone secreted by the pituitary gland, and the actual amount of food that is available to build new cells and tissues. If insufficient food is available, then IGF levels will stay low even if growth-hormone levels are high, and so cell and tissue growth will proceed slowly if at all. Add the necessary food and IGF levels increase, and so will the rate of growth. Unlike insulin, which responds immediately to the appearance of glucose in the bloodstream and so varies considerably from hour to hour, IGF concentrations in the circulation change only slowly over days or weeks, and thus better reflect the *long-term* availability of food in the environment.

Since the mid-1970s, researchers have identified many of the molecules that play a role in regulating the strength of the growth and proliferation signals that IGF communicates to the cells themselves. There are several different insulin-like growth factors, for instance, and they bind to specific IGF receptors on the surfaces of cells. The more IGF receptors on a cell's surface, the stronger the IGF signal to the cell. If insulin levels are high enough, insulin will stimulate the IGF receptors and send IGF signals into cells as well as insulin signals.*

IGF and its receptors appear to play a critical role in cancer. In mice, functioning IGF receptors are a virtual necessity for cancer growth, a discovery that Renato Baserga of Thomas Jefferson University says he "stumbled" upon in the late 1980s, after nearly forty years spent studying the growth processes of normal and cancerous cells. Shutting down the IGF

---

* Different IGFs have different effects. To keep the following discussion reasonably simple, I'll refer to IGF and IGF receptors as though there were only one species of each, although I'm oversimplifying the science by doing so.

receptor in mice will lead to what Baserga calls "strong inhibition, if not total suppression of [tumor] growth"; it is particularly lethal to those tumors that have already metastasized from a primary site elsewhere in the body.

In the bloodstream, virtually all insulin-like growth factors are attached to small proteins that ferry them around to various tissues where they might be needed. But the IGFs, when attached to these proteins, are too large and unwieldy to pass through the walls of blood vessels and get to the tissues and cells where the IGF might be used. At any one time, only a small percentage of IGF in the circulation is left unbound to stimulate the growth of cells.

These *binding proteins* constitute yet another of the mechanisms used by the body to regulate hormonal signals and growth factors. Insulin appears to depress the concentration of IGF-binding proteins, and so high levels of insulin mean more IGF itself is available to effect cell growth—including that of malignant cells. Anything that increases insulin levels will therefore increase the availability of IGF to the cells, and so increase the strength of the IGF proliferation signals. (Insulin has been shown to affect estrogen this way, too, one way in which elevated levels of insulin may potentially cause breast cancer.)

The role of IGF in cancer appears to be fundamental, albeit still controversial. As is the case with insulin, IGF has been found in the laboratory to enhance the growth and formation of tumor cells directly; IGF signals prompt cells to divide and multiply. (This effect seems to be particularly forceful with breast-cancer cells when IGF and estrogen are acting in concert.) IGF has an advantage over other growth factors that might play a role in cancer because it can reach tumors either through the bloodstream— after being secreted by the liver—or as a result of production by nearby tissue. There's even evidence that tumors can stimulate their own further growth and proliferation by secreting their own insulin-like growth factors. In the early 1980s, cancer researchers discovered that tumor cells also *overexpress* IGF receptors, just as they overexpress insulin receptors. The surfaces of tumor cells have two to three times as many IGF receptors as healthy cells, which makes them all that much more responsive to the IGF in their immediate environment.

This is another way in which cancer cells gain their all-important survival growth advantage, suggests Derek LeRoith, whose laboratory at the National Institute of Diabetes and Digestive and Kidney Diseases did much of this research. The extra insulin receptors will cause cancerous cells to receive more than their share of insulin from the environment, which will convey to the cell more blood sugar for fueling growth and proliferation; the extra IGF receptors will assure that these cells are supplied with particularly forceful commands to proliferate. Another critical role of IGF in the development of cancer may be its ability to inhibit or override the cell suicide program that serves as the ultimate fail-safe mechanism to prevent damaged cells from proliferating.

In the past decade, LeRoith and others have demonstrated that the various molecules involved in the communication of the IGF signal from the bloodstream to the nucleus of cells—the insulin-like growth factors themselves, their receptors, and their binding proteins—work together with insulin to regulate both the growth and metastasis (the spread of tumors to secondary sites) of colon and breast cancer. LeRoith has done a series of experiments with mice genetically engineered so that their livers do not secrete IGF. As a result, these mice have only a quarter as much IGF in their circulation as normal mice. When colon or mammary tumors are transplanted into these mice, both tumor growth and metastasis are significantly slower than when identical tumors are implanted in normal mice with normal IGF levels. When insulin-like growth factor is injected back into these genetically engineered mice, tumor growth and metastasis accelerate. David Cheresh, a cancer researcher at the Scripps Institute in La Jolla, California, has demonstrated that both insulin and insulin-like growth factor will prompt otherwise benign tumors to metastasize and migrate through the bloodstream to secondary sites.

The working hypothesis of cancer researchers who study IGF is not that these molecules initiate cancer, a process that occurs through the accumulation of genetic errors, but, rather, that they accelerate the process by which a cell becomes cancerous, and then they work to keep the cells alive and multiplying. At a 2003 meeting in London to discuss the latest work on IGF, researchers speculated that the development of cancerous cells and even benign tumors is a natural side effect of aging. What's not natural is the progression of these cells and tumors to lethal malignancies. Such a transformation requires the chronically high levels of insulin and IGF induced by modern diets. This hypothesis is supported by epidemiological studies linking hyperinsulinemia and elevated levels of IGF to an increased risk of breast, prostate, colorectal, and endometrial cancer.

This hypothesis, if not refuted, would constitute a significant shift in our understanding of the development of malignant cancer. It would mean that the decisive factor in malignant cancer is not the accumulation of genetic damage in cells, much of which is unavoidable, but how diets change the environment around cells and tissues to promote the survival, growth, and then metastasis of the cancer cells that do appear. "People were thinking a bit too much that diet could be a risk factor for cancer almost exclusively based on the idea that it contained carcinogenic substances," explains Rudolf Kaaks, director of the Hormones and Cancer Group at the International Agency for Cancer Research. "Now the idea is that there is a change in the endocrine and growth-factor environment of cells that pushes cells to proliferate further and grow more easily and skip the programmed cell-death events."

IGF and insulin can be viewed as providing fuel to the incipient fire of cancerous cells and the freedom to grow without limit. The critical factor is not that diet changes the *nature* of cells—the mutations that lead to cancer—but that it changes the *nurturing* of those cells; it changes the environment into one in which cancerous and precancerous cells can flourish. Simply by creating "an environment that favored, even slightly, survival (rather than programmed cell death)," says the McGill University oncologist Michael Pollak, insulin and IGF would increase the number of cells that accumulate some genetic damage, and that would increase the number of their progeny that were likely to incur more damage, and so on, until cancer is eventually achieved. "When applied simultaneously to large numbers of at-risk cells over many years," notes Pollak, "even a small influence in this direction would serve to accelerate carcinogenesis."

All of this leads us back to the spectacular benefits of semi-starvation on the health and longevity of laboratory animals. If we take a young rat and restrict its eating to less than two-thirds the calories of its preferred diet, and if we keep this up for its entire life, our rat will likely live 30 to 50 percent longer than had we let it eat to satiation, and any age-related diseases—cancer in particular—will be delayed in their onset and slowed in their progression. This has been shown to hold true for mice and other rodents, and for yeast, protozoans, fruit flies, and worms (and maybe even monkeys).

Two possibilities for how these diets work are that the animals live longer because they are less encumbered by body fat, or because they're leaner all around and so weigh less. Neither of these can explain the evidence, however. Consider a strain of mice known as ob/ob mice. These have a mutation in a single gene that results in such extreme obesity that a mouse ends up looking like a loaf of bread with fur, eyes, whiskers, and a mouth. Nonetheless, these mice can be kept at a normal weight by restricting their food consumption to half of what they would naturally prefer to eat. They are normally short-lived, which supports the idea that the greater the body fat the shorter the lifespan, but on a lifelong very low-calorie diet they will live as long as or longer than lean mice of a similar genetic inheritance but without the mutation that causes obesity. They will do this even though they still have more than twice the body fat of the lean mice. Indeed, when these experiments were done in the early 1980s by David Harrison of the Jackson Laboratory in Bar Harbor, Maine, these calorically restricted ob/ob mice lived just as long as calorically restricted lean mice, even though the former were nearly four times as fat as the latter. "Longevities," Harrison concluded, "were related to food consumption rather than to the degree of adiposity." This has inevitably been the case, whenever these experiments are done. The calorie-restricted animals live longer because of some metabolic or hormonal consequence of semi-starvation, not because they are necessarily leaner or lighter.

So what does eating less do physiologically that leanness does not? With each new study, researchers have honed their hypothesis of why semi-starvation leads to these anti-aging and disease-delaying processes, and what this says about human aging and disease. This has led to some remarkable revelations about insulin and insulin-like growth factor, and what is likely to happen when these two hormone/growth factors are perturbed by modern diets.

One hypothesis proposes that calorie restriction reduces the creation of toxic reactive oxygen species—free radicals—which are considered to be crucial factors in the aging of cells and tissues. Eat less food and the cells burn less fuel, and so generate fewer free radicals. Oxidative stress proceeds at a slower pace, and we live longer, just as a car will last longer in a dry climate that doesn't promote rust. Certainly, calorie restriction suppresses free-radical production. And if fruit flies are either fed antioxidants or genetically transformed to overproduce their own antioxidants, they will live up to 50 percent longer. But similar experimental interventions seem to do nothing for rodents. The genetic evidence suggests that something more profound is happening, although this reduction in oxidative stress likely plays some role.

The characteristics that all these long-lived organisms seem to share definitively are reduced insulin resistance, and abnormally low levels of blood sugar, insulin, and insulin-like growth factor. As a result, the current thinking is that a lifelong reduction in blood sugar, insulin, and IGF bestows a longer and healthier life. The reduction in blood sugar also leads to reduced oxidative stress and to a decrease in glycation, the haphazard binding of sugars to proteins, and glycation end-products and all the toxic sequelae that follow. The decrease in insulin and IGF also apparently bestows on the organism an enhanced ability to protect against oxidative stress and to ward off other pathogens.

The most compelling evidence now supporting this hypothesis has emerged since the early 1990s from genetic studies of yeast, worms, and fruit flies, and it has recently been confirmed in mice. In all four cases, the mutations that bestow extreme longevity on these organisms are mutations in the genes that control both insulin and IGF signaling.

Geneticists and developmental biologists refer to yeast, worms, fruit flies, and mice as *model* organisms because they're easy to study in the laboratory and what we learn from them about genetics will almost assuredly apply to humans as well. This is considered the fundamental principle underlying modern genetic research: once evolution comes upon a genetic mechanism that works, it reuses it again and again. Those genes that regulate the development and the existence of any single living organism will likely be used in some similar fashion in *all* of them. "When reduced to essentials," as the cancer researcher J. Michael Bishop suggested in his 1989 Nobel Prize lecture, "the fruit fly and *Homo sapiens* are not very different."

Consider, for instance, the mutations that control longevity in nematodes, which are the particular type of microscopic worms favored by modern researchers. These mutations, as Cynthia Kenyon and her colleagues from the University of California, San Francisco, reported in *Nature* in 1993, are in a gene that was known to regulate the passage of young worms into a state known as *dauer* that is similar to hibernation in mammals. The worms will enter this dauer state, explains Kenyon, only if they have insufficient food to survive. "The way these worms work," she explains, "is that the worm hatches from the egg, and if there's not a lot of food around, it goes through various larval stages and ends up in this dauer state. . . . It doesn't eat or do anything else. Then, if you give it food, it will exit the state and reproduce and have a normal lifespan." The particular genetic mutation that Kenyon discovered resulted in worms that lived twice as long as normal worms, and this was, at the time, the longest lifespan extension ever reported in an organism. Kenyon then demonstrated that this increased longevity was not simply a consequence of some kind of developmental arrest—as though the mutation had somehow trapped a young worm in a dauerlike limbo—but was actually the result of the mutation's triggering a lifespan-extension mechanism in adult worms. In other words, this mutation was keyed into a genetic program that actually regulates longevity, and does it in a way that would be evolutionarily advantageous.

In 1997, the Harvard geneticist Gary Ruvkun reported that the gene in question was the single worm-equivalent of a trio of insulin-related genes in humans. In retrospect, this wasn't surprising, noted Ruvkun, because here was a gene in worms that regulated a process—dauer—that depended on the presence or absence of food in the environment, and insulin and IGF are the genes in more sophisticated organisms that respond specifically to food availability. As it turns out, particularly long-lived fruit-fly mutants have also been found to have defects in this same insulin-like gene pathway, which serves to regulate in the fly a condition very similar to dauer and hibernation.

The ultimate evidence, at least so far, that insulin and insulin-like growth factor affect longevity and disease comes from a type of transgenic animal experiment known as a *knockout*. The working assumption of such experiments is that the function of a gene can be elucidated by creating an animal that lacks the gene entirely—the gene has been *knocked out*—or has only one copy instead of the usual two. In January 2003, Martin Holzenberger and his colleagues from the Institut National de la Santé et de la Recherche Médicale in Paris reported that they had created mice with only a single copy of the gene for the IGF receptor, which meant that the cells of such mice would be comparatively unresponsive to any IGF that might be available in the circulation. The result was that these mice lived 25 percent longer than their littermates who had both copies of the gene, despite the fact that their weights were effectively identical. That same month, C. Ronald Kahn and his colleagues at the Joslin Diabetes Center published the results of their research on mice that they had genetically engineered to lack the insulin receptor only on their fat cells. With their fat tissue immune to the effect of insulin, Kahn's mice weighed 25 percent less than normal mice. These mice remained lean, even when forced to overeat. They were simply incapable of putting on fat. As Kahn later explained, this wasn't surprising, since fat cells require insulin for fat synthesis. If they have no receptor to detect the insulin that's present, then no fat can accumulate. The transgenic mice lived almost 20 percent longer than normal mice.

These experiments have led to the working hypothesis that insulin and insulin-like growth factor emerged in simple organisms in part to promote the survival of the species when food is hard to come by. These hormone/growth factors regulate metabolism and fat storage and reproduction. The IGF regulates cell division and growth, while the insulin regulates metabolism by apportioning or partitioning the food we consume into those calories that will be used immediately for fuel and those that will be stored for use at a later time. When food is plentiful, activity in the insulin and IGF pathways increases and pushes the animal to grow, mature, and reproduce. When food is scarce, activity in these pathways is reduced, and this shifts the organism into a mode that favors long-term survival over immediate reproduction. As Cynthia Kenyon explains:

> When food becomes limiting, an animal lacking this system would either die of starvation, or produce progeny that die of starvation. In contrast, with this food-sensing system in place, as food declines, the animal begins to build up fat and/or glycogen [the molecular storage form of glucose] reserves, elaborates stress-resistance mechanisms, and delays or suspends reproduction until food is restored. It also activates pathways that extend lifespan, which increases the organism's chance of being alive and still youthful enough to reproduce if it takes a long time for conditions to improve.

If we accept the evolutionary argument that genetic mechanisms are con-served from simple organisms to humans, then we have at least to con-template the implications: if a regulatory system as fundamental as that of insulin and IGF is capable of influencing longevity and susceptibility to disease in flies, worms, and mice, then it is likely to do so in humans as well. This research supports the hypothesis that elevations of insulin and IGF will increase the risk of disease and shorten life, and so any diet or lifestyle that elevates insulin and makes IGF more available to the cells and tissues is likely to be detrimental.

To accept these implications at face value, however, we have to be capa-ble of dismissing the conventional wisdom on diet and chronic disease—that an excess of saturated fat, all fat, or perhaps all calories is responsible. Few researchers are willing to take this approach. One who has is Cynthia Kenyon. Once it became clear that the mutations that prolonged longevity in worms were those that reduced the level of activity in the worms' insulin-IGF pathway, Kenyon began a series of experiments based on a single question: what would happen if she fed worms glucose, in addition to their preferred diet of bacteria? Kenyon added 2 percent glucose to the bacterial medium in which the worms lived, and the lifespan of the worms was reduced by a quarter. Kenyon is still working to establish the nature of this adverse effect of glucose. Her hypothesis: just as mutations increase lifespan in worms by decreasing activity in their insulin-IGF pathway, glucose *shortens* the lifespan of worms by *increasing* activity in the same pathway. In October 2004, when Kenyon presented the results of these experiments at a conference on the molecular genetics of aging, she concluded her presentation with a simple, albeit radical question: "Could a low-carb (i.e., low-glycemic-index) diet lengthen lifespan in humans?"

Kenyon is unusual in this kind of laboratory research in that she had already interpreted the results of her research as relevant to her own life. As Kenyon tells it, the day she realized that glucose shortened the lives of her worms, she decided to restrict her own consumption of carbohydrates to a bare minimum. She lost thirty pounds, she says; her blood pressure, triglycerides, and blood-sugar levels all dropped; and her HDL increased. Kenyon recognizes her experience as anecdotal, but it certainly influenced her suspicion that carbohydrates would also cause chronic disease in humans through their effect on insulin and insulin-like growth factor.

A more common approach to this research implicating insulin and IGF in the causation of chronic disease is to avoid any possible dietary implications and focus solely on the connotations for drug or gene therapies. This was the approach used by Dennis Selkoe and Rudolph Tanzi, who concluded their April 2004 report on insulin and Alzheimer's by suggesting that the results "have attendant therapeutic implications." The only therapeutic implication they discussed was the possibility of creating "compounds" that increase the activity of insulin-degrading enzyme—the equivalent of reducing insulin levels—and so inhibiting the accumulation of Alzheimer's plaques in the brain.

This same approach was used by Ronald Kahn and his collaborators when they discussed the lean, long-lived transgenic mice they had created by knocking out the insulin receptors on the fat cells of the mice. The publication of the research in *Science* was accompanied by a press release from the Joslin Diabetes Center, of which Kahn is president, focused almost exclusively on the "dream of 60 million overweight American adults," which it described as the desire to "throw away those diet books and eat whatever you want without becoming fat, and—as a bonus—not develop diabetes and live longer as well." The press release implied that this dream might be accomplished by the insights gleaned from these transgenic mice, and Kahn was quoted discussing therapeutic implications, although once again diet was not one of them. "Perhaps one day if we are able to find a drug to reduce or block insulin action in fat cells in humans, we might be able to prevent obesity, as well as Type 2 diabetes and other metabolic diseases," Kahn wrote. "And who knows, they might also live longer too." Diabetologists implicitly take the same tack whenever they discuss the need for their diabetic patients to "normalize" blood sugar, while recommending that this be accomplished primarily with "intensive insulin therapy" rather than restricting the carbohydrate content of their diets.

Another common approach today is to accept the chronic elevation of insulin, and so IGF, as a likely cause of chronic disease, but then assume that the hyperinsulinemia is caused by insulin resistance, which in turn is induced by a combination of high-fat, energy-dense, high-calorie diets, physical inactivity, and excess weight. By this logic, any research that implicates increased insulin activity in disease only confirms that too much food and too little exercise are the true banes of our existence. This approach is the one employed by those clinicians and public-health authorities who now acknowledge that hyperinsulinemia, insulin resistance, and the associated physiological abnormalities of metabolic syndrome are important risk factors for heart disease, but then blame the syndrome itself on excess weight or, if the patient happens to be lean, on physical inactivity. The guidelines from the National Cholesterol Education Program manage to merge both of the latter two approaches, by first enumerating the causes of metabolic syndrome as overweight, physical inactivity, and an "atherogenic diet"—defined as a diet high in saturated fat and calories—and then suggesting that "pharmacological modification of the associated risk factors" is the most effective treatment.

In this approach, high-calorie, high-fat diets and sedentary lifestyles are seen as the causes of all the diseases of civilization. The causal link in this chain from diet and lifestyle to disease is excess weight. "Weight sits like a spider at the center of an intricate, tangled web of health and disease," as the Harvard epidemiologist Walter Willett has described it in *Eat, Drink, and Be Healthy: The Harvard Medical School Guide to Healthy Eating*. Or, as Jeremiah Stamler suggested back in 1961, about heart disease in particular, "Excess weight and the common American pattern of gain in weight from young adulthood into middle age are highly prevalent and serious risk factors. . . . The problem is not the severe, marked, huge, circus-type of obesity, but rather the 25 or 40 pounds put on gradually over the years— the moderate, creeping obesity so common among middle-aged American men."

That excess weight is accompanied by an elevated risk of chronic disease is a given. The questionable assumption is that it is an excess of calories of *all* types, and the dense calories of dietary fat in particular, combined with a relative lack of physical activity, that causes weight gain. In the prevailing wisdom, a simple caloric imbalance is the culprit: we get fat because we consume more calories than we expend.

The alternative is that excess weight and obesity, like all diseases of civilization, are caused by the singular hormonal effects of a diet rich in refined and easily digestible carbohydrates. The fattening of our adult years, after all, is not just *associated* with chronic diseases of civilization, it *is* a disease of civilization, and so it, too, may be a symptom of an underlying disorder. In this hypothesis, it is the *quality* of the calories consumed that regulates weight, and the *quantity*—more calories consumed than expended—is a secondary phenomenon. Whatever causes weight gain is at the heart of this tangled web, and that is the question we must now address.

Part Three

OBESITY AND THE REGULATION OF WEIGHT

How may the medical profession regain its proper role in the treatment of obesity? We can begin by looking at the situation as it exists and not as we would like it to be . . . If we do not feel obliged to excuse our failures we may be able to investigate them.

> ALBERT STUNKARD AND MAVIS MCCLAREN-HUME,
> in "The results of treatment for Obesity: A Review of
> the Literature and Report of a Series," 1959

To cultivate the faculty of observation must then be the first duty of those who would excel in any scientific pursuit, and to none is this study more necessary than to the student of medicine. Without the habit of correct observation, no one can ever excel or be successful in his profession. Observation does not consist in the mere habitual sight of objects—in a kind of vague looking-on, so to speak—but in the power of comparing the known with the unknown, of contrasting the similar and dissimilar, in justly appreciating the connection between cause and effect, the sequence of events and in estimating at their correct value established facts.

> THOMAS HAWKES TANNER, *A Manual of Clinical Medicine
> and Physical Diagnosis,* 1869

Chapter Fourteen

THE MYTHOLOGY OF OBESITY

A colleague once defined an academic discipline as a group of scholars who had agreed not to ask certain embarrassing questions about key assumptions.
MARK NATHAN COHEN, *Health and the Rise of Civilization*, 1989

CRITICAL TO THE SUCCESS OF any scientific enterprise is the ability to make accurate and unbiased observations. "To have our first idea of things, we must see those things," is how Claude Bernard explained this in 1865; "to have an idea about a natural phenomenon, we must, first of all, observe it. . . . All human knowledge is limited to working back from observed effects to their cause." But if the initial observations are incorrect or incomplete, then we will distort what it is we're trying to explain. If we make the observations with preconceived notions of what the truth is, if we believe we know the cause before we observe the effect, we will almost assuredly see what we want to see, which is not the same as seeing things clearly.

The trouble with the science of obesity as it has been practiced for the last sixty years is that it begins with a hypothesis—that "overweight and obesity result from excess calorie consumption and/or inadequate physical activity," as the Surgeon General's Office recently phrased it—and then tries and *fails* to explain the evidence and the observations. The hypothesis nonetheless has come to be perceived as indisputable, a fact of life or perhaps the laws of physics, and its copious contradictions with the actual observations are considered irrelevant to the question of its validity. Fat people are fat because they eat too much or exercise too little, and nothing more ultimately need be said.

The more closely we look at the evidence and at obesity itself, the more problematic the science becomes. Lean people will often insist that the secret to their success is eating in moderation, but many fat people insist that they eat no more than the lean—surprising as it seems, the evidence backs this up—and yet are fat nonetheless. As the National Academy of Sciences report *Diet and Health* phrased it, "Most studies comparing normal and overweight people suggest that those who are overweight eat fewer calories than those of normal weight." Researchers and public-health officials nonetheless insist that obesity is caused by overeating, without attempting to explain how these two notions can be reconciled. This situation is not improved by the prevailing attitude of many nutritionists, obesity researchers, and public-health authorities that it is evidence of untoward skepticism to raise such issues, or to ask questions that lead others into contemplating the contradictions themselves.

For the past decade, public-health authorities have tried to explain the obesity epidemic in the United States and elsewhere. In 1960, government researchers began surveying Americans about their health and nutrition status. The first of these surveys was known as the National Health Examination Survey. It was followed by an ongoing series of National Health and Nutrition Examination Surveys (NHANES), of which there have been four so far. According to these surveys, through the 1960s and early 1970s, 12–14 percent of Americans were obese. This figure rose by 8 percent in the 1980s and early 1990s, and another 10 percent by the turn of this century.

This doubling of the *proportion* of obese Americans is consistent through all segments of American society, although obesity remains more common among African Americans and Hispanics than among whites and other ethnic groups, and most common among those in the lowest income brackets and poorly educated. Children were not exempt from this trend. The prevalence of overweight in children six to eleven years old more than doubled between 1980 and 2000; it tripled in children aged eleven to nineteen.*

Some factor of diet and/or lifestyle must be driving weight upward, because human biology and our underlying genetic code cannot change in such a short time. The standard explanation is that in the 1970s we began consuming more calories than we expended and so as a society we began getting fatter, and this tendency has been particularly exacerbated since the early 1980s.

Authorities phrase this concept differently, but the idea is invariably the same. The psychologist Kelly Brownell, director of the Yale Center for Eating and Weight Disorders, coined the term "toxic environment" to

---

* The apparent severity of this epidemic is inflated by the way in which obesity is defined. The use of a threshold for establishing whether or not you're obese—a body mass index (BMI) of 30—means that one can move from the overweight category to the obese category by virtue of gaining a few pounds. As a result, the 10-percent rise in obesity between 1991 and 2000 actually represented an increase in the average BMI of Americans from 26.7 to 28.1, an average weight gain of seven to ten pounds.

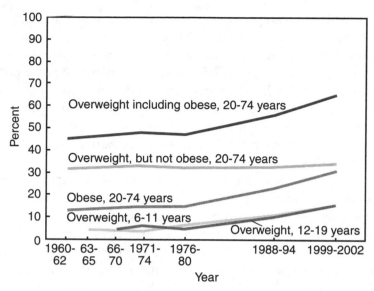

Obesity rates in America, according to the Centers for Disease Control,
with the upturn of the obesity epidemic apparently beginning
in the 1970s or early 1980s.

describe an American culture "that encourages overeating and physical
inactivity" and so encourages obesity as a consequence. "Cheeseburgers
and french fries, drive-in windows and supersizes, soft drinks and candy,
potato chips and cheese curls, once unusual, are as much our background
as trees, grass, and clouds," says Brownell. "Few children walk or bike to
school; there is little physical education; computers, video games, and tele-
visions keep children inside and inactive; and parents are reluctant to let
children roam free to play." In an editorial entitled "The Ironic Politics of
Obesity" published by *Science* in 2003, the New York University nutritionist Marion Nestle summed up this hypothesis of obesity and the obesity
epidemic in two words: "improved prosperity." Nestle, like Brownell, considered the food and entertainment industries culpable: "They turn people
with expendable income into consumers of aggressively marketed foods
that are high in energy but low in nutritional value, and of cars, television
sets, and computers that promote sedentary behavior. Gaining weight is
good for business," Nestle wrote.

More than one billion adults worldwide are overweight, according to the
World Health Organization; three hundred million are obese and obesity
rates have "risen three-fold or more since 1980 in some areas of North

America, the United Kingdom, Eastern Europe, the Middle East, the Pacific Islands, Australasia and China." In these regions, too, prosperity is seen as the problem. "As incomes rise and populations become more urban," the WHO said, "diets high in complex carbohydrates give way to more varied diets with a higher proportion of fats, saturated fats and sugars. At the same time, large shifts towards less physically demanding work have been observed worldwide. Moves towards less physical activity are also found in the increasing use of automated transport, technology in the home, and more passive leisure pursuits."

It all sounds reasonable, but there are so many other variables, so many other possibilities—including the fact that the consumption of refined carbohydrates and sugars has also been increasing dramatically. To determine which hypothesis is most likely to be correct, it's useful to focus on the United States, because it offers a starting point for the epidemic—between the late 1970s and mid-1980s*—and resonably consistent data with which to work.

The question of how much we eat, whether in a population or as an individual, is difficult to assess, but the evidence suggests that we consumed more calories on average in the 1990s than we did in the 1970s. According to NHANES, American men increased their calorie consumption from 1971 to 2000 by an average of 150 calories per day, while women increased their consumption by over 350 calories. This increase in energy intake, according to a 2004 report published by the Centers for Disease Control, was "attributable primarily to an increase in carbohydrate intake." Though the *percentage* of fat in the diet decreased for both sexes, the *absolute* amount of dietary fat decreased only for men. On average, women ate fifty calories more fat each day in 2000 than they did in 1971, and men ate fifty calories less. The NHANES data suggest that either calories or carbohydrates could account for the increase in weight in the United States during this time; it would be difficult to implicate dietary fat.

The identical conclusion could be drawn from the evidence gathered by the U.S. Department of Agriculture and published in a report entitled *Nutrient Content of the U.S. Food Supply, 1909–1997*. The USDA says that the American food supply offered up thirty-three hundred calories a day per capita between 1971 and 1982. By 1993, it had climbed to thirty-eight hundred calories, and it remained at that level through 1997. This increased availability, and so perhaps consumption, of five hundred calories each day could conceivably explain the obesity epidemic. But carbohydrate con-

* Between the second and third National Health and Nutrition Examination Study.

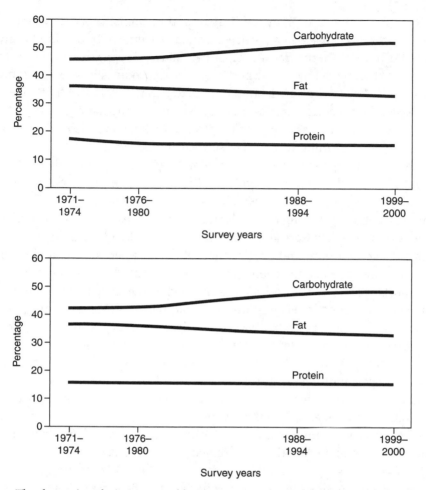

The change in calories consumed between 1971 and 2000 in women ages 20–74
years (upper chart) and in men ages 20–74 (lower chart), according to
the National Health and Nutrition Examination Surveys.

sumption also turned upward after 1982. Of the five hundred additional calories available for consumption each day, 90 percent came from carbohydrates. The remaining 10 percent were from protein and fat, in descending order. Saturated fat dropped from fifty-three grams a day in 1971 to fifty in 1997.*

* The USDA has a variety of mechanisms for estimating macronutrient intake—i.e., protein, carbohydrates, and fat—and has published a variety of reports on the subject. Not all are consistent, but the findings on fat consumption are. For instance, in April 1998, the USDA published an article entitled "Is Total Fat Consumption Really Decreasing?" This article reported that average total fat consumption for men aged nineteen to fifty, for instance, dropped from 113 grams per day in 1977–78 to ninety-six in 1989, the period that encompasses the beginning of the obesity

In 1997, the University of Alabama nutritionist Roland Weinsier reviewed this evidence in an article entitled "Divergent Trends in Obesity and Fat Intake Patterns: The American Paradox." "It appears that efforts to promote the use of low-calorie and low-fat food products have been highly successful," Weinsier noted, but the reduction in fat intake did "not appear to have prevented the progression of obesity in the population."

Population-wide assessments of physical activity are also difficult to make in any meaningful way. Those research agencies that traditionally study such things—the CDC's Behavioral Risk Factor Surveillance System, in particular—have no evidence that would shed light on physical activity during the decade in which the obesity epidemic began. They do have evidence suggesting that Americans were no less active at the end of the 1990s than they were at the beginning of that decade, despite the continued rise in weight and obesity throughout this period. We know, too, that the obesity epidemic coincided with what might be called an exercise or sports epidemic in America, accompanied by the explosion of an entire industry dedicated to leisure-time pursuits. It's worth remembering that in the 1960s Jack La Lanne was the nation's only physical-fitness guru, Gatorade existed solely for the use of University of Florida football players, and skateboarding, in-line skating, snowboarding, mountain biking, power yoga, spinning, aerobics, and a host of other now relatively common physical activities had yet to be invented. To put this in numerical terms, this was an era when the revenues of the health-club industry were estimated at $200 million a year; in 2005, revenues were $16 billion, and nearly forty million Americans belonged to such clubs.*

Press reports also support this version of history. By 1977, the *New York Times* was discussing the "exercise explosion" that had come about because the conventional wisdom of the 1960s that exercise was "bad for you" had been transformed into the "new conventional wisdom—that strenuous exercise is good for you." When the *Washington Post* estimated in 1980 that a hundred million Americans were now partaking in the "new fitness revolution," it also noted that most of them "would have been derided as 'health nuts' " only a decade earlier. "What we are seeing," the *Post* suggested, "is one of the late twentieth century's major sociological events."

---

epidemic. The relevant numbers for women of the same age group are seventy-three grams of fat per day in 1977–78 and sixty-two in 1989.

* According to the Sporting Goods Manufacturers Association, sales of sporting equipment, apparel, and shoes increased from $21.9 billion in 1987 (the earliest year for which they have data) to $52 billion in 2004.

Another apparent contradiction of the notion that either sedentary behavior or prosperity or a toxic food environment is the cause of obesity is that obesity has always been most prevalent among the poorest and thus, presumably, harder-working members of society. In developed nations, the poorer people are, the heavier they're likely to be. The NHANES studies confirmed this observation, first documented more than forty years ago. In 1965, Albert Stunkard and his colleagues at New York Hospital reported that they had surveyed 1,660 New Yorkers and found that obese women were six times more common at the lowest socioeconomic level than at the highest. Thirty percent of the poorest women were obese, compared with 16 percent of those of "middle status" and only 5 percent of the richest. The poor men were twice as likely to be obese as the rich (32 percent to 16 percent). These observations have been confirmed repeatedly throughout the world, in both children and adults. Because poor and immigrant populations are considerably less likely than wealthier, more established populations to own labor-saving devices, and because they are more likely to work in physically demanding occupations, that poverty is a risk factor for obesity is another compelling reason to question the notion that sedentary behavior is a cause.

There is a tendency among public-health authorities, obesity researchers, and health writers to discuss obesity as though the problem on a societal scale were only twenty or thirty years old, but this confuses the problem of obesity with the current obesity epidemic. Because these last few decades also coincide with the spread of McDonald's and other global purveyors of high-fat fast foods, obesity can conveniently be blamed on fast food by virtue of this association. (It has also, by this same logic, been popularly blamed on high-fructose corn syrup.) But the research literature of obesity dates back further than the epidemic, and by including *all* of the relevant observations over the years, we can begin to rule out competing hypotheses. Any hypothesis that purports to explain how obesity is caused, after all, should explain the emergence of obesity in any population and at any time, not just the increasing obesity in the past few decades.

The Pima Indians of southwestern Arizona are now infamous for having the highest rates of obesity and diabetes in the United States. Today the standard explanation for obesity among the Pima is that they have succumbed, as we all have, to prosperity and the toxic environment of American life. Over the last century, the Pima supposedly experienced a nutrition transition—an exaggerated version of the changing-American-diet story. Farmers and hunters became relatively sedentary wage-earners, while their diet changed from one very low in fat and high in fiber-rich carbohydrates and vegetables to a modern high-fat, high-sugar American diet. "As the typical American diet became more available on the reservation after the [Second World] war," according to an NIH report entitled *The Pima Indians: Pathfinders for Health*, "people became more overweight." "If the Pima Indians could return to some of their traditions," explained one NIH authority, "including a high degree of physical activity and a diet with less fat and more starch, we might be able to reduce the rate, and surely the severity, of unhealthy weight in most of the population."

The problem with this version of the Pima history is that obesity and overweight had been evident a century ago, when the relevant nutrition transition was from relative abundance to extreme poverty. From November 1901 to June 1902, the Harvard anthropologist Frank Russell lived on the Pima reservation south of Phoenix studying the tribe and its culture. Many of the older Pima, Russell noted in a report of the Bureau of American Ethnology, "exhibit a degree of obesity that is in striking contrast with the 'tall and sinewy' Indian conventionalized in popular thought."

Russell's assessment of the Pima's relative corpulence was then confirmed by the anthropologist and physician Aleš Hrdlička, who visited the Pima reservation in 1902 and 1905. "Especially well-nourished individu-

Obesity among the Pima is not a new phenomenon,
as demonstrated by this photo of "Fat Louisa"
taken in 1901 or 1902 by the Harvard
anthropologist Frank Russell.

als, females and also males, occur in every tribe and at all ages," Hrdlička reported, "but real obesity is found almost exclusively among the Indians on reservations."

For perhaps two millennia, the Pima had lived as both hunter-gatherers and agriculturalists. Game was abundant in the region, as were fish and clams in the Gila River. When the Jesuit missionary Eusebio Kino arrived among the Pima in 1787, the tribe was already raising corn and beans on fields irrigated with Gila River water. In the decades that followed, they took to raising cattle, poultry, wheat, melons, and figs. They also ate mesquite beans, the fruit of the saguaro cactus, and a mush of what Russell later called "unidentified worms." In 1846, when a U.S. Army battalion passed through Pima lands, the battalion's surgeon John Griffin described the Pima as "sprightly" and in "fine health." He also noted that the Pima had "the greatest abundance of food, and take care of it well, as we saw many of their storehouses full of pumpkins, melons, corn &c."

Life began to change dramatically the following year, when a wagon route was opened to California "by way of Tucson and the Pima villages." This became the southernmost overland route for the California gold rush that began in 1849; tens of thousands of travelers passed through the Pima villages on the way west over the next decade. They relied on the Pima for food and supplies.

With the arrival of Anglo-American and Mexican settlers in the late 1860s, the prosperity of the Pima came to an end, replaced by what the tribe referred to as "the years of famine." Over the next quarter-century, these newcomers hunted the local game almost to extinction, and the Gila River water, on which the Pima depended for fishing and irrigating their own fields, was "entirely absorbed by the Anglo settlements upstream." By the mid-1890s, the Pima were relying on government rations to avoid starvation, and this was still the situation when Hrdlička and Russell arrived in the early 1900s.

Both Hrdlička and Russell struggled with the dilemma of poverty coincident with obesity. Russell knew that the life of these Indians was arduous; sedentary behavior could not be a cause of obesity in the Pima. Instead, he proposed that a dietary factor was responsible. "Certain articles of their diet appear to be markedly flesh producing," Russell wrote. Hrdlička suggested that "the role played by food in the production of obesity among the Indians is apparently indirect." He suggested that life on the reservation might be relatively sedentary and this could play a role— "the change from their past active life to the present state of not a little indolence"—but he did not appear particularly confident about it. After all, he wrote, obesity was quite rare among the Pueblo, "who have been of sedentary habits since ancient times." And obesity among the Pima was found "largely but not exclusively" in the women, and the women of the tribe worked considerably harder than the men, spending their days harvesting the crops, grinding corn, wheat, and mesquite beans and carrying whatever burdens were not carried by pack animals.

Hrdlička also noted that by 1905 the Pima diet already included "everything obtainable that enters into the dietary of the white man," which raises the possibility that this might have been responsible for the obesity. At the half-dozen trading posts that opened on the Pima reservation after 1850, the Indians took to buying "sugar, coffee and canned goods to replace traditional foodstuffs lost ever since whites had settled in their territories."

Neither Hrdlička nor Russell suggested that the U.S. government rations might be the cause of obesity. But if the Pima diet on government rations was anything like that of tribes reduced to similar situations at the time on which data exist—including the Sioux on the Standing Rock Reservation in the Dakotas—then almost 50 percent of their calories came from sugar and flour.

Obesity in association with "widespread poverty" was documented again on the Pima reservation in the early 1950s by Bertram Kraus, a University of Arizona anthropologist working with the Bureau of Indian Affairs. According to Kraus, more than 50 percent of the children on the Pima reservation could legitimately be described as obese by their eleventh birthday. The local Anglos, Kraus wrote, got leaner as they got older (at the time, at least); this was not the case with the Pima. Kraus lamented the absence of dietary data to assess the nutritional state of the tribe, but this situation was remedied a few years later by Frank Hesse, a physician at the Public Health Service Indian Hospital on the Gila River Reservation. Hesse noted that the Pima diet of the mid-1950s was remarkably consistent from family to family and consisted of "mainly beans, tortillas, chili peppers and coffee, while oatmeal and eggs are occasionally eaten for breakfast. Meat and vegetables are eaten only once or twice a week." Hesse neglected to assess sugar consumption, but he did note that "a large amount of soft drinks of all types is consumed between meals." Hesse then concluded that 24 percent of the calories consumed by the Pima (the soft drinks not included) were from fat, which is certainly low by modern standards.*

---

* If each Pima drank two eight-ounce soft drinks a day, this would add roughly two hundred calories a day to Hesse's estimate of both carbohydrate and calorie consumption, and so would drop the fat in the diet to 22 percent.

Over the next twenty years, the prevalence of obesity and diabetes among the Pima continued to rise, now coincident with a change in the foods distributed by government agencies and sold in the reservation trading posts. By the late 1950s, according to the Indian Health Service in Tucson, "large quantities of refined flour, sugar, and canned fruits high in sugar" were being distributed widely on the reservations, courtesy of a surplus commodity food program run by the U.S. Department of Agriculture. When mechanization of the local agriculture industry brought a cash economy to the Pima, the local stores and trading posts "started to carry high caloric pre-packed sweets, such as carbonated beverages (i.e., 'soda pop'), candy, potato chips, and cakes." "Soda pop is used in immense amounts," as one 1962 study described it.

In April 1973, when the evils of dietary fat were still widely considered hypothetical, the NIH epidemiologist Peter Bennett appeared before George McGovern's Senate Select Committee on Nutrition and Human Needs to discuss diabetes and obesity on the Pima reservation. The simplest explanation for why half of all adult Pima were diabetic, said Bennett, was the amount of sugar consumed, which represented 20 percent of the calories in the Pima diet. "The only question that I would have," Bennett had said, "is whether we can implicate sugar specifically or whether the important factor is not calories in general, which in fact turns out to be really excessive amounts of carbohydrates." Bennett's opinion was consistent with that of Henry Dobyns of the D'Arcy McNickle Center for the History of the American Indian, who is considered the foremost authority on Pima history. In 1989, Dobyns described obesity and diabetes in the tribe as being "to some extent a result of inadequate nutrition" and added that this inadequate nutrition had come about because "many of the poorer individuals subsist on a diet of potatoes, bread, and other starchy foods. Their traditional diet is beyond their reach, for they cannot catch fish in a dry riverbed and they cannot afford to buy much meat or many fresh fruits and vegetables."

Studies of the Sioux of the South Dakota Crow Creek Reservation in the 1920s, Arizona Apaches in the late 1950s, North Carolina Cherokees in the early 1960s, and Oklahoma tribes in the 1970s all reported levels of obesity comparable to that in the United States today, but in populations living in extreme poverty. "Men are very fat, women are even fatter," as the University of Oklahoma epidemiologist Kelly West said of the local tribes of the 1970s. "Typically, their lifetime maximum weight has been 185 percent of standard."

The early study of the Sioux, by two investigators from the University of Chicago, is particularly interesting, because it was one of the few published studies of diet, health, and living conditions in such a population, and it appeared the same year that the U.S. Department of the Interior released the results of a lengthy investigation of Native American living conditions. "An overwhelming majority of the Indians are poor, even extremely poor," the Interior Department reported, "living on lands from which a trained and experienced white man could scarcely wrest a reasonable living." The University of Chicago report said most of the Sioux lived in one- or two-room shacks; 40 percent of the children lived in homes without toilet facilities; water had to be hauled from the river. Little milk was consumed, although canned milk was included in the government rations. Butter, green vegetables, and eggs were almost never eaten. No fruit was consumed.* Twenty-five to forty pounds of beef were issued per person as government rations each month, but this was "not an indication of the amount consumed by each person," the report noted, "for the families who receive rations are not left alone to eat them. Issue day is visiting day for the families not on the ration roll, and often the visit lasts until the friends' or relatives' rations of meat are gone. The ration family, therefore, may be compelled to live on bread and coffee for the remainder of the month."

The staple of the Sioux diet on the reservation was "grease bread," fried in fat and made from white flour, supplemented by oatmeal, potatoes, and beans, some squash and canned tomatoes, black coffee, canned milk, and sugar. "Almost two-thirds of the families, including 138 children, were receiving distinctly inadequate diets," the report concluded. Fifteen families, with thirty-two children among them, "were living chiefly on bread and coffee." Nonetheless, 40 percent of the adult women, over 25 percent of the men, and 10 percent of the children "would be termed distinctly fat," the University of Chicago investigators reported, whereas 20 percent of the women, 25 percent of the men, and a slightly greater percentage of the children were "extremely thin."

By the 1970s, when studies of obesity in populations began in earnest, the general attitude was that obesity was simply a fact of life in developed nations. "Even a brief visit to Czechoslovakia," reported a Prague epidemiologist at the first International Conference on Obesity, in 1974, "would reveal that obesity is extremely common and that, as in other industrial countries, it is probably the most widespread form of malnutrition."

---

* The Sioux were "essentially carnivorous" prior to their reservation life, the report noted, and so they "were never in the habit of eating much fruit and vegetables."

The observation that this was also true in poor populations in nonin-
dustrialized countries, that obesity frequently coexists side-by-side with
malnutrition and undernutrition, shows up with surprising consistency.
In a 1959 study of African Americans living in Charleston, South Carolina,
nearly 30 percent of the adult women and 20 percent of the adult men
were obese although living on family incomes of from $9 to $53 a week. In
Chile in the early 1960s, a study of factory workers, most of whom were
engaged in "heavy labor," revealed that 30 percent were obese and 10 percent suffered from "undernourishment." Nearly half the women over
forty-five were obese. In Trinidad, a team of nutritionists from the United
States reported in 1966 that one-third of the women older than twenty-five
were obese, and they achieved this condition eating fewer than two thousand calories a day—an amount lower than the United Nations' Food and
Agriculture Organization recommendation to avoid malnutrition. Only
21 percent of the calories in the diet came from fat, compared with 65 percent from carbohydrates.

In Jamaica, high rates of obesity, again among adult women in particular, were first reported in the early 1960s by a British Medical Research
Council diabetes survey. By 1973, according to Rolf Richards of the University of the West Indies, Kingston, 10 percent of all Jamaican men and
nearly two-thirds of the women were obese in a society in which "malnutrition in infancy and early childhood remains one of the most important
disorders contributing to infant and childhood mortality."

Similar observations were made in the South Pacific and throughout
Africa. In Rarotonga in the South Pacific, for instance, in the mid-1960s,
on a diet of only 25 percent fat, over 40 percent of the women were obese
and 25 percent were "grossly obese." Among Zulus living in Durban,
South Africa, according to a 1960 report, 40 percent of adult females
were obese. Women in their forties averaged 175 pounds. In a population
of urban Bantu "pensioners," the mean weight of women over the age of
sixty was reported in the mid-1960s to be 165 pounds. "Although dietary
habits vary widely amongst the African countries, tribes and villages,"
wrote B. K. Adadevoh from Nigeria's University of Ibadan in 1974, "it
is generally established that the African diet is rich in carbohydrates.
Caloric intake for most is low and protein falls short of the recommended
allowance."

It seems fair to assume that the lives of market women in West Africa
in the 1960s or poor Jamaicans of the same era were nontoxic by any of
the definitions that are commonly associated with the current obesity epidemic. The Sioux of the mid-1920s, or the Pima of the 1900s or 1950s,
living on reservations and relying on government rations to survive,

Obesity in Africa is not associated with prosperity.
These photos from Nigeria, of market women and an
obese eleven-year-old, date to the early 1970s.

clearly lived in a state of poverty that most of us today would find almost unimaginable.

So why were they fat? "It is difficult to explain the high frequency of obesity seen in a relatively impecunious society such as exists in the West Indies, when compared to the standard of living enjoyed in the more developed countries," Rolf Richards wrote about Jamaica in the 1970s. "Malnutrition and subnutrition are common disorders in the first two years of life in these areas, and account for almost 25 per cent of all admissions to pediatric wards in Jamaica. Subnutrition continues in early childhood to the early teens. Obesity begins to manifest itself in the female population from the 25th year of life and reaches enormous proportions from 30 onwards."

The question of what causes obesity in these impoverished populations has typically been ignored by obesity researchers, other than to suggest that there is something unique about given groups of people that exacerbates the problem of obesity. The assumption, as *The New Yorker* writer Malcolm Gladwell wrote about the Pima in 1998, is that they are "different only in degree, not in kind."

The idea of specific populations predisposed to obesity is encapsulated in a notion now known as the thrifty gene—technically, the thrifty-genotype hypothesis—that is now commonly invoked to explain the existence of the obesity epidemic and why we might *all* gain weight easily during periods of prosperity but have such difficulty losing it. The idea, initially proposed in 1962 by the University of Michigan geneticist James Neel, is that we are programmed by our genes to survive in the paleolithic hunter-gatherer era that encompassed the two million years of human evolution before the adoption of agriculture—a mode of life still lived by many isolated populations before extensive contact with Western societies. "Such genes would be advantageous under the conditions of unpredictably alternating feast and famine that characterized the traditional human lifestyle," explained the UCLA anthropologist Jared Diamond in 2003, "but they would lead to obesity and diabetes in the modern world when the same individuals stop exercising, begin foraging for food only in supermarkets and consume three high-calorie meals day in, and day out." In other words, the human body evolved to be what Kelly Brownell has called an "exquisitely efficient calorie conservation machine." And so, by this hypothesis, we suck up calories when they are abundant and store them as fat until they are called upon in a time of need. "Your genes match nicely with a scarce food supply," Brownell explains, "but not with modern living." Such populations as the Pima and the descendants of African tribes, according to this logic, were until very recently still trapped in this cycle of feast and famine and scarce food in general, and so their thrifty genes have yet to evolve to deal with times of continual plenty. The NIH researchers who study the Pima, as Gladwell reported, "are trying to find these genes, on the theory that they may be the same genes that contribute to obesity in the rest of us."

For the first few decades of its existence, this notion that we have evolved "thrifty mechanisms to defend energy stores during times of privation" was invariably referred to as a hypothesis. That qualification is now often dropped, but the thrifty gene remains only a hypothesis, and one that rests on many assumptions that seem unjustifiable.

James Neel initially proposed the idea of a "thrifty genotype rendered detrimental by progress" to explain why diabetes was so prevalent in Western societies and yet apparently absent in primitive tribes, including the Yanomamo of the Brazilian rain forest, who were then the subject of Neel's research. Neel was addressing the diseases of civilization and the kind of observations that led Peter Cleave to propose his saccharine-disease hypothesis. (Neel was unaware of Cleave's work at the time.) The enigma of Type 2 diabetes, Neel observed, is that it bestows significant evolutionary disadvantages upon anyone who has it. Diabetic women are more likely to die in childbirth and more likely to have stillbirths than healthy women; their children are more likely to be diabetic than those of healthy women. This implies that any genes that might predispose someone to become diabetic would evolve out of the population quickly, but this did not seem to have happened. One way to reconcile these observations is to imagine a scenario in which having a genetic predisposition to become diabetic is advantageous in some circumstances. (In a similar way, having the gene for sickle-cell anemia, normally a disadvantage, provides protection against malaria, a major advantage in malarial areas, as Neel himself reported.)

Since diabetic mothers are known to give birth to heavier children, Neel speculated that these diabetic genes bestowed an exceptional ability to use food efficiently, and thus an exceptional ability to convert calories into fat. Those with such thrifty genes, Neel explained, "might have, during a period of starvation, an extra pound of adipose reserve" that would keep them alive when those who failed to fatten easily would die of starvation. So it would be beneficial to have such genes in the event of famine or prolonged food deprivation, which Neel now assumed must have been the case throughout our evolutionary history. Those same genes would lead to obesity and diabetes in an environment in which food was plentiful.

"If the considerable frequency of the disease is of relatively long duration in the history of our species," Neel had asked to begin his discussion, "how can this be accounted for in the face of the obvious and strong genetic selection against the condition? If, on the other hand, this frequency is a relatively recent phenomenon, what changes in the environment are responsible for the increase?"

The thrifty gene could be the answer only if diabetes was of long duration in the species—and there is no evidence of that. The disease seems to appear only after populations have access to sugar and other refined carbohydrates. In the Pima, diabetes appeared to be "a relatively recent phenomenon," as Neel himself later noted. When Russell and Hrdlička discussed the health of the Pima in the early 1900s, they made no mention of diabetes, even while noting the presence of such "rare" diseases as lupus, epilepsy, and elephantiasis.* As late as 1940, when Elliott Joslin reviewed the medical records of the hospitals and physicians in Arizona, he concluded that the prevalence of diabetes was no higher among the Pima and other local tribes than anywhere else in the United States. Only in the 1950s, in studies from the Bureau of Indian Affairs, was there compelling reason to believe that diabetes had become common. When Neel tested

---

* Hrdlička did publish a list of diseases treated by the local agency physician, which did include one case of diabetes.

adolescent Yanomamo for the condition known as glucose intolerance, which might indicate a predisposition to diabetes, he found none, so had no reason to believe that diabetes existed before such isolated populations began eating Western foods. The same was true of an isolated tribe of Pima, discovered living in the Sierra Madre Mountains of northern Mexico. "The high frequency of [Type 2 diabetes] in reservation Amerindians," Neel later explained, "must predominantly reflect lifestyle changes."

By 1982, Neel had come to side with Peter Cleave in believing that the most likely explanation for the high rates of obesity and diabetes in populations like the Pima that had only recently become Westernized was their opportunity to "overindulge in high sugar content foods."

This left open the question of what biological factors or genes might determine who got obese and diabetic and who didn't in the presence of such foods, but it eliminated any reason to suggest that thrifty genes had ever bestowed some evolutionary advantage. "The data on which that (rather soft) hypothesis was based has now largely collapsed," Neel observed. He now suggested that either a tendency for the pancreas to oversecrete insulin and so cause hyperinsulinemia, or a tendency toward insulin resistance, which in turn would result in hyperinsulinemia, was the problem, which is consistent with the carbohydrate hypothesis of chronic disease. Both of these, Neel suggested, would be triggered by the "composition of the diet, and more specifically the use of highly refined carbohydrates."

It wasn't until the late 1970s, just a few years before Neel himself publicly rejected his hypothesis, that obesity researchers began invoking thrifty genes as the reason why putting on weight seems so much easier than losing it. Jules Hirsch of Rockefeller University was among the first to do so, and his logic is noteworthy, because his primary goal was to establish that humans, like every other species of animal, had apparently evolved a homeostatic system to regulate weight, and one that would do so successfully against fluctuations in food availability. We eat during the day, and yet have to supply nutrients to our cells all night long, while we sleep, for example, so we must have evolved a fuel storage system that takes this into account. "To me, it would be most unthinkable if we did not have a complex, integrated system to assure that a fraction of what we eat is put aside and stored," Hirsch wrote in 1977. To explain why these components might cause obesity so often in modern societies, he assumed as fact something that Neel had never considered more than speculation. "The biggest segment of man's history is covered by times when food was scarce and was acquired in unpredictable amounts and by dint of tremendous caloric expenditure," Hirsch suggested. "The long history of food scarcity and its persistence in much of the world could not have gone unnoticed by such an adaptive organism as man. Hoarding and caloric miserliness are built into our fabric."

This was one of the first public statements of the notion that would evolve into the kind of unconditional proclamation made by Kelly Brownell a quarter century later, that the human body is an "exquisitely efficient calorie conservation machine." But it depended now on an assumption about human evolution that was contradicted by the anthropologic evidence itself—that human history was dominated by what Jared Diamond had called the "conditions of unpredictably alternating feast and famine that characterized the traditional human lifestyle." Reasonable as this may seem, we have no evidence that food was ever any harder to come by for humans than for any other organisms on the planet, at least not until our ancestors began radically reshaping their environment ten thousand years ago, with the invention of agriculture.

Both the anthropological remains and the eyewitness testimony of early European explorers suggest that much of the planet, prior to the last century or two, was a "paradise for hunting," in the words of the Emory University anthropologist Melvin Konner and his collaborators, with a diversity of game, both large and small, "present in almost unimaginable numbers."* Though famines have certainly been documented among hunter-gatherer populations more recently, there's little reason to believe that this happened prior to the industrial revolution. Those isolated populations that managed to survive as hunter-gatherers well into the twentieth century, as the anthropologist Mark Nathan Cohen has written, were "conspicuously well-nourished in qualitative terms and at least adequately nourished in quantitative terms."

Hunter-gatherers lived in equilibrium with their environment just as every other species does. The oft-cited example is the !Kung Bushmen of the semi-arid Kalahari desert, who were studied by Richard Lee of the University of Toronto and a team of anthropologists in the mid-1960s. Their observations, Lee noted, were made during "the third year of one of the most severe droughts in South Africa's history." The United Nations had

---

* In 1804 and 1805, when the Corps of Discovery under Meriwether Lewis and William Clark made their historic overland expedition to the Pacific Ocean, they described game so plentiful in places that they literally had to club it out of their way to make progress.

instituted a famine-relief program for the local agriculturalists and pastoralists, and yet the Bushmen still survived easily on "some relatively abundant high-quality foods," and they did not "have to walk very far or work very hard to get them." The !Kung women would gather enough food in one day to feed their families for the next three, Lee and his colleagues reported; they would spend the remaining time resting, visiting, or entertaining visitors from other camps.

The prevailing opinion among anthropologists, not to be confused with that of nutritionists and public-health authorities, is that hunting and gathering allow for such a varied and extensive diet, including not just roots and berries but large and small game, insects, scavenged meat (often eaten at "levels of decay that would horrify a European"), and even occasionally other humans, that the likelihood of the simultaneous failure of all nutritional resources is vanishingly small. When hunting failed, these populations could still rely on foraging of plant food and insects, and when gathering failed "during long-continued drought," as the missionary explorer David Livingstone noted of a South African tribe in the mid-nineteenth century, they could relocate to the local water holes, where "very great numbers of the large game" also congregated by necessity. This resiliency of hunting and gathering is now thought to explain why it survived for two million years before giving way to agriculture. In those areas where human remains span the transition from hunter-gatherer societies to farmers, anthropologists have reported that both nutrition and health declined, rather than improved, with the adoption of agriculture. (It was this observation that led Jared Diamond to describe agriculture as "the worst mistake in the history of the human race.")

Although famines were both common and severe in Europe until the nineteenth century, this would suggest that those with European ancestry should be the most likely to have thrifty genes, and the most susceptible to obesity and diabetes in our modern toxic environments. Rather, among Europeans there is "a uniquely low occurrence of Type 2 diabetes," as Diamond puts it, more evidence that the thrifty-gene hypothesis is incorrect.

Species adapt to their environment over successive generations. Those that don't, die off. When food is abundant, species multiply; they don't get obese and diabetic.

When earlier generations of obesity researchers discussed the storage of fat in humans and animals, they assumed that avoiding excessive fat is as important to the survival of any species as avoiding starvation. Since the average 150-pound man with a body fat percentage of only 10 percent is still carrying enough fat calories to survive one month or more of total starvation, it seems superfluous to carry around more if it might have negative consequences. "Survival of the species must have depended many times both on the ability to store adequate yet *not excessive amounts of energy in the form of fat* [my italics], and on the ability of being able to mobilize these stores always at a sufficient rate to meet the body's needs," observed George Cahill and Albert Renold, considered two of the leading authorities on the regulation of fat metabolism, in 1965. The total amount of fat stored, they suggested, "should be kept sufficiently large to allow for periods of fasting to which a given species in a given environment is customarily exposed, yet sufficiently small to preserve maximum mobility."

The thrifty-gene hypothesis, on the other hand, implies that we (at least some of us) are evolutionarily adapted to survive extreme periods of famine, but assigns to humans the unique concession of having evolved in an environment in which excess fat accumulation would not be a burden or lead to untimely death—by inhibiting our ability to escape from predators or enemies, for instance, or our ability to hunt or perhaps even gather. It presupposes that we remain lean, or at least some of us do, only as long as we remain hungry or simply lack sufficient food to indulge our evolutionary drive to get fat—an explanation for leanness that the British metabolism researchers Nancy Rothwell and Michael Stock described in 1981 as "facile and unlikely," a kind way of putting it. The "major objection" to the thrifty-genotype hypothesis, noted Rothwell and Stock, "must be based on the observation that most wild animals are in fact very lean" and that this leanness persists "even when adequate food is supplied," just as we've seen in hunter-gatherers. If the thrifty-gene hypothesis were true of any species, it would suggest that all we had to do was put them in a cage with plentiful food available and they would fatten up and become diabetic, and this is simply not the case.

Proponents of the thrifty-gene hypothesis, however, will invoke a single laboratory model—the Israeli sand rat—to support the notion that at least some wild animals will get fat and diabetic if caged with sufficient food. "When this animal is removed from the sparse diet of its natural environment and given an abundant, high-calorie diet," wrote Australian diabetologist Paul Zimmet in a 2001 article in *Nature*, "it develops all of the components of the metabolic syndrome, including diabetes and obesity."

But the sand-rat experiments themselves, carried out in the early 1960s at Duke University by the comparative physiologist Knut Schmidt-Nielsen, suggested that the abundance of food was *not* the relevant factor.

Schmidt-Nielsen was trying to establish what aspect of the laboratory diet might be responsible for the obesity and diabetes that appeared in his sand rats. He had taken two groups of rats freshly trapped in Egypt and raised one on Purina Laboratory Chow—"49.4% digestible carbohydrates, 23.4% protein and 3.8% fat"—supplemented with "fresh mixed vegetables," and the other on the fresh vegetables alone. Both had access to as much food as they desired, but only the chow-eating rats got diabetic and obese. This suggested that something about Purina Chow was the determining factor. Perhaps the rats liked it better than vegetables, and so they ate more, although that, too, could be a physiological effect related to the nutrient composition of the chow. It might have been the density of calories in the rat chow, which have less water content than vegetables and so more calories per gram. It was also possible that the cause of the diabetes and obesity in these rats, as Schmidt-Nielsen suggested, was "a carbohydrate intake that is greater than that occuring in the natural diet."*

Depending on the researchers' preconceptions, the Israeli sand rats could have been considered an animal model of the carbohydrate hypothesis, rather than the thrifty-gene hypothesis. Monkeys in captivity, by the way, will also get obese and diabetic on high-carbohydrate chow diets. One of the first reports of this phenomenon was in 1965, by John Brobeck of Yale, whose rhesus monkeys got fat and mildly diabetic on Purina Monkey Chow—15 percent protein, 6 percent fat, and 59 percent digestible carbohydrates. According to Barbara Hansen, who studies diabetes and obesity and runs a primate-research laboratory at the University of Maryland, perhaps 60 percent of middle-aged monkeys in captivity are obese by monkey standards. "This is on the kind of diet recommended by the American Heart Association," she says, "high-fiber, low-fat, no-cholesterol chow."

The world is full of species that do fatten regularly, always to serve a purpose—long-distance migrations, reproduction, or survival during periods when food is either unavailable or too risky to procure. Hibernators seem to be an obvious choice to shed light on the assumptions underlying the thrifty-gene hypothesis. These animals accumulate enormous fat

---

* This conclusion was supported four years later, when German researchers published their protocol for keeping these desert sand rats healthy in captivity. "It is well known that these animals will develop diabetes mellitus soon after their natural vegetative diet is removed and replaced with standard laboratory rations," they noted. But both diabetes and obesity can be avoided if the animals are reared on a suitable diet: in this case, fruits, vegetables, and herbs, supplemented by an *unlimited* supply of insects, shrimps, worms, and grasshoppers.

deposits in response to an environment that offers up periods of feast—spring, summer, and fall—and famine in the winter. Yet this accumulation goes unaccompanied by the chronic ills, such as diabetes, that appear in obese humans. Hibernating ground squirrels, for instance, will double their weight and body fat in a few weeks of late summer. Dissecting such squirrels at their peak weight is akin to "opening a can of Crisco oil," as the University of California biologist Irving Zucker, a pioneer of this research, has described it, "enormous gobs of fat, all over the place."

Investigators who study hibernators, like Nicholas Mrosovsky, a University of Toronto zoologist, point out that weight gain, maintenance, and loss in these animals, and so perhaps in all species, is genetically pre-programmed and particularly resilient to variations in food availability. This program is characterized by its ability to adjust readily to changing circumstances and the unpredictability of the environment. Ground squirrels will gain weight through the summer at the same rate whether they're in the wild or in the laboratory. They will lose it at the same rate during the winter whether they are kept awake in a warm laboratory or are in full hibernation, eating not a bite, and surviving solely off their fat supplies. "It is very hard to prevent them gaining and losing weight" on schedule, explains Mrosovsky, who did much of the original research in this area. When researchers surgically remove a sizable portion of fat from experimental animals—a procedure known as a *lipectomy*—the animals will restore the lost fat so that within months of the surgery they will be just as fat as they would have been without the surgery.*

Even the type of fat found in animals and humans is regulated in a way that accommodates differing internal and external environments. The fat in our limbs, for instance, is less saturated than the fat around our organs, and so is less likely to stiffen in cold weather. We will also change the fatty-acid composition of our subcutaneous fat with temperature—the colder it gets, the more unsaturated the fats. This same phenomenon, independent of the type of fat consumed, has been observed in pigs, rats, and hibernators. Another example of the evolutionary specificity of fat deposits can be seen in those desert animals that do not store fat subcutaneously, as humans and most animals do, apparently because it would inhibit heat loss and cooling. So there are fat-rumped and fat-tailed sheep, and fat-tailed marsupial mice, all desert-dwellers that carry their fat almost exclusively in the so-named locations.

---

* If the surgery is done to rodents during hibernation, they will somehow slow the rate at which they draw on their fat supplies for fuel so as to compensate for the loss.

The storage of fat, it seems clear, like all evolutionary adaptations, tends to be exquisitely well suited to the environment—both internal and external—in a way that maximizes benefits while minimizing risks. This is why most investigators who considered these issues in the 1970s and 1980s assumed that a tendency to gain any excessive weight during periods of abundance would be the kind of obvious liability that evolution would work to select *out* of the species rather than select *in*. The thrifty-gene hypothesis does not hold up. But without a thrifty gene, rendered detrimental by the abundance of food in modern societies and the absence of physical labor needed to procure it, how do we explain why gaining weight in modern societies still seems so much easier than losing it?

Chapter Fifteen

HUNGER

Khrushchev, too, looks like the kind of man his physicians must continually try to diet, and historians will some day correlate these sporadic deprivations, to which he submits "for his own good," with his public tantrums. If there is to be a world cataclysm, it will probably be set off by skim milk, Melba toast, and mineral oil on the salad.

A.J. LIEBLING, The Earl of Louisiana, 1961

IN OCTOBER 1917, FRANCIS BENEDICT, director of the Carnegie Institution of Washington's Nutrition Laboratory (located, as it happens, in Boston), put twelve young men on diets of roughly fourteen hundred to twenty-one hundred calories a day with the intention of lowering their body weights by 10 percent in a month. Their diets would then be adjusted as necessary to maintain their reduced weights for another two months, while Benedict and his colleagues meticulously recorded their psychological and physiological responses. A second squad of twelve men was studied as a comparison and then they were put on similar calorie-restricted diets. The results were published a year later in a seven-hundred-page report entitled *Human Vitality and Efficiency Under Prolonged Restricted Diet.*

Benedict hoped to establish whether humans could adjust to this lower nutritional level and thrive. His subjects lost the expected weight, but they complained constantly of hunger—"a continuous gnawing sensation in the stomach," as described by the Carnegie report—and of being cold to the extent that several found it "almost impossible to keep warm, even with an excessive amount of clothing." They also experienced a 30-percent decrease in metabolism. Indeed, Benedict's subjects reduced their energy expenditure so dramatically that if they consumed more than twenty-one hundred calories a day—a third to a half less than they had been eating prior to the experiment—they would begin to regain the weight they had lost. The men also experienced significant decreases in blood pressure and pulse rate; they suffered from anemia, the inability to concentrate, and marked weakness during physical activity. They also experienced "a decrease in sexual interest and expression, which, according to some of the men, reached the point of obliteration." That these phenomena were caused by the diet itself rather than the subsequent weight loss was demonstrated by the experience of the second squad of men, who manifested, according to the Carnegie report, "the whole picture . . . with striking clearness" after only a few days of dieting.

"One general feature of the post-experimental history," the Carnegie researchers reported, "is the excess eating immediately indulged in by the men." Despite repeated cautions about the dangers of overindulgence after such a strict diet, the men "almost invariably over-ate." As the Carnegie report put it, "the circumstances militated against" any acquisition of "new dietetic habits." In particular, the cravings for "sweets and accessory foods of all kinds,"—i.e., snacks—were now free to be indulged, and so they were. Perhaps for this reason, Benedict's young subjects managed to regain all the lost weight and body fat in less than two weeks. Within another three weeks, they had gained, on average, eight pounds more, and came out of this exercise in calorie restriction considerably heavier than they went in. "In practically every instance the weight prior to the beginning of the experiment was reached almost immediately and was usually materially exceeded," Benedict and his collagues wrote.

In 1944, Ancel Keys and his colleagues at the University of Minnesota set out to replicate Benedict's experiment, although with more restrictive diets and for a greater duration. Their goal was to reproduce and then study the physiological and psychological effects of starvation of the kind that Allied troops would likely confront throughout Europe as the continent was liberated. Thirty-two young male conscientious objectors would serve as "guinea pigs," the phrase Keys used in this context. These volunteers would eventually spend twenty-four weeks on a "semi-starvation diet," followed by another twelve to twenty weeks of rehabilitation.

The subjects consumed an average of 1,570 calories each day, split between two meals designed to represent the daily fare of European famine areas. "The major food items served," the researchers noted, "were whole-wheat bread, potatoes, cereals, and considerable amounts of turnips and cabbage. Only token amounts of meats and dairy products were provided."* This diet provided roughly half the calories that the subjects had been consuming to maintain their weight. It was expected to induce an

---

* The diet constituted roughly 400 calories a day of protein, 270 calories of fat, and 900 calories of carbohydrates.

average weight loss of 20 percent—or forty pounds in a two-hundred-pounder—aided by a routine that required the subjects to walk five to six miles each day, which would burn off another two to three hundred calories.

Keys's conscientious objectors lost, on average, a dozen pounds of fat in the first twelve weeks of semi-starvation, which constituted more than half of their original fat tissue, and they lost three more pounds of body fat by the end of twenty-four weeks. But weight loss, once again, was not the only physiological response to the diet. Nails grew slowly, and hair fell out. If the men cut themselves shaving, they would bleed less than expected, and take longer to heal. Pulse rates were markedly reduced, as was the resting or basal metabolism, which is the energy expended by the body at rest, twelve to eighteen hours after the last meal. Reflexes slowed, as did most voluntary movements: "As starvation progressed, fewer and fewer things could stimulate the men to overt action. They described their increasing weakness, loss of ambition, narrowing of interests, depression, irritability, and loss of libido as a pattern characteristic of 'growing old.' " And, like Benedict's subjects, the young men of the Minnesota experiment complained persistently of being cold. Keys's conscientious objectors reduced their total energy expenditure by over half in response to a diet that gave them only half as many calories as they would have preferred. This was a reasonable response to calorie deprivation, as Keys and his colleagues explained, "in the sense that a wise man reduces his expenditure when his income is cut."

More than fifty pages of the two-volume final report by Keys and his colleagues, *The Biology of Human Starvation*, document the "behavior and complaints" induced by the constant and ravenous hunger that obsessed the subjects. Food quickly became the subject of conversations and daydreams. The men compulsively collected recipes and studied cookbooks. They chewed gum and drank coffee and water to excess; they watered down their soups to make them last. The anticipation of being fed made the hunger worse. The subjects came to dread waiting in line for their meals and threw tantrums when the cafeteria staff seemed slow. Two months into the semi-starvation period, a buddy system was initiated, because the subjects could no longer be trusted to leave the laboratory without breaking their diets.

Eventually, five of the subjects succumbed to what Keys and his colleagues called "character neurosis," to be distinguished from the "semi-starvation neurosis" that all the subjects experienced; in two cases, it "bordered on a psychosis." One subject failed to lose weight at the expected rate, and by week three was suspected of cheating on the diet. In week eight, he binged on sundaes, milk shakes, and penny candies, broke down "weeping, [with] talk of suicide and threats of violence," and was committed to the psychiatric ward at the University Hospital. Another subject lasted until week seven, when "he suffered a sudden 'complete loss of willpower' and ate several cookies, a bag of popcorn, and two overripe bananas before he could 'regain control' of himself." A third subject took to chewing forty packs of gum a day. Since his weight failed to drop significantly "in spite of drastic cuts in his diet," he was dropped from the study. For months afterward, "his neurotic manifestations continued in full force." A fifth subject also failed to lose weight, was suspected of cheating, and was dropped from the study.

With the relaxation of dietary restriction, Keys avoided the dietary overindulgence problem that had beset Benedict's subjects by restricting the rehabilitation diets to less than three thousand calories. Hunger remained unappeased, however. For many of the subjects, the depression deepened during this rehabilitation period. It was in the very first week of rehabilitation, for instance, that yet another subject cracked—his "personality deterioration culminated in two attempts at self-mutilation."

Even during the last weeks of the Minnesota experiment, when the subjects were finally allowed to eat to their hearts' content, they remained perversely unsatisfied. Their food intake rose to "the prodigious level of 8,000 calories a day." But many subjects insisted that they were still hungry, "though incapable of ingesting more food." And, once again, the men regained weight and body fat with remarkable rapidity. By the end of the rehabilitation period, the subjects had added an average of ten pounds of fat to their pre-experiment levels. They weighed 5 percent more than they had when they arrived in Minneapolis the year before; they had 50 percent more body fat.

These two experiments were the most meticulous ever performed on the effects on body and mind of long-term low-calorie diets and weight reduction. The subjects were selected to represent a range of physiological types from lean to overweight (albeit all young, male, and Caucasian). They were also chosen for a certain strength of character, suggesting they could be trusted to follow the diets and remain dedicated to the scientific goals at hand.

The diets may seem severe in the retelling, but, in fact, fourteen to sixteen hundred calories a day for weight loss could be considered generous compared with the eight-to-twelve-hundred-calorie diets that are now commonly prescribed, what the 1998 *Handbook of Obesity* refers to as "conventional reducing diets." Nonetheless, such diets were traditionally known as semi-starvation diets, a term that has fallen out of use, perhaps because it implies an unnatural and uncomfortable condition that few individuals could be expected to endure for long.

In both experiments, even after the subjects lost weight and were merely trying to maintain that loss, they were still required to eat considerably fewer calories than they would have preferred, and were still beset by what Keys and his colleagues had called the "persistent clamor of hunger." Of equal importance, simply restraining their appetites, independent of weight loss, resulted in a dramatic reduction in energy expenditure. This could be reversed by adding calories back into the diet, but then any weight or fat lost returned as well. One lesson learned was that, for the weight reduction to be permanent, some degree of semi-starvation has to be permanent. These experiments indicated that would never be easy.

Obese patients also get hungry on semi-starvation diets. If they have to restrict their calories to lose weight, then by definition they are forcing themselves to eat less than they would otherwise prefer. Their hunger is not being satisfied. As with lean subjects, their energy expenditure on a semi-starvation diet also "diminishes proportionately much more than the weight," as the Pittsburgh clinician Frank Evans reported in 1929 of his obese subjects. This same observation was reported in 1969 by George Bray, who was then at the Tufts University School of Medicine in Boston, and who entitled his article, for just this reason, "The Myth of Diet in the Management of Obesity." "There is no investigator who has looked for this effect and failed to find it," the British obesity researcher John Garrow wrote in 1978.

The latest reiteration of these experiments, using obese subjects, was conducted by Jules Hirsch at Rockefeller University, and the results were published in *The New England Journal of Medicine* in 1995. Calorie restriction in Hirsch's experiment resulted in disproportionate reductions in energy expenditure and metabolic activity. Increasing calorie consumption resulted in disproportionate increases in metabolic activity.

Hirsch and his colleagues interpreted their observations to mean that the human body seems surprisingly intent on maintaining its weight— resisting both weight gain and weight loss—so that the obese remain obese and the lean remain lean. As Hirsch explained it, the obese individual appears to be somehow metabolically normal in the obese state, just as Keys's and Benedict's young men were metabolically normal in their lean or overweight states before their semi-starvation diets. Once Hirsch's obese subjects took to restricting their calories, however, they experienced what he called "all the physiological and psychological concomitants of starvation."

A semi-starvation diet induces precisely that—semi-starvation—whether the subject is obese or lean. "Of all the damn unsuccessful treatments," Hirsch later said, "the treatment of weight reduction by diet for obese people just doesn't seem to work."

Over the course of a century, a paradox has emerged. Obesity, it has been said, is caused, with rare exceptions, by an inability to eat in moderation combined with a sedentary lifestyle. Those of us who gain excessive weight consume more calories than we expend, creating a *positive caloric balance* or a *positive energy balance,* and the difference accumulates as excessive pounds of flesh. But if this reconciles with the equally "indisputable" notion that "eating fewer calories while increasing physical activity are the keys to controlling body weight," as the 2005 USDA *Dietary Guidelines for Americans* suggest, then the problems of obesity and the obesity epidemic should be easy to solve. Those few individuals for whom obesity is a preferred condition, such as sumo wrestlers, would remain obese through their voluntary program of overeating, and the rest would create a *negative energy balance,* lose the excess weight, and return to leanness. The catch, as Hirsch pointed out, is that this doesn't happen.

The documented failure of semi-starvation diets for the obese dates back at least half a century. It begins with Albert Stunkard's analysis of the relevant research in the mid-1950s, motivated by his desire to resolve what he called the "paradox" between his own failure to reduce obese patients successfully by diet at New York Hospital and "the widespread assumption that such treatment was easy and effective." Stunkard managed to locate eight reports in the literature that allowed for an accurate assessment of whether semi-starvation diets worked. In 1959, he reported that the existing evidence confirmed his own failures: semi-starvation diets were "remarkably ineffective" as a treatment for obesity. Only 25 percent of the subjects discussed in these articles had lost as much as twenty pounds on their semi-starvation diets, "a small weight loss for the grossly overweight persons who are the subjects of these reports." Only 5 percent successfully lost forty pounds. As for Stunkard's own experience with a hundred obese patients, all prescribed "balanced" diets of eight to fifteen hundred calories a day, "only 12% were able to lose 20 lb., and only 1 patient lost 40 lb. . . . Two years after the end of treatment only two patients had maintained their weight loss."*

* Though Stunkard's analysis has widely been perceived as a condemnation of all methods of dietary treatment of obesity, the studies he reviewed included only semi-starvation, calorie-restricted diets.

A decade later, when Stunkard was invited to discuss obesity at Richard Nixon's White House Conference on Food, Nutrition, and Health, he had come to believe that the adverse effects caused by semi-starvation diets as a treatment for obesity often outweighed any benefits. "Attempts at weight reduction are often accompanied by anxiety and depression, at times severe enough to warrant discontinuation," he said. "Many obese persons today might well be better off if they learned to live with their condition and stopped subjecting themselves over and over to painful and frustrating attempts to lose weight."

More recent assessments of the efficacy of semi-starvation diets tend to be studies that set out to evaluate the efficacy of low-fat, calorie-restricted diets, but because they do so by comparing these diets with more balanced calorie-restricted diets, they provide evidence for the efficacy of the latter as well. In 2002, a Cochrane Collaboration review of the evidence concluded that low-fat diets induced no more weight loss than calorie-restricted diets, and in both cases the weight loss achieved "was so small as to be clinically insignificant." A similar analysis was published in 2001 by the U.S. Department of Agriculture. In this case, the authors identified twenty-eight relevant trials of low-fat diets, of which at least twenty were also calorie-restricted. The overweight subjects consumed, on average, less than seventeen hundred calories a day for an average weight loss of not quite nine pounds over six months. Only one of these studies tracked its participants for more than a year, and in that case the subjects reportedly reduced their caloric intake to thirteen hundred calories for eighteen months. In other words, these subjects reportedly consumed fewer calories per day than had Keys's conscientious objectors, they maintained this semi-starvation regimen for three times as long—and they emerged from the trial having *gained*, on average, a pound. In the Women's Health Initiative, discussed earlier (see pages 74–5) twenty thousand women were prescribed a low-fat diet and reportedly reduced their calorie consumption by an average of 360 calories a day. After almost eight years of this regimen, they weighed only two pounds less than when they started, and their average waist circumference, which is a measure of abdominal fat, had *increased*.

The evidence for the failure of semi-starvation as a treatment of obesity hasn't stopped obesity researchers from recommending the approach. The *Handbook of Obesity*, published in 1998 and edited by three of the most prominent authorities in the field—George Bray, Claude Bouchard, and W.P.T. James—says that "dietary therapy remains the cornerstone of treatment and the reduction of energy intake continues to be the basis of successful weight reduction programs." It also notes, in contradiction, that the results of such calorie-restricted diets "are known to be poor and not long-lasting." The chapter on obesity in the latest edition of *Joslin's Diabetes Mellitus*, written by two clinical investigators from Harvard Medical School, also describes "reduction of caloric intake" as "the cornerstone of any therapy for obesity." It then notes that reducing energy intake to a level substantially below that of energy expenditure "is difficult to accomplish despite a wide variety of specific dietary approaches." A deficit of seventy-five hundred calories, the authors explain, "is predicted to produce a weight loss of [2.2 pounds]," and so a reduction in food intake of a hundred calories a day "should bring about [an eleven-pound] weight loss over 1 year." But this doesn't seem to happen. "It is clear from common experience, however, that attempts at dieting that rely on such small reduction in food intake are rarely successful. Thus, more severe reductions in energy intake are typically prescribed," the Harvard physicians write. These more severe regimens include total starvation, but "the extreme nature of the therapy," the loss of muscle rather than fat tissue, and the many complications "have led to the virtual disappearance of this approach." They also include very low-calorie diets of two to six hundred calories a day, which will inevitably lead to weight loss, but the weight loss diminishes as the diet progresses, once again because metabolism and energy expenditure both decrease, and when the patients go off the diet, they regain the weight lost. Finally, there are the "many different diets" that provide eight hundred to a thousand calories and are in common use, all of which "should result in weight loss." "None of these approaches," the authors say, "has any proven merit."

In response to these pessimistic assessments, it is commonly suggested that the obese would ameliorate their problem, or prevent it, if they merely exercised—perhaps sixty or ninety minutes a day, as now prescribed by the USDA *Dietary Guidelines*. A negative energy balance can be created, according to this logic, by increasing energy expenditure as well as by eating less. Advice to engage in daily physical activity is now ever-present in public-health messages and popular writing on the problems of obesity and overweight. It's reinforced by the existence of the ubiquitous electronic displays on stair-climbers, treadmills, and other exercise apparatus that tell us how many calories we allegedly expended in our latest workout.

The belief in physical activity as a method of weight control is relatively new, however, and it has long been contradicted by the evidence. When Russell Wilder of the Mayo Clinic lectured on obesity in 1932, he noted that his patients tended to lose more weight with bed rest, "while unusually strenuous physical exercise slows the rate of loss." "The patient reasons quite correctly," Wilder said, "that the more exercise he takes the more fat should be burned and that loss of weight should be in proportion, and he is discouraged to find that the scales reveal no progress."

Until the 1960s, clinical investigators routinely pointed out that moderate exercise would lead only to insignificant increases in energy expenditure, and these could be easily matched by slight and comparatively effortless changes in diet. A 250-pound man will expend three extra calories climbing a flight of stairs, as Louis Newburgh of the University of Michigan calculated in 1942, and this in turn is the equivalent of depriving himself of one-fourth of a teaspoon of sugar or a tenth of an ounce of butter. "He will have to climb twenty flights of stairs to rid himself of the energy contained in one slice of bread!" Newburgh observed.

Though more strenuous exercise would burn more calories, it would also lead to a significant increase in appetite. This is the implication of the phrase "working up an appetite." "Vigorous muscle exercise usually results in immediate demand for a large meal," noted the Northwestern University endocrinologist Hugo Rony in 1940. "Consistently high or low energy expenditures result in consistently high or low levels of appetite. Thus men doing heavy physical work spontaneously eat more than men engaged in sedentary occupations. Statistics show that the average daily caloric intake of lumberjacks is more than 5,000 calories while that of tailors is only about 2,500 calories. Persons who change their occupation from light to heavy work or *vice versa* soon develop corresponding changes in their appetite."* If a tailor became a lumberjack and, by doing so, took to eating like one, there was little reason to think that the same wouldn't happen, albeit on a lesser scale, to an obese tailor who chose to work out like a lumberjack for an hour a day. In 1960, when the epidemiologist Alvan Feinstein examined the efficacy of various obesity treatments in a lengthy review in the *Journal of Chronic Diseases,* he dismissed exercise in a single paragraph. "There has been ample demonstration that exercise is an ineffective method of increasing energy output," Feinstein noted, "since it takes far too much activity to burn up enough calories for a significant weight loss. In addition, physical exertion may evoke a desire for food

---

* Physical activity is the primary determinant of the *variation* in energy intake in human populations, as Walter Willett and his Harvard colleague Meir Stampfer note in the 1998 textbook *Nutritional Epidemiology:* "Indeed, in most instances, energy intake can be interpreted as a crude measure of physical activity. . . ."

so that the subsequent intake of calories may exceed what was lost during the exercise."

By this time, though, exercise had a profoundly influential proponent: the Harvard nutritionist Jean Mayer, who would almost single-handedly overturn a century of clinical evidence and anecdotal experience. In the 1950s, when Mayer established himself as *the* leading authority on obesity in the United States, he did so based more on the romance of his background than his expertise as a clinical scientist: he was the son of the famous French physiologist André Mayer, and he had fought in the French resistance during World War II.

Mayer represented a new breed of obesity authority, of a kind that would now come to dominate the field. His predecessors—among them Louis Newburgh, Hugo Rony, Hilde Bruch, Frank Evans, Julius Bauer, and Russell Wilder—had all been physicians who worked closely with obese patients. Collectively, they had treated thousands of them. Their views on the cause of obesity differed, often radically, but their firsthand experience was unquestionable. Mayer was not a clinician. His training was in physiological chemistry; he had obtained a doctorate at Yale on the interrelationship of vitamins A and C in rats. In the ensuing decades, he would publish hundreds of papers on different aspects of nutrition, including obesity, but he never treated obese patients, so his hypotheses were less fettered by any anecdotal or real-life experience.

As early as 1953, after just two years of research on genetically obese mice, Mayer was extolling the virtues of exercise for weight control. By the end of the decade, he was getting credit from the *New York Times* for having "debunked" the "popular theories," argued by clinicians and their obese patients, that exercise had little influence on weight. Mayer knew that the obese often eat no more than the lean, and often even less. This seemed to exclude overeating, which meant the obese *had* to be less physically active. Otherwise, how could they have achieved positive energy balance and become obese? Mayer himself first reported this phenomenon in a strain of laboratory mice that were prone to both obesity and diabetes. They ate little more than their lean littermates, he noted, but their activity was "almost nil"; this sedentary behavior could explain their propensity to grow fat.

Through the 1960s, Mayer documented this relationship between energy intake, inactivity, and obesity in a series of human studies. He noted that high-school girls who were overweight ate "several hundred calories *less*" than those who weren't. "The laws of thermodynamics, however, were not flouted by this finding," he said, because the obese girls spent only a third as much time in physical activity as the lean girls; they spent four times as many hours watching television. Mayer studied adolescent girls at summer camp and reported that the obese girls expended "far less energy," even during scheduled exercise periods, than their non-obese counterparts. He also studied infants. "The striking phenomenon is that the fatter babies were quiet, placid babies that had moderate intake," Mayer reported, "whereas the babies who had the highest intake tended to be very thin babies, cried a lot, moved a lot, and became very tense." Thus, Mayer concluded, "some individuals are born very quiet, inactive, and placid and with moderate intake get fat, and some individuals from the very beginning are very active and do not get particularly fat even with high intake."

Mayer also believed that this link between physical inactivity and overweight explained another troubling conflict in the evidence. How could the prevalence of obesity and overweight be increasing in the 1950s if calorie consumption, according to USDA estimates, had dropped significantly since the turn of the century? (Recall the changing-American-diet story.)

Descriptions of typical meals in the nineteenth century, as Mayer noted, suggest they were enormous compared with what we eat today. Breakfasts of the British gentry of the late nineteenth century "frequently assumed prodigious proportions," according to the anthropologist Eric Ross. In a typical country house, wrote one British authority in the late 1880s, breakfasts consisted of "fish, poultry, or game, if in season; sausages, and one meat of some sort, such as mutton cutlets, or filets of beef; omelettes, and eggs served in a variety of ways; bread of both kinds, white and brown, and fancy bread of as many kinds as can conveniently be served; two or three kinds of jam, orange marmalade, and fruits when in season; and on the side table, cold meats such as ham, tongue, cold game, or game pie, galantines, and in winter a round of spiced beef." In the United States, according to the historian Hillel Schwartz, such enormous meals were also the norm: "The 75-cent special at Fred Harvey restaurants in the late 1870s included tomato purée, stuffed whitefish with potatoes, a choice of mutton or beef or pork or turkey, chicken turnovers, shrimp salad, rice pudding and apple pie, cheese with crackers, and coffee. . . . When life insurance medical directors sat down to their banquet in 1895, they had clams, cream soup, kingfish with new potatoes, filet mignon with string beans, sweetbreads and green peas, squabs and asparagus, petits fours, cheese with coffee, and liqueurs to follow. . . ." Incredibly, Schwartz noted, these gargantuan repasts "were two or more courses and thirty to sixty minutes shorter than formal dinners of the previous era, and their portions were smaller."

Having concluded that caloric intake had actually fallen since the nineteenth century, Mayer pioneered the practice of implicating the sedentary nature of our lives as the "most important factor" in obesity and the chronic diseases that accompany it. Americans in the mid-twentieth century, as Mayer perceived it, were more inert than their "pioneer forebears," who were "constantly engaged in hard physical labor." Every modern convenience, from the car to the extensions on our telephones and even the electric toothbrush, only served to make our lives ever more sedentary. "The development of obesity (and of heart disease as well as a number of other pathologic conditions)," Mayer wrote in 1968, "is to a large extent the result of the lack of foresight of a civilization which spends tens of billions annually on cars, but is unwilling to include a swimming pool and tennis courts in the plans of every high school."

But Mayer's hypothesis always had shortcomings. First, the association between reduced physical activity and obesity doesn't tell us what is cause and what is effect. "It is a common observation," noted Hugo Rony, "that many obese persons are lazy, i.e., show decreased impulse to muscle activity. This may be, in part, an effect that excess weight would have on the activity impulse of any normal person." It's also possible that both obesity *and* physical inactivity are the symptoms of the same underlying cause. This was a likely explanation for the inactivity and obesity that Mayer had observed in his laboratory mice. The same genetic mutation that rendered these mice sedentary could also have induced obesity (and perhaps diabetes).

Another problem, as we discussed in the last chapter, is that obesity is also associated with poverty, and even extreme poverty, and that should be a compelling argument against physical inactivity as a cause of the disease. Those who earn their living through manual labor tend to be the less advantaged members of societies in developed nations, and yet they will have the greatest obesity rates.

A third problem was the observation that exercise accomplishes little in the way of tilting the caloric balance when compared with a very modest restriction of intake—walking a few miles as opposed to eating one less slice of bread—and that increasing activity will increase appetite. Mayer ignored the comparison of intake and expenditure by focusing on expenditure alone. "For a long period the role of exercise in weight control was disregarded, if not actually ridiculed," he wrote in a 1965 *New York Times Magazine* article. "One reason often advanced for this neglect is that 'exercise consumes very little energy.' . . . Somehow the impression was given that any such exercise had to be accomplished in a single uninterrupted session. Actually, exercise does correspond to a caloric expenditure that can be considerable, and this expenditure will take place in a day or a decade." And so the expenditure of calories by exercise, no matter how small, according to Mayer, would accumulate, leading to long-term weight reduction. This, of course, would be true only if the excess expenditure went unaccompanied by a compensatory increase in appetite and intake.

Mayer acknowledged that exercise could increase food intake, but said it wasn't "necessarily" the case. This was the heart of Mayer's hypothesis—a purported loophole in the relationship between appetite and physical activity. "If exercise is decreased below a certain point," Mayer told the *New York Times* in 1961, "food intake no longer decreases. In other words, walking one-half hour a day may be equivalent to only four slices of bread, but if you don't walk the half hour, you still want to eat the four slices. . . ." Mayer based this conclusion on two of his own studies from the mid-1950s.* The first was with laboratory rats and purported to demonstrate that rats that are exercised for one to two hours every day will actually eat *less* than rats that don't exercise at all. The second was a study of mill workers in West Bengal, India, and stands as a reminder that dreadful science can pass for seminal research in the field of obesity.

Mayer worked with the dietician and chief medical officer of the company that owned the Bengali mill and an accompanying bazaar, and it was these Indian colleagues who assessed the physical activity and diet of the resident workers. These men, as Mayer reported, ranged from "extraordinarily inert" stall holders "who sat at their shop all day long," to those engaged in intense physical activity who "shoveled ashes and coal in tending furnaces all day long."

The evidence reported in Mayer's paper could have been used to demonstrate any point. The more active workers in the mill, for example, both weighed more *and* ate more. As for the sedentary workers, the more sedentary they were, the *more* they ate and the *less* they weighed. The twenty-two clerks who lived on the premises and sat all day long weighed ten to fifteen pounds *less* and were reported to have eaten four hundred calories *more* on average than the twenty-three clerks who had to walk three to six miles to work, or even than those five clerks who walked to work and played soccer every day.

Nonetheless, Mayer claimed that the study confirmed the findings of

---

* When Mayer wrote about this research, or when he spoke to reporters about it, he would often give the impression that it included multiple studies in animals and humans—"J. Mayer has since demonstrated, in both animal and human studies . . . ," as he would write in *Science* in 1967. This was technically true, in that he had performed studies of both humans and animals— one study of each.

his rat experiment. He based his conclusion exclusively on the relative girth of thirteen stallholders and eight supervisors. These men weighed, respectively, fifty to sixty pounds and thirty to forty pounds more than the clerks who worked for them, and yet, according to Mayer's data, consumed the same amount of calories. Mayer implied that they added this extra weight because they were *somehow* even less active than those employees whose jobs entailed sitting all day, but he had no evidence for it. It's also possible that their relative wealth introduced other dietary factors that could have explained the dramatic differences in weight. Either way, as John Garrow noted, these findings would never be replicated, which is why such authorities as the Institute of Medicine of the National Academies of Science still cite Mayer's study today as the only evidence for the proposition that "too little" exercise can disrupt the mechanisms that normally regulate food intake.

Mayer's advocacy of exercise for weight control did not go unchallenged. After his 1965 *New York Times Magazine* article, entitled "The Best Diet Is Exercise," physicians working with obese patients wrote to the newspaper saying that Mayer's faith in exercise was unreasonable and flouted common sense. "As much as Dr. Mayer minimizes the thirst and appetite increases after exercise, my patients all seem to be thirstier after tennis and find it difficult to stick to plain water," wrote Morton Glenn of New York University College of Medicine; "and who hasn't heard someone say: 'This walk all the way home sure gave me an appetite!' Exercise can and does increase thirst and appetite, in most persons, in most situations, and most people respond to these sensations accordingly!"

Despite these commonsensical objections, Mayer's hypothesis won out. It helped that Mayer—like Ancel Keys and Dennis Burkitt—perceived the process of convincing the public and the medical-research community to be akin to a crusade. This served to absolve him, apparently, of the obligation to remain strictly accurate about what the research, including his own, had or had not demonstrated. In the popular press, Mayer would unleash his less scientific impulses. He wrote about the "false idea which continues to have broad and pernicious acceptance" that exercise would increase appetite, and he insisted that the "facts overwhelmingly demonstrate" that this was "*not* necessarily" the case.

As Mayer's political influence grew through the 1960s, his prominence and his proselytizing contributed to the belief that his hypothesis had both been proven true and was widely accepted. In 1966, when the U.S. Public Health Service advocated increased physical activity and diet as the best ways for us to lose weight, Mayer was the primary author of the report.

Three years later, Mayer chaired Richard Nixon's White House Conference on Food, Nutrition, and Health. "The successful treatment of obesity must involve far reaching changes in life style," the conference report concluded. "These changes include alterations of dietary patterns and physical activity. . . ." In 1972, Mayer began writing a syndicated newspaper column on nutrition that clearly did not hold to the standards of a serious scientific publication. Sounding suspiciously like a diet doctor selling a patent claim, Mayer said that exercise "makes weight melt away faster." "Contrary to popular belief," Mayer asserted, "exercise won't stimulate your appetite."

The current culture of physical exercise in the United States emerged in the late 1960s, coincident with Mayer's crusade and accompanied by a media debate about whether exercise is or is not good for us. "While it is generally agreed that exercise programs can improve strength, stamina, coordination and flexibility and provide an overall sense of well-being, two crucial questions remain," a 1977 *New York Times Magazine* article observed: "(1) Does exercise prolong life? and (2) does it give any protection against the modern scourge, heart disease?" A handful of observational studies had linked exercise to greater longevity—the most famous being a study of seventeen thousand Harvard alumni published by Ralph Paffenbarger in 1978—but these didn't reveal whether this effect was due to the health benefits of exercise or the fact that healthier people are more likely to exercise. Those who exercised regularly also tended to smoke less and pay more attention to their diets.

Nonetheless, the view of exercise as a panacea for excess weight soon became conventional wisdom. "Diligent exercisers tend to lose weight," was how a *Washington Post* article on the fitness revolution phrased it in 1980. No source for this claim was deemed necessary. All doubts about whether the weight-reducing benefits of exercise actually existed were left behind. In 1983, Jane Brody of the *New York Times* was counting the numerous ways in which exercise was "the key" to successful weight loss. Exercise, she explained, increases metabolism for hours afterward, which further increases caloric expenditure. It is also "an appetite suppressant, sometimes delaying the return of hunger for hours." Exercise builds up muscle tissue, Brody said, which in turn burns more calories than fat. And muscle tissue is denser than fat, Brody concluded, "so even if you do not lose any weight, exercise will make you trimmer." By the end of the decade, as *Newsweek* observed, exercise was now considered "essential" to any weight-loss program. In 1989 the *New York Times* counseled readers that, on those infrequent occasions "when exercise isn't enough" to induce sufficient weight loss, "you must also make sure you don't overeat."

The press may have been convinced, but the scientific evidence never supported Mayer's hypothesis. In October 1973, when the National Institutes of Health hosted its first conference on obesity, Per Björntorp, a Swedish investigator, reported about his own clinical trials on obesity and exercise. After six months of a thrice-weekly exercise program, his seven obese subjects remained both as heavy *and* as fat as ever. Four years later, when the NIH again hosted a conference on obesity, the conference report concluded that "the importance of exercise in weight control is less than might be believed, because increases in energy expenditure due to exercise also tend to increase food consumption, and it is not possible to predict whether the increased caloric output will be outweighed by the greater food intake." In 1989, when Xavier Pi-Sunyer, director of the Obesity Research Clinic at St. Luke's–Roosevelt Hospital Center in New York, reviewed the evidence that exercise "without caloric restriction" could lead to weight loss, he still found little reason for optimism, despite what the press was now claiming as gospel. "Decreases, increases, and no changes in body weight and body composition have been observed," Pi-Sunyer noted. That same year, Danish investigators reported that they had indeed trained previously sedentary individuals to run marathons (26.2 miles). At the end of this eighteen-month training period—a time of almost fanatic exercise—the eighteen men in the study had lost an average of five pounds of body fat. "No change in body composition was observed" among the nine female subjects.

Throughout this period, the research in laboratory animals was equally unsupportive of Mayer's hypothesis. Male rats might actually limit their food intake after running for *hours* on a running wheel, as Mayer had suggested was possible, but they ate more on days when they didn't exercise. They also made up for the exercise by moving less at other times. Moreover, these rats had to be forced to exercise to suppress hunger even temporarily; it did not happen voluntarily. In Mayer's experiments, the rats were put on a motorized treadmill; they ran because they had no choice. This suggested that any decrease in appetite observed in these less-than-voluntary exercise experiments might have been induced by either stress or exhaustion rather than the exercise itself, and particularly by the use of what are technically known as *shock grids* to "motivate" the rats. In those experiments that relied on voluntary physical activity, the more the rats ran, the more the rats ate, and weights remained unchanged. When the rats were retired from forced-exercise programs, they ate more than ever and gained weight "more rapidly" than those rats that had been allowed to remain sedentary. With hamsters and gerbils, voluntary running activity produced "permanent increases" in body weight and adiposity—exercising made these rodents fatter, not leaner.

If Mayer's hypothesis was true, if physical activity played a meaningful role in weight regulation, then researchers' increasing interest in demonstrating this fact should have led, over the decades, to an unambiguous demonstration that this was the case. On the contrary. "When surveying the scientific literature on the treatment of obesity one cannot help but come away . . . underwhelmed by the minor contribution of exercise to most weight-loss programs," University of California, Davis, nutritionist Judith Stern, who had obtained her doctorate at Harvard with Mayer, wrote in 1986.

In the past few years, a series of authoritative reports have advocated ever more physical activity for adults—now up to ninety minutes a day of moderate-intensity exercise—but they have done so precisely because the evidence in support of the hypothesis is so unimpressive. No substantial evidence in fact supports this recommendation for weight loss or maintenance.

These reports, from the USDA and others, rely for their conclusions on a handful of systematic reviews of the medical literature that have been published over the past decade. The most comprehensive of these, and the one cited most frequently by these authoritative reports, is a 2000 analysis by two Finnish investigators. The Finnish review reveals that only a dozen or so clinical trials exist that test the benefits of exercise to maintain weight. The great proportion of the studies are observational studies, which survey the amount of physical activity reported by individuals in various populations and then compare this with how much weight these people gain over a certain period of time. These studies—like the famous Framingham Heart Study—are capable only of identifying associations, not cause and effect, and even these associations are inconsistent. Some studies imply that physical activity might inhibit weight gain, the Finnish investigators report; some that it might accelerate weight gain; and some that it has no effect whatsoever. The clinical trials were equally inconsistent. When the Finnish investigators tried to quantify the results of the dozen trials that addressed the effect of an exercise program on weight maintenance, or what the USDA describes as preventing "unhealthy weight gain," they concluded, depending on the type of trial, that it either led to a decrease of 90 grams (3.2 ounces) per month in weight gained or regained, or to an increase of 50 grams (1.8 ounces). Because "the more rigorous study designs (randomized trials)" yielded the least impressive results, the authors noted, the association between physical activity and weight change, even if it existed, was "more complex" than they might otherwise have assumed. This last point is crucial.

If we consider the last forty years of research as a test of Mayer's hypothesis that physical activity induces weight loss or even inhibits weight gain, it's clear the hypothesis leads nowhere meaningful. What Mayer initially insisted *had* to be true, so much so that he publicly accused the "enemies of exercise" of propagating "pseudo-science," had devolved over the intervening decades into an analysis of whether the prescription of an exercise program would inhibit weight gain by three ounces each month or accelerate it by two.

The fact that appetite and thus calories consumed will increase to compensate for physical activity, however, was lost along the way. Clinicians, public-health authorities, and even exercise physiologists had taken to thinking and talking about hunger as though it were a phenomenon that was exclusive to the brain, a question of willpower rather than the natural consequence of a physiological drive to replace whatever calories may have been expended. When we are physically active, we work up an appetite. Hunger increases in proportion to the calories we expend, just as restricting the calories in our diet will leave us hungry until we eventually make good the deficit, if not more. The evidence suggests that this is true for both the fat and the lean. It is one of the fundamental observations we have to explain if we're to understand why we gain weight and how to lose it.

Chapter Sixteen

PARADOXES

The literature on obesity is not only voluminous, it is also full of conflicting and confusing reports and opinions. One might well apply to it the words of Artemus Ward: "The researches of so many eminent scientific men have thrown so much darkness upon the subject that if they continue their researches we shall soon know nothing."

HILDE BRUCH, *The Importance of Overweight*, 1957

L ET'S ASK A FEW MORE SIMPLE QUESTIONS about the nature of obesity and weight regulation. Even if we accept—just for the moment—that obesity is caused by a positive energy balance and thus some combination of overeating and sedentary behavior, why would anyone willingly continue to overeat or remain sedentary if obesity is the undesirable state it certainly appears to be? Why would energy balance remain positive when there are so many compelling reasons and so much time to stop the process and maybe reverse it? If a positive energy balance can be turned into a negative energy balance with reasonable facility by exercise and calorie-restricted diets, why is it so difficult to lose weight?

This is the paradox that haunts a century of obesity research. As Marian Burros wrote in the *New York Times* in 2004: "Those who consume more calories than they expend in energy will gain weight. There is no getting around the laws of thermodynamics." This was the "very old and immutable scientific message," she explained. And yet the great majority of those who attempt to expend more calories than they consume don't lose weight. Those who do, lose only a little, and for short periods of time. This suggests that obesity is a disease, "a chronic condition," as Albert Stunkard described it over thirty years ago, "resistant to treatment, prone to relapse, for which we have no cure."

In 1983, Jules Hirsch of Rockefeller University framed this enigma in the form of two alternative hypotheses. One was the common belief "that obesity is the result of a willful descent into self-gratification." The other was the "alternative hypothesis that there is something 'biologic' about obesity, some alteration of hormones, enzymes or other biochemical control systems which leads to obesity." Because no such biologic abnormality had been unambiguously identified, Hirsch believed, "it is perhaps better to maintain the illusion that obesity is not an illness. It is more pleasant to believe that it is no more than an error of good judgment and that better judgments and choices will eventually lead" to a better outcome.

Here is another apparent contradiction: it may be true that, "for the vast majority of individuals, overweight and obesity result from excess calorie consumption and/or inadequate physical activity," as the Surgeon General's Office says, but it also seems that the accumulation of fat on humans and animals is determined to a large extent by factors that have little to do with how much we eat or exercise, that it has a biologic component.

The deposition of fat in men and women is distinctly different. Men tend to store fat above the waist—hence the beer belly—and women below it. Women put on fat in puberty, at least in the breasts and hips, and men lose it. Women gain weight (particularly fat) in pregnancy and after menopause. This suggests that sex hormones are involved, as much as or more than eating behavior and physical activity. "The energy conception can certainly not be applied to this realm," as the German clinician Erich Grafe observed in 1933 about this anatomical distribution of fat deposits and how it differs by sex.

Fat, or lack of it, runs in families and even does so, noted the pediatrician-turned-psychiatrist Hilde Bruch in 1957, with such characteristic shapes or body types that "this similarity may be as striking as facial resemblance." And if girth has a genetic component, then that means it is regulated by biological factors—perhaps tilted in one direction for those who gain weight easily, and tilted in another for those who don't. "It is genetics, and not the environment, that accounts for a large proportion of the marked differences in individual body weight in our population today," wrote the Rockefeller University molecular biologist Jeffrey Friedman in 2004. If obesity does have such a significant genetic factor—"equivalent to that of height, and greater than that of almost every other condition that has been studied," according to Friedman—then how does this figure into the equation of overeating and sedentary behavior?

The same could be asked about metabolic or hormonal factors, which also contribute to excessive adiposity, as Jerome Knittle of Rockefeller University explained in 1976, when he testified before George McGovern's Senate Select Committee on Nutrition and Human Needs. "Infants born to diabetic mothers are heavier at birth, are relatively fatter and have a higher rate of subsequent obesity than infants of non-diabetic mothers of equal gestational age," Knittle said. But if these physiologic factors make for fatter babies and subsequently fatter adults, couldn't the same be true for those of us without diabetic mothers, too?

Some of us simply seem predisposed, if not fated, to put on weight from infancy onward. Some of us lie further along what Friedman described as the distribution of adiposity than others. In the early 1940s, the Harvard psychologist William Sheldon was referring to what he called the "morphology" of body types when he commented, "It does not take a science to tell that no two human beings are identically alike." According to Sheldon, every human body could be described by some combination of three basic physical types: ectomorphs, who tend to be long and lean; mesomorphs, who are broad and muscular; and endomorphs, who are round and fat. You could starve endomorphs, Sheldon said, and they might lose weight and even appear emaciated, "but they do not change into mesomorphs or ectomorphs any more than a starved mastiff will change into a spaniel or a collie. They become simply emaciated endormorphs."

In 1977, when McGovern's committee held a hearing on obesity, Oklahoma Senator Henry Bellmon captured this dilemma perfectly. The committee had spent the day listening to leading authorities discuss the cause and prevention of obesity, and the experience had left Bellmon confused. "I want to be sure we don't oversimplify . . . ," Bellmon said. "We make it sound like there is no problem for those of us who are overweight except to push back from the table sooner. But I watched Senator [Robert] Dole in the Senate dining room, a double dip of ice cream, a piece of blueberry pie, meat and potatoes, yet he stays as lean as a west Kansas coyote. Some of the rest of us who live on lettuce, cottage cheese and Ry-Krisp don't do nearly as well. Is there a difference in individuals as to how they utilize fuel?" The assembled experts acknowledged that they "constantly hear anecdotes of this type," but said the research was ambiguous. In fact, the evidence was clear, but it was difficult to reconcile with the assembled experts' preconceived notion—the dogma—that obesity is caused by gluttony and/or sloth.

Over the past century, numerous studies have addressed this issue of how much more easily some of us fatten than others. In these studies, volunteers are induced to overeat to considerable excess for months at a time. The most famous such study was conducted by the University of Vermont endocrinologist Ethan Sims beginning in the late 1960s. Sims first used students for his experiments, but found it difficult to get them to gain significant weight. He then used convicts at the Vermont State Prison, who initially raised their food consumption to four thousand calories a day. They gained a few pounds, but then their weights stabilized. So they ate five thousand calories a day, then seven thousand (five full meals a day), then ten thousand, while remaining sedentary.

There were "marked differences between individuals in ability to gain weight," Sims reported. Of his eight subjects that went two hundred days on this mildly heroic regimen, two gained weight easily and six did not. One convict managed to gain less than ten pounds after thirty weeks of forced gluttony (going from 134 pounds to 143). When the experiment ended, all the subjects "lost weight readily," Sims said, "with the same alacrity," in fact, as that with which obese patents typically return to their usual weights after semi-starvation diets. Sims concluded that we're all endowed with the ability to adopt our metabolism and energy expenditure "in response to both over- and undernutrition," but some of us, as with any physiological trait, do it better than others.

Another overfeeding study, led by Claude Bouchard, who is now head of the Pennington Biomedical Research Center in Louisiana, was published in 1990. Bouchard and his colleagues overfed twenty-four young men—twelve pairs of identical twins—by a thousand calories a day, six days a week, for twelve weeks. The subsequent weight gain varied from nine to thirty pounds. The amount of body fat gained also varied by a factor of three. In 1999, James Levine from the Mayo Clinic reported that he had overfed sixteen healthy volunteers by a thousand calories a day, seven days a week, for eight weeks. The amount of fat these subjects managed to put on ranged from less than a single pound to almost nine; "fat gain varied ten-fold among our volunteers," Levine reported.

None of these experiments could explain what happened to the extra calories in those subjects who did not fatten easily, and why some of these subjects fattened more than others. Why is it that when two people eat a thousand calories a day more than they need to otherwise maintain their weight, and this overfeeding continues for weeks on end, one barely adds a pound of fat while the other puts on nearly ten? Bouchard and his colleagues used identical twins for their study to determine whether genetics contributed to this ability to fatten, and they reported that, indeed, pairs of twins gained similar amounts of weight and fat. "Genetic factors are involved" was all they could say. "These may govern the tendency to store energy as either fat or lean tissue and the various determinants of the resting expenditure of energy."

Those engaged in the practice of animal husbandry have always been implicitly aware of the genetic, constitutional component of fatness. This is why they breed livestock to be more or less fatty, just as they breed dairy cattle to increase milk production, racehorses for speed and endurance, or dogs for hunting or herding ability. It's conceivable, as the logic of overeating and sedentary behavior might suggest, that breeders of fat cattle or pigs have merely identified genetic traits that determine the will to eat in moderation and a propensity to exercise, but it strains the imagination that these are the relevant factors.

Much of the laboratory research on both obesity and diabetes is carried out on strains of rats and mice that grow reliably obese (sometimes monstrously so) eating no more than others that remain lean. The German physiologist Ingrid Schmidt says that when she first saw an example of an obese Zucker rat her immediate response was disbelief. "Up until that moment," Schmidt recalls, "I thought if someone is too fat he should eat less. Then I saw that animal and thought, That's incredible, one gene is broken, and this is the result. And once they get fat, you have the same problem you do with fat humans: everything is changed, and you have no idea what's the cause and what's secondary to this underlying defect."

When Jean Mayer began studying a strain of obese mice in 1950, he observed that if he starved them sufficiently he could reduce their weight beneath that of normal rats, but they'd "still contain more fat than the normal ones, while their muscles have melted away," which made them sound suspiciously like rodent versions of Sheldon's emaciated endomorphs. For centuries, fat men and women have been complaining that virtually everything they eat turns to fat, and this was precisely what was happening with Mayer's obese mice. "These mice will make fat out of their food under the most unlikely circumstances," he wrote, "even when half starved."

Something more is going on than mere immoderation in lifestyle—metabolic or hormonal factors in particular. Yet the accepted definitions of the cause of obesity do not allow for such a possibility. Why?

An obese Zucker rat will be fatter than a lean one,
even if it's semi-starved from birth onward.
(Photo courtesy of Charles River Laboratories.)

The answer dates back to the birth of modern nutrition research in the late nineteenth century. Until then, obesity had been considered no more likely to be cured by any facile prescription than was any other debilitating disease. As early as 1811, one French physician's list of the curative agents promoted for obesity included several that might, naïvely, be considered the last resorts of desperate individuals: bleeding from the jugular vein, for example, and leeches to the anus. In the 1869 edition of *The Practice of Medicine*, the British physician Thomas Hawkes Tanner added to these "ridiculous" prescriptions, those of Thomas King Chambers, whose 1850 book, *Corpulence; or Excess of Fat in the Human Body*, recommended eating "very light meals of substances that can be easily digested" and devoting "many hours daily to walking or riding." "All these plans," wrote Tanner, "however perseveringly carried out, fail to accomplish the object desired; and the same must be said of simple sobriety in eating and drinking."*

The paradox developed with the understanding of the energy content of foods—the calorie—and the development of a technology, known as calorimetry, that could measure the heat production and respiration of living organisms and so equate the caloric content of foods to the calories expended as energy in the process of living. This was the culmination of a hundred years of science, beginning in the mid-eighteenth century with the Frenchman Antoine-Laurent Lavoisier, who demonstrated that the heat generated by an animal (literally a guinea pig in his experiments) was directly related to how much oxygen it consumed and carbon dioxide it exhaled. Living organisms are burning or combusting just as any other fire or flame does, which is why both will expire without sufficient oxygen. By 1900, a succession of legendary German chemists—Justus von Liebig, his students Max von Pettenkofer and Carl von Voit, and their student Max Rubner, among others—had worked out how organisms burn protein, fat, and carbohydrates and the basics of both metabolism and nutrition science. "The amount of information [the Germans] acquired within a comparatively few years past is remarkable," wrote Wilbur Atwater the pioneer of nutrition research in the United States, in 1888.

It was Rubner who discovered that fat had more than twice as many calories per gram as did protein or carbohydrates. He also demonstrated, in 1878, what he originally called the *isodynamic law*, which has since been

---

* Tanner did believe that William Banting's French predecessor Jean-François Dancel had finally provided a "more sure basis" for the treatment of obesity, and that Banting himself deserved credit for "bringing the subject before the public in a plain and sensible manner."

distilled by nutritionists to the phrase "a calorie is a calorie." A calorie of protein provides the same amount of energy to the body as a calorie of fat or carbohydrate. Lost in this distillation is the fact that the effects of these different nutrients on metabolism and hormone secretion are so radically different, as is the manner in which the body employs the nutrients, that the energetic equivalence of the calories themselves is largely irrelevant to why we gain weight. As Rubner suggested more than a century ago, "the effect of specific nutritional substances upon the glands" may be the more relevant factor.

Rubner gets credit for being the first to demonstrate that the law of conservation of energy holds in living organisms. Rubner studied the heat expenditure and respiration of a dog for forty-five days and published his findings in 1891. Eight years later, Francis Benedict and Wilbur Atwater confirmed the observation in humans: the calories we consume will indeed either be burned as fuel—metabolized or *oxidized*—or they'll be stored or excreted. The research of Rubner, Benedict, and Atwater is the origin of the pronouncement often made by nutritionists with regard to weight-reducing diets that "calories in are equal to calories out." As Marian Burros of the *New York Times* observed, there's no violating the laws of thermodynamics.

It was with the application of these laws to the problem of human obesity that the paradoxes emerged. This work was done in the first years of the twentieth century by Carl von Noorden, the leading German authority on diabetes, the author-editor of several multivolume medical texts, and author of one 1900 monograph on obesity entitled, in the original German, *Die Fettsucht*. "His work contains many ideas which have become so incorporated, and in such a matter of fact way, into medical thinking, that his name is no longer mentioned with them," noted Hilde Bruch fifty years ago. The same is still true today.

Von Noorden proposed three hypotheses for the cause of obesity. One of these, what he called diabetogenous obesity, was remarkably prescient, but so far ahead of its time that it had no influence on how the science evolved. (We will discuss this hypothesis later, in Chapter 22.) Von Noorden's other two hypotheses, however, which he called exogenous and endogenous obesity, though simplistic in comparison, have dominated thinking and research on obesity ever since.

Von Noorden worked directly from the law of energy conservation: "The ingestion of a quantity of food greater than that required by the body," he wrote, "leads to an accumulation of fat, and to obesity should the disproportion be continued over a considerable period." This left open the question of what would cause such a positive energy balance,* and von Noorden suggested that it was due either to an immoderate lifestyle (exogenous obesity, driven by forces external to the body) or to the fact that some people seemed predestined to grow fat and stay fat, regardless of how much they ate or exercised (endogenous obesity, driven by internal forces, not external).

In the cases where immoderate lifestyle was to blame—by "far the most common" of the two, von Noorden believed—the metabolism and physiology of the obese individual are normal, but "the mode of living" is defective, marked by that now familiar combination of "overeating or deficient physical exercise." In endogenous obesity, the lifestyle is normal, and the weight gain is caused by an abnormally slow metabolism. These unfortunate individuals might eat no more than anyone else, but their metabolisms use a smaller proportion of the calories they consume, and so a greater proportion is stored as fat.

Just as heart-disease researchers came to blame cholesterol because it seemed to be an obvious culprit and they could measure it easily, von Noorden and the clinical investigators who came after him implicated metabolism and the energy balance because that's what they could measure and that, too, seemed obvious. In 1892, a German chemist named Nathan Zuntz had developed a portable device to measure an individual's oxygen consumption and carbon-dioxide respiration. This, in turn, allowed for the calculation, albeit indirectly, of energy expenditure and the metabolism of anyone who had the patience to remain immobile for an hour while breathing into a face mask. Within a year, Adolf Magnus-Levy, a colleague of von Noorden, had taken this calorimeter to the hospital bedside and begun a series of measurements of what later became known as basal metabolism, the energy we expend when we're at "complete muscular repose," twelve to eighteen hours after our last meal. By the end of World War I, calorimetric technology had been refined to the point where measuring metabolism had become "an extremely popular, almost fashionable field."

Von Noorden's focus on metabolic expenditure set the science of obesity on the path we still find it. The evolution of this research, however, proceeded like a magician's sleight-of-hand. By the 1940s, common sense, logic, and science had parted ways.

---

* It also assumes that the ingestion of food greater than that required by the body won't lead to a compensatory increase in energy expenditure, which is a point we'll discuss at length in the next chapter.

The most obvious difficulty with the notion that a retarded metabolism explains the idiosyncratic nature of fattening is that it never had any evidence to support it. Before von Noorden proposed his hypothesis, Magnus-Levy had reported that the metabolism of fat patients seemed to run as fast if not faster than anyone else's.* This observation would be confirmed repeatedly: The obese tend to expend more energy than lean people of comparable height, sex, and bone structure, which means their metabolism is typically burning off more calories rather than less. When people grow fat, their lean body mass also increases. They put on muscle and connective tissue and fat, and these will increase total metabolism (although not by the same amount).

The tendency of the obese to expend more energy than do the lean (of comparable height, age, and sex) led to the natural assumption that they *must* eat more than the lean do. Otherwise, they would have to lose weight. Researchers from Magnus-Levy onward avoided this conclusion by calculating energy expenditure as a *metabolic rate*—the total metabolism divided by weight, for instance, or by the skin-surface area of the subject. The obese could then be said to have a metabolic rate that seemed, on average, to run slower than that of the lean. That was beside the point, though, at least when it came to the amount of calories that must be consumed either to cause obesity or to reverse it. The factor of interest, noted the British physiologists Michael Stock and Nancy Rothwell in 1982, is "the metabolism of the individual and not a unit fraction of that individual."

One of the most telling observations that emerged from these studies of metabolic rate was how greatly it might differ between any two individuals of equal weight, or how similar it might be among individuals of vastly different weights. In 1915, Francis Benedict published his studies of the basal metabolism translated into the minimal amount of energy expended over the course of a day, as measured in eighty-nine men and sixty-eight women. Though men expended more energy than women on average, and large men more than small, there were huge variations. For men who weighed roughly 175 pounds, the minimal energy expenditure daily ranged from sixteen to twenty-one hundred calories. This implies that one 175-pounder could eat five hundred calories a day more than another 175-pounder each day—a quarter-pounder with cheese from McDonald's—and yet would

---

* This left von Noorden explaining that the detection of a retarded metabolism seemed to require "special knowledge and acumen on the part of the observer," and he acknowledged that even he lacked sufficient expertise. Hence, the only way to diagnose a retarded metabolism was by implication: if the patient's weight could not be "brought under control through intelligent regulation of diet and exercise," then the patient probably had a retarded metabolism. The circularity of this argument was evidently not apparent to him.

gain no more weight by doing so, even if the amount of physical activity in their lives was identical. Heavier women also tended to expend more energy, but the variations were striking. One of Benedict's female subjects weighed 106 pounds, whereas another weighed 176, and yet both had a basal metabolism of 1,475 calories.

The idea that obesity can be preordained by a constitutional predisposition to grow fat, what von Noorden had called endogenous obesity, would ultimately be rejected by the medical community, based largely on the efforts of Hilde Bruch, who did the actual research, and Louis Newburgh, who shaped the way it would come to be interpreted. Bruch was a German pediatrician who in 1934 had immigrated to New York, where she established a clinic to treat childhood obesity at Columbia University's College of Physicians and Surgeons. She began her career testing what she called the "fashion" of the day: that obese children must suffer from a hormonal or endocrine disorder. How else to reconcile their claims to eat like birds, as obese adults often claim? Bruch failed to find evidence for this hypothesis and so set out to study in exhaustive detail the lives and diet of her young obese patients.

In 1939, Bruch published the first of a series of lengthy articles reporting what she had learned from treating nearly two hundred obese pediatric patients at her clinic. All of these children, upon close investigation, reported Bruch, ate significant quantities of food. "Overeating was often vigorously denied and it took some detective work, with visits to the home to obtain an accurate picture," Bruch wrote. For whatever reason, the mothers tended to be more candid about their children's eating habits at home than at the clinic. "The terms used for depicting the amounts eaten varied a good deal," Bruch reported; "they ranged from 'good appetite' and 'he eats very well' to 'most tremendous appetite,' 'he eats voraciously' and 'food is the only interest she has.' "

Bruch's conclusion was that "excessive eating and avoidance of muscular exercise represent the most obvious factors in the mechanisms of a disturbed energy balance." And this was either caused or exacerbated by psychological factors of the mother-child relationship. A mother will substitute food for affection, Bruch said, and by doing so overfeed the child. She may compound the damage by being overprotective, which leads her to "keep the child from activities with peers lest the child be hurt."* For the fat children themselves, she wrote, giving up food means "giving up [their]

---

* This notion has survived in the suggestion that weight gain in children is exacerbated by the refusal of parents to allow their children to walk or ride to school, for fear they will be kidnapped or abused by strangers.

only source of pleasure and enjoyment. The very size itself, although resented because it is being constantly ridiculed, nevertheless gives the fat child, who has no basic security in his interpersonal relations, a certain sense of strength and security."

It was Newburgh, a professor of medicine at the University of Michigan, who then killed off von Noorden's hypothesis of endogenous obesity once and for all, and with it any explanation for obesity that didn't blame it on simple gluttony and sloth. Unlike Bruch, Newburgh had easily been convinced that obesity was the result of what he called a "perverted appetite." "All obese persons are alike in one fundamental respect—they literally overeat," he was insisting as early as 1930. The obese were responsible for their condition, Newburgh argued, regardless of whether or not their metabolism was somehow retarded. If it was, then the obese were culpable because they were unwilling to rein in their appetites to match their "lessened outflow of energy." If their metabolisms ran at normal speed, they were even more culpable, guilty of "various human weaknesses such as overindulgence and ignorance."

In 1942, Newburgh published a sixty-three-page article in the *Archives of Internal Medicine* meticulously documenting the evidence against von Noorden's endogenous-obesity hypothesis. He rejected the role of any "endocrine disorder" in fattening—a pituitary tumor, for instance, or the particularly slow secretion of thyroid hormones, which were the two leading candidates—on the basis that these could explain, at best, only a tiny percentage of cases. The great majority of the obese had perfectly normal thyroid glands, Newburgh wrote, and there were considerable cases of pituitary tumors that were not accompanied by obesity. He scoffed at the notion that "retarded metabolism" could play a role in obesity, because the obese expend as much energy as the lean, or more. And Bruch's research, Newburgh went on, constituted the definitive proof that even the most obese children earned their condition by eating too much. If obese children could no longer hide behind the excuse of a constitutional predisposition, then neither could obese adults, Newburgh said. Thus, the only obstacle standing between obesity and leanness was insufficient willpower. As proof, Newburgh offered up a case study of a patient who lost 286 pounds in a year on a diet of three hundred calories a day, and then another eighty pounds the following year while eating six hundred calories a day. By then this patient had returned to his normal weight; "his gluttonous habits had been abolished," Newburgh wrote, and he had subsequently maintained his weight "without any effort to restrict his food intake." This may have been true, but if so, Newburgh's patient was virtually unique in the annals of obesity research.

By the end of Newburgh's review, he had dismissed any possibility of a constitutional predisposition as a factor in the etiology of obesity. If genes had anything to do with obesity, which Newburgh did not believe, "it might be true that a good or poor appetite is an inherited feature." If obesity ran in families, "a more realistic explanation is the continuation of the familial tradition of the groaning board and the savory dish." If women became matronly after menopause, it had nothing to do with hormones—that "the secretions of the sex glands, now in abeyance, formerly had the power to restrain the growth of the adipose tissue,"—but, rather, that the postmenopausal woman now had the time and the inclination to indulge herself. "She does not resist gain in weight, since the friends in whom she has the greatest confidence have assured her that nature intends her to lay on weight at this time of life," Newburgh wrote.

To the generation of physicians who took up the treatment of obesity in the decade following World War II, Newburgh's 1942 review was the seminal article on human obesity. "The work of Newburgh showed clearly . . . ," these physicians would say, or "Newburgh answered that . . . ," they would respond to any evidence suggesting that obesity was caused by anything other than what Newburgh had called a "perverted appetite"—overeating, or the consumption of more calories than are expended.

But this simple concept had a fundamental flaw, which dated back to von Noorden's original conception of exogenous obesity. The statement that obesity is accompanied by an imbalance between energy intake and energy output—calories in over calories out—is a tautology. As Marian Burros said, it has to be true, because it is implied by the law of energy conservation. So, then, what causes this imbalance? Von Noorden's proposition that the imbalance is caused by "overeating and deficient physical exercise" (or "excess calorie consumption and/or inadequate physical activity," as the Surgeon General's Office put it) is both an assumption (unproved) and a tautology. The assumption is that something that accompanies the process of becoming obese—overeating and deficient physical activity—causes it. The tautology is that these terms are defined in such a way that they have to be true.

The terms "overeating" and "deficient physical exercise" are applied only to the overweight and obese. "If eating behavior did not produce deposits of body fat we could not call it overeating," is how this phenomenon was phrased in 1986 by William Bennett, then editor of the *Harvard Medical School Health Letter* and one of the rare investigators interested in obesity ever to make this point publicly. If someone is fat, then he has overeaten by definition. If he's lean, the amount of food he consumes is not considered relevant to his weight, nor is the amount of physical activity in his life. This is why lean individuals who consume comparatively large quantities of food are said to have a healthy appetite or are big eaters. No one suggests that they are suffering from excess calorie consumption.

Von Noorden's proposition, which still obtains today, is the equivalent of saying that "alcoholism is caused by chronic overdrinking" or "chronic fatigue syndrome is caused by excessive lethargy and/or deficient energy." These propositions are true, but meaningless. And they confuse an association with cause and effect. They tell us nothing about why one person becomes obese (or alcoholic or chronically fatigued) and another person doesn't. Moreover, as Bennett noted, even if fat people did eat more and/or expend less energy than most or all lean people—something that has never been shown to be true—it would still beg what should be the salient question in all obesity research: why wasn't intake adjusted downward to match expenditure, or vice versa? Nor does it explain why reversing this caloric imbalance fails to reverse the weight gain reliably.*

Those who are overweight or obese, with exceedingly rare exceptions, do not continue to gain weight year in and year out. Rather, they gain weight over long periods of time and then stabilize at a weight that is higher than ideal, remaining there for a long period of time, if not indefinitely. Why, as Bennett asked, "is energy balance achieved at a particular level of fat storage and not some other?" This is another question that any reasonable hypothesis of obesity must address. In 1940, the Northwestern University endocrinologist Hugo Rony described the problem in a way that brings to mind Hirsch's comment of fifty years later: "An obese person who maintains his weight at 300 pounds indefinitely, is in caloric equilibrium the same as any person of normal weight. The conception that his obesity is due to positive caloric balance might be useful in explaining how he reached this excessive weight, but cannot inform us why he maintains it, why he resists attempts to reduce it to normal, why he tends to regain it after successful reduction."

It's tempting to suggest that one reason why the obesity-research community has paid little attention to the logical and scientific deficiencies of the overeating/sedentary-behavior hypothesis is that it becomes difficult even to discuss the subject without constantly tripping over the solecisms it

---

* The common response to confronting this dilemma, as Bennett noted, "is to ignore it," which is what happened to Bennett's commentary, even though he discussed this issue at a 1986 obesity conference hosted by the New York Academy of Sciences and attended by many of the prominent authorities in the field.

engenders. To say someone "overeats" or "eats a lot" immediately raises the question, Compared with whom? One of the most reproducible findings in obesity research, as I've said, is that fat people, on average, eat no more than lean people. They may not eat as little as they say or think they do, but they don't necessarily eat any more than anyone else. "On the few occasions when the food intake of a group of obese persons has been measured with an approved technique," wrote the British physiologists J.V.G.A. Durnin and Reginald Passmore in 1967, "it has been found to be no greater than that of a control group of persons of normal weight. Fat people are not necessarily gluttons: some indeed are truly abstemious." Passmore and Durnin neglected to ask then how such an abstemious individual becomes fat. Rather, they insisted that there was "not a shred of evidence" to support the belief of the obese and "also their friends and sometimes regretfully their medical attendants that they are 'mysterious engines' and can conserve energy in an unknown manner." A mysterious conservation of energy does, however, seem to be the only explanation. Why do they remain fat when others would remain effortlessly lean on the same diet? What does it mean to overeat, if that's the case?

James Boswell and Samuel Johnson struggled with the same paradox in the late eighteenth century, as Boswell reported in *The Life of Samuel Johnson:*

> Talking of a man who was growing very fat, so as to be incommoded with corpulency; [Johnson] said, "He eats too much, Sir." Boswell. "I don't know, Sir; you will see one man fat who eats moderately, and another lean who eats a great deal." Johnson. "Nay, Sir, whatever may be the quantity that a man eats, it is plain that if he is too fat, he has eaten more than he should have done."

But to clarify, as Johnson did, that obesity is caused by eating more than one should have, is not a satisfying answer. We're still left asking why.

This question is built into the logic of the overeating/sedentary-behavior hypothesis. Why do people overeat, or why are they so sedentary, if the inevitable result is obesity? And because both overeating and deficient physical activity are, after all, behavioral conditions, not physiological ones, the only answer allowed by the hypothesis is a judgment on the behavior of the obese. To say that the obese eat more than they should, as Johnson phrased it, or are less active than they should be—thus, inducing their positive caloric balance—implies only two possibilities. Either it's beyond their control, in which case there is another, more profound cause of their condition—perhaps a metabolic or hormonal disorder for which we should still be searching—or it is within their control, and so we are led to the judgment that the obese are weaker of will than the lean. It may be true, as von Noorden noted, that their appetite is unable to regulate their energy consumption, but why, then, do they not consciously adjust? The logic keeps taking us in circles.

We arrive at the same conclusion if we ask why semi-starvation diets fail to cure obesity reliably, inducing only short-term weight loss by creating a negative caloric balance. Again, there are two possibilities. The first is that the obese stay on the diet but the weight loss eventually stops or even reverses itself. If this is the case, then whatever physiological mechanism is at work may be the cause of the obesity as well. If so, obesity may be caused not by overeating, whatever that means, and sedentary behavior—i.e., by positive caloric balance—but by some more profound underlying disorder. Since a metabolic disorder is not an option in the overeating/sedentary-behavior hypothesis (if it were, then we might be discussing the *metabolic-disorder* hypothesis), the only allowable answer is the second possibility: the obese lack the willpower to remain on the diet—a character defect.

The closer we look at the overeating hypothesis, the more counterintuitive its logic becomes. Consider a thought experiment. The subjects are two middle-aged men of similar height and age. One eats three thousand calories a day and is lean. The other eats three thousand calories a day and is obese. (The epidemiologic and metabolic studies of the past century make clear that we could find two such men with little difficulty.) Let's cut the calorie intake of our obese subject in half and semi-starve him on fifteen hundred calories a day. He will lose weight, although, if Albert Stunkard's 1959 analysis holds true, there's only one chance in eight he'll lose even as much as twenty pounds. Our lean subject will lose weight on this diet as well, as Keys demonstrated with his conscientious objectors in 1944. That's what the law of energy conservation implies. But they would both be hungry continuously, making it likely they would fall off the diet given time. That's what common sense, the history of obesity research, and the Carnegie, Minnesota, and Rockefeller experiments tell us. And after some amount of weight loss, their weight will plateau, because their metabolism and energy expenditure will adjust to this new level of calorie intake. "Eventually, calorie balance is re-established at a new (low) plateau of body weight and the calorie deficit is zero," as Keys explained.

Our intuition is that our obese subject will lose more weight because he has more to lose, but we have little evidence to that effect, one way or the other. And yet, if both our obese and lean subjects fall off the diet and

return to eating three thousand calories a day, the obese individual will
return to obesity, perhaps even fatter than ever, and thus will satisfy our
diagnostic criterion for a character defect; our lean subject will also put
back the weight he lost, and perhaps a little more, but will still be lean, and
will not have to think of himself as possessed of a perverted appetite or
some other character defect.

The same conclusion will be reached if our obese subject undergoes
bariatric surgery. "This procedure alters gastrointestinal anatomy to
reduce caloric intake beyond what could be achieved volitionally," explains
Jeff Friedman of Rockefeller University in a recent issue of *Nature Medicine*. "Although people who undergo bariatric surgery lose a significant
amount of weight, nearly all remain clinically obese." We will now have
two individuals of more similar size and weight, one of whom needs a surgically altered gastrointestinal tract to reduce calorie intake so much that
he can stay at that weight, and the other who doesn't and can eat to his
heart's content. Our surgical patient is perceived as defective in character,
having had to rely on surgery to curb his appetite. Our naturally lean subject is not, despite the possession of an identical appetite. "The implication," as Friedman noted, "is that something metabolically different about
morbidly obese individuals results in obesity independently of their caloric
intake."

Whatever the accepted wisdom, making obesity a behavioral issue is
endlessly problematic. "Theories that diseases are caused by mental states
and can be cured by will power," as Susan Sontag observed in her 1978
essay *Illness as Metaphor*, "are always an index of how much is not understood about the physical terrain of a disease." This is certainly the case
with obesity. One goal of any discussion of the cause of obesity must be a
way to think about it that escapes the facile and circular reasoning of the
overeating/sedentary-behavior hypothesis and permits us to proceed in a
direction that leads to real progress, to find a way of discussing the condition, as the philosopher of science Thomas Kuhn might have put it, that
allows for a "playable game."

Obesity researchers over the last century have struggled with this dilemma,
but they failed to escape it, which is the inevitable consequence of circular
logic. Von Noorden, for instance, sought to absolve the obese of character
defects by suggesting that weight was gained so imperceptibly as to go
unnoticed. He inaugurated the practice, ubiquitous today, of enumerating
the subtle ways in which excess calories creep into our diet, or fail to be
expended in our sedentary lives. Two hundred calories a day, he suggested,

the content of five pats of butter or twelve ounces of beer, could easily slip into the diet unobserved and result in a weight increase, by his calculation, of nearly seventeen pounds a year. "These 200 calories represent such a small amount of food," he explained, "that neither eyesight nor appetite afford any indication of it, and therefore the person can say to the best of his knowledge that his food-supply has not been altered, although he has obviously become corpulent." Any such claims that obesity is caused by the slow and imperceptible accumulation of excess calories inevitably serves to blame obesity on the behaviors of overeating and inactivity, while avoiding the explicit accusation of a character defect. Such explanations also beg the question of how the victim managed to make the transition from lean through overweight to obese without noticing and then choosing to reverse the process.

The hypothesis that the currently rising tide of obesity is caused by a toxic food environment, as Yale's Kelly Brownell has proposed, is another example of an attempt to blame obesity on the behavior of overeating, even while sympathizing with the sufferers. "As long as we have the food environment we do," says Brownell, "the epidemic of obesity is predictable, inevitable, and an understandable consequence." That environment, in his view, is the fault of the food industry, aided and abetted by the makers of computer games and television shows that encourage sedentary entertainment. Following this argument, severely obese people have sued fast-food chains, the inventors of supersizing, which supposedly pushes extra calories on unsuspecting bargain-conscious Americans. "Our culture's apparent obsession with 'getting the best value' may underlie the increased offering and selection of larger portions and the attendant risk of obesity," as James Hill of the University of Colorado and his colleague John Peters of Procter & Gamble suggested in *Science* in 1998.

But if the environment is so toxic, as the Mayo Clinic diabetologist Russell Wilder asked seventy years ago, "why then do we not all grow fat?" After all, Wilder observed, "we continue to be protected against obesity, most of us, even though we hoodwink our appetite by various tricks, such as cocktails and wines with our meals. The whole artistry of cookery, in fact, is developed with the prime object of inducing us to eat more than we ought." This brings us right back to the character issue. Some misbehave in this toxic environment and become obese. Some do not.*

* That the toxic-environment hypothesis is deeply immersed in moral and class judgments is evidenced by the observation that few or none of the condemnations of fast-food restaurants include a coffee chain such as Starbucks, despite the copious excess calories it peddles. A "grande" (sixteen ounces) Tazo® Chai Crème Frappuccino,® for instance, with whipped cream has roughly 510 calories, equivalent to a quarter-pounder with cheese at McDonald's. The same judgments are

Albert Stunkard and Jean Mayer are among those investigators who argued that it was wrong to blame obesity on character defects, and yet still failed to extricate themselves from the circular logic of the overeating/sedentary-behavior hypothesis. In his 1959 analysis of semi-starvation diets, Stunkard wrote that obesity research went astray once investigators concluded that "excessive body fat results from an excess of caloric intake over caloric expenditure" and then enshrined this thinking as the dictum that "all obesity comes from overeating." After that, wrote Stunkard, the physician's job became nothing more than to explain that "semi-starvation reduces fat stores, to prescribe a diet for this purpose," and then to sit by and await the result. "If the patient lost weight as predicted, this merely confirmed the comfortable feeling that treatment of obesity was really a pretty simple matter," wrote Stunkard. "However, if, as so often happened, the patient failed to lose weight, he was dismissed as uncooperative or chastised as gluttonous." Mayer also ridiculed the logic that obesity was caused by gluttony or whatever was meant by the term *overeating*. "Obesity," he wrote in *The Atlantic* in 1955, "it is flatly stated, comes from eating too much and that is all there is to it. Any attempt to search for causes deeper than self-indulgence can only give support to patients already seeking every possible means to evade their own responsibility."

But the trap that Stunkard and Mayer had identified is built into the logic of the positive-caloric-balance hypothesis; there is no escaping it. Mayer, as we've discussed, proceeded to insist in his book *Overweight*, as in all his writing, that obesity was the result of sedentary behavior, which simply implicated sloth rather than gluttony and still left the issue defined as a behavioral one. Although Mayer gets credit for convincing his peers that obesity has a genetic component, he implied that the only role of these genes was to make us want to be more or less sedentary. By the end of *Overweight*, Mayer was insisting not only that the obese must exercise more, but that they must also try harder to eat less. "Obesity is not a sin," he wrote. "At most, it is the consequence of errors of omission, the result of not having kept up the life-long battle against an inherited predisposition and against an environment which combines constant exposure to food with the removal of any need to work for it physically. In the pilgrim's progress of the constitutionally plump, salvation demands more than the shunning of temptation. It requires . . . the adoption of an attitude almost

---

made when discussing physical activity: If we sit around all day watching television, we're condemned as couch potatoes, and our obesity is only a matter of time. If we sit around studying or reading books, this same accusation is rarely voiced.

stoic in its asceticism and in the deliberate daily setting aside of time for what will be often lonely walking and exercising."

Stunkard became a leading authority in the study of behavioral therapy for obesity, which can be defined as a system of behavioral techniques by which obese patients might come to endure semi-starvation, while avoiding the explicit judgment that they achieved their obesity because they lacked willpower or had a defect of character. For instance, they eat too fast, or they are overly responsive to the external cues of their environment that tell them to eat, while being unresponsive to the internal cues of satiation, as one popular theory of the early 1970s had it. "Fat Americans: They Don't Know When They're Hungry, They Don't Know When They're Full," as a *New York Times* headline suggested in 1974. By that time, obesity, like anorexia, was categorized as an eating disorder, and the field of obesity therapy had become a subdiscipline of psychiatry and psychology. All these behavioral therapies, call them what you may, were in fact aimed at correcting failures of will. Every attempt to treat obesity by inducing the obese to eat less or exercise more is a behavioral treatment of obesity, and implies a behavioral-psychological cause of the condition.

Even if we accept that obese individuals are possessed of a defective character, then we're still left in the dark. Why doesn't the same defect—"the combination of weak will and a pleasure seeking outlook upon life," said Louis Newburgh—cause obesity in everyone? "It exists in many non-obese individuals as well," observed Hugo Rony; "in some of these it leads to chronic alcoholism, or drug addiction, others may become gamblers, playboys, prostitutes, petty criminals, etc. Evidently, such mental makeup, in itself, is not conducive to obesity. Those who do become obese apparently have something additional to and independent from this mental makeup: an intrinsic tendency to obesity."

If we can believe that people become obese because they simply ignored the fact that they were getting increasingly fatter, year in and year out, with the passive accumulation of excess calories, and that by the time they noticed it was either too late to do anything about it or they *really* didn't care (despite claims they might make to the contrary); if we can believe that obese individuals fail to survive indefinitely on semi-starvation diets because they are gluttonously unwilling to forgo temptation and so prefer, consciously or unconsciously, obesity to a life of moderation, then, as Stunkard observed in 1959, the matter is settled. Our job is done. But, of course, it isn't.

The more thoughtful analyses of obesity over the years have inevitably taken a more empathic view of those who suffer from it. They posit that there is no scientifically justifiable reason—or evidence—to assume that the obese are any more defective in character or behavior than you or I. Eric Ravussin, a diabetologist and metabolism researcher who began studying obesity among the Pima in 1984, has reported that Pima men who gained excessive weight—more than twenty pounds—over the course of a three-year study had a significantly lower basal-metabolic rate before their weight gain than men who remained relatively lean. (This same observation, as Ravussin points out, was made in infants: those who are heavier at one year of life have abnormally low daily energy expenditures when they are three months old.)* This suggests a *constitutional* difference in these individuals; it would be difficult to explain it in terms of sloth and a weak character. As a result, Ravussin questioned the logic and implications of the positive-caloric-balance hypothesis. "If obesity was only caused by an excessive appeal for food," Ravussin asked in a 1993 article in the journal *Diabetes Care*, "how can we explain the complete failure of treating it with behavioral therapies? Can we really believe that so many obese patients are liars and are cheating their doctors? How many more times do we need to demonstrate the high rate of recidivism among obese patients after weight loss to persuade others that unwanted metabolic forces contribute significantly to the causes of obesity in man?"

When Ravussin was interviewed more recently, he insisted that overeating and sedentary behavior could not explain the prevalence of obesity and diabetes in modern societies, and particularly not in the Pima. "I was shocked when I went to work with the Pima and I saw the amount of suffering in this population," he said. "It's not fun to see your mother [having a limb] amputated when she's thirty-two or thirty-five because she's had poorly controlled diabetes for twenty years. There's not a population in the world as aware as the Pima of the damages of diabetes and obesity. They know that. They are told from the age of two to avoid it, and still they cannot make it."

Hilde Bruch is now given credit for initiating a "revolution in thinking about childhood obesity"—doing "the first systematic investigation of the inner compulsions of the fat person," as the *New York Times* reported in 1950—and so purportedly demonstrating that its roots are not physiologi-

---

* These observations do not contradict Magnus-Levy's. Magnus-Levy compared lean and obese subjects. These latter observations compare those who gain weight to those who don't; this difference, as we'll see, is critical.

cal but behavioral. Indeed, Bruch may be the person most responsible for initiating the belief that obesity is an "eating disorder," and thus sending several generations of psychiatrists and psychologists off to work with obese patients. Yet, ironically, Bruch never embraced this conclusion herself and always considered the primary underlying cause of obesity to be metabolic and/or hormonal.

Despite Bruch's research linking childhood obesity to overeating and pathologies in the mother-child relationship, she was all too aware that her own research had failed to establish what was cause and what was effect. Her research had been uncontrolled, she noted, because she had studied *only* obese children and their families. "The literature on behavior disorders in childhood abounds with references to maternal rejection and overprotection," she explained. There was no way to know whether what she had discovered about her obese subjects actually played a major role in the development of obesity. It was also possible that the children had a predisposition to fatten and that this affected the children's desire to eat to excess, which in turn affected the family dynamics and how the families treated the children. What appeared to be a cause could in fact be an effect. "Life situations and emotional experiences of this kind," Bruch wrote, "provoke increased desire for food only in a certain type of person and result in obesity only when such a person has a special tendency to store fat in larger amounts than others and does not increase the energy expenditure correspondingly."

After publishing her observations on childhood obesity, Bruch put aside her clinical practice temporarily to study psychiatry, in the hope of helping these children. Through the early 1960s, she practiced psychiatry in New York, and then took a position as professor of psychiatry at Baylor College of Medicine in Houston. Throughout this period, she continued to specialize in anorexia and obesity. In 1957, with the publication of *The Importance of Overweight*, she was still questioning the role of psychological factors in obesity. (She described the book as a "critical re-evaluation and reintegration" of the obesity literature, including her own research.) Bruch couldn't escape the fact that restricting calories failed to bring obesity under control for any extended period of time, and she was simply unwilling to blame such consistent failure on her patients or their upbringing. "The efficacy of any treatment of obesity can be appraised only by the permanence of the result," Bruch stated. "When I began to work with obese children," she wrote, "I was impressed by the seeming ease with which some were able to lose weight once I had gained their co-operation. Having followed such cases over twenty years, I am today even more impressed by the speed with which they will regain the lost weight and by the tenacity with which they maintain their weight at an individually characteristic high level. It is possible to force the weight below this individual level, but such efforts are usually short-lived."

That obese people overeat, at least during periods of weight increase, Bruch said, had been "adequately established." What she disagreed with was what had now become, thanks in good part to her own research, the conventional interpretation of this observation: that overeating is the cause of obesity, and that the logical treatment is underfeeding. "In the course of my observations," she noted, "studying many obese people in great detail and following them over a long period of time, I have come to the conclusion that . . . overeating, though it is observed with great regularity, is not the cause of obesity; it is a symptom of an underlying disturbance. . . . Food, of course, is essential for obesity—but so is it for the maintenance of life in general. The *need* for overeating and the *changes* in weight regulation and fat storage are the essential disturbances."

In 1973, when Bruch published *Eating Disorders: Obesity, Anorexia Nervosa, and the Person Within*, she was still struggling with this conflict between psychological and physiological factors in the development of obesity. She acknowledged the need to prescribe reducing diets, and much of her analysis focused on those interpersonal and familial relationships that might contribute to obesity and dietary failure. Yet, she could not escape the suspicion, implied by a growing body of research, that the cause of obesity is a "primary metabolic or enzymatic disorder." And she acknowledged that it was still up to researchers to unambiguously identify the nature of the disorder. "Studies of human obesity," she wrote, "are not yet able to differentiate between factors that are the cause of obesity, or the result of it."

Chapter Seventeen

CONSERVATION OF ENERGY

The complicated mechanism of the body must be taken into consideration, and the ways it takes to reach its goals are not always the straight paths envisioned in our calculations.

MAX RUBNER, *The Laws of Energy Conservation in Nutrition*, 1902

B EFORE WORLD WAR II, the proposition that obesity was caused by overeating—the positive-caloric-balance hypothesis—was one of several competing hypotheses to explain the condition. After Hilde Bruch reported that obese children ate immoderately, and Louis Newburgh insisted that a perverted appetite was the fundamental cause of obesity, the positive-caloric-balance hypothesis became the conventional wisdom, and the treatment of obesity, as Jean Mayer observed, became the provenance of psychiatrists, psychologists, and moralists whose primary goal was to rectify our dietary misbehavior. Any attempt to dispute the accepted wisdom was treated, as it still is, as an attempt to absolve the obese and overweight of the necessity to exercise and restrain their appetites, or to sell something, and often both.

This conviction that positive caloric balance causes weight gain is founded on the belief that this proposition is an incontrovertible implication of the first law of thermodynamics. "The fact remains that no matter what people eat, it is calories that ultimately count," as Jane Brody explained in the *New York Times*. "Eat more calories than your body uses and you will gain weight. Eat fewer calories and you will lose weight. The body, which is after all nothing more than a biochemical machine, knows no other arithmetic."

For fifty years, clinicians, nutritionists, researchers, and public health officials have used this logic as the starting point for virtually every discussion of obesity. Anyone who challenges this view is seen as willfully disregarding a scientific truth. "Let me state," said the Columbia University physiologist John Taggart in his introduction to an obesity symposium in the early 1950s, "that we have implicit faith in the validity of the first law of thermodynamics." "A calorie is a calorie," and "Calories in equals calories out," and that's that.

But it isn't. This faith in the laws of thermodynamics is founded on two misinterpretations of thermodynamic law, and not in the law itself. When these misconceptions are corrected, they alter our perceptions of weight regulation and the forces at work.

The first misconception is the assumption that an association implies cause and effect. Here the context is the first law of thermodynamics, the law of energy conservation. This law says that energy is neither created nor destroyed, and so the calories we consume will be either stored, expended, or excreted. This in turn implies that any change in body weight must equal the difference between the calories we consume and the calories we expend, and thus the positive or negative energy balance. Known as the energy-balance equation, it looks like this:

Change in energy stores = Energy intake − Energy expenditure

The first law of thermodynamics dictates that weight gain—the increase in energy stored as fat and lean-tissue mass—will be *accompanied by* or *associated with* positive energy balance, but it does not say that it is *caused* by a positive energy balance—by "a plethora of calories," as Russell Cecil and Robert Loeb's 1951 *Textbook of Medicine* put it. There is no arrow of causality in the equation. It is equally possible, without violating this fundamental truth, for a change in energy *stores,* the left side of the above equation, to be the driving force in cause and effect; some regulatory phenomenon could drive us to gain weight, which would in turn cause a positive energy balance—and thus overeating and/or sedentary behavior. Either way, the calories in will equal the calories out, as they must, but what is cause in one case is effect in the other.

All those who have insisted (and still do) that overeating and/or sedentary behavior *must* be the cause of obesity have done so on the basis of this same fundamental error: they will observe correctly that positive caloric balance must be *associated* with weight gain, but then they will assume without justification that positive caloric balance is the *cause* of weight gain. This simple misconception has led to a century of misguided obesity research.

When the law of energy conservation is interpreted correctly, either of two possibilities is allowed. It may be true that overeating and/or physical inactivity (positive caloric balance) can cause overweight and obesity, but the evidence and the observations, as we've discussed, argue otherwise.

The alternative hypothesis reverses the causality: we are driven to get fat by "primary metabolic or enzymatic defects," as Hilde Bruch phrased it, and this fattening process induces the compensatory responses of overeating and/or physical inactivity. We eat more, move less, and have less energy to expend because we are metabolically or hormonally driven to get fat.

In 1940, Hugo Rony, former chief of the endocrinology clinic at Northwestern University's medical school, discussed this reverse-causation problem in a monograph entitled *Obesity and Leanness*, which is easily the most thoughtful analysis ever written in English on weight regulation in humans.* Rony's goal, as he explained it, was to "separate recognized facts from suggestive evidence, and reasonable working hypothesis from mere speculations." This set Rony apart from Louis Newburgh, Jean Mayer, and others who were more interested in convincing their colleagues in the field that their speculations were correct.

When Rony discussed positive energy balance, he compared the situation with what happens in growing children. "The caloric balance is known to be positive in growing children," he observed. But children do not grow because they eat voraciously; rather, they eat voraciously because they are growing. They require the excess calories to satisfy the requirements of growth; the result is positive energy balance. The growth is induced by hormones and, in particular, by growth hormone. This is the same path of cause and effect that would be taken by anyone who is driven to put on fat by a metabolic or hormonal disorder. The disorder will cause the excess growth—horizontal, in effect, rather than vertical. For every calorie stored as fat or lean tissue, the body will require that an extra calorie either be consumed or conserved. As a result, anyone driven to put on fat by such a metabolic or hormonal defect would be driven to excessive eating, physical inactivity, or some combination. Hunger and indolence would be side effects of such a hormonal defect, merely facilitating the drive to fatten. They would not be the fundamental cause. "Positive caloric balance may be regarded as the cause of fatness," Rony explained, "when fatness is artificially produced in a normal person or animal by forced excessive feeding or forced rest, or both. But obesity ordinarily develops spontaneously; some intrinsic abnormality seems to induce the body to establish positive

---

* *Obesity and Leanness* was the first serious book on obesity published after 1900, when von Noorden published *Die Fettsucht*. In the years since, there have been only half a dozen similar attempts (out of the innumerable professional texts and proceedings now available) to present a comprehensive and balanced analysis of the evidence, and only three come close to *Obesity and Leanness* in critical analysis—the chapters on obesity and undernutrition in the 1933 English translation of Eric Grafe's *Metabolic Diseases and Their Treatment*, Hilde Bruch's *Importance of Overweight*, and, a distant fourth, John Garrow's *Energy Balance and Obesity in Man*.

caloric balance leading to fat accumulation. Positive caloric balance would be, then, a result rather than a cause of the condition."

An obvious example of this reverse causation would be pregnant women, who are driven to fatten by hormonal changes. This hormonal drive induces hunger and lethargy as a result. In the context of evolution, these expanded fat stores would assure the availability of the necessary calories to nurse the infants after birth and assure the viability of the off-spring. The mother's weight loss after birth may also be regulated by hormonal changes, just as it appears to be in animals.

What may be the single most incomprehensible aspect of the last half-century of obesity research is the failure of those involved to grasp the fact that both hunger and sedentary behavior can be driven by a metabolic-hormonal disposition to grow fat, just as a lack of hunger and the impulse to engage in physical activity can be driven by a metabolic-hormonal disposition to burn calories rather than store them. Obesity researchers will immediately acknowledge that height, and thus the growth of skeletal bones and muscle tissue, is determined by genetic inheritance and driven by hormonal regulation, and that this growth will induce the necessary positive caloric balance to fuel it. But they see no reason to believe that a similar process drives the growth of fat tissue. What they believe is what they were taught in medical school, which was and is the conventional wisdom: the growth of skeletal muscle and bones, and thus our height, is driven by the secretion of growth hormone from the pituitary gland; the growth of fat tissue, and thus our girth, is driven by eating too much or physical inactivity.

This notion that fattening is the cause and overeating the effect, and not vice versa, also explains why a century of researchers have made so little progress, and why they keep repeating the same experiments over and over again. By this logic, those who become obese have a constitutional tendency to fatten, whereas those who remain lean have a constitutional tendency to resist the accumulation of fat. This tendency is the manifestation of very subtle deviations in metabolism and hormonal state. The obese have a constitutional predisposition to accumulate slight excesses of fat in their adipose tissue, which in turn induces compensatory tendencies to consume slightly more calories than the lean or expend slightly less. Obese individuals will put on fat until they have counterbalanced the influence of this underlying disorder. Eventually, these individuals achieve energy balance—everyone does—but only at an excessive weight and with an excessive amount of body fat.

The essential question, then, is: what are the metabolic and hormonal deviation that drives this fattening process? When we have that answer, we will know what causes obesity.

For the past half-century, obesity researchers have focused on a different question: establishing the characteristics that distinguish fat people from lean. Do fat people expend less energy? Do they consume more? Are they aware of how much they're eating? Are they less physically active? Is their metabolism slower? Are they more or less insulin-sensitive? All of these address factors that may be *associated* with the condition of being obese, but none address the question of what causes it initially.

Even if it could be established that all obese individuals eat more than do the lean—which they don't—that only tells us that eating more is associated with being obese. It tells us nothing about what causes obesity, because it doesn't tell us why the obese don't respond to an increase in food intake by expending more energy. After all, this must be the case when a lean person has a healthy appetite. "The statement that primary increase of appetite may be a cause of obesity does not lead us very far," Rony explained, "unless it is supplemented with some information concerning the origin of the primarily increased appetite. . . . What is wrong with the mechanism that normally adjusts appetite to caloric output? What part of this mechanism is primarily disturbed . . . ?"

Slightly more relevant are *prospective* studies, in which a population of individuals is observed to determine what distinguishes those who go on to become obese from those who don't. These studies, however, also fail to establish cause and effect. Such studies have repeatedly demonstrated that those who are *pre-obese* expend less energy—even at the age of three months—than those who will remain lean, which means that the low energy expenditure is a *risk factor* for obesity. This suggests that the pre-obese do indeed have a retarded metabolism, as von Noorden suggested, but it does not imply that relatively low energy expenditure causes obesity, only that it is associated with the condition of being pre-obese, and perhaps facilitates the drive to become obese.

As we've discussed, obesity is associated with all the physiological abnormalities of metabolic syndrome and all the attendant chronic diseases of civilization. For this reason, public-health authorities now assume that obesity causes or exacerbates these conditions. The alternative logic, with the causality reversed, implies a different conclusion: that the same metabolic-hormonal disorder that drives us to fatten also causes metabolic syndrome and the attendant chronic diseases of civilization.

The second misinterpretation of the law of energy conservation inevitably accompanies the first and is equally unjustifiable. The idea that obesity is caused by the slow accumulation of excess calories, day in and day out, over years or decades, and the associated idea that it can be prevented by reductions in caloric intake and/or increases in physical activity, are both based on an assumption about how the three variables in the energy-balance equation—energy storage, energy intake, and energy expenditure—relate to each other. They assume that energy intake and energy expenditure are what mathematicians call *independent variables*; we can change one without affecting the other. "We cannot get away from the fact that, *given no change in physical activity* [my italics], increased food means increased weight," as John Yudkin phrased it in 1959. "Yet this simple expression of the laws of conservation of mass and of energy is still received with indignation by very many people." But Yudkin's purportedly inescapable truth included an assumption that may not be physiologically plausible: "given no change in physical activity." The question is whether one can actually change energy intake in a living organism without prompting compensatory changes in energy expenditure.

When Carl von Noorden suggested in 1900 that obesity could be caused by eating one extra slice of bread every day or climbing fewer flights of stairs, so that a few extra dozen calories each day would accumulate over a decade into tens of pounds, and when the USDA *Dietary Guidelines*, over a century later, evoked the same concept with the suggestion that "for most adults a reduction of 50 to 100 calories per day may prevent gradual weight gain," they were treating human beings as though they are simple machines. "There is only one trouble," as Hilde Bruch commented about von Noorden's logic—"human beings do not function this way."

If we consume an average of twenty-seven hundred calories a day, that's almost a million calories a year; almost twenty million calories consumed over the course of two decades—more than twenty-five tons of food. Maintaining our weight within a few pounds for twenty years requires that we adapt our food intake to our expenditure over that period with remarkable accuracy. It's all too easy, therefore, to imagine how a metabolic or hormonal defect might lead to obesity by inducing the slightest compensatory inclination to consume more calories than we expend, and why it would be so subtle as to go undetected by virtually any imaginable diagnostic technology. "It is conceivable," as Eugene Du Bois of Cornell University suggested seventy years ago in his classic textbook *Basal Metabolism in Health and Disease*, "that common obesity is the only manifestation of an endocrine disturbance . . . so slight that it upsets the balance of intake and output by less than 0.1 of 1 percent."

Less easy to imagine, though, is how anyone avoids this fate, particularly if we believe that the balancing of intake and expenditure is maintained not by some finely tuned regulatory system, one honed over a few million years of evolution to accomplish its task under any circumstances, but, rather, by our conscious behavior and our perspicacity at judging the caloric value of the foods we eat. Looked at this way, as Du Bois suggested, "there is no stranger phenomenon than the maintenance of a constant body weight under marked variation in bodily activity and food consumption."

In 1961, the Cambridge University physiologist Gordon Kennedy discussed the paradoxes of obesity and weight regulation in the context of two propositions that he described as "common sense rather than physiology." The first was that "there must be long-term regulation of energy balance." The second was that "there is no a priori reason why this balance should be maintained by control of appetite alone, since it depends as much on calorie expenditure as on calorie intake."

Like Kennedy, most researchers who studied metabolism and the science of bioenergetics and growth through most of the twentieth century assumed that energy balance must be regulated involuntarily, without conscious intent, and that the mechanisms that do so adapt both intake to expenditure and expenditure to intake. Our bodies work to minimize long-term fluctuations in energy reserves and maintain a stable body weight, and they do so, as with all our homeostatic systems, via what George Cahill of Harvard and Albert Renold of the University of Geneva in 1965 called "multiple metabolic control mechanisms." This idea evolved in the 1970s into the popular set-point hypothesis, that our bodies will defend a certain preferred amount of body fat against either an excess or a deficit of calories. It fell out of favor because it implied that neither calorie-restricted diets nor exercise would lead to long-term weight loss.

The fundamental assumption of this idea that body weight is regulated homeostatically is that energy intake and expenditure are very much *dependent* variables—that they are physiologically linked so that a change in one forces a corresponding change in the other—and it is energy storage that is determined biologically within a certain range set by the interaction between genetics and the environment. Now the same law of energy conservation that decrees that calories in equal calories out, tells us that any increase in energy expenditure will have to induce a compensatory increase in intake, and so hunger has to be a consequence. And any enforced decrease in intake will have to induce a compensatory decrease in expenditure—a slowing of the metabolism and/or a reduction in physical activity.

In the nineteenth century, Carl von Voit, Max Rubner, and their contemporaries demonstrated that this was indeed what happened, at least in animals. Francis Benedict, Ancel Keys, George Bray, Jules Hirsch, and others have demonstrated this in humans, showing that neither eating less nor exercising more will lead to long-term weight loss, as the body naturally compensates. We get hungry, and if we can't satisfy that hunger, we'll get lethargic and our metabolism will slow down to balance our intake. This happens whether we're lean or obese, and it confounds those authorities who recommend exercise and calorie restriction for weight loss. They operate on the assumption that the only adjustment to the caloric deficit created by either dieting or exercise will be a unilateral reduction in fat tissue. This would be convenient, but the evidence argues against it.

Among researchers who study malnutrition, as opposed to those whose specialty is obesity, these compensatory effects to caloric deprivation are taken for granted, as is the fact that hormones regulate this process. "Changes in . . . hormones such as insulin and glucagon* play an important role in this metabolic response to energy restriction," explains Prakash Shetty, director of the Nutrition Planning, Assessment and Evaluation Service of the United Nations' Food and Agriculture Organization. "These physiological changes may be considered as metabolic adaptations which occur in a previously well-nourished individual and are aimed at increasing the 'metabolic efficiency' and fuel supply of the tissues at a time of energy deficit." We should not be surprised that "dieting is difficult," as Keith Frayn of Oxford University says in his 1996 textbook, *Metabolic Regulation*. "It is a fight against mechanisms which have evolved over many millions of years precisely to minimize its effects. . . . As food intake drops, the level of thyroid hormone falls and metabolic rate is lowered. Food intake has to be reduced yet further to drop below the level of energy expenditure. Hunger mechanisms, including the feeling of an empty stomach, lead us to search for food. . . ."

Though the traditional response to the failure of semi-starvation diets to produce long-term weight loss has been to blame the fat person for a lack of willpower, Bruch, Rony, and others have argued that this failure is precisely the evidence that tells us positive caloric balance or overeating is not the underlying disorder in obesity. No matter what technique is used to achieve a caloric deficit, whether eating less or exercising more, it will only serve to induce hunger and/or a compensatory decrease in energy expenditure. These are the "usual symptoms resulting from reduced food

---

* A hormone also secreted by the pancreas that tends to counteract the effects of insulin.

intake," as Ancel Keys and his collaborators described them, and anyone will experience them, regardless of weight.

Obese patients who try to reduce their weight by semi-starvation, as Rony noted, will always be fighting what he called their "spontaneous impulses of eating and activity." Once they give in to these impulses, which is effectively preordained, they will get fat again. This is exactly what we would expect to see if obesity were merely a consequence of an underlying disorder, much as high blood sugar and glycosuria—i.e., sugar in the urine—are symptoms and consequences of diabetes. Consuming fewer calories can serve only to address the symptoms temporarily, just as with diabetes. It does not remove the underlying abnormality.

This is why the long-term failure of semi-starvation diets is significantly more informative about the true nature of obesity than is the short-term weight loss. This failure is an important "clue to the puzzle," as Bruch suggested in 1955. The obese, Bruch noted, "react exactly like normal people after starvation. They continue overeating." This drive to become fat can be inhibited or even temporarily reversed by restricting calories—just as a child's growth can be stunted by starvation or malnutrition—but in neither case will the caloric deprivation address the metabolic and hormonal forces at work.

Just as we will decrease energy expenditure in response to caloric deprivation, we will also increase expenditure in response to caloric surplus. This compensatory effect of overeating was also demonstrated in the late nineteenth century by Carl von Voit and Max Rubner, although they disagreed about the mechanisms at work. It has since been encapsulated in a German word, *Luxuskonsumption,* which means a spendthrift metabolism that wastes excess calories as heat or superfluous physical activity. The term was first used in this context in 1902 by the German physiologist R. O. Neumann, who spent three years studying how his own body weight responded to extended fluctuations in caloric intake. *Luxuskonsumption* was Neumann's explanation for the apparent disassociation between the calories he consumed and the ease with which he maintained his weight.

Through the first half of the twentieth century, this capacity for *Luxuskonsumption* was assumed to be a critical factor in the genesis of obesity or leanness. To borrow Gordon Kennedy's phrase, this seemed like common sense rather than physiology. "Food in excess of immediate requirements and not needed to replenish stores can be readily disposed of, being burnt up and dissipated as heat," wrote David Lyon and Sir Derrick Dunlop, clinicians at the Royal Infirmary of Edinburgh, in 1932. "Did this capacity not exist, obesity would be almost universal." And so the ability to burn up small excesses, they observed, on the order of a few hundred calories a day, is "well within the capacity of the ordinary person, but in the obese individual the power of flexibility is much less evident."

Investigators studying obesity argued about the same handful of studies on *Luxuskonsumption,* and then the subject went out of fashion with the general acceptance of Newburgh's argument that obesity is caused by a perverted appetite. "The idea that people burned off excess energy when overfed was regarded with great disfavor by respectable nutritionists," as the British clinician John Garrow later noted. "It was a story put about by charlatans to justify magic cures, or by self-indulgent obese people as a justification for their obesity." It experienced a renaissance in the 1960s, sparked by the British physiologist Derek Miller, who reported that young pigs fed a low-protein diet would consume five times as many calories as those fed a high-protein diet, and yet could burn off the excess so as not to gain weight. This led Miller to speculate that the pigs would eat until they satisfied their protein requirements, and while doing so would stay lean through this process of *Luxuskonsumption.** It was thought that the ability to burn off excess calories would be of particular survival advantage when confronted with a poor-quality diet, when excessive amounts of food had to be consumed to achieve a requisite amount of protein or essential vitamins or minerals. Miller's observations prompted the renewed interest in overfeeding experiments of the kind we discussed in the last chapter (page 272). The one consistent finding in these studies has been that individuals vary dramatically in response to prolonged and enforced gluttony. Some will fatten easily, and some will not. The conclusion, seemingly unavoidable, is that a critical variable in the facility with which we gain weight is whether we respond to superfluous calories by storing them away as fat and/or muscle or by converting them to heat and physical activity—i.e., *Luxuskonsumption.*

At least some of these excess calories are lost in the various chemical reactions required to digest and store the nutrients. Rubner referred to this as the heat generated by the "thermochemical tangle of breakdowns"

---

* This phenomenon led to the notion of low-protein diets for weight loss. Regrettably, the ability to burn off excess calories when consuming a protein-deficient diet appears to be specific to young animals, and maybe even young pigs. When researchers tried to replicate this result in other animals—rats, sheep, cattle, or even older pigs—they noted that the animals eating the lower-protein diet got considerably fatter. They had more fat and less muscle, even if they weighed the same as the control animals.

that occur during the process of digestion. Physicians measure basal or resting metabolism after a twelve-to-eighteen-hour fast because by then this *diet-induced thermogenesis* has played itself out. The protein in the diet, as Rubner discovered, dominates this effect. The more protein digested over the amount necessary to maintain tissues and organs, the greater the heat generation. It's what Rubner called the *specific dynamic effect* of protein that is usually invoked as the rationale to eat high-protein diets for weight loss; excessive calories lost as heat in the process of digesting and utilizing protein can't then be stored as fat or used for fuel.

As the external environment changes, though, our bodies change the manner in which they utilize this heat. Maintaining our bodies at a constant temperature (roughly 98.6°F) requires more energy when it's cold than when it's warm. More of the heat from this thermochemical tangle of breakdowns, as Rubner reported, will go to that purpose when it's cold, as it will when our energy reserves are low—when we're undernourished—and we need to conserve the biologically useful energy for other purposes. In short, we will put this heat to use when we need to conserve energy, and we will waste it when it might be to our benefit to avoid the accumulation of excess calories as fat.

The primary source of controversy today remains the question that Rubner and Voit disputed a hundred years ago: whether the excess calories consumed have to be dissipated entirely as heat, or whether they can also be used biologically. Rubner argued that the energy requirements of our cells are basically constant. Under some set requirement, determined by temperature among other factors, our cells will adjust by conserving energy. Anything greater, and the energy is wasted as heat. Voit believed that the metabolic rate of our cells responds to the fuel available. The more fuel, the more energy generated. According to Voit, overeating leads to an increase in the available energy for cells, tissues, and muscles, and so perhaps to what the clinical investigators studying obesity in the first half of the century called the "impulse to physical activity" or the "impulse to move." That feeling of restlessness, they believed, is the manifestation of cells and tissues, literally, having energy to burn.

Both interpretations suggest the same fundamental conclusion about how our bodies work. We have thrifty metabolisms when we are undernourished and so need to use efficiently every calorie we consume, and we have spendthrift metabolisms when we're overnourished, so as to avoid excessive weight gain and obesity. Our cells may have a certain maximal or ideal capacity for metabolizing nutrients, but the amount that they actually metabolize is ultimately determined by the quantity and perhaps the quality of the nutrients delivered in the circulation. This determination is made on a cellular and hormonal level, not a cognitive or conscious one.

This idea that energy expenditure increases to match consumption, and that the ability to do this differs among individuals, also serves to reverse the cause-and-effect relationship between weight and physical activity or inactivity. Lean people are more active than obese people, or they have, pound for pound, a higher expenditure of energy,* because a greater proportion of the energy they consume is made available to their cells and tissues for energy. By this conception, lean people become marathon runners because they have more energy to burn for physical activity; their cells have access to a greater proportion of the calories they consume to use for energy. Less goes to making fat. That's why they're lean. Running marathons, however, will not make fat people lean, even if they can get themselves to do it, because their bodies will adjust to the extra expenditure of energy, just as they would adjust to calorie-restricted diets.

Our propensity to alter our behavior in response to physiological needs is what the Johns Hopkins physiologist Curt Richter called, in a heralded 1942 lecture, "total self-regulatory functions." Behavioral adaptation is one of the fundamental mechanisms by which animals and humans maintain homeostasis. Our responses to hunger and thirst are manifestations of this, replenishing calories or essential nutrients or fluids. Physical activity, as Richter suggested, is another example of this behavioral regulation, in response to an excess or dearth of calories. "We may regard the great physical activity of many normal individuals, the play activity of children, and perhaps even the excessive activity of many manic patients, as efforts to maintain a constant internal balance by expending excessive amounts of energy," he explained. "On the other hand, the low level of activity seen in some apparently normal people, the almost total inactivity seen in depressed patients, again may be regarded as an effort to conserve enough energy to maintain a constant internal balance."

In 1936, when Eugene Du Bois published the third edition of his metabolism textbook, *Basal Metabolism in Health and Disease*, he described the system that accomplished the regulation of a stable body weight as it was then understood. How much we want to eat on any given day, Du Bois

---

* Although, as we discussed in Chapter 16, the *total* energy expenditure of obese individuals is likely to be greater, because they have, simply put, more pounds to expend energy and generate heat.

explained, is determined by how much we've depleted whatever our body considers the necessary reserves of protein, fat, and carbohydrates. If we then consume more calories than we need, the excess will either be burned off as heat or induce physical activity: "When well nourished, the individual tends to become more energetic and it is quite possible that he will soon burn up his stored fat by extra work or exercise which would not have been undertaken had it not been for the overfeeding." If we consume less food than we might require to replenish our reserves, then the amount of heat generated in response to a meal is minimized, and the stores of carbohydrates (glycogen), fat, and protein are used to make up the difference. Should the caloric deficit continue, the result is "a gradual lowering of metabolism and a tendency toward restriction of activities, due to a lack of energy and initiative."

However this homeostatic system works to balance energy intake and output and thus maintain a steady supply of fuel to the cells and a stable body weight, it is extraordinarily complex and involves the entire body. Rony discussed this: "The appetite mechanism, which is but a part, although the most important one, of body weight regulation is in itself a highly complex mechanism involving [the central nervous system], endocrine glands, the gastric neuro-muscular apparatus, and the organs of the glycogen, protein, and fat reserves." This notion was supported by a host of experimental and clinical studies, as we'll discuss in Chapter 21, which demonstrated that disturbances in body-weight regulation—like obesity—could be caused by "pathological changes in certain parts of the nervous system, endocrine system and depot organs."

It is also vital to understand that it's our cells and tissues that require and expend the energy we consume, so this adjustment of intake to expenditure is occurring first and foremost on a cellular level. "Whatever may be the mechanisms controlling food intake," as the University of California, Berkeley, nutritionist Samuel Lepkovsky wrote in 1948, "the chief site of their action must be the cell." A fundamental requirement of any living organism is to provide a steady and reliable supply of fuel to its cells, regardless of the circumstances. We apparently evolved an intricate and extraordinarily robust regulatory system of hormones, enzymes, and the nervous system to accomplish this task. If the necessary fuel fails to reach the cells, the body compensates. The crucial factor is not how much is eaten—how many calories are consumed—or how much is expended, but how those nutrients or the energy they contain is ultimately distributed, how those calories are utilized and made available when needed. It's not the energy balance that is driving this system, but the distribution of that energy, the demand for energy at the cellular level.

Chapter Eighteen

FATTENING DIETS

Oversupply of food does not necessarily produce excessive nutrition. The appropriation depends in part on the character of the food and the ease or difficulty with which it is converted into a condition suitable for absorption, in part on such extrinsic and intrinsic influences as heredity, age, sexual and psychical habits, exercise and sleep; but to a great extent also on personal peculiarities of the metabolic processes. . . .

JAMES FRENCH, The Practice of Medicine, *1907*

I N 1857, JOHN HANNNING SPEKE AND Richard Burton set off through West Africa to search for the source of the Nile River. After Burton fell ill, Speke discovered the river's origin on his own. When he returned to the region five years later, according to his memoirs, he heard about the custom of local Abyssinian nobility to fatten up their wives to "such an extent that they could not stand upright." He went to see for himself. "There was no mistake about it," he recalled. "On entering the hut I found the old man and his chief wife sitting side by side on a bench. . . . I was struck with the extraordinary dimensions, yet pleasing beauty, of the immoderately fat fair one his wife. She could not rise; and so large were her arms that, between the joints, the flesh hung down like large, loose stuffed puddings." Two weeks later, when Speke visited "another one of those wonders of obesity," he took the opportunity to measure her. Her chest was fifty-two inches around. Her arms were nearly two feet in circumference and her thighs over two and a half feet.

With the notable exception of the current prevalence of obesity in Western societies, there is little reason to believe that fattening up the constitutionally lean is any easier than inducing leanness in the obese. For successful fattening, the excess calories consumed have to be stored as fat, rather than expended in metabolism or physical activity or stored as muscle. This isn't a given, considering these alternative uses for the calories. Continuing to consume excess calories is necessary, too—the person being fattened has to continue eating long after becoming sated—and these calories also have to be stored as fat.

In the early 1970s, the British physician John Garrow attempted to add twelve hundred calories a day to his daily diet, hoping to sustain it for a hundred days. After failing with several methods, he found that he could accomplish his goal by keeping chocolate biscuits on hand and, "whenever the prospect didn't seem too revolting, eating however many of these biscuits that I could." He managed to gain fifteen pounds in sixty days and then gave up the experiment and lost the weight in fifty days. "I learned that for me it is difficult to move my weight at all rapidly in any direction," he said, "and I saw absolutely no reason to suppose that obese people would find it easier than I did."

Various foods have been used to induce extensive fattening. The tribes that Speke visited relied on milk to fatten their women. In the mid-1970s, the French ethnologist Igor de Garine documented two male fattening sessions of the Massa tribe of northern Cameroon. In an individual ritual, the man ingests both milk and a porridge made from sorghum, a cornlike grain that provides, like sugarcane, a sweet syrup from the stalk. In 1976, Garine reported, one Massa tribesman gained seventy-five pounds on this ceremonial binge, apparently averaging ten thousand calories a day throughout. In a group fattening ritual, the men consume thirty-five hundred calories a day, rather than their usual twenty-five hundred, the excess consiting of milk and porridge. The weight gain tends to be fifteen to twenty pounds. The Massa are cattle herders, and their staple diet is primarily milk. This fattening, therefore, is accomplished by the addition of carbohydrates almost exclusively—one thousand to seventy-five hundred calories a day of sorghum.

The sumo wrestlers of Japan, whose weight commonly exceeds three hundred pounds, typically reach that level by their early twenties. In 1976, a University of Tokyo collaboration, led by Tsuneo Nishizawa, published an article in *The American Journal of Clinical Nutrition* that still constitutes the most comprehensive analysis in the English medical literature of the sumo diet, body composition, and health. The world of professional sumo wrestling, according to Nishizawa, is divided into an "upper group," constituting the best wrestlers in the country, and a "lower group." The members of the upper group consumed on average some fifty-five hundred calories' worth of *chanko nabe* (a pork stew) a day, out of which 780 grams were carbohydrates, 100 grams fat and 365 grams protein. This constituted more than twice the calories and carbohydrates of the typical Japanese diet of the era,* slightly less than twice the fat, and four and a half

---

* "The mean diet for Japanese people," Nishizawa et al. reported, citing a 1972 survey by the Ministry of Health and Welfare, "consists of 359 g of carbohydrate, 50.1 g of fat, 82.9 g of protein and a total of 2,279 calories."

times the amount of protein. The sumo diet was very high in carbohydrate by our standards—57 percent of the calories—and very low in fat—16 percent—considerably beneath what most public-health authorities in America consider a feasible low-fat target. The lower group of sumo weighed as much as their more accomplished colleagues, but were significantly fatter and less muscular. They consumed, on average, only 5,120 calories of *chanko nabe* a day, consisting of 1,000 grams of carbohydrates, 165 grams of protein, and only 50 grams of fat; these lesser sumo attained and maintained their corpulence on a diet of nearly 80 percent carbohydrate calories and 9 percent fat.

It seems that if we wanted to design a diet capable of inducing pathological obesity in young men in their prime, we might start with just such a very low-fat, high-carbohydrate diet. The diet would provide an enormous amount of calories, which might be the salient factor, but we would have to wonder what it is about this dietary composition that allows for such extraordinary overconsumption, not just for a few days, but for years or perhaps decades.

For the past quarter century, public-health authorities and obesity researchers have insisted that it is dietary fat, not carbohydrates, that fattens most effectively and causes obesity. This is why low-fat, low-calorie diets are recommended for weight loss as well as prevention of heart disease. This notion is based on four pieces of evidence, all of which are easily challenged.

The one that has been most influential is the association between heart disease, obesity, and diabetes. If heart disease is caused by high-fat diets, as is commonly believed, then so are obesity and diabetes, since these diseases appear together in both individuals and populations. But there is no evidence linking obesity to dietary-fat consumption, neither between populations nor in the same populations.* And, of course, if dietary fat is *not* responsible for heart disease, then it's unlikely that it plays a role in obesity and diabetes.

Second, laboratory rats will become obese on a high-fat diet. This is the evidence that convinced George Bray that excessive dietary fat would cause obesity in humans, too, and Bray has been among the most influential obesity authorities and the foremost proponent of this dietary-fat/obesity hypothesis. According to Bray, the rats used in his laboratory experiments

---

* "Obesity itself," as the National Academy of Sciences noted in 1989, "has not been found to be associated with dietary fat in either inter- or intra-population studies."

would grow reliably obese on high-fat diets. "I could feed them any kind of composition of carbohydrates I want," Bray said, "and in the presence of low fat, they don't get fat. If I raised the fat content, particularly saturated fat, in *susceptible* [my italics] strains I would get obesity regularly."

But some strains of rats, perhaps most of them, will *not* grow obese on high-fat diets, and even those that do will grow fatter on a high-fat, high-carbohydrate diet than a high-fat, low-carbohydrate diet. Moreover, to induce obesity even in susceptible rodents, the percentage of fat in the diet has to be greater than 30 percent, and usually closer to 40 or even 60 percent (which still makes only some strains of rats fat). Though 30 percent sounds like a low-fat or moderate-fat diet for humans, it's far greater than anything rats would normally consume, either in the wild or in the laboratory. It's what researchers will call a *pharmaceutical* dosage of fat. Rat chow is typically 2–6 percent fat calories. Rats will also fatten when fed large amounts of carbohydrates in the form of sugar. Moreover, other animals fatten on carbohydrates, including pigs—whose digestive apparatus is most like that of humans among experimental animals—cattle, and monkeys.

In the 1970s, Anthony Sclafani of Brooklyn College demonstrated that rats get "super obese" if allowed to freely consume a selection of foods from the local supermarket. This made their eating habits and subsequent obesity seem particularly like ours in character. But, as Sclafani explained, his rats fattened preferentially on sweetened condensed milk, chocolate-chip cookies, and bananas. Among the foods they didn't eat to excess were cheese, pastrami, and peanut butter—the items that were high in fat and low in carbohydrates.

The third supporting leg of the hypothesis that fat is particularly fattening is an assumption that the density of fat calories fools people into eating too many. Density was originally invoked to explain why some rats would eat fat to excess and become obese. Because the fat used in these experiments is typically an oil—Crisco cooking oil poured over the rat chow—it was hard to imagine that palatability was the deciding factor. As a result, researchers suggested that the density of the fat calories—nine calories per gram, compared with four for protein and carbohydrates—fooled the rats into consuming too many.

This was in line with the belief that we match our intake to expenditure by simple mechanisms such as those that limit the volume of food consumed in a single meal. It also led to the notion that eating fiber-rich, leafy vegetables will prevent weight gain by filling our stomachs with fewer digestible calories than if we consumed the densely packed calories of fat or refined carbohydrates. The more rigorous experiments with laboratory animals, however, suggest otherwise. The seminal experiments on this question were done by the University of Rochester physiologist Edward Adolph back in the 1940s. Adolph diluted the diets of his rats with water, fiber, and even clay, and noted that the rats would continue to eat these adulterated diets until they consumed the same amount of calories they had been eating when he had fed them unadulterated rat chow. The more Adolph diluted the chow with water, the more the rats consumed—until the meals were more than 97 percent water. At these very low dilutions, the rats apparently expended so much energy drinking that they couldn't consume enough calories to balance the expenditure. When Adolph put 90 percent of their daily calories directly into the rats' stomachs, "other food was practically refused for the remainder of the twenty-four hour period." Putting water directly into their stomachs had no such effect. Adolph's conclusion was that rats adjust their intake in response to caloric content, not volume, mass, or even taste, and this is presumably true of humans as well.

The fourth piece of evidence is thermodynamic. The idea dates to the late nineteenth century and its revival by the University of Massachusetts nutritionist J. P. Flatt was coincident with the rise of the dietary-fat/heart-disease hypothesis in the 1970s. According to Flatt's calculations, the "metabolic cost" of storing the calories we consume in adipose tissue—the proportion of energy dissipated in the conversion-and-storage process—is only 7 percent for fat, compared with 28 percent for carbohydrates. For this reason, when carbohydrates are consumed in excess, as the University of Vermont obesity researchers Ethan Sims and Elliot Danforth explained in 1987, the considerable calories expended in converting them to fat will "blunt the effect on weight gain of high-carbohydrate, high-caloric diets." High-fat diets, on the other hand, would lead "to a metabolically efficient and uncompensated growth of the fat stores." Flatt's analysis omitted all hormonal regulation of fuel utilization and fat metabolism (as well as a half-century's worth of physiological and biochemical research that we will discuss shortly) but it has nonetheless been invoked often during the last twenty years to make the point, as Sims and Danforth did, that obesity is yet another "penalty for living off the fat of the land rather than the carbohydrate."

Like much of the established wisdom on diet and health, this conclusion was based on very little experimental evidence. In this case, its only supporting evidence came from Sims's overfeeding studies. These began in the mid-1960s with four small trials that led to the observation that some people will gain weight easily and others won't, even when consuming the same quantity of excess calories. Another half dozen trials followed, each with only a handful of subjects, intending to shed light on what Sims and his collaborators called the "obvious question" of whether a carbohydrate-rich diet, independent of the calories consumed, could raise insulin levels, cause obesity, and induce hyperinsulinemia and insulin resistance. Sims and his collaborators varied the composition of the diets that their volunteers would then eat to excess. Some diets were "fixed carbohydrate" regimens, in which the amount of fat was increased as much as possible but the carbohydrates were limited to what the subjects would have normally consumed in their pre-experiment lives; others were "variable carbohydrate" regimens, in which both fat and carbohydrates were added to excess.

In the mid-1970s shortly after finishing their research, Sims and Danforth believed that obesity was most likely caused by chronically elevated levels of insulin, and that the elevated levels of insulin were likely the product of carbohydrate-rich diets. In the 1980s, their opinions changed and fell into step with the prevailing consensus on the evils of dietary fat. Sims and Dansforth now found in their decade-old results an observation that supported Flatt's argument that it was thermodynamically more efficient to fatten on fat than on carbohydrates. When excess calories were provided in the form of fat alone, they now explained, the subjects converted a greater proportion of the excess into body fat than when the excess calories included both fat and carbohydrates. "Simply stated, when taken in excess, fat is more fattening than carbohydrate," Danforth wrote in 1985. "Therefore, if one is destined to overeat and desires to suffer the least obesity, overindulgence in carbohydrate rather than fat should be recommended." "In view of these considerations and the tendency toward overnutrition in most affluent societies," he added, "main attention should be toward reducing both caloric and fat intake."

What the Vermont investigators failed to take into account, however, was their own previous observation that the nutrient composition of the diet seemed to affect profoundly the desire to consume calories to excess. One potentially relevant observation that Sims and his colleagues neglected to publish, for example, was that it seemed impossible to fatten up their subjects on high-fat, high-protein diets, in which the food to be eaten in excess was meat. According to Sims's collaborator Edward Horton, now a professor of medicine at Harvard and director of clinical research at the Joslin Diabetes Center, the volunteers would sit staring at "plates of pork chops a mile high," and they would refuse to eat enough of this meat to constitute the excess thousand calories a day that the Vermont investigators were asking of them. Danforth later described this regimen as the experimental equivalent of the diet prescribed by Robert Atkins in his 1973 diet book, *Dr. Atkins' Diet Revolution.* "The bottom line," Danforth said, "is that you cannot gain weight on the Atkins diet. It's just too hard. I challenge anyone to do an overfeeding study with just meat. You can't do it. I think it's a physical impossibility."

Getting their volunteers to add a thousand calories of fat to their daily diet also proved surprisingly difficult. Throughout their numerous publications, Sims and his colleagues comment on the "difficult assignment of gaining weight by increasing only the fat." Those fattening upon both carbohydrates and fat, on the other hand, easily added two thousand calories a day to their typical diet. Indeed, subjects in some of his studies, Sims and his colleagues reported, experienced "hunger late in the day . . . while taking much greater caloric excesses of a mixed diet"—as much as ten thousand calories a day.

Sims and his collaborators evidently did not wonder why anyone would lose appetite—develop "marked anorexia," as they put it—on a diet that includes eight hundred to a thousand excess fat calories a day, and yet feel "hunger late in the day" on a diet that includes six to seven thousand excess calories of fat and carbohydrates together. It would seem there is something about carbohydrates that allows the consumption of such enormous quantities of food and yet still induces hunger as the night approaches.

By perceiving obesity as an eating disorder, a defect of behavior rather than physiology, and by perceiving excessive hunger as the cause of obesity, rather than a symptom that accompanies the drive to gain weight, those investigators concerned with human obesity had managed to dissociate the perception of hunger and satiety from any underlying metabolic conditions. They rarely considered the possibility that hunger, satiety, and level of physical activity might be symptomatic of underlying physiological conditions. Imagine if diabetologists had perceived the ravenous hunger that accompanies uncontrolled diabetes as a behavioral disorder, to be treated by years of psychotherapy or behavioral modification rather than injections of insulin. These researchers simply never confronted the possibility that the nutrient composition of the diet might have a fundamental effect on eating behavior and energy expenditure, and thus on the long-term regulation of weight.

There is one way to test this latter notion, and, in fact, such tests were done from the 1930s onward. Alter the proportion of fats and carbohydrates in experimental diets and see what happens. Test low-fat diets versus low-carbohydrate diets, keeping in mind that a diet low in fat must be high in carbohydrates and vice versa. This would test the notion that these nutrients have unique metabolic and hormonal effects that influence weight, hunger or satiety, and energy expenditure. Such trials provide the means of answering these fundamental questions: What happens when we eat a diet restricted in carbohydrates, but not calories? Do we lose or gain weight? Are we as hungry as we are when calories are restricted? Do we eat more or less? Do we expend more or less energy? And what about when fat is restricted, but carbohydrates or calories are not? What are the effects on hunger, energy expenditure, and weight?

Chapter Nineteen

REDUCING DIETS

Concentrated carbohydrates, such as sugars and breadstuffs, and fats must be restricted. Diets, therefore, should exclude or minimize the use of rice, bread, potato, macaroni, pies, cakes, sweet desserts, free sugar, candy, cream, etc. They should consist of moderate amounts of meat, fish, fowl, eggs, cheese, coarse grains and skimmed milk.

> ROBERT MELCHIONNA of Cornell University,
> describing the reducing diet prescribed at
> New York Hospital in the early 1950s

THE AMERICAN HEART ASSOCIATION TODAY insists that severe carbohydrate restriction in a weight-loss diet constitutes a "fad diet," to be taken no more seriously than the grapefruit diet or the ice-cream diet. But this isn't the case. After the publication of Banting's *Letter on Corpulence* in 1863, physicians would routinely advise their fat patients to avoid carbohydrates, particularly sweets, starches, and refined carbohydrates, and this practice continued as the standard treatment of obesity and overweight through the better part of the twentieth century. Only after the AHA itself started recommending fat-restricted, carbohydrate-rich diets for heart disease in the 1960s and this low-fat prescription was then applied to obesity as well, was carbohydrate restriction forced to the margins. "In the instruction of an obese patient," as Louis Newburgh of the University of Michigan explained in 1942, "it is a simple matter to teach him to omit sugar because sweet flavors are not easily disguised. It is also relatively simple to teach him to limit the use of foods high in starch."

Those early weight-loss diets were meant to eliminate fat tissue while preserving muscle or lean-tissue mass. The protein content of the diet would be maximized and calories reduced. Only a minimal amount of carbohydrates and added fats—butter and oils—would be allowed in the diet, because these were considered the nonessential, i.e., nonprotein, elements. When physicians from the Stanford University School of Medicine described the diet they prescribed for obesity in 1943, it was effectively identical to the diet prescribed at Harvard Medical School and described in

1948, at Children's Memorial Hospital in Chicago in 1950, and at Cornell Medical School and New York Hospital in 1952. According to the Chicago clinicians, the "general rules" of a successful reducing diet were as follows:

1. Do not use sugar, honey, syrup, jam, jelly or candy.
2. Do not use fruits canned with sugar.
3. Do not use cake, cookies, pie, puddings, ice cream or ices.
4. Do not use foods which have cornstarch or flour added such as gravy or cream sauce.
5. Do not use potatoes (sweet or Irish), macaroni, spaghetti, noodles, dried beans or peas.
6. Do not use fried foods prepared with butter, lard, oil or butter substitutes.
7. Do not use drinks such as Coca-Cola, ginger ale, pop or root beer.
8. Do not use any foods not allowed on the diet and [for other foods use] only as much as the diet allows.

With the carbohydrates and added fats minimized in these diets, meat was inevitably the primary constituent. This would provide the protein necessary to ensure that weight loss came mostly from the patient's fat and not the muscle. The idea was to keep the body in what is called nitrogen equilibrium, with the nitrogen consumed from the protein in the diet balancing out the nitrogen being excreted in the urine from the breakdown of muscle protein.

When these clinicians discussed what plant foods they would allow in their diets, they typically did so on the basis of the carbohydrate content: potatoes are nearly 20 percent carbohydrate by weight (the rest is mostly water), so they were known as 20-percent vegetables. Green peas and artichokes are 15-percent vegetables. Onions, carrots, beets, and okra are 10-percent vegetables. Most of the green vegetables—including lettuce, cucumbers, spinach, asparagus, broccoli, and kale—are 5-percent, which means carbohydrates constitute at most 5 percent of their weight. These weight-loss diets allowed only 5-percent vegetables, which ruled out all starchy vegetables, like potatoes. Because a one-cup serving of a 5-percent vegetable will yield only twenty to thirty calories, as the University of Toronto physician Walter Campbell wrote in 1936, "the inclusion of an extra portion or omission of an undesired portion is of little moment in the [dietary] scheme as a whole." Some of these diets did allow an ounce or two of bread—usually whole-grain, because white bread had too few vitamins to make it worth including. But most did not. "All forms of bread contain a large proportion of carbohydrate, varying from 45–65 percent," noted H. Gardiner-Hill of London's St Thomas's Hospital Medical School in 1925, "and the percentage in toast may be as high as 60. It should thus be condemned."

When these physicians talked about lean meat as the basis of a weight-reducing diet, they did not mean a chicken breast without the skin, as has been the iconic example for the past twenty years. They meant any meat, fish, or poultry (bacon, salt pork, sausage, and duck occasionally excepted) in which the visible fat had been trimmed away.

Once weight was satisfactorily lost, weight-maintenance diets were also restricted in carbohydrates, although not so drastically. For maintaining a reduced weight, as described by the Pittsburgh physician Frank Evans in the 1947 edition of the textbook *Diseases of Metabolism*, the daily diet should include at least one egg, a glass of skimmed milk, a portion of raw fruit, "a generous portion of any cut of lean fresh lamb, beef, poultry or fish," and a portion of each of three 5-percent vegetables. Individuals trying to maintain their weight loss could then eat anything else they wanted, Evans wrote, but they could do so only as long as they maintained a stable weight and were sufficiently "sparing with" alcohol, added fats and oils, "concentrated carbohydrate foods," "starches," "mealy vegetables, which are potatoes, beans, peas," and "cereals, used as vegetables, which are: macaroni, spaghetti, rice, corn."

Evans provided one of the few variations on this regime that caught on as an obesity therapy in the years before World War II. This was a very low-calorie diet, of 360 to 600 calories a day, rather than the common prescription of 1,200 to 1,500 calories, then considered the minimal amount that a patient would tolerate and that would produce a safe and consistent weight loss. Evan's diet could induce a loss of up to five pounds a week, rather than the two pounds predicted for the more typical semi-starvation diets. The daily menu, explained Evans in 1929, was "composed of fresh meat and egg white. Approximately 100 [grams] of lean steak was the backbone of each of the two largest meals. When necessary, fresh fish was given at intervals." No starches or sugars were allowed, but the patient could eat a few ounces of 5-percent vegetables and one ounce of bread each day. These minimal carbohydrates—perhaps twenty grams—were included to "spare" the protein in the diet, so that it would be utilized for balancing out nitrogen losses rather than having some of it converted to glucose to fuel the brain and central nervous system. "The secret of the success of this procedure depends, almost certainly, on giving enough protein," Russell Wilder of the Mayo Clinic wrote after first prescribing the diet for his patients in 1931. Evans's very low-calorie diet may also have been popular because it appealed to the puritanical sense of those clinicians like Louis

Newburgh, who believed that gluttony had to be vigorously curbed in obese patients. One of the fundamental rules of Evans's diet was: "No concession to gustatory sensualism is permitted."

In the century before the medical community began prescribing fat-restricted, carbohydrate-rich diets for weight loss, one point of controversy was whether carbohydrates should be avoided because they are uniquely fattening or perhaps even cause obesity—as Jean Anthelme Brillat-Savarin and William Banting would have suggested—in which case they would be the only nutrient restricted, or because they constitute superfluous calories, in which case dietary fat was restricted as well, by avoiding oils, lard, and butter. "The next question to decide," wrote the Chicago physician Alfred Croftan in the *Journal of the American Medical Association* in 1906, "is whether the carbohydrates or the fats are to be chiefly restricted."

One observation made repeatedly through the 1960s was that the obese favor carbohydrates, and that these constitute the great proportion of all calories they consume. Though the obese did not appear to eat more calories, on average, than the lean, they did consume more carbohydrates. Such a dietary assessment was inevitably difficult to make with any accuracy, explained Sir Derrick Dunlop of the Royal Infirmary in Edinburgh, when he reported in 1931 on the lessons he had learned from treating 523 obese patients. Nonetheless, Dunlop believed that "obesity does occur in persons without showing any direct relationship to food intake, and that a certain group of patients do become overweight on an apparently normal well-balanced diet," and, second, "that an outstanding dietetic abnormality was an excessive intake of carbohydrate." "In some extreme cases," he noted, "the diet had consisted almost exclusively of sweet tea, white bread and scones."

This observation was echoed in *The Lancet* in 1935 by the British physician John Anderson, and in the 1940s by Hilde Bruch, Hugo Rony, and the Harvard physician Robert Williams and his colleagues, all of whom had questioned their fat patients extensively about their diets. Their common finding was an excessive consumption of starches and sweets. Rony reported that the craving for sweets and starches among his patients was so common that it suggested an underlying physiological mechanism at work, possibly related to a greater need for or reduced availability of glucose. "It is easier to induce the gluttonous obese to control his general appetite than to control his craving for sweets," Rony noted. One common rationale for restricting carbohydrates in weight-reducing diets was that it eliminated a disproportionately large share of the calories that the obese would normally eat.*

When carbohydrates are restricted, however, calories may also be cut— and the reverse is nearly always true. One of the revolutionary aspects of Frank Evans's very low-calorie diet was that it also restricted carbohydrates almost entirely.† When Louis Newburgh subsequently concluded that all obese patients can sustain a significant rate of weight loss for months or years if their diet is sufficiently draconian—as was the case with his patient who lost over 360 pounds—he was using Evans's very low-calorie, very low-carbohydrate diet to generate this weight loss. His patient lost the weight while eating at most a hundred calories of carbohydrates daily. It could have been the restriction of carbohydrates that was responsible for the weight loss. It could also have been the calorie restriction.

This same confounding of calories and carbohydrates might also ex-plain the success stories attributed to low-calorie diets—Albert Stunkard's one patient in a hundred, as he reported in 1959, who lost as much as forty pounds and managed to keep it off. It's effectively impossible to restrict calories significantly without also reducing the carbohydrates. Any calorie-restricted diet that restricts all calories equally, restricts carbohydrates, too. Even diets that preferentially reduce fat will have to reduce carbohydrates to achieve a significant reduction in calories (unless the dieters are willing to sacrifice protein in fish and meat, for instance, in order to avoid the fat that accompanies it). If dieters avoid sweets and snacks, and if they drink sugar-free soda but not regular soda, they're reducing their carbohydrate consumption significantly, and they're changing the type of carbohydrates they consume. Any benefit may be due to the calories reduced, or the car-bohydrates, or even just the relative absence of sugar.

Another issue that complicates this issue of calorie versus carbohydrate restriction is that the effect of weight-loss diets changes over time. The modest benefits of semi-starvation slowly diminish with time, as the calo-rie restriction induces a compensatory inhibition of energy expenditure. Moreover, much of the initial weight loss comes from losing water, not fat (see page 148). Because of this "tendency to retain water on a carbohydrate

* The Duke University pediatrician James Sidbury, Jr., who would go on to become director of the National Institute of Child Health and Human Development, made the same observation about the obese children he treated in the early 1970s: "A pattern of constant nibbling was consistently found. Most common snack foods are predominantly carbohydrate: crackers, potato chips, french fries, cookies, soft drinks, and the like."

† Evans's first test diets "called for no carbohydrate whatever"; only later did he settle on twenty grams of carbohydrates to address nitrogen balance.

diet and to give it out on a rich fat diet," as Dunlop described it, restricting carbohydrate calories specifically will induce a more dramatic and immediate loss of water. Testing diets for only a few weeks will demonstrate that carbohydrate-restricted diets induce weight loss at a greater rate than calorie- or fat-restricted diets, but whether they induce *fat* loss at a greater rate is a different question. "Changes in body-weight are to be taken, therefore, as of significance only when the experiment continues for a period of several weeks," as Francis Benedict cautioned in 1910. "Certainly, for short experiments, body-weight is for the most part wholly without significance."

For this reason, the first meaningful report on the efficacy of carbohydrate restriction for weight loss was one published in 1936, by Per Hanssen of the Steno Memorial Hospital in Copenhagen. Hanssen reported treating twenty-one obese patients over two years with an 1,850-calorie diet that contained only 450 calories of carbohydrates, or a little less than 25 percent. Nearly 60 percent of the calories came from fat: 65 grams of cream, 65 grams of butter, and 25 grams of olive oil every day, along with two eggs, cheese, and a liberal portion of meat or fish. Some of his patients, Hanssen reported, were so fat initially that they "could scarcely move when they arrived at the hospital, and were unable to work." On the diet from one to four months, the patients lost an average of two pounds a week. "During the stay in the hospital the patients never felt hungry," he reported. "The fatigue, a prominent and disturbing symptom, improved often very rapidly, and before the occurrence of any considerable reduction in weight." Hanssen compared his results with those reported five years earlier by physicians at the nearby University Clinic using a diet consisting of half the calories but twice the proportion of carbohydrates (over 50 percent). "At Steno Memorial Hospital," Hanssen noted, "a diet of 1,850 calories will reduce weight as quickly as a diet of 950 calories at the University Clinic of Copenhagen."

If obese individuals can lose weight and keep it off, without hunger, on a diet of 1,850 calories, it's a reasonable assumption that they will find it easier to sustain such a diet than one that allows only 950 calories, or even less, and assumes, as Evans put it, that the obese "should be hungry most of the time as this is normal." A diet "relatively poor in carbohydrates," Hanssen suggested, might "not be so difficult to adhere to as the diets commonly used."

What additionally complicates any assessment of the role of carbohydrate restriction in reducing diets is that the composition of a diet is never quite so simple as merely being high or low in carbohydrates or refined carbohydrates. Proteins, fats, and calories assume different roles depending on the diet. Also, carbohydrates in these diets can be restricted, but the standard thinking is that they have to remain sufficiently high so that the brain and central nervous system derive all their necessary fuel from this dietary source of glucose. Nutritionists will often insist that 130 grams a day of carbohydrates are the minimal safe amount in a human diet.

Though glucose is a primary fuel for the brain, it is not, however, the only fuel, and dietary carbohydrates are not the only source of that glucose. If the diet includes less than 130 grams of carbohydrates, the liver increases its synthesis of molecules called ketone bodies, and these supply the necessary fuel for the brain and central nervous system. If the diet includes no carbohydrates at all, ketone bodies supply three-quarters of the energy to the brain. The rest comes from glucose synthesized from the amino acids in protein, either from the diet or from the breakdown of muscle, and from a compound called glycerol that is released when triglycerides in the fat tissue are broken down into their component fatty acids. In these cases, the body is technically in a state called ketosis, and the diet is often referred to as a ketogenic diet. Whether the diet is ketogenic or anti-ketogenic—representing a difference of a few tens of grams of carbohydrates each day—might influence the response to the diet, complicating the question of whether carbohydrates are responsible for some effect or whether there is another explanation. (Ketosis is often incorrectly described by nutritionists as "pathological." This confuses ketosis with the ketoacidosis of uncontrolled diabetes. The former is a normal condition; the latter is not. The ketone-body level in diabetic ketoacidosis typically exceeds 200 mg/dl, compared with the 5 mg/dl ketone levels that are typically experienced after an overnight fast—twelve hours after dinner and before eating breakfast—and the 5–20 mg/dl ketone levels of a severely carbohydrate-restricted diet with only 5–10 percent carbohydrates.)

For fifty years after William Banting publicized William Harvey's prescription for a carbohydrate-restricted diet in 1863, the primary clinical disagreements were on the role of fat in the diet. Banting's original prescription was a high-fat diet, but then it was modified by Harvey himself and by the German clinicians Felix von Niemeyer and Max Oertel into lower-fat, higher-protein versions, and by Wilhelm Ebstein into a version featuring still more fat. "The fat of ham, pork or lamb is not only harmless but useful," Ebstein wrote.

The notion of a carbohydrate-restricted diet based exclusively on fatty meat was publicized after World War I by the Harvard anthropologist-turned-Arctic-explorer Vilhjalmur Stefansson, who was concerned with the overall healthfulness of the diet, rather than its potential for weight loss. Stefansson had spent a decade eating nothing but meat among the Inuit of northern Canada and Alaska. The Inuit, he insisted, as well as the visiting explorers and traders who lived on this diet, were among the healthiest if not the most vigorous populations imaginable.

Among the tribes with whom Stefansson lived and traveled, the diet was primarily caribou meat, "with perhaps 30 percent fish, 10 percent seal meat, and 5 or 10 percent made up of polar bear, rabbits, birds and eggs." The Inuit considered vegetables and fruit "not proper human food," Stefansson wrote, but they occasionally ate the roots of the knotweed plant in times of dire necessity.

The Inuit paid little attention to the plants in their environment "because they added nothing to their food supply," noted the Canadian anthropologist Diamond Jenness, who spent the years 1914–16 living in the Coronation Gulf region of Canada's Arctic coast. Jenness described their typical diet during one three-month stretch as "no fruit, no vegetables; morning and night nothing but seal meat washed down with ice-cold water or hot broth." (The ability to thrive on such a vegetable- and fruit-free diet was also noted by the lawyer and abolitionist Richard Henry Dana, Jr., in his 1840 memoirs of life on a sailing ship, Two Years Before the Mast. For sixteen months, Dana wrote, "we lived upon almost nothing but fresh beef; fried beefsteaks, three times a day . . . [in] perfect health, and without ailings and failings."

None of Stefansson's observations would have been controversial had not the conventional wisdom at the time been—as it is still—that a varied diet is essential for good health. A healthy diet, it is said, must contain protein, fats, and carbohydrates, the latter because of the misconception that the brain and central nervous system require dietary glucose to function, and the debatable assumption that fresh vegetables and fruit, which contain carbohydrates, are essential to prevent deficiency diseases.

Because it is still common to assume that a meat-rich, plant-poor diet will result in nutritional deficiencies, it's worth pausing to investigate this issue. The assumption dates to the early decades of the twentieth century, the golden era of research on vitamins and vitamin-deficiency diseases, as one disease after another—scurvy, pellagra, beriberi, rickets, anemia—was found to be caused by a lack of essential vitamins and minerals. This was The Newer Knowledge of Nutrition, as it was called by the Johns Hopkins nutritionist Elmer McCollum; it dictated that the only way to ensure all the essential elements for health was to eat as many types of foods as possible, and nutritionists still hold by this logic today. "A safe rule of thumb," as it was recently described, "is that the more components there are in a dietary, the greater the probability of balanced intake."

This philosophy, however, was based almost exclusively on studies of deficiency diseases, all of which were induced by diets high in refined carbohydrates and low in meat, fish, eggs, and dairy products. When the Scottish naval surgeon James Lind demonstrated in 1753 that scurvy could be prevented and cured by the consumption of citrus juice, for example, he did so with British sailors who had been eating the typical naval fare "of water gruel sweetened with sugar in the morning, fresh mutton broth, light puddings, boiled biscuit with sugar, barley and raisins, rice and currants." Pellagra was associated almost exclusively with corn-rich diets, and beriberi with the eating of white rice rather than brown. When beriberi broke out in the Japanese navy in the late 1870s, it was only after the naval fare had been switched from vegetables and fish to vegetables, fish, and white polished rice. The outbreak was brought under control by replacing the white rice with barley and adding meat and evaporated milk. Pellagra, too, could be cured or ameliorated, as Carl Voegtlin demonstrated in 1914, by adding fresh meat, milk, and eggs to a pellagra-causing diet, which in Voegtlin's experiments constituted primarily wheat bread, cabbage, cornmeal and corn syrup, turnips, potatoes, and sugar. Nutritionists working with lab animals also found that they could induce deficiency diseases by feeding diets rich in refined grains and sugar. Guinea pigs were given scurvy in a series of laboratory experiments in the 1940s when they were fed diets of mostly *crushed* barley and chickpeas.

This research informed the conventional wisdom of the era that fresh meat, milk, and eggs were what the Scottish nutritionist Robert McCarrison called "protective foods" (which is how they were known before Ancel Keys and his contemporaries established them as the fat-rich agents of coronary disease), but it also bolstered the logic that a "balanced" diet, with copious vegetables, fruits, and grains, was necessary for health. Because diets of mostly grains and starches, or diets of refined grains, fish, and vegetables, such as the Japanese sailors consumed, might be deficient in a vitamin or vitamins essential for health, nutritionists considered it a reasonable assumption that this might be true of any such "unbalanced" diets, including those that were made up exclusively of animal products.

What the nutritionists of the 1920s and 1930s didn't then know is that animal foods contain all of the essential amino acids (the basic structural building blocks of proteins), and they do so in the ratios that maximize their utility to humans.* They also contain twelve of the thirteen essential vitamins in large quantities. Meat is a particularly concentrated source of vitamins A, E, and the entire complex of B vitamins. Vitamins D and B$_{12}$ are found *only* in animal products (although we can usually get sufficient vitamin D from the effect of sunlight on our skin).

The thirteenth vitamin, vitamin C, ascorbic acid, has long been the point of contention. It is contained in animal foods in such small quantities that nutritionists have considered it insufficient and the question is whether this quantity is indeed sufficient for good health. Once James Lind demonstrated that scurvy could be prevented and cured by eating fresh fruits and vegetables, nutritionists assumed that these foods are an absolutely essential dietary source of vitamin C. What had been demonstrated, they will say, is that scurvy is "a dietary deficiency resulting from lack of fresh fruit and vegetables." To be technically accurate, however, Lind and the nutritionists who followed him in the study of scurvy demonstrated only that the disease is a dietary deficiency that can be cured by the addition of fresh fruits and vegetables. As a matter of logic, though, this doesn't necessarily imply that the lack of vitamin C is caused by the lack of fresh fruits and vegetables. Scurvy can be ameliorated by adding these to the diet, but the original lack of vitamin C might be caused by other factors. In fact, given that the Inuit and those Westerners living on the Inuit's vegetable- and fruit-free diet never suffered from scurvy, as Stefansson observed, then other factors *must* be involved. This suggested another way of defining a balanced diet. It's possible that eating easily digestible carbohydrates and sugars increases our need for vitamins that we would otherwise derive from animal products in sufficient quantities.

This was the issue that Stefansson was raising in the early 1920s. If the Inuit thrived in the harshest of environments without eating carbohydrates and whatever nutrients exist in fruits and vegetables, they, by definition, were consuming a balanced, healthy diet. If they did so solely because they had become evolutionarily adapted to such a diet, which was a typical rejoinder to Stefansson's argument, then how can one explain those traders and explorers, like Stefansson himself and the members of

* "Wheat contains all of the essential amino acids," explained the Columbia University nutritional anthropologist Marvin Harris, "but to get enough of the ones that are in scarce supply a man weighing 176 pounds (80 kilos) would have to stuff himself with 3.3 pounds (1.5 kilos) of whole wheat bread a day. To reach the same safe level of protein, he would need only .75 pounds (340 grams) of meat."

his expeditions, who also lived happily and healthfully for years at a time on this diet?

Nutritionists of the era assumed that all-meat diets were unhealthy because (1) excessive meat consumption was alleged to raise blood pressure and cause gout; (2) the monotony of eating only meat—or any other single food—was said to induce a physical sense of revulsion; (3) the absence of fresh fruit and vegetables in these diets would cause scurvy and other deficiency diseases, and (4) protein-rich diets were thought to induce chronic kidney damage, a belief based largely on early research by Louis Newburgh.

None of these claims were based on compelling evidence. Newburgh, for instance, had based his conclusions largely on experiments in which he fed excessive quantities of soybean, egg whites, and beef protein to rabbits, which, as critics would later observe, happen to be herbivores. Their natural diet is buds and bark, not their fellow animals, and so there was little scientific value in force-feeding them meat or animal protein. Nonetheless, the dangers of an all-meat diet were considered sufficiently likely that even Francis Benedict, as Stefansson told it, claimed that it was "easier to believe" that Stefansson and all the various members of his expeditions "were lying, than to concede that [they] had remained in good health for several years on an exclusive meat regimen."

In the winter of 1928, Stefansson and Karsten Anderson, a thirty-eight-year-old Danish explorer, became the subjects in a yearlong experiment that was intended to settle the meat-diet controversy. The experiment was planned and supervised by a committee of a dozen respected nutritionists, anthropologists, and physicians.* Eugene Du Bois and ten of his colleagues from Cornell and the Russell Sage Institute of Pathology would oversee the day-to-day details of the experiment.

For three weeks, Stefansson and Anderson were fed a typical mixed diet of fruits, cereals, vegetables, and meat while being subjected to a battery of tests and examinations. Then they began living exclusively on meat, at which point they moved into Bellevue Hospital in New York and were put under twenty-four-hour observation. Stefansson remained at Bellevue for three weeks, Anderson for thirteen weeks. After they were released, they continued to eat only meat for the remainder of one year. If they cheated on the diet, according to Du Bois, the experimenters would know it from

---

* These included Graham Lusk and Eugene Du Bois from Cornell and the Russell Sage Institute of Pathology; Russell Pearl and William McCallum from Johns Hopkins; the Harvard anthropologist Earnest Hooton; and Clark Wissler of the American Museum of Natural History.

regular examinations of Stefansson's and Anderson's urine. "In every individual specimen of urine which was tested during the intervals when they were living at home," Du Bois wrote, "acetone [ketone] bodies were present in amounts so constant that fluctuations in the carbohydrate intake were practically ruled out."

The experimental diet included many types of meat. To test the argument that the vitamins necessary in such a diet to avoid scurvy and remain healthy could be obtained only by eating raw meat, as was incorrectly assumed to be the practice of the Inuit, all of the meat was cooked. (In fact, the Inuit only occasionally ate raw meat.) Stefansson and Anderson each consumed an average of almost two pounds of meat per day, or twenty-six hundred calories: 79 percent from fat, 19 percent protein, and roughly 2 percent from carbohydrates (a maximum of fifty calories a day), which came from glycogen contained in the muscle meat. (Glycogen is the compound that stores glucose, a carbohydrate, in the liver and the muscle.)

"The only dramatic part of the study was the surprisingly undramatic nature of the findings," wrote Du Bois, when he later summarized the results. "Both men were in good physical condition at the end of the observation," he reported in 1930, in one of the nine articles he and his colleagues published on the study. "There was no subjective or objective evidence of any loss of physical or mental vigor." Stefansson lost six pounds over the course of the year, and Anderson three, even though "the men led somewhat sedentary lives." Anderson's blood pressure dropped from 140/80 to 120/80; Stefansson's remained low (105/70) throughout. The researchers detected no evidence of kidney damage or diminished function, and "vitamin deficiencies did not appear." Nor did mineral deficiencies, although the diet contained only a quarter of the calcium usually found in mixed diets, and the acidic nature of a meat-rich diet was supposed to increase calcium excretion and so deplete the body of calcium. Among the minor health issues reported by Du Bois and his colleagues was the observation that Stefansson began the experiment with mild gingivitis (inflammation of the gums), but this "cleared up entirely, after the meat diet was taken."

When Stefansson published Not by Bread Alone, a popular treatise on fat-and-protein diets, in 1946, a New York Times reviewer wrote, "Mr. Stefansson makes the mixed-diet technicians and the nuts-and-fruits addicts look terribly silly." Du Bois, who supervised the experiments, wrote an introduction to Stefansson's book. After Stefansson and Anderson were living exclusively on meat, he said, "a great many dire predictions and brilliant theories faded into nothingness." A diet that should have left Stefansson and Anderson deathly ill from scurvy had left them as healthy as or healthier than the balanced diet they had been eating in the years immediately preceding the study. "Quite evidently we must revise some of our text book statements," Du Bois concluded.

The textbook statements on vitamins would go unrevised, however, despite laboratory research that has confirmed Stefansson's speculations. Nutritionists would establish by the late 1930s that B vitamins are depleted from the body by the consumption of carbohydrates. "There is an increased need for these vitamins when more carbohydrate in the diet is consumed," as Theodore Van Itallie of Columbia University testified to McGovern's Select Committee in 1973. A similar argument can now be made for vitamin C. Type 2 diabetics have roughly 30 percent lower levels of vitamin C in their circulation than do nondiabetics. Metabolic syndrome is also associated with "significantly" reduced levels of circulating vitamin C, which suggests that vitamin-C deficiency might be another disorder of civilization. One explanation for these observations—described in 1997 by the nutritionists Julie Will and Tim Byers, of the Centers for Disease Control and the University of Colorado respectively, as both "biologically plausible and empirically evident"—is that high blood sugar and/or high levels of insulin work to increase the body's requirements for vitamin C.

The vitamin-C molecule is similar in configuration to glucose and other sugars in the body. It is shuttled from the bloodstream into the cells by the same insulin-dependent transport system used by glucose. Glucose and vitamin C compete in this *cellular-uptake* process, like strangers trying to flag down the same taxicab simultaneously. Because glucose is greatly favored in the contest, the uptake of vitamin C by cells is "globally inhibited" when blood-sugar levels are elevated. In effect, glucose regulates how much vitamin C is taken up by the cells, according to the University of Massachusetts nutritionist John Cunningham. If we increase blood-sugar levels, the cellular uptake of vitamin C will drop accordingly. Glucose also impairs the reabsorption of vitamin C by the kidney, and so, the higher the blood sugar, the more vitamin C will be lost in the urine. Infusing insulin into experimental subjects has been shown to cause a "marked fall" in vitamin-C levels in the circulation.

In other words, there is significant reason to believe that the key factor determining the level of vitamin C in our cells and tissues is not how much or little we happen to be consuming in our diet, but whether the starches and refined carbohydrates in our diet serve to flush vitamin C out of our system, while simultaneously inhibiting the use of what vitamin C we do have. We might get scurvy because we don't faithfully eat our fruits and vegetables, but it's not the absence of fruits and vegetables that causes the scurvy; it's the presence of the refined carbohydrates.* This hypothesis has not been proven, but, as Will and Byers suggested, it is both biologically plausible and empirically evident.

When we discuss the long-term effects of diets that might reverse or prevent obesity, we must not let our preconceptions about the nature of a healthy diet bias the science and the interpretation of the evidence itself.

* There are only four experiments in the medical literature, not including Stefansson and Anderson's, in which the goal was to induce scurvy in human subjects—in one, four, twenty, and four subjects respectively. In each case, the goal was accomplished and the diets were carbohydrate- and/or sugar-rich.

Chapter Twenty

UNCONVENTIONAL DIETS

Here was a treatment, that, in its encouragement to eat plentifully, to the full satisfaction of the appetite, seemed to oppose not only the prevailing theory of obesity but, in addition, principles basic to the biological sciences and other sciences as well. It produced a sense of puzzlement that was a mighty stimulant to thought on the matter.

ALFRED PENNINGTON, *talking about a high-fat,
high-protein diet, unrestricted in calories, in the
American Journal of Digestive Diseases,* 1954

Does it help people lose weight? Of course it does. If you cannot eat bread, bagels, cake, cookies, ice cream, candy, crackers, muffins, sugary soft drinks, pasta, rice, most fruits and many vegetables, you will almost certainly consume fewer calories. Any diet will result in weight loss if it eliminates calories that previously were overconsumed.

JANE BRODY, talking about a high-fat,
high-protein diet, unrestricted in calories,
in the *New York Times,* 2002

A. J. LIEBLING, THE CELEBRATED AUTHOR of *The New Yorker's* "On Press" column, once wrote that he had enunciated a journalistic truth with such clarity that it was suitable for framing. "There are three kinds of writers of news in our generation," Liebling wrote. "In inverse order of worldly consideration, they are:

1. The reporter, who writes what he sees.
2. The interpretive reporter, who writes what he sees and what he construes to be its meaning.
3. The expert, who writes what he construes to be the meaning of what he hasn't seen.

"To combat an old human prejudice in favor of eyewitness testimony," Liebling wrote, "the expert must intimate that he has access to some occult source or science not available to either reporter or reader. He is the Priest of Eleusis, the man with the big picture. . . . All is manifest to him, since his conclusions are not limited by his powers of observation."

Leibling was talking about journalism, but a similar ranking holds true in medicine. In fact, the medical experts have the further advantage that they can disseminate their opinions with considerably greater influence. They can make their case with the imprimatur of the institutions that employ them—the American Medical Association, for instance, or Harvard University. They can easily attract the media's attention. Physicians' case reports and the patients' anecdotal experience have a fundamental role in medicine, but if these conflict with what the experts believe to be true, the experts' opinions win out.

This conflict between expertise and observational evidence has had a significant influence in the science of obesity. Reliable eyewitness testimony has come only from those who have weight problems themselves, or the clinicians who regularly treat obese patients, and neither group has ever garnered much credibility in the field. (The very assumption that obesity is a psychological disorder implies that the obese cannot be trusted as reliable witnesses to their own condition.) But it is these individuals who have the firsthand experience. When Hilde Bruch reported in 1957 that a fine-boned girl in her teens, "literally disappearing in mountains of fat," lost nearly fifty pounds over a single summer eating "three large portions of meat" a day, it was easier for the experts to ignore the testimony as a freakish phenomenon than to contemplate how such a thing was possible. But the process of discovery in science, as the philosopher of science Thomas Kuhn has put it, only begins with the awareness that nature has violated our expectations. Often it is the unconventional events—the anomalous data, as these are called in science—that reveal the true nature of the universe.

In 1920, while Vilhjalmur Stefansson was just beginning his campaign to convince nutritionists that an all-meat diet was a uniquely healthy diet, it was already making the transition into a reducing diet courtesy of a New York internist named Blake Donaldson. Donaldson, as he wrote in his 1962 memoirs, began treating obese patients in 1919, when he worked with the cardiologist Robert Halsey, one of four founding officers of the American Heart Association. After a year of futility in trying to reduce these patients ("fat cardiacs," he called them) with semi-starvation diets, he spoke with the resident anthropologists at the American Museum of Natural History, who told him that prehistoric humans lived almost exclusively on "the fattest meat they could kill," perhaps supplemented by roots and berries. This led Donaldson to conclude that fatty meat should be "the

essential part of any reducing routine," and this is what he began prescribing to his obese patients. Through the 1920s, Donaldson honed his diet by
trial and error, eventually settling on a half-pound of fatty meat—three
parts fat to one part lean by calories, the same proportion used in Stefansson's Bellevue experiment—for each of three meals a day. After cooking,
this works out to six ounces of lean meat with two ounces of attached fat
at each meal. Donaldson's diet prohibited all sugar, flour, alcohol, and
starches, with the exception of a "hotel portion" once a day of raw fruit or a
potato, which substituted for the roots and berries that primitive man
might have been eating as well. Donaldson also prescribed a half-hour
walk before breakfast.

Over the course of four decades, as Donaldson told it, he treated seventeen thousand patients for their weight problems. Most of them lost two to
three pounds a week on his diet, without experiencing hunger. Donaldson
claimed that the only patients who didn't lose weight on the diet were those
who cheated, a common assumption that physicians also make about
calorie-restricted diets. These patients had a "bread addiction," Donaldson
wrote, in that they could no more tolerate living without their starches,
flour, and sugar than could a smoker without cigarettes. As a result, he
spent considerable effort trying to persuade his patients to break their
habit. "Remember that grapefruit and all other raw fruit is starch. You can't
have any," he would tell them. "No breadstuff means any kind of bread. . . .
They must go out of your life, now and forever." (His advice to diabetics
was equally frank: "You are out of your mind when you take insulin in
order to eat Danish pastry.")

Had Donaldson published details of his diet and its efficacy through the
1920s and 1930s, as Frank Evans did about his very low-calorie diet, he
might have convinced mainstream investigators at least to consider the
possibility that it is the quality of the nutrients in a diet and not the quantity of calories that causes obesity. As it is, he discussed his approach only
at in-house conferences at New York Hospital. Among those who heard of
his treatment, however, was Alfred Pennington, a local internist who tried
the diet himself in 1944—and then began prescribing it to his patients.

After the war, Pennington worked for the industrial-medicine division
of E. I. du Pont de Nemours & Company, and specifically for George
Gehrmann, the company's medical director and a pioneer in the field of
occupational health.* Gehrmann founded and was the first president,

---

* According to Lewis Finn, then president of the Delaware Academy of Medicine, Gehrmann's
department at DuPont was "one of the most outstanding industrial medical departments in the
country."

from 1946 to 1949, of the American Academy of Occupational Medicine, an organization that has since merged and evolved into the American College of Occupational and Environmental Medicine. By 1948, according to Gehrmann, DuPont as a corporation had become anxious about the apparent epidemic of heart disease in America. Just as Ancel Keys said he was prompted to pursue dietary means to prevent heart disease after perusing the obituaries, Gehrmann said he was prompted by the heart attack of a DuPont executive. Gehrmann decided to attack overweight and obesity, hoping heart-disease risk would diminish as a result.

"We had urged our overweight employees to cut down on the size of the portions they ate," Gehrmann said, "to count their calories, to limit the amounts of fats and carbohydrates in their meals, to get more exercise. None of those things had worked." These failures led Gehrmann and Pennington to test Donaldson's meat diet on overweight DuPont executives.

In June 1949, Pennington published an account of the DuPont experience in the journal *Industrial Medicine*. He had prescribed Donaldson's regimen to twenty executives, and they lost between nine and fifty-four pounds, averaging nearly two pounds a week. "Notable was a lack of hunger between meals," Pennington wrote, "increased physical energy and sense of well being." All of this seemed paradoxical: the DuPont executives lost weight on a diet that did not restrict calories. The subjects ate a minimum of twenty-four hundred calories every day, according to Pennington: eighteen ounces of lean meat and six ounces of fat divided over three meals. They averaged over three thousand calories. Carbohydrates were restricted in their diet—no more than eighty calories at each meal. "In a few cases," Pennington reported, "even this much carbohydrate prevented weight loss, though an ad-libitum [unrestricted] intake of protein and fat, more exclusively, was successful."*

In June 1950, *Holiday* magazine called Pennington's diet a "believe it or not diet development" and "an eat-all-you-want reducing diet." Two years later, Pennington discussed his diet at a small obesity symposium hosted by the Harvard department of nutrition and chaired by Mark Hegsted. "Many of us feel that Dr. Pennington may be on the right track in the practical treatment of obesity," Hegsted said afterward. "His high percentage of favorable results is impressive and calls for more extensive and for impartial comparative trials by others"—although, Hegsted concluded,

---

* One DuPont executive, discussed by Pennington in a later report, lost sixty-two pounds on the diet and kept it off for more than two years, while averaging thirty-three hundred calories of meat a day. If he ate *any* carbohydrates, "even an apple," Pennington wrote, his weight would climb upward.

"any method of [obesity] treatment other than caloric restriction still requires study by all methods that can be brought to bear on the problem."

The Harvard symposium led to the publication of Pennington's presentation in *The New England Journal of Medicine,* and this, along with the *Vogue* article, prompted the competing medical journals to address it. In a scathing editorial called "Freak Diets!" *The Journal of the American Medical Association (JAMA)* took the position that calorie restriction was the only legitimate way to induce weight loss, and that what Hegsted had called "impartial comparative trials by others" were not necessary. "The proposed high-fat diet will probably add unduly to the patient's weight and thus, in addition to the other harmful effects of obesity, increase the hazard of atherosclerosis," wrote *JAMA.* In Britain, *The Lancet* wrote, "A low calorie intake is the best way to restore the composition of the body to normal, and this is most easily arranged by eliminating fat from the diet." If Pennington's diet worked, according to *The Lancet,* it did so only because "any monotonous diet leads to a loss of weight."

Clinicians—doctors who actually treated obese patients—pushed back against the experts. After *The Lancet*'s editorial, local clinicians wrote that the diet was successful in "a surprisingly large proportion of cases," as one Devonport physician put it. "Results so far certainly seem to support the work of Pennington which you rather lightly dismiss." "Pennington's idea of cutting out the carbohydrate but allowing plenty of protein and fat works excellently . . . ," wrote the prominent British endocrinologist Raymond Greene, "and allows of a higher caloric intake than a proportionate reduction of protein, fat and carbohydrate. . . . The diet need not be monotonous. Many patients come to prefer it." By early 1954, *The Lancet*'s editors were backpedaling, just as they had with Banting a century earlier. "Pennington has hardly proved his case," the journal argued, but it accepted the possibility that his diet worked, and perhaps not through the usual method of restricting calories.

The challenge to *JAMA* came from a physician within the American Medical Association itself—from George Thorpe, a Kansas doctor who both treated obese patients and chaired the AMA's Section on General Practice. At the AMA annual meeting in 1957, Thorpe charged that semi-starvation diets would inevitably fail, because they work "not by selective reduction of adipose deposits, but by wasting of all body tissues," and "there- fore any success obtained must be maintained by chronic undernourishment." Thorpe had tried Pennington's diet, he said, after "considering a personal problem of excess weight." He then began prescribing the diet to his patients, who experienced "rapid loss of weight, without hunger, weakness, lethargy or constipation." Even with small portions of salad and vegetables included, Thorpe said, weight losses of six to eight pounds a month could be obtained. "Evidence from widely different sources," he concluded, "seems to justify the use of high-protein, high-fat, low-carbohydrate diets for successful loss of excess weight."

In response to Thorpe's testimonial, *JAMA* could no longer claim outright that a high-fat, carbohydrate-restricted diet would actually *increase* weight, as it had asserted five years earlier, but it still insisted in a 1958 editorial that the diet would endangered health, whatever else it might accomplish.* Pennington's diet failed to fulfill the criterion of being "adequate in all essential nutrients," *JAMA* wrote. Thus, "the most reasonable diet to employ for weight reduction is one that maintains normal proportions of fat, proteins, and carbohydrates and simply limits the total quantity of the mixture." As it would do for the next fifty years, *JAMA* disregarded firsthand testimony from clinicians and trivialized the scientific issues; it promoted diets not because they were effective, but because they were supposedly "least harmful"—invariably basing its notion of harm on ideas that had been and would be strongly challenged for decades.

All the while, the DuPont experience would be confirmed in the literature repeatedly. The first confirmation came from two dietitians, Margaret Ohlson and Charlotte Young, who published their observations in the *Journal of the American Dietetic Association* in 1952. Ohlson was chair of the food and nutrition department at Michigan State University. Young had studied with Ohlson in the 1940s and then moved to Ithaca, New York, to become a nutritionist at Cornell. Young also worked with Cornell's Student Medical Clinic, and it was in this capacity, along with struggles to control her own weight (she was five ten and weighed 260 pounds), that she had become dissatisfied with calorie-restricted diets.

Ohlson began her research by testing Pennington's diet on members of her own laboratory. "The edibility of the food mixture, the feeling of well-being of the subjects and the ease with which meal pattern could be fitted into a daily schedule involving business and social engagements, suggested a further trial with patients," Ohlson reported. She then prepared a version of Pennington's diet that restricted both carbohydrates and calories, on the mistaken assumption that the diet must work by restricting calories. This was the diet that Young would also use at Cornell. It allowed

---

* These critiques were written by anonymous "competent authorities." In this case, the likely authority was Philip White, formerly at Harvard, now beginning his job as secretary of the AMA's Council on Foods and Nutrition and a columnist for *JAMA*. He would write a similar dismissal of high-fat, carbohydrate-restricted diets under his own name in 1962, and then edit another anonymous version in 1973.

only fourteen to fifteen hundred calories a day, out of which 24 percent was protein, 54 percent was fat, and 22 percent was carbohydrates.* Because the diet was also calorie-restricted, it did not actually test Pennington's observation that weight would be lost even without such a calorie limitation. Nor did Ohlson or Young address the question of why their subjects never reported feeling hungry even though it provided no more calories than a typical semi-starvation diet. Still, their observations are relevant, particularly because they came in an era when high-fat diets were not yet widely considered deadly, so that researchers were not biased by this perception.

Ohlson initially tested a twelve-hundred-calorie low-fat diet on four overweight young women. This was eight hundred to a thousand calories less than these women normally ate to maintain their weight, Ohlson reported, so they should have lost at least twenty-two pounds each over the fifteen weeks of the trial. Rather, the four women lost zero, six, seven, and seventeen pounds. The "subjects reported lack of 'pep' throughout . . . [and] they were discouraged because they were always conscious of being hungry."

Ohlson then tested her calorie-restricted version of Pennington's diet on seven women who ranged from mildly overweight to obese. These women followed the diet for sixteen weeks and lost between nineteen and thirty-seven pounds. In a comparison of the low-fat diet of twelve hundred calories with the carbohydrate-restricted diet of fourteen to fifteen hundred calories, the former resulted in an average weight loss of a half-pound a week, whereas the latter diet, higher in calories, induced an average weight loss of almost three pounds weekly. "Without exception, the low-carbohydrate reducing diet resulted in satisfactory weight losses," Ohlson wrote. "The subjects reported a feeling of well-being and satisfaction. Hunger between meals was not a problem."

Over a ten-year period, Ohlson's laboratory tested a range of dietary compositions on nearly 150 women, including between 50 and 60 women on her version of Pennington's diet. She also tested low-protein diets and diets low in fat (only 180 calories, or less than 15 percent fat) but high in carbohydrates. Her subjects considered these low-fat diets to be "dry, uninteresting, [and] hard to eat," no more satisfying than those regimens of turnips, bread, and cabbage that Ancel Keys had fed his conscientious

* Ohlson worried that "the large servings of meat" could get monotonous and that the diet did not meet the recommended daily allowances for essential vitamins recently introduced by the Food and Nutrition Board of the National Research Council. She therefore included in her diet more milk, cheese, and eggs than Pennington had recommended, and expanded the choice of fruits and vegetables.

objectors. Diets with 360 calories of fat proved "sufficient to provide acceptability," she added, but her subjects "uniformly" preferred the high-fat diets, with seven to eight hundred calories of fat. At that level, the women "did not appear to give as much thought to forbidden foods," and "they also appeared to be more successful in controlling appetite during college vacations." Simply put, Ohlson's subjects were not as hungry on the high-fat, low-carbohydrate diet as they were on the low-fat, high-carbohydrate regimens.

On these high-fat, high-protein diets, according to Ohlson, her subjects appeared to add muscle or lean-tissue mass, rather than losing it, which she believed to happen inevitably with both balanced semi-starvation diets and low-protein diets. On Ohlson's version of Pennington's diet, her subjects stored nitrogen while losing one to three pounds of weight a week. This "can only mean that replenishment of the lean muscle mass is taking place," Ohlson said, an observation reinforced in some of her subjects by "a reduction in dress size [that] appeared to be greater than seemed reasonable on the basis of pounds lost."

Meanwhile, Charlotte Young at Cornell first tested Ohlson's version of Pennington's diet on sixteen overweight women, who lost between nine and twenty-six pounds in ten weeks, averaging nearly two pounds per week. They were "unanimous in saying that they had not been hungry," Young wrote. She reported that her subjects seemed unexpectedly healthy while on the diets, "despite an unusually heavy siege of colds and 'flu' on the campus," and that several "reported that their skins had never looked better than during the reducing regimen." "No excessive fatigue was evident; there was a sense of well-being unusual during weight reduction." In 1957, Young published the results of a second trial with eight overweight male students, and the results were comparable. Young fed these men an eighteen-hundred-calorie version of Ohlson's diet. After nine weeks, the men had lost between thirteen and twenty-eight pounds, averaging almost three pounds each week. Their weight loss, Young said, "in every case" actually exceeded that expected purely from the reduction in calories. Ohlson's and Young's journal articles were ignored.

As with virtually all weight-loss diet studies until the last decade, these were not the kind of randomized, well-controlled trials necessary to establish whether a particular diet actually extends life or prevents chronic disease. Subjects were not randomly selected to follow a low-carbohydrate diet, or a low-calorie diet, or no diet at all, and then followed for months or years to compare the treatments and their respective risks and benefits. Rather, the logic behind them was that obese patients were themselves the controls because they had tried calorie-restricted diets and they hadn't been successful.

For an obese person, it's a reasonable assumption that they have tried to weigh less by eating less—i.e., calorie restriction. If that approach had worked, as Hilde Bruch noted, that person would not be obese. When Bruch described a fifty-pound weight loss in a young patient eating Pennington's diet, she also reported that the woman had described her life, as Bruch's obese patients often did, as a constant, ongoing failure to control her appetite and restrict her calories to a level that would maintain or reduce her weight.

In 1961, William Leith of McGill University reported his clinical experience with forty-eight patients on Pennington's diet all of whom had previously tried low-calorie diets "without measurable success." Half had used appetite-suppressant drugs ("anorectic agents," as Leith called them), seven had taken "bulk substitutes," and "eight had participated in group psychotherapy for a period of eight months," and yet "none of them showed a sustained loss of weight." Twenty-eight, by contrast, lost a significant portion of their excess weight on Pennington's diet—between ten and forty pounds, averaging one and a half pounds each week. "Our results do show that satisfactory weight loss may be accomplished by a full caloric, low carbohydrate diet," Leith concluded. "The patients ingested protein and fat as desired." For the successful dieters, a significant success had followed a lifetime of failure.

Neither the individuals who wish to lose weight nor the clinicians who prescribe the diet need a randomized trial to tell them if it works. Such a trial is necessary only to establish that the diet works better than some other diet, and whether it leads to sustained benefits in health and longevity.

Until recently, few nutritionists or clinicians considered it worth their time and effort to test weight-reducing diets. Instead, they spent their careers studying the physiological and psychological abnormalities associated with the condition of obesity, comparing food consumption and physical activity in obese and lean individuals, and studying obesity in animals. They tried to induce fat people to endure semi-starvation by behavioral modification; they studied pharmacological methods of suppressing hunger, or surgical methods of reducing the amount of food that could be consumed or digested.* Testing diets or even treating obese patients was regarded as lesser work. "To be honest, obesity treatment is extremely boring," said Per Björntorp, who was among the most prominent European authorities on obesity in the 1970s and 1980s. "It's very difficult and unrewarding." When obese individuals came to his biochemistry laboratory at the University of Göteborg, they were referred to the local nutritionists to be taught how to count and restrict calories. Since everyone knew that obesity was caused by overeating, why bother with diet trials? "There's no point wasting your time on them," George Bray, considered one of the world's leading authorities, said in a recent interview. "If you get restriction of energy you will lose weight, unequivocally. It's not an issue."

When clinical investigators did test the efficacy of high-fat, carbohydrate-restricted diets, however, the results were remarkably consistent. Every investigator reported weight losses of between one and five pounds a week even when the investigators running the trial seemed more concerned with establishing that the diets caused deleterious side effects. Every investigator who discussed the subjective experiences of the test subjects reported that they suffered none of the symptoms of semi-starvation or food deprivation—"excessive fatigue, irritability, mental depression and extreme hunger," as Margaret Ohlson described them.

The last of these symptoms may be the most telling. The diets induced significant weight loss without hunger even when the patients ate only a few hundred calories a day, as Russell Wilder's did at the Mayo Clinic in the early 1930s, or 650–800 calories per day, as was the case with the patients treated by George Blackburn and Bruce Bistrian of MIT's department of nutrition and food science and the Harvard Medical School in the 1970s. Wilder was treating his obese patients with the very low-calorie diet developed by Frank Evans, principally meat, fish, and egg white, with 80–100 calories' worth of green vegetables. "The absence of complaints of hunger has been remarkable," Wilder wrote. Bistrian and Blackburn reported in 1985 that they had prescribed their diet of lean meat, fish, and fowl—almost 50 percent protein calories and 50 percent fat—to seven hundred patients. On average, the patients lost forty-seven pounds over a period of four months; nearly three pounds a week. "People loved it," said Blackburn.†

---

* Drug studies were encouraged by the relative ease of obtaining money and resources from the pharmaceutical industry, and the absence of funding for dietary treatments.

† In 1989, William Dietz, who now serves as director of the Division of Nutrition and Physical Activity at the Centers for Disease Control, reported that Bistrian and Blackburn's diet was "especially successful" on obese patients with a genetic disorder called Prader-Willi syndrome, "whose characteristic ravenous appetites appeared to be suppressed."

Significant weight loss without hunger was also reported when the diet was prescribed at 1,000 calories, as the University of Würzburg clinicians Heinrich Kasper and Udo Rabast did in a series of trials through the 1970s; at 1,200 calories, as the University of Iowa nutritionist Willard Krehl reported in 1967; at 1,320 calories, as Edgar Gordon of the University of Wisconsin reported in *JAMA* in 1963; at 1,400 or 1,800 calories, as Young and Ohlson did; at 2,200 calories, as the Swedish clinician Bertil Sjövall reported in 1957, and even when the diet provided more than 2,700 calories a day, as reported also in 1957 by Weldon Walker, who would later become chief of cardiology at the Walter Reed Army Medical Center in Washington. The same has invariably been the case even when patients are simply "encouraged to eat as much as [is] necessary to avoid feeling hungry," but to avoid carbohydrates in doing so, as John LaRosa, now president of the State University of New York Downstate Medical Center, reported in 1980.

Every investigator who compared these carbohydrate-restricted diets with more balanced low-calorie diets also reported that the carbohydrate-restricted diet performed at least as well, and usually better, even when the caloric content of the carbohydrate-restricted diet was significantly greater—say, 1,850 calories versus 950 calories, as Per Hanssen reported in 1936; or 2,200 calories versus 1,200 calories, as Bertil Sjövall reported in 1957; or even an "eat as much as you like" diet compared with a 1,000-calorie diet, as Trevor Silverstone of St. Bartholomew's Hospital in London reported in 1963 in a study of obese diabetics. The same held true for children, too. In 1979, L. Peña and his colleagues from the Higher Institute of Medical Sciences in Havana reported that they had randomized 104 obese children to either an "eat as much as you like" high-fat, high-protein diet with only 80 calories of carbohydrates, or an 1,100-calorie diet of which half the calories came from carbohydrates. The children on the carbohydrate-restricted diet lost almost twice as much weight as those who were semistarved on the balanced diet.

Between 1963 and 1973, Robert Kemp, a physician at Walton Hospital in Liverpool, published three articles reporting his clinical experience with a low-carbohydrate, unrestricted-calorie diet. Kemp reported that his obese patients craved carbohydrates and were invariably puzzled and frustrated by two aspects of their condition: "that other people can eat just the same diet and remain thin," and "that they themselves in earlier life may well have been thin on the same amount and type of food on which they subsequently became fat." These observations led Kemp to formulate "a working hypothesis that the degree of tolerance for carbohydrate varies from patient to patient and indeed in the same patient at different periods of life."

He then translated this hypothesis into a carbohydrate-restricted, calorie-unrestricted diet. Doing so, he said, made it "possible for the first time in [his] experience to produce worth-while results in obesity treatment."

Beginning in 1956, Kemp prescribed this diet to 1,450 overweight and obese patients. More than seven hundred (49 percent) were "successfully reduced" in his practice, which Kemp defined as having lost more than 60 percent of their excess weight. These patients averaged twenty-five pounds of weight loss after a year on the diet. Another 550 patients (38 percent) defaulted, which means they stopped appearing at Kemp's monthly counseling sessions. Nearly two hundred patients (13 percent) failed to lose significant weight while apparently following through with the treatment. This failure suggested to Kemp that the diet may not work on everyone, despite some claims by popular diet books that it can.

Still, even if we assume that all of Kemp's patients who defaulted on the diet also failed to lose significant weight, Kemp's track record would still suggest that his carbohydrate-restricted diet was at least four times more effective than the balanced semi-starvation diet that Albert Stunkard used when reporting on his clinical experiences in 1959.

The last decade has witnessed a renewed interest in testing carbohydrate-restricted diets as obesity levels have risen and a new generation of clinicians have come to question the prevailing wisdom on weight loss. Six independent teams of investigators set out to test low-fat semi-starvation diets of the kind recommended by the American Heart Association in randomized control trials against "eat as much as you like" Pennington-type diets, now known commonly as the Atkins diet, after Robert Atkins and *Dr. Atkins' Diet Revolution.* Five of these trials tested the diet on obese adults, one on adolescents. Together they included considerably more than six hundred obese subjects. In every case, the weight loss after three to six months was two to three times greater on the low-carbohydrate diet—unrestricted in calories—than on the calorie-restricted, low-fat diet.

In 2003, seven physicians from the Yale and Stanford medical schools published an article in *JAMA* that claimed to be the "first published synthesis of the evidence" in the English-language medical literature on the efficacy and safety of carbohydrate-restricted diets. These doctors concluded that the evidence was "insufficient to recommend or condemn the use of these diets," partly because there had been no long-term randomized controlled trials that established the safety of the diets. Nonetheless, they did report the average weight loss from the trials that the authors had culled from the last forty years of medical research. "Of the 34 of 38 lower-carbohydrate diets for which weight change after diet was calculated," they noted, "these lower-carbohydrate diets were found to produce greater weight loss than higher-carbohydrate diets"—an average of thirty-seven pounds when carbohydrates were restricted to less than sixty grams a day, as Pennington had prescribed, compared with four pounds when they were not.*

Accepting that high-calorie diets can lead to greater weight loss than semi-starvation diets requires overturning certain common assumptions. One is that a calorie is a calorie, which is typically said to be all we need to know about the relationship between eating and weight. "Calories are all alike," said the Harvard nutritionist Fred Stare, "whether they come from beef or bourbon, from sugar or starch, or from cheese and crackers. Too many calories are too many calories." But if a calorie is a calorie, why is it that a diet restricted in carbohydrates—eat cheese, but not crackers—leads to weight loss, largely if not completely independent of calories? If significant weight can be lost on all these carbohydrate-restricted diets, even when subjects eat twenty-seven hundred or more calories a day, how important can calories be to weight regulation? Wouldn't this imply that the quantity of carbohydrates is at least a critical factor, in which case there must be something unique about these nutrients that affects weight but falls outside the context of energy content? Isn't it possible, as Max Rubner suggested a century ago, that "the effect of specific nutritional substances upon the glands" might be a factor when it comes to weight regulation, and perhaps the more relevant one?

Look at this another way. When Bruce Bistrian and George Blackburn instructed their patients to eat nothing but lean meat, fish, and fowl—650 to 800 calories a day of fat and protein—half of them lost at least forty pounds each. That success rate held true for "thousands of patients" from the 1970s on, Bistrian said. "It's an extraordinarily effective and safe way to get large amounts of weight loss." But had they chosen to *balance* these very low-calorie diets of fat and protein with carbohydrates—say, by adding another 400 calories of "wonderful fruits and vegetables," as Bistrian phrased it—they would then be consuming the kind of semi-starvation diet that inevitably fails: 1,200 calories evenly balanced between protein, fat, and carbohydrates. "The likelihood of losing forty pounds on that diet is *one* percent," Bistrian said.

---

* When the authors included only randomized control trials in their calculations, they identified seven relevant studies of this severe carbohydrate restriction and seventy-five of higher-carbohydrate diets. The average weight loss was eight pounds for the carbohydrate-restricted diets and four for the higher-carbohydrate diets.

The bottom line: If we add 400 calories of fat and protein to 800 calories of fat and protein, we have a 1,200-calorie high-fat, carbohydrate-restricted diet that will still result in considerable weight loss. If we add 400 calories of carbohydrates to the 800 calories of fat and protein, we have a *balanced* semi-starvation diet of the kind commonly recommended to treat obesity—and we reduce the efficacy by a factor of fifty. We now have a diet that will induce forty pounds of weight loss in perhaps one in a hundred patients rather than one in two.

This striking contrast also relates to hunger. One obvious explanation for the failure of balanced semi-starvation diets is hunger. (Another, as I noted earlier, is that our bodies adjust to caloric deprivation by reducing energy expenditure.) We're semi-starved, and so we eventually break the diet. We cannot withstand the "nagging discomfort," as William Leith put it. This is why clinicians like Peña and Leith believed that the carbohydrate-restricted diets were more successful: their obese patients could eat whenever they got hungry and would sustain the diet longer. It's why Per Hanssen in 1936 suggested that the 1,800-calorie carbohydrate-restricted diet was likely to make weight maintenance easier than a 900-calorie balanced diet. But, as Willard Krehl noted, the diet at 1,200 calories also abated hunger: the desire for food, he wrote, was "more than amply satisfied." Bistrian and Blackburn were able to reduce or eliminate hunger even at 650 to 800 calories. Had hunger remained acute, as Bistrian said, it's likely that the patients would have eventually cheated, and this would have thwarted the weight loss *if they cheated with carbohydrates*. If the cheaters reached daily for a few hundred calories of carbohydrates—say, a bagel or a couple of a sodas—they would be eating a balanced semi-starvation diet with its 1-percent success rate. The 50-percent success rate on the half-protein, half-fat diet suggests that these dieters do not feel hunger, or certainly do not feel it as acutely as they would had they been eating a diet that came with carbohydrates as well. "Isn't the proof of the pudding in the eating?" asked Bistrian.

These observations would suggest that we can *add* 400 calories to a diet of 800 calories—400 calories of fruits and vegetables on top of our 800 calories of meat, fish, and fowl—and be *less* satisfied. But, again, this will happen only if the initial diet is protein and fat and the added calories come from carbohydrates. If we add more fat and protein, we have a 1,200-calorie carbohydrate-restricted diet that will satisfy our hunger. So is the amount of calories consumed the critical variable, or is there something vitally important about the presence or absence of carbohydrates? The implication is that there is a direct connection between carbohydrates and our experience of hunger, or between fat and protein and our experience of satiety, which is precisely what Ethan Sims's overfeeding experiments had suggested—that it's possible to eat up to 10,000 calories of mostly carbohydrates and be hungry at the end of the day, whereas eating a third as many calories of mostly fat and protein will more than satiate us.

Now take into account the experience of prolonged starvation. In 1963, Walter Bloom, then director of research at Atlanta's Piedmont Hospital, published a series of articles on starvation therapy for obesity, noting that total starvation—i.e., fasting, or eating nothing at all—and carbohydrate restriction had much in common. In both cases, our carbohydrate reserves are used up quickly, and we have to rely on protein and fat for fuel. When we fast, the protein and fat come from our muscle and fat tissue; when we restrict carbohydrates, they're provided by the diet as well. "At a cellular level, the major characteristic of fasting is limitation of available carbohydrate as an energy source," Bloom wrote. "Since fat and protein are the energy sources in fasting, there should be little difference in cellular metabolism whether the fat and protein come from endogenous [internal] or exogenous [external] sources." And this turns out to be the case. The metabolic responses of the body are virtually identical.

And, once again, there is "little hunger" during prolonged starvation. "In total starvation," Keys wrote in *The Biology of Human Starvation,* "the sensation of hunger disappears in a matter of days." This assessment was confirmed in the early 1960s by Ernst Drenick at UCLA, when he starved eleven obese patients for periods of twelve to 117 days. "The most astonishing aspect of this study," wrote Drenick and his colleagues in *JAMA,* "was the ease with which prolonged starvation was tolerated. This experience contrasted most dramatically with the hunger and suffering described by individuals who, over a prolonged period, consume a calorically inadequate diet." As the editors of *JAMA* suggested in an accompanying editorial, this absence of hunger made starvation seem to be a viable weight-loss therapy for severely obese patients: "The gratifying weight loss without hunger may bring about the desired immediate results and help establish a normal eating pattern where other dietary restrictions may fail."

The implication is that we will experience no hunger if we eat nothing at all—zero calories—and our cells are fueled by the protein and fat from our muscle and fat tissue. If we break our fast with any amount of dietary protein and fat, we'll still feel no hunger. But if we add carbohydrates, as Drenick noted, we'll be overwhelmed with hunger and will now suffer all the symptoms of food deprivation. So why is it when we add carbohydrates to the diet we get hungry, if not irritable, lethargic, and depressed, but this will not happen when we add only protein and fat? How can the amount of calories possibly be the critical factor?

In the early 1950s, Alfred Pennington noted the paradoxes engendered by a diet restricted in carbohydrates and relatively rich in fat and protein, and described them as a "mighty stimulant to thought on the matter." But this is not how the medical-research establishment has perceived them. Rather, the accepted explanation for the success of carbohydrate-restricted diets is that they work via the same mechanism as calorie-restricted diets—they restrict calories, creating a negative energy balance. Either they so limit the choices of food that dieters simply find it too difficult to consume as many calories as they might otherwise prefer, or they bore the dieters into eating less, or both. "Many individuals spontaneously and unconsciously reduce their energy intakes by as much as 30% when placed on low carb diets," Johanna Dwyer, a Tufts University nutritionist, explained in 1985. They do this "because there is insufficient carbohydrate permitted for them to eat many common and highly palatable foods in which they might otherwise indulge." So where's the paradox?

"The fact remains that some patients have lost weight on the low-carbohydrate diet 'unrestricted in calories,'" the AMA Council on Food and Nutrition conceded in 1973 in a critique of such diets. "When obese patients reduce their carbohydrate intake drastically, they are apparently unable to make up the ensuing deficit by means of an appreciable increase in protein and fat." By this logic, weight loss on a diet "unrestricted in calories" does not represent a refutation of the hypothesis that calorie restriction itself—creating a negative energy balance—is the only way to lose weight, because it suggests that a carbohydrate-restricted diet *is* a calorie-restricted diet in disguise. And the sensation of hunger isn't an issue, because it can apparently be ignored.*

This rationale, which has been invoked frequently over the past four decades, is curious on many levels. First of all, it seems to contradict the underlying principle of low-fat diets for weight control and the notion that we get obese because we overeat on the dense calories of fat in our diets. One reason that bread has always been considered the ideal staple of a low-fat reducing diet, as Jean Mayer noted, is that it only has about sixty calo-

---

* Indeed, the AMA's 1973 critique escaped the issue of hunger by including "anorexia" as one of the "untoward side effects" of the diet. Since anorexia, in this context, is the technical term for loss of appetite, it seemed a peculiar criticism to make of a weight-loss diet.

ries a slice. "If you put a restaurant-size pat of butter on your toast, for example, you triple the calories," Mayer said. If we avoid the dense calories of fat in the butter, the argument goes, we will naturally eat fewer total calories and lose weight accordingly. (This was the fallback position in 1984 for the official NIH recommendation of a low-fat diet for heart disease: if nothing else, we'd lose weight on such a diet, and so that would reduce heart-disease risk.) To explain the peculiar efficacy of carbohydrate-restricted reducing diets, the circuitous reasoning is that if we avoid the not-nearly-so-dense calories of bread and potatoes, we will also not consume the dense calories in the butter. We could still eat the dense calories of meat, cheese, and eggs, and we could certainly increase the portion sizes to compensate for the now absent butter, but apparently we won't want to do that, or somehow won't be able to, if we don't have the bread, potatoes, and pasta to eat as well.

Ironically, this argument is based almost exclusively on the research efforts of John Yudkin. "Yudkin showed that a long time ago," as George Bray recently said. "We don't generally slice butter off a dish and put in our mouth to eat. We like to put it on bread. That's why lowering carbohydrates lowers calorie intake." Yudkin was ridiculed for his advocacy of the hypothesis that sugar causes heart disease. Yet he is considered the essential source for the rationale that reconciles carbohydrate-restricted diets with the conventional wisdom of calories and weight, based on two papers, a decade apart, discussing the experience of seventeen subjects over two weeks of dieting.

Yudkin was the most prominent advocate of carbohydrate-restricted diets among nutritionists through the 1970s. He also had unconditional faith, however, in the popular misinterpretation of the law of conservation of energy. "The irrefutable, unarguable fact is that overweight comes from taking in more calories than you need," Yudkin explained in a 1958 diet book entitled *This Slimming Business*. He reconciled this belief with his advocacy of carbohydrate-restricted diets because he also believed that "much of the extra fat today" in the diet "comes together with carbohydrate in cakes, biscuits, ice cream, and sweetmeats of various sorts." If we remove the carbohydrates, Yudkin proposed, the fat calories will come down, too.

In 1960, Yudkin provided experimental evidence to support this statement in a *Lancet* article entitled "The Treatment of Obesity by the 'High-Fat' Diet." He had asked four women and two men to consume a carbohydrate-restricted diet for two weeks. They all lost weight, he reported, by consuming significantly less carbohydrates and no more fat than they typically ate on a balanced diet. The two men ate roughly twenty-nine hundred and thirty-five hundred calories normally, but reported consuming only fifteen to sixteen hundred when they abstained from carbohydrates. Their fat consumption dropped by two hundred calories a day as well. This led Yudkin to the "unequivocal" conclusion that "the high-fat diet leads to weight-loss because, in spite of its unrestricted allowance of fat and protein, it is in fact a low-calorie diet. . . ." Weight is lost by restricting calories, even if calorie restriction is not required by the diet.

Here again, however, Yudkin was confusing an association with cause and effect. Even if Yudkin's subjects had reduced their calorie consumption on the carbohydrate-restricted diet, which is a common finding in these studies, it does not mean that the reduction in calories *caused* the weight loss, only that the diet was associated with a reduction in calories as well as a reduction in weight. The diet could have worked by some other mechanism entirely, but both weight loss and decreased appetite were consequences. The fact that a reduction of appetite associates with weight loss does not mean that it is the fundamental cause.

And, of course, what may have been true, on average, for Yudkin's seventeen subjects—six in his 1960 study and eleven a decade later—is not necessarily the case for everyone who loses weight on such diets.* Even before Yudkin published *This Slimming Business*, Weldon Walker and the Columbia University physician Sidney Werner had both reported that their subjects lost significant weight while consuming at least twenty-seven hundred and twenty-eight hundred calories a day respectively. In 1954, when the Swiss clinician B. Rilliet discussed his experiences using Pennington's diet to treat obese patients at the County Hospital of Geneva, he reported that his successes were "numerous and encouraging" with both a twenty-two-hundred-calorie version of the diet and a three-thousand-calorie version. It's hard to avoid the observation that at least some individuals lose weight on carbohydrate-restricted diets while eating considerably more calories than would normally be consumed in a semi-starvation diet. This is why Werner speculated that his obese subjects must have typically been eating four to five thousand calories a day before he set about experimentally reducing them. But if that is true, why don't obese patients regularly lose weight on twenty-seven or twenty-eight-hundred calorie balanced diets, and why have clinicians always believed it necessary to semi-starve

---

* The auxiliary "may" is critical here, because Yudkin based his conclusions on three-day dietary records, which are notoriously inaccurate. He then assumed that these three-day records could be extrapolated to the entire two weeks of the study, and from there to what would happen over months or years on the diet.

them with twelve to fifteen hundred calories, or even feed them very low-calorie diets of eight hundred calories or less, to achieve any significant weight loss? Something else is going on here, and it has nothing to do with calories.

The argument that carbohydrate-restricted diets work by the same mechanism as calorie-restricted diets only changes the nature of the dilemma we have to unravel. It does not make it disappear. Even if we accept Yudkin's notion that all people who lose weight while abstaining from carbohydrates do so because they spontaneously feel the compulsion to eat less, we must then explain why anyone would willingly suffer the symptoms of semi-starvation—hunger, irritability, depression, and lethargy—rather than simply eat another piece of cheese, or steak, or lamb. The standard explanations are that it's simply too much trouble to do so, or that "all-you-can-eat-diet[s]," as Jane Brody wrote in the *New York Times* in 1981, "so restrict the dieter's choices that boredom and distaste automatically produce a calorie cutback." But these are unacceptably facile. If the obese end up eating less on this diet, the most likely explanation is that they're less hungry, in the same way that, if we don't drink water when water is there for the drinking, we're probably not thirsty. If we don't feel or act semi-starved, it's a reasonable bet that we're not. "The best definition of food deficiency," as Ancel Keys and his colleagues wrote in *The Biology of Human Starvation*, "is to be found in the consequence of it."

Keys's starvation studies suggest where the "no bread, no butter" logic will take us. We know from these studies that if we feed people a carbohydrate-rich diet of fifteen or sixteen hundred calories a day, they will be obsessed with the "persistent clamor of hunger," so much so that they might be willing to mutilate themselves to escape the ordeal. Meanwhile, if those same people were allowed to consume unlimited calories of only meat, cheese, and eggs, this school of thought dictates, they will voluntarily restrict their consumption to the same fifteen or sixteen hundred calories—or at least they will if they're obese or need to lose ten or twenty pounds—because in this case, as Harvard endocrinologist George Cahill suggested, the "nonappetizing nature" of this meat-egg-and-cheese diet will overcome the urge to amply satisfy their desire for food. Our subjects will voluntarily starve themselves, as though hunger itself, and all its regrettable side effects, have been rendered impotent in the face of monotony, which is to say, a diet that these experts define as unappetizing because it does not allow consumption of starches, flour, sugar, or beer.

But Keys had also severely restricted the choice of foods he fed his subjects. Remember, he had wanted to simulate the foods available during wartime in Eastern Europe and so had allowed his conscientious objectors only bread, potatoes, cereals, turnips, cabbages, and "token" amounts of meat and dairy products. Yet, in the entire fourteen hundred pages of his *Biology of Human Starvation*, there is not the slightest hint that his semi-starved subjects, or those starving populations he discusses in his comprehensive history of famine, would have turned down more cabbages, bread, or turnips had they been available, not to mention meat, cheese, fish, or eggs. The notion that hunger can be relieved or eliminated simply by limiting the choice of food is exceedingly difficult to embrace.

Over the years, a common way to avoid thinking about the paradox of a diet that allegedly restricts calories but does not induce hunger is to attribute the suppression of appetite to a factor that these authorities consider irrelevant to the bigger picture of weight and health—to ketosis, the condition produced when the liver increases its production of ketone bodies to replace glucose as a fuel for the brain and nervous system. Once ketone bodies are produced, "their appetite-depressing activity takes effect," as Richard Spark of Harvard Medical School claimed in 1973. "Substances called ketones will accumulate in your bloodstream [during carbohydrate restriction] and can make you slightly nauseated and light-headed and cause bad breath," wrote Jane Brody in the *New York Times* in 1996. "This state is not exactly conducive to a hearty appetite, so chances are you will eat less than you might otherwise have of the high-protein, high-fat foods permitted on the diet."

But this, too, fails as a viable explanation. The liver increases ketone-body synthesis only when carbohydrates are unavailable and the body is relying predominantly on stored fat for its fuel. Ketone bodies could be responsible for appetite suppression, as Spark and Brody suggested, but so could the absence of carbohydrates or the burning of fat, or something else entirely. All of these are *associated* with the absence of hunger. In fact, the existing research argues against the claim that ketone bodies suppress appetite. Individuals with uncontrolled diabetes, for example, will suffer from ketoacidosis, during which ketone-body levels can be tenfold or even forty-fold higher than the mild ketosis of carbohydrate restriction, and yet these people are ravenous. "It is not clear why the sensation of hunger subsides [in starvation studies], but the disappearance is apparently not related to ketosis," wrote Ernst Drenick in 1964 about his fasting studies at UCLA. Hunger sensations often disappeared in his subjects before ketone bodies could be detected in their blood or urine, "and it did not reappear" in those periods when ketone body levels were low. The same dissociation between ketone bodies and hunger was reported in 1975 by

Duke University pediatrician James Sidbury, Jr., in the treatment of obese children.

Another common explanation for the absence of hunger on carbohydrate-restricted diets is that fat and protein are particularly satiating—"these foods digest slowly, making you feel satisfied longer," as Brody has explained in the *Times*. (Even those investigators who published studies supporting Yudkin's idea that carbohydrate-restricted diets work by restricting calories would invariably comment that high-protein, high-fat diets still induced the least hunger and the greatest feeling of satiation. "There is a good reason to believe that the satiety value of such diets is superior to diets high in carbohydrate and low in fat, and hence, may be associated with better dietary adherence," the metabolism researcher Laurance Kinsell wrote in an influential 1964 article entitled "Calories Do Count.") But this is also unsatisfying as an explanation. The statement that fat and protein satisfy us longer is equivalent to the statement that carbohydrates are less satisfying—they either make us experience hunger sooner than fat and protein or perhaps induce hunger, whereas fat and protein suppress it. This leads us back to the now familiar question: what is it about carbohydrates, or about the speed with which we digest them, that accelerates or exacerbates our sensation of hunger and our desire to eat?

Even Yudkin had struggled with the question of why people would willingly semi-starve themselves on a carbohydrate-restricted diet. "For reasons I do not clearly understand," he wrote, there must be something unique about carbohydrates that either stimulates our appetites or fails to satiate us. "It would seem from this that carbohydrate does not satisfy the appetite," he noted; "it may even increase it. . . ."

This conclusion is simply hard to avoid, considering the half century of experimental observations on these diets. It leaves us with two seemingly paradoxical observations. The first is that weight loss can be largely independent of calories. The second is that hunger can also be. Even if we could establish that weight loss on these diets is universally attended by a decrease in calories consumed—no bread, no butter—we then have to explain why the subjects of these diets don't manifest the symptoms of semi-starvation. If they eat less on the diets, why aren't they hungry? And if they don't eat less, why do they lose weight?

"It is better to know nothing," wrote Claude Bernard in *An Introduction to the Study of Experimental Medicine*, "than to keep in mind fixed ideas based on theories whose confirmation we constantly seek, neglecting meanwhile everything that fails to agree with them." In the study of human obesity, that fixed idea has been what Yudkin called "the inevitability of calories," which in turn is based on the ubiquitous misconception of the law of energy conservation. If we believe that conservation of energy—calories in equal calories out—implies cause and effect, then we will refuse to believe that obese patients can lose significant weight without restricting their energy intake beneath some minimal expenditure. Any reports to the contrary will be rejected on the basis that they cannot possibly be true. "Claims that weight loss occurs even with high-caloric intake, but no carbohydrate, are absurd," as the American Medical Association insisted in 1974. "Although authors of popular diet books frequently say that loss of body fat can occur regardless of high-calorie intake, this is not supported by evidence and, in fact, is refuted by the laws of thermodynamics."

Because such a possibility is not refuted by the laws of thermodynamics, we should take such claims seriously, as Alfred Pennington did. Although several of Pennington's articles appeared in journals that were widely read, including *The New England Journal of Medicine,* they would have little influence on the thinking about obesity. A few practicing physicians took his work seriously—George Thorpe and Herman Taller, a Brooklyn obstetrician who published a 1961 best-seller based on Pennington's science entitled *Calories Don't Count*—but they only lost professional credibility by doing so. The great majority of clinicians and nutritionists would not go against the conventional wisdom.

Nonetheless, Pennington was on to something. He set out to understand why his DuPont patients lost weight on a calorie-unrestricted diet that they enjoyed. He knew it contradicted the conventional wisdom but was determined to pursue the evidence. First he read what he called the "voluminous experimental literature on obesity." He concluded that only "meager and conflicting" evidence existed to support the popular contention that calorie restriction would induce long-term weight loss, or even that it should induce long-term weight loss. He came to believe that experts who invoked the first law of thermodynamics to defend their beliefs did great damage. "These tended to distract the general attention from examination of the evidence on the real question, whether or not common obesity arises from a metabolic defect," he wrote.

Pennington based his analysis of the obesity problem on one fundamental premise that he adopted from the research on homeostasis in the 1930s and early 1940s: Because fuel is ultimately used by the cells themselves, the relationship between fuel supply and demand at this cellular level determines both hunger and energy expenditure. The less fuel available to supply the metabolic demands of our cells, the greater the hunger and the less energy we will expend. The greater the fuel available to the cells, the greater the metabolic activity and perhaps physical activity also. This was something Francis Benedict had suggested in the 1920s and Eugene Du Bois believed. Energy expenditure, wrote Pennington, is an "index of calorie nutrition at the cellular level."

Pennington considered two facts about obesity to be particularly revealing. One was Hugo Rony's observation that an obese individual will spend much of his life in energy balance—in the "static phase" of obesity, to use Rony's term—just as the lean do. "His caloric intake, like that of people of normal weight, is dictated by the energy needs of his body," Pennington wrote. "His appetite, far from being uncontrolled, is precisely and delicately regulated."

The second fact was that when obese individuals try consciously to eat less—when they go on a low-calorie diet—their metabolism and energy expenditure inevitably decrease, just as they do when lean individuals are semi-starved. Benedict had observed this diet-induced decrease in energy expenditure in his lean subjects in his 1917–18 semi-starvation studies. Frank Evans and Margaret Ohlson had made the same observation of the obese. Pennington believed, as Benedict, the Cornell nutritionist Graham Lusk, and others had suggested, that this was the natural response to a diminished supply of energy. Less energy is available to the cells, and so they expend less. On a calorie-restricted diet, Pennington suggested, the obese and the lean become hungry and lethargic for identical reasons— "their tissues are not receiving enough nutriment."

This presented a dilemma. That the tissues of the lean are semi-starved by calorie restriction is easy to imagine; they don't have a lot of excess calories to spare. But why would this happen with the obese, who do? Pennington found his answer in a 1943 article by the Columbia University biochemist DeWitt Stetten, who reported that the rate at which fatty acids were released from the fat deposits of congenitally obese mice was significantly slower than it was in lean mice. Stetten had suggested that obesity in these animals was caused by a suppression of the flow of fat from the adipose tissue back into the circulation and its subsequent use by the tissues for fuel.

Pennington proposed that the same thing causes obesity in humans. The adipose tissue amasses fat calories in a normal manner after meals, but it doesn't release those calories fast enough, for whatever reason, to satisfy the needs of the cells between meals. This was the metabolic defect that causes obesity, he said, and it could apparently be corrected or minimized by removing carbohydrates from the diet.

By hypothesizing the existence of such a defect, Pennington was able to explain the entire spectrum of observations about obesity in humans and animals simply by applying the same law of energy conservation that other obesity researchers had misinterpreted. The law applies to the fat tissue, Pennington noted, just as it does to the entire human body. If energy goes into the fat tissue faster than it comes out, the energy stored in the fat tissue has to increase. Any metabolic phenomenon that slows down the release of fat from the fat tissue—that retards the "energy out" variable of the equation—will have this effect, as long as the rate at which fat enters the adipose tissue (the energy in) remains unchanged, or at least does not decrease by an equal or a greater amount. Fat calories accumulating in the adipose tissue wouldn't be available to the cells for fuel. We would have to eat more to compensate, or expend less energy, or both. We'd be hungrier or more lethargic than individuals without such a defect.

Pennington suggested that as the adipose tissue accumulates fat its expansion will increase the rate at which fat calories are released back into the bloodstream (just as inflating a balloon will increase the air pressure inside the balloon and the rate at which air is expelled out of the balloon if the air is allowed to escape), and this could eventually compensate for the initial defect itself. We will continue to accumulate fat—and so continue to be in positive energy balance—until we reach a new equilibrium and the flow of fat calories out of the adipose tissue once again matches the flow of calories in. At this point, Pennington said, "the size of the adipose deposits, though larger than formerly, remains constant: the weight curve strikes a plateau, and the food intake is, again, balanced to the caloric output."

By Pennington's logic obesity is simply the body's way of compensating for a defect in the storage and metabolism of fat. The compensation, he said, occurs homeostatically, without any conscious intervention. It works by a negative feedback loop. By expanding with fat, the adipose tissue "provides for a more effective release of fat for the energy needs of the body." Meanwhile, the conditions at the cellular level remain constant; the cells and tissues continue to function normally, and they do so even if we have to become obese to make this happen.

This notion of obesity as a compensatory expansion of the fat tissue came as a revelation to Pennington: "It dawned on me with such clarity that I felt stupid for not having seen it before." By working through the further consequences of this compensatory process, Pennington said, all the seemingly contradictory findings in the field suddenly fit together "like clockwork."

This defect in fat metabolism would explain the sedentary behavior typically associated with obesity, and why all of us, fat or lean, will become easily fatigued when we restrict calories for any length of time. Rather than drawing on the fat stores for more energy, the body would compensate by expending less energy. Any attempt to create a negative energy balance, even by exercise, would be expected to have the same effect.

Clinicians who treat obese patients invariably assume that the energy or caloric requirement of these individuals is the amount of calories they can consume without gaining weight. They then treat this number as though it were fixed by some innate facet of the patients' metabolism. Pennington explained that this wasn't the case. As long as obese individuals have this metabolic defect and their cells are not receiving the full benefit of the calories they consume, their tissues will always be conserving energy and so expending less than they otherwise might. The cells will be semi-starved, even if the person does not appear to be. Indeed, if these individuals are restraining their desire to eat in an effort to curb, if possible, still further weight gain, this inhibition of energy expenditure will be exacerbated.

Consider the kind of young, active men Ancel Keys had employed in his starvation experiments. These men might normally expend thirty-five hundred calories a day, and this was what they would eat from day to day to maintain their weight. In a healthy state, the supply of fuel to their cells would be unimpeded by any metabolic defects, and so the cells would have plenty of energy to burn, and their metabolism would run unimpeded. Every day, the calories temporarily stored in their fat deposits would be mobilized and burned for fuel. But imagine that one of these men develops a metabolic defect that retards the release of fat from the adipose tissue. Now more energy enters his fat tissue than exits. If this amounts to a hundred calories a day, he'll gain roughly one pound every month. After a while, he's likely to go on a diet to rid himself of this excess fat. He might try to reduce his consumption to three thousand calories. In a healthy state, this would have worked, but now he is dogged by a defect in fat metabolism. Fat still accumulates in his fat tissue. Rather than remedy the imbalance between the calories coming to and going from the adipose tissue, this self-imposed calorie restriction further decreases the fuel available to the cells, because now fewer calories have been consumed. He's even hungrier, and if he doesn't give in to the hunger, his body has to get by on even less fuel than before. His metabolic rate slows in response, and he finds himself lacking the desire to expend energy in physical activity. If he wants to inhibit this accumulation of fat in his adipose tissue, he might further restrict his diet. If he does, however, this will further diminish the amount of calories his cells can expend.

To Pennington, this explained the observation that some obese patients can maintain their weight consuming as little as seventeen hundred calories a day, as Keys had reported. It would also explain why malnutrition and obesity could coexist in the same populations and even the same families, as we discussed earlier (see page 240). The chronic, long-term effect of such a defect in fat metabolism, combined with a diet that continues to exacerbate the problem, would so constrain the energy expenditure of adults that they could conceivably gain weight and grow obese on a caloric intake that would still be inadequate for their children.

"What happens when low calorie diets are applied is that the starved tissues of the obese are starved further," Pennington wrote. Since the consequences of this food deprivation are likely to be the same in the obese as in the lean, they had already been adequately described by the semi-starvation experiments of Benedict and Keys. "The first noticeable effect of such a calorie shortage is limitation of the voluntary activities of leisure hours," Pennington wrote. "The various avenues of caloric expenditure are all contracted in adjustment to the diminished food intake . . . and thus deflect the purpose for which low calorie diets are prescribed."

"A more rational form of treatment," Pennington suggested, would be one that makes fat once again flow readily out of the fat cells, that directs "measures primarily toward an *increased* mobilization and utilization of fuel" by the muscles and organs. Pennington believed that this is what carbohydrate restriction accomplished and this was why the diets worked. The cells would respond to this increased supply of fuel by accelerating the rate of metabolism—utilizing the fuel. Now the body would have to establish a new equilibrium between the three variables of the energy-balance equation—energy storage, intake, and expenditure. This new equilibrium, however, would be commensurate with a healthy—i.e., uninhibited—flow of fat from the adipose tissue.

If Pennington was right, a high-protein, high-fat diet that was restricted in carbohydrates but not calories would correct the metabolic fault. The adipose tissue (i.e., energy storage) would shrink, because fat would no longer be trapped in the fat tissue. It would flow out at an accelerated rate, and this would continue until a healthy equilibrium was reestablished between fat storage and fat release. Appetite (i.e., energy in) would adjust downward to compensate for the increased availability of fuel from the fat tissue. Edward Adolph of the University of Rochester and Curt Richter of Johns Hopkins had repeatedly demonstrated that laboratory animals will increase or decrease their food intake in response to the available calories. Slip nutrients into their drinking water or deposit them through a tube directly into their stomachs, and the animals compensate by eating less.

Dilute their food with water or indigestible fiber, and the animals compensate by consuming a greater volume to get the same amount of calories. There is no reason to think that this adjustment in caloric intake will not occur if the increase in available nutrients comes from the internal fat stores, rather than external manipulations—no reason to think that the body or its cells and tissues could tell the difference. "Mobilization of increased quantity of utilizable fat, then, would be the limiting factor on the appetite, effecting the disproportion between caloric intake and expenditure which is necessary for weight reduction," Pennington wrote.

If the fat can be mobilized from the adipose tissue with "sufficient effectiveness," Pennington suggested, "no calorie restriction would be necessary" on a carbohydrate-restricted diet. A greater share of the energy needs would be supplied by the calories from the fat tissue, and the appetite would naturally adjust. "Weight would be lost, but a normal caloric production would be maintained." A person would be eating less because his appetite would be reduced by the increased availability of fat calories in his circulation, not because the diet somehow bored, restricted, or revolted him. He would be eating less because his fat tissue was shrinking; his fat tissue would not be shrinking because he was eating less. "The result would appear to be a 'negative energy balance,'" Pennington said, "because so much of the energy needs would be supplied from stored amounts."

Energy expenditure would also increase on such a diet. The now unconstrained flow of fat calories from the adipose tissue would increase the fuel available for cellular metabolism. The cells would no longer be undersupplied, as though living in a constant state of semi-starvation, and their metabolism would no longer be inhibited. Metabolic rate would increase, as would the impulse to physical activity—the urge to expend some of the energy now freely available. That such an effect is possible in humans, Pennington said, had been one of the observations reported by Du Bois and his colleagues in their yearlong all-meat-diet experiment with Stefansson and his colleague Anderson. These investigators had measured Stefansson's and Anderson's metabolism on a balanced diet and then measured their metabolism repeatedly during the yearlong trial. Both men lost some weight while eating the meat diet; both increased their basal-metabolic rate—7 percent for Stefansson and 5 percent for Anderson. Such an increase in energy expenditure could account for a weight loss of twenty pounds or more over the course of a year. If this change in expenditure went in the other direction when the diet included carbohydrates, it could easily account for the slow development of obesity.

When the obese or overweight go on a carbohydrate-restricted diet, Pennington theorized, there will be an increase in metabolic and physical activity as their bodies expend this newly available energy, and an attendant weight loss. The naïve assumption would be that the physical activity caused the weight loss, and it would be wrong. They will finally be burning off their accumulated fat stores and putting that energy to use.

Under these conditions, the energy expenditure of the obese individual might rise to what it otherwise would have been in a healthy state. It was not out of the question, as Frank Evans had reported and Sidney Werner had speculated, that this might be more than four thousand calories a day for someone who was definitively obese. Such an individual might easily eat over three thousand calories a day and still lose a pound or two a week.

This brings us back to the questions we asked earlier: If people eat less on carbohydrate-restricted diets, why aren't they hungry. And if they don't eat less, why do they lose weight? If the restriction of carbohydrates works to ameliorate this defect in fat metabolism, as Pennington speculated, then weight will be lost, hunger will be absent, and calorie consumption may decrease, while energy expenditure will increase. This is no more than the consequences of the law of energy conservation applied to a biological system that works to conserve body composition and maintain a healthy flow of fuel to the cells and tissues.

In an ideal world, Pennington's metabolic-defect hypothesis of obesity would have been tested directly. Instead, it was ignored. Pennington made this easier by speculating that the root cause of obesity was an inability to metabolize properly a compound called pyruvic acid. This made physiological sense, but further research quickly refuted it. Pennington's error allowed his contemporaries in nutrition and obesity research to dismiss him as just another renegade who refused to accept the reality of energy conservation. He deserved far better, as it wouldn't be long before researchers pinned down the precise nature of the metabolic-hormonal defect that appears to be the driving force in the accumulation of excess fat.

Chapter Twenty-one

THE CARBOHYDRATE HYPOTHESIS, I: FAT METABOLISM

Looking at obesity without preconceived ideas, one would assume that the main trend of research should be directed toward an examination of abnormalities of the fat metabolism, since by definition excessive accumulation of fat is the underlying abnormality. It so happens that this is the area in which the least work has been done.

HILDE BRUCH, *The Importance of Overweight,* 1957

IN JUNE 1962, EDWIN ASTWOOD OF Tufts University gave the presidential address to the annual meeting of the Endocrinology Society in Chicago. Although Astwood was not known as an obesity researcher, he nonetheless took the opportunity to present what he considered the obvious explanation for its cause. A physician who had spent thirty years studying and treating hormone-related disorders, Astwood had discovered the reproductive hormone luteotropin (now known as luteinizing hormone), and he had created the standard technique for purifying pituitary hormones. He had performed what *The New England Journal of Medicine* would call a "brilliant series of experiments" to demonstrate that hyperthyroidism could be controlled with anti-thyroid drugs. By 1976, when Astwood died, three dozen of his former students had become full professors; eight were department chairmen—"a record perhaps unequaled in medicine," according to his obituary in the journal *Endocrinology.* He was a man who knew what he was talking about, even when he was speculating, as he was in his 1962 address, entitled "The Heritage of Corpulence."

Astwood believed that obesity and a disposition to fatten are genetic disorders. If genes determine stature and hair color, the size of our feet, and a "growing list of metabolic derangements," he asked, then "why can't heredity be credited with determining one's shape?" Although it's possible to fatten animals by stuffing them, "and doubtless we could do the same thing to ourselves if we put our minds to it," Astwood did not consider this a cause of overweight. "Not many people try to get into the circus this way," he said—"they become candidates spontaneously." He also considered inactivity to be of dubious importance. "Many of our moderately fat patients sit like bumps on a log," he said, but that could be an effect, not a cause. "It would be interesting to know whether adiposity and inertia go together for some reason common to both. If fatty acid is needed for energy, a deficit could indeed promote lethargy and indolence."

Astwood then described what had been learned over the past thirty years about the hormonal regulation of fat metabolism. "To turn what is eaten into fat, to move it and to burn it requires dozens of enzymes and the processes are strongly influenced by a variety of hormones," he explained. Sex hormones, for instance, determine where fat is stored, as evidenced by the differences in fat distribution between men and women. Thyroid hormones, adrenaline, and growth hormone accelerate the release of fatty acids from fat depots, as does a hormone known as glucagon, secreted by the pancreas. "The reverse process," Astwood said, "the reincorporation of fat into the depots and the conversion of other food to fat, tends to be reduced by these hormones, but to be strongly promoted by insulin." All of this demonstrated "what a complex role the endocrine system plays in the regulation of fat."

Finally, Astwood speculated on what he considered the simplest possible explanation for obesity, and here he echoed Alfred Pennington, although, if he had read Pennington's work, he neglected to mention it. "Now just suppose that any one of these (or other unlisted) regulatory processes were to go awry," Astwood said.

> Suppose that the release of fat or its combustion was somewhat impeded, or that the deposition or synthesis of fat was promoted; what would happen? Lack of food is the cause of hunger and, to most of the body, [fat] is the food; it is easy to imagine that a minor derangement could be responsible for a voracious appetite. It seems likely to me that hunger in the obese might be so ravaging and ravenous that skinny physicians do not understand it.
>
> There is no reason to suppose that only one of these mechanisms ever goes wrong. . . . There are so many possibilities here that I am willing to give odds that obesity is caused by a metabolic defect. I would not want to wager about how many enzymes determine the shape of voluminous pulchritude.
>
> This theory would explain why dieting is so seldom effective and why most fat people are miserable when they fast. It would also take care of our friends, the psychiatrists, who find all kinds of preoccupation with food, which pervades dreams among patients who are obese. Which of us would not be preoccupied with thoughts of food if we were suffering from internal starvation? Hunger is such an awful thing that it is classically cited with pestilence and war as one of our three worst burdens. Add to the physical discomfort the emotional stresses of being fat, the taunts and teasing from

the thin, the constant criticism, the accusations of gluttony and lack of "will power," and the constant guilt feelings, and we ha*e reasons enough for the emotional disturbances which preoccupy the psychiatrists.

For the past century, the conspicuous alternative to the positive-caloric-balance hypothesis has always been, as Pennington, Astwood, and Hilde Bruch suggested, that obesity is caused by a defect in the regulation of fat metabolism. At the risk of repetition, it is important to say this is, by definition, a disorder of fat accumulation, not a disorder of overeating. For whatever reason, the release of fat or its combustion is impeded, or the deposition or synthesis of fat is promoted, as Astwood said, and the result is obesity. That in turn will cause a deficit of calories elsewhere in the body—Astwood's "internal starvation"—and thus a compensatory hunger and sedentary behavior.

This alternative hypothesis differs in virtually every respect from the positive-caloric-balance/overeating hypothesis. It implies a cause of weight gain and a treatment that stand in contradiction to virtually everything we have come to believe over the past fifty years. For this reason, it's a good idea to compare the basic propositions of these two competing hypotheses before we continue.

The positive-caloric-balance/overeating hypothesis dictates that the primary defect is in the brain, in the "regulation of ingestive behaviors, particularly at the cognitive level," as it was described by the University of California, Santa Cruz, biologist M.R.C. Greenwood in 1985. This defect purportedly causes us to consume more calories than we expend, and thus induces weight gain. Overeating and sedentary behavior are defined (tautologically) as the causes of obesity. The treatment is to create a caloric deficit by eating less and/or expending more. This hypothesis presupposes that excess calories accumulate in the body and thus are effectively "pushed" into the fat cells, which play a passive role in the process. And the calories remain bound up as fat only because we never expend sufficient energy to require their use.

Implicit in this hypothesis is the assumption that energy expenditure and energy intake are *independent* variables. Because they are independent, one of these variables can be manipulated, consciously or unconsciously, so that the primary result will be an increase or a reduction in energy stores—i.e., the amount of fat we carry—without the other responding. It is almost impossible to overstate the extent to which this hypothesis now pervades all thinking and research on obesity and weight, and underlies every accepted method of treatment and prevention. As

Greenwood observed, "The vast majority of the notoriously unsuccessful weight control programs are predicated on this assumption."

By contrast, the alternative hypothesis proposes that the primary defect is hormonal and metabolic—in fat storage and/or the burning of fat for fuel (oxidation)—and is in the body, not the brain. This defect causes the excessive accumulation of calories as fat and compensatory urges to eat more and expend less energy. In this hypothesis, overeating and inactivity (hunger and lethargy) are side effects of this underlying metabolic defect; they are not causes. The hypothesis presupposes that calories are effectively "pulled" into the fat cells, rather than pushed, with our fat tissue playing a very active role in this process. It assumes that energy intake and expenditure are *dependent* variables: a change in one induces a compensatory change in the other, because the body constantly works to maintain a healthy body composition and a dependable flow of energy to the cells. Immoderate eating and physical inactivity do not induce obesity, because the body adjusts intake to expenditure and expenditure to intake. Neither eating less nor exercising more addresses the cause of the problem, and that's why these approaches fail. The only effective treatments, according to this hypothesis, would be those that remedy the fundamental regulatory defect.

The only thing missing from this hypothesis as it was originally conceived a century ago, or as reconceived by Pennington and then Bruch and Astwood, was an explanation for the epidemiological observations. In other words, obesity may be caused by a hormonal or metabolic defect determined primarily by genetic inheritance, but the epidemiology tells us that this defect is triggered by environmental factors. Genetics determines our propensity to put on weight, but those genes (nature) have to be triggered by an agent of diet or lifestyle (nurture) to explain the association of obesity with poverty, the present obesity epidemic, and the emergence of obesity in recently Westernized populations. A change in the environment is also necessary to explain why man alone seems to grow chronically obese, not other species of animals. "Something has happened in the past twenty, thirty, forty years in the incidence of obesity, and that has to be environmental," as George Cahill has said about the present obesity epidemic.

The likely explanation is the effect of diet on this regulation of fat metabolism and energy balance. Since insulin, as Astwood noted, is the hormone responsible for promoting the incorporation of fat into our adipose tissue and the conversion of carbohydrates into fat, the obvious suspects are refined carbohydrates and easily digestible starches, which have well-documented effects on insulin. This is what Peter Cleave argued, albeit without understanding the underlying hormonal mechanisms at work, and what the geneticist James Neel, father of the thrifty-gene hypothesis, came to believe as well. And it's the effect of these carbohydrates on insulin that would explain the dietary observations—the futility of calorie restriction, the relative ease of weight loss when carbohydrates are restricted, and perhaps two centuries of anecdotal observations that sweets, starches, bread, and beer are uniquely fattening.

In this hypothesis, obesity is another variation on the theme of insulin dysfunction and diabetes. In Type 1 diabetes, the cause is a lack of insulin. The result is an inability to use glucose for fuel and to retain fat in the fat tissue, leading to internal starvation, as Astwood put it, excessive hunger, and weight loss. In obesity, the cause is an excess of insulin or an inordinate sensitivity to insulin by the fat cells; the result is an overstock of fuel in the adipose tissue and so, once again, internal starvation. But now the symptoms are weight gain and hunger. In obesity, the weight gain occurs with or without satisfying the hunger; in Type 1 diabetes, the weight loss occurs irrespective of the food consumed.

This alternative hypothesis of obesity ultimately vanished in the 1980s, a casualty of the official consensus that fat was the dietary evil and carbohydrates were the cure. Ironically, it disappeared just as all the relevant physiological mechanisms had been worked out and a causal path established from the carbohydrates in the diet through insulin to the regulatory enzymes and molecular receptors in the adipose tissue itself.

This alternative hypothesis of obesity constitutes three distinct propositions. First, as I've said, is the basic proposition that obesity is caused by a regulatory defect in fat metabolism, and so a defect in the distribution of energy rather than an imbalance of intake and expenditure. The second is that insulin plays the primary role in this fattening process, and the compensatory behaviors of hunger and lethargy. The third is that carbohydrates, and particularly refined carbohydrates—and perhaps the fructose content as well, and thus the amount of sugars consumed—are the prime suspects in the chronic elevation of insulin; hence, they are the ultimate cause of common obesity. These latter two propositions—that insulin regulates fat deposition and carbohydrates regulate insulin—have never been controversial, but they've been dismissed as irrelevant to obesity, given the ubiquitous belief that obesity is caused by overeating. That, I will argue, was a mistake.

Through the beginning of World War II, the notion that a defect in fat metabolism causes obesity was known as the *lipophilia* hypothesis. "Lipophilia" means "love of fat." The term was invoked in 1908 by the German internist Gustav von Bergmann to explain why areas of the body differ in their affinity for accumulating fat—a vitally important phenomenon, one would think, since obesity is a malady of fat accumulation. Bergmann considered the energy-balance hypothesis of obesity to be nonsensical: "It seems just as illogical," he wrote, "to say: Child, you shoot up in height because you eat too much or you exercise too little—or you have remained small because you play sports too much. What the body needs to grow, it always finds, and what it needs to become fat, even if it's ten times as much, the body will save for itself from the annual balance."

Just as we grow hair in some places and not typically in others, Bergmann noted, there are places more or less prone to fatten, and some biological factor must regulate that. Some regions of the body are more or less lipophilic than others. This is the kind of observation that can obsess us individually: Why do we have love handles or a double chin? Why fat ankles, thighs, or buttocks? Why is it that some men accumulate excessive fat in the abdomen (a beer belly) and yet are lean elsewhere? Why do some women have significant fat deposits in their breasts and so are considered voluptuous, whereas other women have little or none? These are all variations on the question of which biological factors determine the regional and local distribution of fat.

The example commonly cited in discussions of the nature of this localized lipophilia was that of a twelve-year-old girl in the early 1900s who burned the back of her hand. Her doctors used skin from her abdomen as a graft over the burn. By the time this girl turned thirty, she had grown fat, and the skin that had been transplanted to the back of her hand had grown fat as well. "A second operation was necessary for the removal of the big fat pads which had developed in the grafted skin," explained the University of Vienna endocrinologist and geneticist Julius Bauer, "exactly as fatty tissue had developed in the skin of the lower part of the abdomen." Some biological factor must regulate this, Bauer believed.

Several clinical conditions also demonstrate this phenomenon of local lipophilia. Benign fat masses a few inches in diameter characterize a condition known as *lipomatosis,* and there are fatty tumors known as *lipomas.* In both cases, these masses of fat appear unaffected by any weight loss by the patients themselves; whatever it is that causes fat to accumulate in localized masses seems to be independent of the fat content of the body itself. There's also a rare condition known as *lipodystrophy,* characterized by the inability to store fat in subcutaneous tissue. Those who suffer from it appear abnormally emaciated; lipodystrophy, too, can be localized, and even progressive. In one case reported in 1913, a ten-year-old girl first lost fat from her face; then, over the next three years, this emaciation gradually extended down her trunk and arms. "Adiposity of the lower body," as the report described it, began at age fifteen and eventually became "lower body obesity." By the time she was twenty-four, the patient, who was five foot four and weighed 185 pounds, had effectively all of her body fat localized below her waist.

Bergmann and Julius Bauer, the "noted Vienna authority on internal diseases," as the *New York Times* called him, were the two most prominent proponents of the lipophilia hypothesis, but only Bauer wrote about the hypothesis in English, attempting to influence how obesity would be perceived by physicians in the United States. Bauer's expertise was in the

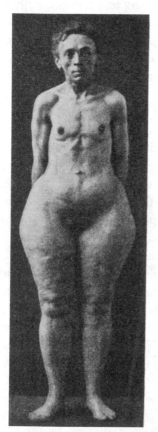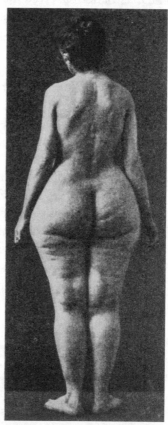

A case of progressive lipodystrophy with lower-body obesity. If emaciation above the waist is followed by obesity below it, can the quantity of calories consumed have anything to do with it?

application of genetics and endocrinology to clinical medicine, a field he arguably pioneered in a 1917 monograph entitled *Constitution and Disease*. Bauer had taken case histories from 275 obese patients and reported that nearly 75 percent had one or both parents who were also obese. He considered this compelling evidence that the condition had a genetic component, which in turn implied the existence of genetically determined hormonal and metabolic factors that would bestow a constitutional disposition to put on excessive fat. "The genes responsible for obesity," Bauer wrote, "act upon the local tendency of the adipose tissue to accumulate fat (lipophilia) as well as upon the endocrine glands and those nervous centers which regulate lipophilia and dominate metabolic functions and the general feelings ruling the intake of food and the expenditure of energy. Only a broader conception such as this can satisfactorily explain the facts."

Lipophilia, as Bauer observed, has nothing to do with energy balance. Where we accumulate fat is regulated by something other than how much we eat or how little we exercise. Someone who has a double chin, fat ankles, or large breasts but is lean elsewhere, or the women of African tribes who have the characteristic fat deposits of the buttocks known as *steatopygia*, did not develop these fat deposits by eating too much. Rather, as Bauer wrote, "A local factor must exist which influences the fat deposition in particular regions independently of the general energy balance or imbalance." If a person becomes emaciated above the waist and then, a few years later, obese below it, as in these cases of progressive lipodystrophy, how can the obese half be blamed on overeating? And, if not, why does overeating become the cause when the obesity exists above the waist as well? The difference between local lipophilia and generalized obesity, Bauer observed, is one of distribution and not quantity.

Whatever mechanisms lead some parts of the human body to be more or less lipophilic, Bauer argued, exist to different extents in individuals as well. Those of us who seem constitutionally predisposed to fatten simply have adipose tissue that is generally more lipophilic than that of lean individuals; our adipose tissue may be more apt to store fat or less willing to give it up when the body needs it. And if our adipose tissue is so predisposed to accumulate excessive calories as fat, this will deprive other organs and cells of nutrients, and will lead to excessive hunger or lethargy. "Like a malignant tumor or like the fetus, the uterus or the breasts of a pregnant woman, the abnormal lipophilic tissue seizes on foodstuffs, even in the case of undernutrition," wrote Bauer in 1929. "It maintains its stock, and may increase it independent of the requirements of the organism. A sort of anarchy exists; the adipose tissue lives for itself and does not fit into the precisely regulated management of the whole organism."

In 1941, when Bauer turned to the question of which biological factors might determine or regulate this lipophilia, the understanding of the function of hormones and enzymes in regulating metabolism was still in its infancy. Bauer based his understanding, as Astwood would twenty years later, largely on clinical observations. Local factors in the adipose tissue itself have to be involved, he thought. How else to explain the lipophilic skin graft? Surely something attached to the skin and the adipose tissue determines how much fat it will hold. Hormonal factors have to be involved. Male sex hormones seem to inhibit the kind of fat formation typically seen in women—men who are castrated or whose testicles are destroyed by disease often develop a fat distribution that is more typically feminine. This type of fat distribution, Bauer wrote, is also present in "obese boys in whom the physiologic production of the testicular hormone is not yet sufficient to prevent the accumulation of adipose tissue of the female type. The larger the quantity of fat deposited, the more striking is the resemblance to the female type. . . ." Female sex hormones do not appear to play a major role in determining *where* fat appears on the body— women who have their ovaries removed put on fat very much like other women. These hormones do, however, seem to affect the *quantity* of fat, which would explain the tendency of women to gain weight after menopause. Bauer also suggested that insulin plays a role, by enhancing the deposition of glucose in the adipose tissue, a phenomenon first demonstrated in the 1920s, and by increasing the general affinity of the adipose tissue for accumulating fat. The nervous system plays a role as well, Bauer said: researchers had demonstrated that they could increase the amount of fat in fat deposits by severing the nerve fibers that run to the relevant tissue.

Through the 1920s, discussions of the lipophilia hypothesis were confined to the German and Austrian research communities. The relevant research appeared almost exclusively in the German medical literature. Clinicians in the United States began to take notice only in 1933, after Eugene Du Bois convinced Erich Grafe, director of the Clinic of Medicine and Neurology at the University of Würzburg in Germany, that the American medical community could benefit from an English translation of Grafe's textbook, *Metabolic Diseases and Their Treatment*. By that time, as Hugo Rony noted, the hypothesis was "more or less fully accepted" in Europe. "It seems to me this conception deserves attentive consideration," Russell Wilder of the Mayo Clinic wrote in 1938. "The effect after meals of withdrawing from the circulation even a little more fat than usual might well account both for the delayed sense of satiety and for the frequently abnormal taste for carbohydrate encountered in obese persons. . . . A slight tendency in this direction would have a profound effect in the course of time."

Knowledge and research on the hypothesis, though, remained largely confined to the German and Austrian research community. When this school of research evaporated with the rise of Hitler and World War II, the notion of lipophilia evaporated with it. Anti-German sentiments in the postwar era may have contributed as well to the disappearance.* In 1955, the year Bergmann died, the primary German textbook on endocrinology and internal medicine included a lengthy discussion of the lipophilia hypothesis in its chapter on obesity, but it was never translated into English. By that time, English had become the international language of science, and the belief that researchers had at least to read German to keep up with the latest advances no longer held sway. (This disappearance of the German and Austrian influence on obesity research is conspicuous in the literature itself. In Rony's *Obesity and Leanness*, published in 1940, 191 of 587 references are from German publications; in the 1949 manual *Obesity . . .* , written by the Mayo Clinic physicians Edward Rynearson and Clifford Gastineau, only thirteen of 422 references are from the German literature, compared with a dozen from Louis Newburgh alone. By the 1970s, when George Bray, John Garrow, and Albert Stunkard wrote and edited the next generation of obesity textbooks and clinical manuals, this German research was treated as ancient history and entirely absent.)

Bauer published three articles on lipophilia in English: in 1931 (with Solomon Silver, an endocrinologist at New York's Mount Sinai Hospital), 1940, and 1941, the latter two after he fled to the United States following the German annexation of Austria. By then, however, Bauer was a scholar without an institution. He eventually took a position with the College of Medical Evangelists in Los Angeles, which was affiliated with the Seventh-day Adventist Church, and he became a senior attending physician at Los Angeles County General Hospital. But these were not institutions that bestowed credibility. Meanwhile, Newburgh's seminal paper establishing a perverted appetite as the definitive cause of obesity was published in 1942, and Newburgh rejected the lipophilia hypothesis with the alacrity with which he rejected any explanation that didn't implicate gluttony as the primary cause.

What made the disappearance of the lipophilia hypothesis so remark-

---

* When Ted Van Itallie, who worked with Jean Mayer in the 1950s, was asked why Mayer paid so little attention to the prewar German literature on obesity, he said, "Mayer hated the Germans. He shot a few of them in World War II."

able is that it could easily be tested in the laboratory, in animal models. These experiments should have settled the issue. Instead, they generated two distinct interpretations of the same evidence. The scientists who study weight regulation in animals came to conclude that obesity is caused by a defect in the regulation of fat metabolism, just as Bauer would have predicted. Their interpretation influenced Pennington and informed his metabolic-defect hypothesis of obesity. The clinicians, nutritionists, and psychologists concerned with human obesity, however, concluded from this same work that the cause of obesity is overeating, as Newburgh would have predicted, or sedentary behavior, as Mayer would, although they had to ignore considerable contrary evidence to do so. When these latter researchers were confronted by results inconsistent with their beliefs, the matter was reconciled by rejecting the relevance of obesity in animals to obesity in humans. As George Cahill explained in 1978, it was "indubitable" that animals had evolved a regulatory system of fat metabolism and energy balance that had to be crippled or *dysregulated* before these animals could gain an unhealthy amount of weight. Such a system "is also probably present in man," Cahill acknowledged, "but markedly suppressed by his intellectual processes."

The value of these animal models of obesity, ideally, is to see if they can refute or exclude one of the two competing hypotheses. For instance, these models can be used to test the hypothesis that obesity is caused by eating too many calories. We have only to ask a simple question: when laboratory animals grow obese, do they require more food to do so than lean animals would normally eat? If they grow excessively fat even when their calorie intake is restricted, then that refutes the notion that obesity (at least in these animals) is *caused* by consuming too many calories. The restriction *controls* for overeating. The explanation we'd be left with is that they're redistributing the calories they do eat. The fundamental defect would seem to be in the body, not in the brain. Overeating would be a side effect of the fattening process. And this might well apply to humans.

In 1934, the Harvard physiologist Milton Lee reported that when rats had their pituitary glands removed and were injected with growth hormone (a product of the pituitary gland), they gained "significantly more weight" than their untreated littermates, even when eating identical quantities of food. The implication was that the weight gain was caused by the effect of growth hormone, independent of calorie consumption. The treated rats grew heavier, larger, and more muscular, Lee reported; the rats found the calories to do so by consuming what fat they had and by expending less energy in physical activity.

As for genetically obese mice, it is invariably the case, as Jean Mayer discovered in the early 1950s, that these animals will fatten excessively regardless of how much they eat. Their obesity is not dependent on the number of calories they consume, although allowing them to consume excessive calories may speed up the fattening process. "These mice will make fat out of their food under the most unlikely circumstances, even when half starved," Mayer had reported. And if starved sufficiently, these animals can be reduced to the same weight as lean mice, but they'll still be fatter. They will consume the protein in their muscles and organs rather than surrender the fat in their adipose tissue. Indeed, when these fat mice are starved, they do not become lean mice; rather, as William Sheldon might have put it, they become emaciated versions of fat mice. Francis Benedict reported this in 1936, when he fasted a strain of obese mice. They lost 60 percent of their body fat before they died of starvation, but still had five times as much body fat as lean mice that were allowed to eat as much as they desired.

In 1981, M.R.C. Greenwood reported that if she restricted the diet of an obese strain of rats known as Zucker rats (or fa/fa rats in the genetic terminology), and did it from birth onward, these rats would actually grow fatter by adulthood than their littermates who were allowed to eat to their hearts' content. Clearly, the number of calories these rats consumed over the course of their life was not the critical factor in their obesity (unless we are prepared to argue that eating *fewer* calories induces greater obesity). What's more, as Greenwood reported, these semi-starved Zucker rats had 50 percent less muscle mass than genetically lean rats, and 30 percent less muscle mass than the Zucker rats that ate as much as they wanted. They, too, were sacrificing their muscles and organs to make fat.

The most dramatic of these animal obesity models is known as *hypothalamic obesity,* and it served as the experimental obesity of choice for researchers from the 1930s onward. It also became another example of the propensity to attribute the cause of obesity to overeating even when the evidence argued otherwise. The interpretation of these experiments became one of a half-dozen critical turning points in obesity research, a point at which the individuals involved in this research chose to accept an interpretation of the evidence that fit their preconceptions rather than the evidence itself and, by so doing, further biased the perception of everything that came afterward.

The hypothalamus sits directly above the pituitary gland, at the base of the brain. It is hard-wired by the nervous system to the endocrine organs, which allows it to regulate the secretion of hormones and thus all physiological functions that themselves are regulated hormonally. Tumors in the hypothalamus have been linked to morbid obesity since 1840, when a German physician discovered such a tumor in a fifty-seven-year-old woman who had become obese in a single year. The manifestation of these tumors can be both grotesque and striking. Stylianos Nicolaidis of the Collège de France recounted the story of being driven to study obesity as a young physician in 1961, when a forty-eight-year-old woman was referred to his hospital for tests after gaining thirty pounds in a single *month*. He never got a chance to do the tests, however, because she literally choked to death over the hospital dinner. "She was eating so fast that she swallowed down the wrong pathway and suffocated," Nicolaidis said. "When I performed the autopsy, I cut the brain in sections and found two very, very tiny metastatic tumors in the hypothalamus."

Because of the proximity of the hypothalamus to the pituitary gland— the two together are known as the hypothalamic-pituitary axis—a question that haunted this research in its early years was which of these two regions played the dominant role in weight regulation. Researchers had managed to induce extreme corpulence in rats, mice, monkeys, chickens, dogs, and cats by puncturing their brains in this pituitary-hypothalamic region. The controversy was definitively resolved in 1939 by Stephen Ranson, who was then director of the Institute of Neurology at Northwestern University and perhaps the leading authority on the neuroanatomy of the brain, and his graduate student Albert Hetherington. The two demonstrated that it was, indeed, the hypothalamus, not the pituitary, that regulated adiposity in the rats; lesions in a region called the *ventromedial hypothalamus* (VMH) would induce corpulence even in those animals that had their pituitary glands removed.

John Brobeck, a Yale researcher who had done his Ph.D. work with Ranson, was the first to propose a mechanistic explanation for the phenomenon. Brobeck had replicated Hetherington's experiments in his Yale laboratory and then read Newburgh's articles arguing a perverted appetite as the cause of obesity. Now Brobeck perceived his research as providing experimental confirmation in laboratory animals of Newburgh's hypothesis. The hypothalamic lesions, Brobeck argued, served to damage what amounted to a center of hunger regulation in the hypothalamus. The lesions made the rats hungry, and so the rats over ate and grew obese. He would later write about his astonishment at how voraciously these surgically lesioned rats ate. Because obesity in most of his rats (but not all) appeared only after the rats began eating ravenously, Brobeck reasoned incorrectly that "the laws of thermodynamics suggest that . . . food intake determines weight gain." Brobeck coined the term *hyperphagia* to describe the extraordinary hunger manifested by these animals, and *hyperphagia* would become the accepted technical term for a perverted appetite that leads to obesity.

The alternative hypothesis, that the obesity in these animals was a disorder of fat metabolism, came from Ranson and Hetherington. Whereas Brobeck interpreted his argument in the context of Newburgh's beliefs, Ranson interpreted his from the context of thirty years of brain research. Some of the lesioned animals ate voraciously, Ranson noted, which might have been due to hunger alone, but others ate normally and still grew obese. (Several of Brobeck's rats also grew obese while eating no more than lean rats did, but Brobeck dismissed their relevance to his overeating hypothesis on the basis that some other effect "related to the feeding habits" of these animals might be responsible.*) Ranson also noted "the tremendously decreased activity of these obese rats."

Ranson argued that Brobeck's hyperphagia hypothesis missed the bigger picture. "Insistence upon the primary importance" of either overeating or inactivity "would in all probability represent oversimplification of the problem, and this for at least two reasons," Ranson wrote.

> In the first place, the two factors are complementary in their effect upon body weight. Both would tend to increase it. A very sedentary life, combined with a high caloric intake would seem to be an ideal combination for building up a thick *panniculus adiposus* [layer of fat]. Secondly, these two factors may be only symptomatic, and not fundamental. It is not difficult to imagine, for example, a condition of hidden cellular semistarvation caused by a lack of easily utilizable energy-producing material, which would soon tend to force the body either to increase its general food intake or to cut down its energy expenditure, or both.

Damage to the ventromedial hypothalamus caused a defect that directed nutrients away from the tissues and organs where they were needed for fuel and into the fat tissue, Ranson argued. It made the animals more

---

* Brobeck paired each of a dozen lesioned rats with a healthy control rat and fed the lesioned rat precisely the same amount of food that the control rat had consumed on the previous day. "In three pairs of animals," Brobeck wrote, "the rat with lesions gained more rapidly than the control when they were fed the same amount of food." Thus overeating could not be the cause of the excessive fattening, because these rats weren't overeating.

lipophilic. This reduced the supply of fuel to the other cells of the body and so caused "hidden cellular semistarvation," or what Astwood later called "internal starvation." That in turn led to the voracious hunger—hyperphagia—that Brobeck had considered the primary defect. As long as nutrients continued to be channeled into fat and away from the cells of other tissues and organs, the animals would remain hungry. If they couldn't satisfy this hunger by eating more—when their food supply was restricted, for instance—they would respond by expending less energy.

Brobeck's scenario—that the primary role of the ventromedial hypothalamus is to regulate food intake—would survive into the modern era of obesity research, but Ranson's insights were far more profound. Only Ranson could explain all the observations, and he did so based on an ongoing revolution in the understanding of the brain, and particularly the role of the hypothalamus. This was Ranson's expertise. The hypothalamus is the "concertmaster" of homeostasis, as *Time* wrote in 1940, reporting on a two-day conference dedicated to discussing the "orchestral effects" of the hypothalamus and paying tribute to Ranson, who had done much of the research.

Just before Ranson and Hetherington took to inducing corpulence in rats, Ranson had studied the hypothalamic regulation of fluid balance. This influenced his interpretation of the later research. Our bodies conserve fluids and water, just as they do fuel. Even our saliva and gastric juices are reabsorbed and reused. Just as damage to the ventromedial hypothalamus can induce obesity, damage elsewhere in the hypothalamus can induce diabetes insipidus. The symptoms of this rare condition are excessive urination and a tremendous and constant thirst. These symptoms appear in uncontrolled diabetes mellitus as well, but in diabetes insipidus, insulin secretion is not impaired, so blood sugar and fat metabolism remain regulated and no sugar appears in the urine.

The similarities between diabetes mellitus and diabetes insipidus had led Ranson and other physiologists to conclude that the homeostatic regulation of fluid balance was akin to that of blood sugar. That both diabetes insipidus and obesity could be caused by hypothalamic lesions informed Ranson's interpretation of the underlying disorders. In the case of diabetes insipidus, the lesions inhibit the ability of the kidneys to conserve water by suppressing the secretion of an anti-diuretic hormone that normally works in the healthy animal to inhibit urination. This failure in the homeostatic regulation of fluids causes the kidney to excrete too much water, and that leads to a compensatory thirst to replace the fluid that's lost.

The same cause and effect are evident in Type 1 diabetes mellitus. The inability of diabetics to utilize the food they eat, and particularly the carbo-hydrates, results in a state of starvation and extreme hunger. Diabetics also urinate more, because the body gets rid of the sugar that accumulates in the bloodstream by allowing it to overflow into the urine, and this is why diabetics will be abnormally thirsty as well.

Lesions to the ventromedial hypothalamus can induce tremendous hunger *and* cause obesity, but now Ranson considered it naïve to assume that the hunger caused the obesity. Rather, the hunger was another conse-quence of a breakdown in homeostasis—the loss of calories into the fat tis-sue. This is why the animals get fat even when they aren't allowed to satisfy their appetite. And this is why these lesioned animals are always hungry, at least until they put on enough fat so that the excess counteracts the dam-age caused by the hypothalamic lesion. Sedentary behavior is another way their bodies compensate for the loss of calories to the fat tissue. As Ranson perceived it, both hunger and physical inactivity are manifestations of the internal starvation of the tissues. These are the ways that the homeostatic regulation of energy balance compensate for the loss of nutrients into the fat tissue.

It's hard to avoid the suggestion that one major factor in how this research played out was the preconceptions of the investigators and their urge to make a unique contribution to the science. Ranson had suggested that all the more obvious manifestations of hypothalamic lesions were the conse-quences of a primary defect in the homeostatic control of energy balance that made the animals accumulate excessive fat in the adipose tissue. Brobeck and the other investigators who took to studying hypothalamic obesity would conclude that whatever phenomenon they happened to find most remarkable in their own postoperative rodents was the critical factor, or at least a critical factor, requiring intensive investigation. By doing so, as Ranson had cautioned in the early 1940s, they oversimplified the physiol-ogy and only directed attention away from the fundamental problem. Jean Mayer, for instance, would discuss hypothalamic obesity in the plural—as the "classic type of experimental obesities"—and he would say that one such obesity was caused by lack of physical activity, as in his mice. Philip Teitelbaum, who did his research as a doctoral student at Johns Hopkins in the early 1950s, observed that VMH-lesioned rodents, at the peak of their obesity, became finicky eaters, and he concluded that this was an obvious manifestation of the *behavior* of taste aversion. This observation established his reputation in the field and also the common belief that the ventromedial hypothalamus controls food preference, too, and the motivation to eat. "Of course they overate," he said of his obese rodents; "that's why they became obese." But he simultaneously acknowledged that they were also so inactive that they would fatten even without overeating.

In 1951, Brobeck and his colleague Bal Anand reported that lesioning a different region of the hypothalamus—the lateral hypothalamus—would induce rats to stop eating and lose weight and even die of starvation. Ranson's lab had reported this phenomenon in rats, cats, and monkeys in the 1930s, but now Brobeck and Anand reinterpreted it to support Brobeck's belief that the hypothalamus regulates eating behavior. Brobeck proposed that the lateral hypothalamus is a "feeding center" that motivates animals to eat, and the ventromedial hypothalamus works as a "satiety center" to inhibit eating.

In August 1942, just three months after Ranson and Hetherington published their research, Ranson died of a heart attack. If there was a single event that derailed the course of obesity research in the United States, this may have been it. With World War II raging and his adviser gone, Hetherington left Northwestern to do research for the U.S. Air Force. This left Brobeck, still a medical student at the time, as the leading authority on these experiments, and so it was Brobeck's emphasis on overeating—hyperphagia—as the cause of the obesity in these brain-damaged animals that dominated thinking in the field, despite its inability to explain the observations. Though later editions of Ranson's textbook *The Anatomy of the Nervous System* would continue to refer to the ventromedial hypothalamus as a regulator of fat metabolism, the investigators writing about human obesity would refer to the VMH as a regulator of hunger and ingestive behavior.

Once human-obesity research became the domain of psychologists and psychiatrists in the 1960s, studies of hypothalamic obesity left behind once and for all the greater context of homeostasis and the use and storage of metabolic fuels, and focused instead on how Brobeck's dual centers of the hypothalamus allegedly regulate eating *behavior*. This served to further the conviction that defects in this region of the brain cause overeating, and overeating causes obesity. Hunger, and the overeating that accompanies it, would be considered exclusively a psychological phenomenon, not a physiological one. (Because these psychologists would consider eating behavior to be the subject of their research, they would often screen their animals after surgery and those that didn't eat voraciously would be "discarded." They would then omit these animals from their subsequent analyses, even if the discarded animals became obese as well.) Hunger was something that occurred only in the head, and so it could be decoupled from the needs of the body, at least with sufficient willpower.

Animal research continued to confirm Ranson's hypothesis, even though its author had died, no matter whether the fattening was induced by hypo-thalamic lesions, genetic defects, or as the naturally occurring seasonal weight gain of hibernators. In 1946, for example, the Johns Hopkins phys-iologist Chandler Brooks reported that his albino mice become "definitely obese" after VMH lesions, and that they gained *six times* as much weight *per calorie* of food consumed as normal mice. In other words, it wasn't how much these mice ate that determined their ultimate weight, or the number of calories, but how these calories were utilized. They were turned into fat, not used for fuel.

Though Brooks reported that he could prevent his albino mice from growing obese, he could do so only by imposing "severe and permanent" food restriction. If he subjected them to "long continued limitation of food," the animals would lose some weight, but they would never lose the drive to fatten or the hunger that went with it. Periods of fasting, Brooks noted, were "followed by an augmentation of appetite and development of a greater degree of obesity than had been attained before fasting." And so Brooks's lesioned mice, as Hilde Bruch might have noted, were acting exactly like normal healthy humans and obese humans after a semi-starvation diet. These VMH lesions also resulted in changes in the repro-ductive cycles of the animals, and in their normal nocturnal eating patterns, which Ranson and Hetherington had also reported; once the ani-mals became obese, they slept more than normal animals, all of which suggested that the VMH lesions had profound effects on the entire home-ostatic system and could not be written off as simply affecting hunger and thus food intake.

When physiologists began studying animal hibernation in the 1960s, they again demonstrated this decoupling of food intake from weight gain. Hibernating ground squirrels will double their body weight in late sum-mer, in preparation for the winter-long hibernation. But these squirrels will get just as fat even when kept in the laboratory and not allowed to eat any more in August and September than they did in April. The seasonal fat deposition is genetically programmed—the animals will accomplish their task whether food is abundant or not. If they didn't, a single bad summer could wipe out the species.

This same decoupling of food intake and weight would also be demon-strated when researchers studied what are now known as dietary models of obesity. Certain strains of rats will grow obese on very high-fat diets, and others on high-sugar diets. In both cases, the animals will get fatter even if they don't consume any more calories than do lean controls eating their usual lab chow. This same decoupling occurs in animals that are regaining weight after lengthy periods of fasting. "It doesn't matter how long you food-deprive the animal," said Irving Faust, who did this work in the 1970s; "the recovery of body weight is not connected to the amount of food eaten during the recovery phase." And this same decoupling of calories and weight has also been made consistently, if not universally, in the recent research on transgenic animals, in which specific genes are manipulated.

What may have been the most enlightening animal experiments were carried out in the 1970s by physiologists studying weight regulation and reproduction. In these experiments, the researchers removed the ovaries from female rats. This procedure effectively serves to shut down production of the female sex hormone estrogen (technically estradiol). Without estrogen, the rats eat voraciously, dramatically decrease physical activity, and quickly grow obese. When the estrogen is replaced by infusing the hormone back into these rats, they lose the excess weight and return to their usual patterns of eating and activity. The critical point is that when researchers remove the ovaries from these rats, but restrict their diets to only what they were eating before the surgery, the rats become just as obese, just as quickly; the number of calories consumed makes little difference.

George Wade, the University of Massachusetts biologist who did much of this research, described it as a "revelation" that obesity could be brought on without overeating, just as Pennington had described it as revelatory that weight could be lost without undereating. "If you keep the animals' food intake constant and manipulate the sex hormones, you still get substantial changes in body weight and fat content," Wade said. Another consequence of removing the ovaries was that the rats hoarded more food in their cages, which is analogous to storing excess calories as fat. Infusing estrogen back into these rats suppressed the food-hoarding, just as it prompted weight loss. "The animals overeat and get fat," said Tim Bartness, who worked on this research as part of his doctoral studies with Wade in the 1970s, "but they are overeating because they're socking all the calories away into adipose tissue and they can't get to those calories. They're not getting fat because they're overeating; they're overeating because they're getting fat. It's not a trivial difference. The causality is quite different."

One critical idea here is that survival of a species is dependent on successful reproduction, and that in turn depends first and foremost on the availability of food. Fat accumulation, energy balance, and reproduction are all intimately linked, and all regulated by the hypothalamus. This is why food deprivation suppresses ovulation, and why the same kind of hormonal control of reproduction ensures that herbivores, such as sheep, tend to give birth in the springtime, when food is available. The link between food availability and reproduction was something that Charles Darwin had also observed: "Hard living . . . retards the period at which animals conceive," he wrote.

The lesson of these animal experiments is that understanding energy balance and weight control requires Claude Bernard's harmonic-ensemble perspective of homeostasis: an appreciation of the entire organism and the entire homeostatic web of hormonal regulation. "Fertility is linked to food supply, physical exercise involved in foraging for food and avoiding predators, and energy expenditure associated with temperature regulation and other physiological processes," Wade explains. These functions are controlled by a tight orchestration of both sex hormones and those hormones that control the "partitioning and utilization of metabolic fuels," and this is accomplished in ways that are "reciprocal, redundant and ubiquitous."

The idea that obesity in humans is caused, as it is in animals, by a defect in the homeostatic maintenance of energy distribution and fat metabolism—that we overeat because we're getting fat, and not vice versa—barely survived into the second half of the twentieth century, although the evidence has always supported it.

This homeostatic hypothesis effectively vanished from the mainstream thinking on human (as opposed to animal) obesity with the coming of World War II. The war destroyed the German and Austrian community of clinical investigators, who had done the most perceptive thinking about the causes of obesity and had a tradition of rigorous scientific research dating back two hundred years. In the United States, it resulted in a suspension of obesity research that lasted for most of a decade. Meanwhile, Stephen Ranson had died, Hugo Rony and Julius Bauer retired. The generation of physiologists who had founded the field of nutrition in the United States and actually studied human metabolism disappeared with them. Francis Benedict's Nutrition Laboratory at the Carnegie Institution did contract work for the armed services during the war and then was shut down in 1946. The Russell Sage Institute of Pathology, where Graham Lusk and Eugene Du Bois did their research, was also gone by the 1950s. Lusk himself died in 1932, Francis Benedict retired in 1937. Du Bois retired four years later.

Among the few investigators whose careers spanned the war years, Louis Newburgh was the most influential and conspicuous. As late as

1948, Newburgh was still promoting his perverted-appetite hypothesis of obesity. The first obesity textbook published after the war, *Obesity* . . . (1949), by Edward Rynearson and Clifford Gastineau, would be considered the standard text on obesity for twenty years. It faithfully communicated Newburgh's belief that obesity is caused by overeating. Any suggestion to the contrary, wrote Rynearson and Gastineau, constituted little more than "an excuse for avoidance of the necessary corrective measures."

An entire generation of young researchers and clinicians effectively started the study of obesity from scratch after the war. They did so with little concern for whatever understanding had been achieved before they arrived, and so they embraced a hypothesis of causation that flew in the face of much of the evidence. The institutionalized skepticism and meticulous attention to experimental detail that are necessary to do good science—"being ruthless in self-criticism and . . . taking pains in verifying facts," as the Nobel laureate chemist Hans Krebs said—had also been left behind.

Chapter Twenty-two

THE CARBOHYDRATE HYPOTHESIS, II: INSULIN

Every woman knows that carbohydrate is fattening.

REGINALD PASSMORE AND YOLA SWINDELLS,
British Journal of Nutrition, 1963

The fact that insulin increases the formation of fat has been obvious ever since the first emaciated dog or diabetic patient demonstrated a fine pad of adipose tissue, made as a result of treatment with the hormone.

REGINALD HAIST AND CHARLES BEST,
The Physiological Basis of Medical Practice, 1966

IN 1929, WHEN LOUIS NEWBURGH FIRST rejected the possibility of an "endocrine abnormality" as the cause of obesity, and insisted instead that all fat people had a perverted appetite, hormones were still widely known as "internal secretions" and endocrine glands as "ductless glands." The first purification of growth hormone had been only nine years earlier, the purification of insulin only eight years before. In 1955, when *The Journal of the American Medical Association* declared unconditionally that those "theories that attributed obesity to an endocrine disturbance have been shown to be erroneous," it was five years before Rosalyn Yalow and Solomon Berson would publish the details of the first method for measuring the insulin level in the blood, and a few more years after that before the ensuing revelations that obesity was associated with the endocrine disturbances and abnormalities of hyperinsulinemia and insulin resistance.

In other words, the editors at *JAMA*—and the clinical investigators they represented—were declaring that hormones, as a rule, play little role in the genesis of obesity, even before the relevant hormones could be measured accurately in the human bloodstream. In fact, it's hard to imagine, as Julius Bauer noted, that hormones *wouldn't* play a role. Here again we have that familiar scenario we first discussed with regard to dietary fat and heart disease. Once the "truth" has been declared, even if it's based on incomplete evidence, the overwhelming tendency is to interpret all future observations in support of that preconception. Those who *know* what the answer is lack the motivation to continue looking for it. Entire fields of science may then be ignored, on the assumption that they can't possibly be relevant.

In 1968, Jean Mayer pointed out that obesity researchers may have "eliminated" hormones "from legitimate consideration" as a cause of obesity, or so they believed, but the evidence continued to accumulate just the same. Researchers had demonstrated that insulin seemed to have a dramatic effect on hunger, that insulin was the primary regulator of fat deposition in the adipose tissue, and that obese patients had chronically high levels of insulin. Other hormones, such as adrenaline, had been shown to increase the mobilization of fat from the fat cells. "It is probable that different concentrations of these hormones in blood are characteristic of different body types and fat contents," Mayer wrote.

> At the beginning of this century, when hormones were first discovered, it was commonly believed that obesity would be found to be due to the absolute excess or deficiency of a single hormone. When this was found to be almost never true, the popular medical position swung to the other extreme: "obesity is almost never due to hormonal disturbances; it is almost always due to overeating." Actually, the reasonable position ought to be: "in order to be obese, you always have to eat more than you expend for a certain period. How often this is due to a slight shift of relative or absolute hormone concentrations, each one of which is in the 'normal' range, we don't know."

Among the hormones that play a role in regulating fat metabolism and thus potentially play a causative role in obesity, insulin was always an obvious choice. Some failure in what clinicians a century ago called the *insular** apparatus* of the pancreas is the fundamental defect in diabetes, and diabetes is intimately associated with obesity in those who develop the disease as adults, and with emaciation, which was the end stage of the disease in the pre-insulin era. In 1905, Carl von Noorden invoked this intimate association between diabetes and weight to formulate the third of his speculative hypotheses of obesity, what he called *diabetogenous obesity*. His ideas were remarkably prescient. They received little attention because insulin had not yet been discovered, let alone the technology to measure it.

Von Noorden suggested that obesity and diabetes are different consequences of the same underlying defects in the mechanisms that regulate carbohydrate and fat metabolism. In severe diabetes (Type I), he noted, the patients are unable either to utilize blood sugar as a source of energy or to

---

* In reference to the islets of Langerhans, the pancreatic cells that secrete insulin.

convert it into fat and store it. This is why the body allows the blood sugar to overflow into the urine, which is a last resort since it wastes potentially valuable fuel. The result is glycosuria, the primary symptom of diabetes. These diabetics must be incapable of storing or maintaining fat, von Noorden noted, because they eventually become emaciated and waste away. In obese patients, on the other hand, the ability to utilize blood sugar is impaired, but not the ability of the body to convert blood sugar into fat and store it. "Obese individuals of this type have already an altered metabolism for sugar," von Noorden wrote, "but instead of excreting the sugar in the urine, they transfer it to the fat-producing parts of the body, whose tissues are still well prepared to receive it." As the ability to burn blood sugar for energy further deteriorates and "the storage of the carbohydrates in the fat masses [also suffers] a moderate and gradually progressing impairment," sugar appears in the urine, and the patient becomes noticeably diabetic. Using the modern terminology, this is the route from obesity to Type 2 diabetes. "The connection between diabetes and obesity," as von Noorden put it, "ceases in the light of my theory to be any longer an enigmatical relation, and becomes a necessary consequence of the relationship discovered in the last few years between carbohydrate transformation and formation of fat."

After the discovery of insulin in 1921, the potential role of insulin as a fattening hormone would become a long-running controversy. Those physicians who believed, as Louis Newburgh did without reservation, that obesity was an eating disorder, rejected the idea that insulin could fatten humans, if for no other reason than that this suggested the existence of a defective hormonal mechanism that could lead to obesity. The evidence, however, suggested exactly that. When insulin was injected into diabetic dogs in the laboratory, or diabetic human patients in the clinic, they put on weight and body fat. As early as 1923, clinicians were reporting that they had successfully used insulin to fatten chronically underweight children— patients who would be diagnosed today as anorexic—and to increase their appetite in the process.

In 1925, Wilhelm Falta, a student of von Noorden and a pioneer of the science of endocrinology in Europe, began using insulin therapy to treat underweight and anorexia in adults as well. Falta had argued, even in the pre-insulin era, that whatever pancreatic hormone was absent or defective in diabetes governed not only the use of carbohydrates for fuel, but also the assimilation of fat in adipose tissue. "A functionally intact pancreas is necessary for fattening," Falta wrote. He also noted that the only way to fatten anyone efficiently was to include "abundant carbohydrates in the diet."

Otherwise, the body would adjust to eating "very much more than the appetite really craves," by either lessening appetite still further or creating "an increased demand for movement." The only way to get around this natural balance of intake and expenditure is by increasing the secretion from the pancreas. "We can conceive," Falta speculated, "that the origin of obesity may receive an impetus through a *primarily strengthened function of the insular apparatus,* in that the assimilation of larger amounts of food goes on abnormally easily, and hence there does not occur the setting free of the reactions that in normal individuals work against an ingestion of food which for a long time supersedes the need." After the discovery of insulin, Falta reported that giving it to patients would increase their appetite for carbohydrates specifically, and the carbohydrates in turn would stimulate the patient's own insulin production. It would create a vicious cycle— although, in the case of anorexic and underweight patients, one that might return them to a normal appetite and normal weight.

By the 1930s, clinicians throughout Europe and the United States had taken to using insulin therapy to fatten their pathologically underweight patients. These patients could gain as much as six pounds a week eating meals "rich in carbohydrates" after receiving injections of small doses of insulin, reported Rony, who used insulin therapy on seven anorexic patients in his own clinic; it worked on five of them. None of these patients had been able to gain weight, but now they added an average of twenty pounds each in three months. "All reported a more or less pronounced increase of appetite," Rony wrote, "and occasional strong feelings of hunger." Until the 1960s, insulin was also used to treat severe depression and schizophrenia. Among the more renowned patients subjected to what was then called insulin-shock therapy was the Princeton mathematician John Nash, made famous by Sylvia Nasar's 1998 biography, *A Beautiful Mind.* Its efficacy for treating mental illness was debatable, but as Nasar observed, "all the patients gained weight." Another memorable recipient was the poet Sylvia Plath, who experienced a "drastic increase in weight" on the treatment. (In her autobiographical novel, *The Bell Jar,* Plath's protagonist, Esther Greenwood, gains twenty pounds on insulin therapy—"I just grew fatter and fatter," she says.)

Insulin's fattening properties have long been particularly obvious to diabetics and the physicians who treat them. Because diabetics will gain weight with insulin therapy, even those who are obese to begin with, clinicians have always had difficulty convincing their patients to continue taking their insulin. When they start to fatten, they naturally want to slack off on the therapy, so the need to control blood sugar competes with the desire to remain lean, or at least relatively so. This is also a clinical dilemma, because the weight gain will also increase the risk of heart disease. In the chapter on insulin therapy in the 1994 edition of *Joslin's Diabetes Mellitus*, the Harvard diabetologist James Rosenzweig portrayed this insulin-induced weight gain as uncontroversial: "In a number of studies of patients treated with insulin for up to 12 months, weight gains of 2.0 to 4.5 kg [roughly four to ten pounds] were reported. . . ." This weight gain, he wrote, then leads to "the often-cited vicious cycle of increased insulin resistance, leading to the need for more exogenous insulin, to further weight gain, which increases the insulin resistance even more."*

If insulin fattens those who receive it, as the evidence suggests, then how does it work? The prewar European clinicians who used insulin therapy to treat anorexics accepted the possibility, as Falta suggested, that the hormone can directly increase the accumulation of fat in the fat tissues. Insulin was "an excellent fattening substance," Erich Grafe wrote in *Metabolic Diseases and Their Treatment*. Grafe believed that the fattening effect of insulin is likely "due to improved combustion of carbohydrate and increased synthesis of glycogen and fat." In the United States, however, the conventional wisdom came from Louis Newburgh and his colleagues at the University of Michigan. When insulin increases weight, Newburgh said, it does so either through the power of suggestion—a placebo effect—or by a reduction of blood sugar to the point where the patient eats to avoid very low blood sugar (hypoglycemia) and the accompanying symptoms of dizziness, weakness, and convulsions.

When Rony reviewed the experimental and clinical reports in 1940, he considered any conclusion to be premature. Because obese individuals tend to have *high* blood sugar, rather than low, Rony said, it was hard to imagine how insulin, which lowered blood sugar, could cause obesity. "Still," he noted, "it might be possible that in obese subjects a latent or conditional form of hyperinsulinism exists which would promote fat deposition without causing hypoglycemia." This was not supported by conclusive evidence, he added, and so it "remains, for the time being, at best a working hypothesis."

Only Newburgh's interpretation of the evidence, however (and only the

---

* "Diet therapy and weight loss are extremely important in reversing this process," Rosenzweig added, "but the long-term results of these therapies have generally been disappointing, even in patients not receiving insulin."

obesity research community in the United States), survived the war years. Afterward, clinical investigators would state unambiguously—as Edward Rynearson and Clifford Gastineau did in their 1949 clinical manual *Obesity* . . .—that insulin puts weight on only by lowering blood sugar to the point where patients *overeat* to remain conscious. This hypoglycemia was considered a rare pathological condition, one with no relevance to everyday life, and so *only* in that condition were elevated insulin levels to be considered a causal agent in weight gain and common obesity.

In 1992, the University of Texas diabetologist Denis McGarry published an article in *Science* with the memorably idiosyncratic title "What If Minkowski Had Been Ageusic? An Alternative Angle on Diabetes." The German physiologist Oskar Minkowski was the first to identify the role of the pancreas in diabetes. The word "ageusic" refers to a condition in which the sense of taste is absent. "Legend has it," McGarry wrote, "that on a momentous day in 1889 Oskar Minkowski noticed that urine collected from his pancreatectomized* dogs attracted an inordinate number of flies. He is said (by some) to have tasted the urine and to have been struck by its sweetness. From this simple but astute observation he established for the first time that the pancreas produced some entity essential for control of the blood sugar concentration, which, when absent, resulted in diabetes mellitus." Some thirty years later, when Frederick Banting and Charles Best in Toronto identified insulin as the relevant pancreatic secretion, McGarry wrote, they naturally did so in the context of Minkowski's observations about blood sugar, and thus "diabetes mellitus has been viewed ever since as a disorder primarily associated with abnormal glucose metabolism." But if Minkowski had been ageusic and so missed the sweet taste of the urine, McGarry speculated, he might have noticed instead the smell of acetone, which is produced in the liver from the conversion of fat into ketone bodies. "He would surely have concluded that removal of the pancreas causes fatty acid metabolism to go awry," McGarry wrote. "Extending this hypothetical scenario, the major conclusion of Banting's work might have been that the preeminent role of insulin is in the control of fat metabolism."

McGarry's parable focused on diabetes, but the point he made extends to virtually everything having to with insulin. Just as diabetes has traditionally been perceived as a disorder of carbohydrate metabolism—even

* In which the pancreas had been removed.

though fat metabolism is also dysfunctional—insulin has always been perceived as a hormone that primarily functions to regulate blood sugar, though, as we've discussed, it regulates the storage and use of fat and protein in the body as well. Because blood sugar could be measured easily through the first half of the twentieth century, but not yet the fats in the blood, the focus of research rested firmly on blood sugar.

From the 1920s through the 1960s, a series of discoveries in the basic science of fat metabolism led to a revolution in the understanding of the role of insulin and the regulation of fat tissue in the human body. This era began with a handful of naïve assumptions: that fat tissue is relatively inert (a "garbage can," in the words of the Swiss physiologist Bernard Jeanrenaud); that carbohydrates are the primary fuel for muscular activity (which is still commonly believed today); and that fat is used for fuel only after being converted in the liver into supposedly toxic ketone bodies. The forty years of research that followed would overturn them all—but it would have effectively no influence on the mainstream thinking about human obesity.

Those who paid attention to this research either had no influence themselves—Alfred Pennington comes to mind—or were so convinced that obesity is caused by overeating that they couldn't imagine why the research would be relevant. From the 1950s onward, clinical investigators studying and treating obese patients, as Hilde Bruch commented, seemed singularly uninterested in this research. "Until recently, knowledge of the synthesis and oxidation of fat was quite rudimentary," Bruch wrote in 1957. "As long as it was not known how the body builds up and breaks down its fat deposit, the ignorance was glossed over by simply stating that food taken in excess of body needs was stored and deposited in the fat cells, the way potatoes are put into a bag. Obviously, this is not so." By 1973, after details of the regulation of fat metabolism and storage had been worked out in fine detail, Bruch found it "amazing how little of this increasing awareness . . . is reflected in the clinical literature on obesity."

There are three distinct phases of the revolution that converged by the mid-1960s to overturn what Bruch called the "the time-honored assumption that fat tissue is metabolically inert," and the accompanying conviction that fat only enters the fat tissue after a meal and only leaves it when the body is in negative energy balance.

The first phase began in the 1920s, when biochemists realized that the cells of adipose tissue have distinct structures and are not, as was previously believed, simply connective tissue stuffed with a droplet of oily fat. Researchers then demonstrated that the adipose tissue is interlaced with blood vessels such that "no marked quantity of fat cells escapes close contact with at least one capillary," and that the fat cells and these blood vessels are regulated by "abundant" nerves running from the central nervous system.

This led to the revelation that the fat in the cells of the adipose tissue is in a continual state of flux. This was initially the work of a German biochemist, Rudolf Schoenheimer. In the early 1930s, while working at the University of Freiburg, Schoenheimer demonstrated that animals continually synthesize and degrade their own cholesterol, independent of the amount of cholesterol in the diet. After Hitler came to power in January 1933, Schoenheimer moved to New York, where he went to work at Columbia University. It was in New York that Schoenheimer collaborated on the development of a technique for measuring serum cholesterol and, by doing so, launched the medical profession's obsession with cholesterol levels. Then, with David Rittenberg, he developed the technique to *label* or *tag* molecules with a heavy form of hydrogen known as deuterium* so that their movement through the metabolic processes of the body could be followed. Schoenheimer and Rittenberg put this technique to work studying the metabolism of fat, protein, and carbohydrates in the body.

Among their discoveries is that both dietary fat *and* a considerable portion of the carbohydrates we consume are stored as fat—or, technically, triglycerides—in the adipose tissue before being used for fuel by the cells. These triglycerides are then continuously broken down into their component fatty acids, released into the bloodstream, moved to and from organs and tissues, regenerated, and merged with fatty acids from the diet to re-form a mixture of triglycerides in the fat cells that is, as Schoenheimer put it, "indistinguishable as to their origin." Fat stored as triglycerides in the adipose tissue, and the fatty acids and triglycerides moving through the bloodstream are both part of the same perpetual cycle of fat metabolism. "Mobilization and deposition of fat go on continuously, without regard to the nutritional state of the animal," as the Israeli biochemist Ernst Wertheimer explained in 1948, in a seminal review of this new science of fat metabolism.[†] "The 'classical theory' that fat is deposited in the adipose tissue only when given in excess of the caloric requirement has been

---

* Schoenheimer and Rittenberg worked in Harold Urey's lab at Columbia. Urey had recently discovered deuterium and won the 1934 Nobel Prize in Chemistry for the discovery.

† Wertheimer began his career at the University of Halle in Germany and was expelled from his position in the same purge that sent Schoenheimer to New York. Wertheimer immigrated to Jerusalem, where in the 1940s he became head of pathophysiology and biochemistry at the Hebrew University.

finally disproved," Wertheimer wrote. Fat accumulates in the adipose tissue when these forces of deposition exceed those of mobilization, he explained, and "the lowering of the fat content of the tissue during hunger is the result of mobilization exceeding deposition."

The controlling factors in this movement of fat to and from the fat tissue have little to do with the amount of fat present in the blood, thus little to do with the quantity of calories consumed at the time. Rather, they must be controlled, Wertheimer wrote, by "a factor acting directly on the cell," the kind of hormonal and neurological factors that Julius Bauer had discussed. Over the next decade, investigators would begin to refer to these factors that increase the synthesis of fat from carbohydrates and the deposition of fat in the adipose tissue as *lipogenic,* and those that induce the breakdown of fat in the adipose tissue and its subsequent release into the circulation as *lipolytic.*

The second phase of this revolution began in the 1930s, with the work of Hans Krebs, who showed how our cells convert nutrients in the bloodstream into usable energy. The Krebs cycle, for which Krebs shared the Nobel Prize in Medicine in 1953, is a series of chemical reactions that generate energy in the mitochondria of cells, which are those compartments commonly referred to as the "power plants" of the cells. The Krebs cycle starts with the breakdown products of fat, carbohydrates, and protein and then transforms them into a molecule known as adenosine triphosphate, or ATP, which can be thought of as a kind of "energy currency," in that it carries energy that can be used at a later time.* This cycle of reactions will generate energy whether the initial fuel is fat, carbohydrates, or protein. Indeed, Krebs had initiated his research assuming, as was common at the time, that carbohydrate was "the main energy source of muscle tissue." But he came to realize that fat and protein also supply fuel for muscle tissue, and that there was no reason why carbohydrates should be the preferred fuel. "All three major constituents of food supply carbon atoms . . . for combustion," he wrote.

By 1950, the addition of the Krebs cycle to the revelations about fat metabolism from Schoenheimer and others provided the foundation for understanding the fundamental mechanisms that assure a constant supply of energy to our tissues and organs, regardless of how the demand might change in response to the environment and over the course of seconds, hours, days, or seasons. It is based on a generator—the Krebs

---

* ATP gives up a phosphate molecule, becoming adenosine diphosphate, or ADP, and releases energy in the process.

cycle—that burns fat, carbohydrates, and protein with equal facility, and then a supply chain from the adipose tissue that ensures the circulation of fuels at a level that will always be more than adequate for the needs at hand. "The high degree of metabolic activities present in the fat tissues," as Hilde Bruch explained, "becomes understandable as necessary for a continuous reserve for energy requirements. Instead of a savings account for unneeded surplus, as fat deposits have commonly been described, a coin purse would be a far closer analogy. Fat tissues contain the ready cash for all the expenditures of the organism. Only when the organism does not or cannot draw on the ready cash for its daily business is it put into depots, and excessive replenishment, through overeating, takes place."

To understand the path of events that leads to obesity, "the big question," as Bruch noted, was "why the metabolism is shifted in the direction of storage away from oxidation?" Why is fat deposited in the adipose tissue to accumulate in excess of its mobilization for fuel use? Once again, this has little to do with calories consumed or expended, but addresses the questions of how the cells utilize these calories and how the body regulates its balance between fat deposition and mobilization, between lipogenesis (the creation of fat) and lipolysis (the breakdown of triglycerides into fatty acids, their escape from the fat tissue, and their subsequent use as fuel). "Since it is now assumed that the genes and enzymes are closely associated," Bruch wrote in 1957, "it is conceivable that people with the propensity for fat accumulation have been born with enzymes that are apt to facilitate the conversion of certain reactions in that direction."

The third phase of this research finally established the dominant role of fatty acids in supplying energy for the body, and the fundamental role of insulin and adipose tissue as the regulators of energy supply. As early as 1907, the German physiologist Adolf Magnus-Levy had noted that during periods of fasting between meals "the fat streams from the depots back again into the blood . . . as if it were necessary for the immediate needs of the combustion processes of the body." A decade later, Francis Benedict reported that blood sugar provides only a "small component" of the fuel we use during fasting, and this drops away to "none at all" if our fast continues for more than a week. In such cases, fat will supply 85 percent of our energy needs, and protein the rest, after its conversion to glucose in the liver. Still, because the brain and central nervous system typically burn 120 to 130 grams of glucose a day, nutritionists insisted (as many still do) that carbohydrates must be our primary fuel, and they remained skeptical of the notion that fat plays any role in energy balance other than as a long-term reserve for emergencies.

Among physiologists and biochemists, any such skepticism began to evaporate after Wertheimer's review of fat metabolism appeared in 1948. It vanished after the 1956 publication of papers by Vincent Dole at Rockefeller University, Robert Gordon at NIH, and Sigfrid Laurell of the University of Lund in Sweden that reported the development of a technique for measuring the concentration of fatty acids in the circulation. All three articles suggested that these fatty acids were the form in which fat is burned for fuel in the body. The concentration of fatty acids in the circulation, they reported, is surprisingly low immediately after a meal, when blood-sugar levels are highest, but then increases steadily in the hours that follow, as the blood sugar ebbs. Injecting either glucose or insulin into the circulation diminishes the level of fatty acids almost immediately. It's as though our cells have the option of using fatty acids or glucose for fuel, but when surplus glucose is available, as signaled by rising insulin or blood-sugar levels, the fatty acids are swept into the fat tissue for later use. The concentration of circulating fatty acids rises and falls in "relation to the need" for fuel, wrote Gordon. And because injections of adrenaline cause a flooding of the circulation with fatty acids, and because adrenaline is naturally released by the adrenal glands as an integral part of the flight-fight response, Gordon suggested that the concentration of fatty acids also rises in relation to "the anticipated need" for fuel.

In 1965, the American Physiological Society published an eight-hundred-page *Handbook of Physiology* dedicated to the latest research on adipose-tissue metabolism. As this volume documented, several fundamental facts about the relationship between fat and carbohydrate metabolism had become clear. First, the body will burn carbohydrates for fuel, as long as blood sugar is elevated and the reserve supply of carbohydrates stored as glycogen in the liver and muscles is not being depleted. As these carbohydrate reserves begin to be tapped, however, or if there's a sudden demand for more energy, then the flow of fatty acids from the fat tissue into the circulation accelerates to take up the slack. Meanwhile, a significant portion of the carbohydrates we consume and all of the fat will be stored as fat in our fat cells before being used for fuel. It's this stored fat, in the form of fatty acids, that will then provide from 50 to 70 percent of all the energy we expend over the course of a day. "Adipose tissue is no longer considered a static tissue," wrote the Swiss physiologist Albert Renold, who coedited the *Handbook of Physiology;* "it is recognized as what it is: the major site of active regulation of energy storage and mobilization, one of the primary control mechanisms responsible for the survival of any given organism."

Since the excessive accumulation of fat in the fat tissue is the problem in obesity, we need to understand this primary control mechanism. This means, first of all, that we have to appreciate the difference between triglycerides and free fatty acids. They're both forms fat takes in the human body, but they play very different roles, and these are tied directly to the way the oxidation and storage of fats and carbohydrates are regulated.

When we talk about the fat stored in the adipose tissue or the fats in our food, we're talking about triglycerides. Oleic acid, the monounsaturated fat of olive oil, is a fatty acid, but it is present in oils and meats in the form of a triglyceride. Each triglyceride molecule is composed of three fatty acids (the "tri"), linked together on a backbone of glycerol (the "glyceride"). Some of the triglycerides in our fat tissue come from fat in our diet. The rest come from carbohydrates, from a process known as de novo lipogenesis, which is Latin for "the new creation of fat," a process that takes place both in the liver and, to a lesser extent, in the fat tissue itself. The more carbohydrates flooding the circulation after a meal, the more will be converted to triglycerides and stored as fat for future use (perhaps 30 percent of the carbohydrates in any one meal). "This lipogenesis is regulated by the state of nutrition," explained Wertheimer in an introductory chapter to the *Handbook of Physiology*: "it is decreased to a minimum in carbohydrate deficiency and accelerated considerably during carbohydrate availability."*

A second critical point is that while the fat is stored as triglycerides it enters and exits the fat cells in the form of fatty acids—actually, free fatty acids, to distinguish them from the fatty acids bound up in triglycerides—and it's these fatty acids that are burned as fuel in the cells. As triglycerides, the fat is locked into the fat cells, because triglycerides are too big to slip through the cell membranes. They have to be broken down into fatty acids—the process technically known as lipolysis—before the fat can escape into the circulation. The triglycerides in the bloodstream must also be broken down into fatty acids before the fat can diffuse into the fat cells. It's only reconstituted into triglycerides, a process called esterification, once the fatty acids have passed through the walls of the blood vessels and the fat-cell membranes and are safely inside. This is true for all triglycerides, whether they originated as fat in the diet or were converted from carbohydrates in the liver.

Inside the fat cells, triglycerides are continuously broken down into their component fatty acids and glycerol (i.e., in lipolysis), and fatty acids and glycerol are continuously reassembled into triglycerides (i.e., esteri-

---

* Synthesis of the enzymes required to convert carbohydrates into fat will also increase and decrease in proportion to the carbohydrate content of the diet.

fied)—a process known as the triglyceride/fatty-acid cycle. Any fatty acids that are not immediately repackaged back into triglycerides will slip out of the fat cell and back into the circulation—"a ceaseless stream of [free fatty acids], a readily transportable source of energy, into the bloodstream," as it was described in the *Handbook of Physiology* by one team of NIH researchers.

Some of these free fatty acids will be taken up by the tissues and organs and used as fuel. Perhaps as much as half of them will not. These will be incorporated in the liver back into triglycerides, loaded on lipoproteins,* and shipped back again to the fat tissue. And so fatty acids are continuously slipping from the fat tissue into the circulation, while those fatty acids that aren't immediately taken up and used for fuel are continuously being reconverted to triglycerides and transported back to the fat tissue for storage. "The storage of triglyceride fat in widely scattered adipose tissue sites is a remarkably dynamic process," explained the University of Wisconsin endocrinologist Edgar Gordon in 1969, "with the stream of fatty acid carbon atoms flowing in widely fluctuating amounts, first in one direction and then the other in a finely adjusted minute by minute response to the fuel requirements of energy metabolism of the whole organism."

This remarkably dynamic process, however, is regulated by a remarkably simple system. The flow of fatty acids out of the fat cells and into the circulation depends on the level of blood sugar available. The burning of this blood sugar by the cells—the oxidation of glucose—depends on the availability of fatty acids to be burned as fuel instead.

A single molecule plays the pivotal role in the system. It goes by a number of names, the simplest being *glycerol phosphate*. This glycerol-phosphate molecule is produced from glucose when it is used for fuel in the fat cells and the liver, and it, too, can be burned as fuel in the cells. But glycerol phosphate is also an essential component of the process that binds three fatty acids into a triglyceride. It provides the glycerol molecule that links the fatty acids together.† In other words, a product of carbohydrate metabolism—i.e., burning glucose for fuel—is an essential component in the regulation of fat metabolism: storing fat in the fat tissue. In fact, the rate at which fatty acids are assembled into triglycerides, and so the rate at which fat accumulates in the fat tissue, depend primarily on the availability of glycerol phosphate. The more glucose that is transported into the fat cells and used to generate energy, the more glycerol phosphate

* The VLDL particles we discussed when we talked about heart disease.

† The addition of a phosphate molecule to glycerol to make glycerol phosphate is said to "activate" the glycerol so that it can now be used in this process.

will be produced. And the more glycerol phosphate produced, the more fatty acids will be assembled into triglycerides. Thus, anything that works to transport more glucose into the fat cells—insulin, for example, or rising blood sugar—will lead to the conversion of more fatty acids into triglycerides, and the storage of more calories as fat.

This brings us to the mechanisms that control and regulate the availability of fat and carbohydrates for fuel and regulate our blood sugar in the process.

The first is the triglyceride/fatty-acid cycle we just discussed. This cycle is regulated by the amount of blood sugar made available to the fat tissue. If blood sugar is ebbing, the amount of glucose transported into the fat cells will decrease; this limits the burning of glucose for energy, which in turn reduces the amount of glycerol phosphate produced. With less glycerol phosphate present, fewer fatty acids are bound up into triglycerides, and more of them remain free to escape into the circulation. As a result, the fatty-acid concentration in the bloodstream increases. The bottom line: as the blood-sugar level decreases, fatty-acid levels rise to compensate.

If blood-sugar levels increase—say, after a meal containing carbohydrates—then more glucose is transported into the fat cells, which increases the use of this glucose for fuel, and so increases the production of glycerol phosphate. This is turn increases the conversion of fatty acids into triglycerides, so that they're unable to escape into the bloodstream at a time when they're not needed. Thus, elevating blood sugar serves to decrease the concentration of fatty acids in the blood, and to increase the accumulated fat in the fat cells.

The second mechanism that works to regulate the availability of fuel and to maintain blood sugar at a healthy level is called the glucose/fatty-acid cycle, or the Randle cycle, after the British biochemist Sir Philip Randle. It works like this: As blood-sugar levels decrease—after a meal has been digested—more fatty acids will be mobilized from the fat cells, as we just discussed, raising the fatty-acid level in the bloodstream. This leads to a series of reactions in the muscle cells that inhibit the use of glucose for fuel and substitute fatty acids instead. Fatty acids generate the necessary cellular energy, and the blood-sugar level in the circulation stabilizes. When the availability of fatty acids in the blood diminishes, as would be the case when blood-sugar levels are rising, the cells compensate by burning more blood sugar. So increasing blood-sugar levels decreases fatty-acids levels in the bloodstream, and decreasing fatty-acid levels in the bloodstream, in turn, increases glucose use in the cells. Blood-sugar levels always remain within safe limits—neither too high nor too low.

These two cycles are the fundamental mechanisms that maintain and ensure a steady fuel supply to our cells. They provide a "metabolic flexibil-ity" that allows us to burn carbohydrates (glucose) when they're present in the diet, and fatty acids when they're not. And it's the cells of the adipose tissue that function as the ultimate control mechanism of this fuel supply.

Regulation by hormones and the nervous system is then layered onto these baseline mechanisms to deal with the vagaries of the external envi-ronment, providing the moment-to-moment and season-to-season fine-tuning necessary for the body to work at maximum efficiency. Hormones modify this flow of fatty acids back and forth across the membranes of the fat cells, and they modify the expenditure of energy by the tissues and organs. Hormones, and particularly insulin—"even in trace amounts," as Ernst Wertheimer explained—"have powerful direct effect on adipose tissue."

With the invention by Rosalyn Yalow and Solomon Berson of their radioimmunoassay to measure insulin levels, it quickly became clear that insulin was what Yalow and Berson called "the principal regulator of fat metabolism." Insulin stimulates the transport of glucose into the fat cells, thereby effectively controlling the production of glycerol phosphate, the fixing of free fatty acids as triglycerides, and all that follows. The one fun-damental requirement to increase the flow of fatty acids out of adipose tis-sue—to increase lipolysis—and so decrease the amount of fat in our fat tissue, is to lower the concentration of insulin in the bloodstream. In other words, the release of fatty acids from the fat cells and their diffusion into the circulation require "only the negative stimulus of insulin deficiency," as Yalow and Berson wrote. By the same token, the one necessary require-ment to shut down the release of fat from the fat cells and increase fat accumulation is the presence of insulin. When insulin is secreted, or the level of insulin in the circulation is abnormally elevated, fat accumulates in the fat tissue. When insulin levels are low, fat escapes from the fat tis-sue, and the fat deposits shrink.

All other hormones will work to release fatty acids from the fat tissue, but the ability of these hormones to accomplish this job is suppressed almost entirely by the effect of insulin and blood sugar. These hormones can mobilize fat from the adipose tissue only when insulin levels are low—during starvation, or when the diet being consumed is lacking in carbohydrates. (If insulin levels are high, that implies that there is plenty of carbohydrate fuel available.) In fact, virtually anything that increases the secretion of insulin will also suppress the secretion of hormones that release fat from the fat tissue. Eating carbohydrates, for example, not only elevates insulin but inhibits growth-hormone secretion; both effects lead to greater fatty-acid storage in the fat tissue.

Hormones that promote fat mobilization	Hormones that promote fat accumulation
Epinephrine Norepinephrine Adrenocorticotropic hormone (ACTH) Glucagon Thyroid-stimulating hormone Melanocyte-stimulating hormone Vasopressin Growth hormone	Insulin

In 1965, hormonal regulation of adipose tissue looked like this: at least eight hormones that worked to release fat from the adipose tissue and one, insulin, that worked to put it there.

That increasing the secretion of insulin can in fact cause obesity (i.e., excess fat accumulation) would be demonstrated conclusively in animal models of obesity, particularly in the line of research we discussed in Chapter 21 on rats and mice with lesions in the area of the brain known as the ventromedial hypothalamus, or VMH. In the 1960s, this research became another beneficiary of Yalow and Berson's new technology to measure circulating levels of insulin. As investigators now reported, insulin secretion in VMH-lesioned animals increases dramatically within seconds of the surgery. The insulin response to eating also goes "off the scale" with the very first meal. The more insulin secreted in the days after the surgery, the greater the ensuing obesity. Obesity in these lesioned animals could be prevented by short-circuiting the exaggerated insulin response—by severing the vagus nerve, for example, that links the hypothalamus with the pancreas.* Similarly, the hypersecretion of insulin was reported to be the earliest detectable abnormality in genetic strains of obesity-prone mice and rats.

By the mid-1970s, it was clear that Stephen Ranson's insights into obesity in these animals had been confirmed. The lesion causes a defect in the part of the hypothalamus that regulates what researchers have come to call fuel partitioning—the result is the hypersecretion of insulin. The insulin forces the accumulation of fat in the fat tissue, and the animal overeats to compensate. This research refuted John Brobeck's notion, which has since become the standard wisdom in the field, that the VMH lesion causes overeating directly and the animals grow fat simply because they eat too much. These studies were neither ambiguous nor controversial. In 1976, University of Washington investigators Stephen Woods and Dan Porte

---

* For this reason, *vagotomy*, as this surgical procedure is known, was later considered a potential treatment for obese humans with various syndromes of hypothalamic obesity.

described as "overwhelming" the evidence that the increased secretion of insulin is the primary effect of VMH lesions, the driving force of obesity in these animals.

This half century of research unequivocally supported the alternative hypothesis of obesity. It established that the relevant energy balance isn't between the calories we consume and the calories we expend, but between the calories—in the form of free fatty acids, glucose, and glycerol—passing in and out of the fat cells. If more and more fatty acids are fixed in the fat tissue than are released from it, obesity will result. And while this is happening, as Edgar Gordon observed, the energy available to the cells is reduced by the "relative unavailability of fatty acids for fuel." The consequence will be what Stephen Ranson called *semi-cellular starvation* and Edwin Astwood, twenty years later, called *internal starvation*. And as this research had now made clear, the critical molecules determining the balance of storage and mobilization of fatty acids, of lipogenesis and lipolysis, are glucose and insulin—i.e., carbohydrates and the insulin response to those carbohydrates.

Just a few more details are necessary to understand why we get fat. The first is that the amount of glycerol phosphate available to the fat cells to accumulate fat—to bind the fatty acids together into triglycerides and lock them into the adipose tissue—also depends directly on the carbohydrates in the diet. Dietary glucose is the primary source of glycerol phosphate. The more carbohydrates consumed, the more glycerol phosphate available, and so the more fat can accumulate. For this reason alone, it may be impossible to store excess body fat without at least some carbohydrates in the diet and without the ongoing metabolism of these dietary carbohydrates to provide glucose and the necessary glycerol phosphate.

"It may be stated categorically," the University of Wisconsin endocrinologist Edgar Gordon wrote in *JAMA* in 1963, "that the storage of fat, and therefore the production and maintenance of obesity, cannot take place unless glucose is being metabolized. Since glucose cannot be used by most tissues without the presence of insulin, it also may be stated categorically that obesity is impossible in the absence of adequate tissue concentrations of insulin. . . . Thus an abundant supply of carbohydrate food exerts a powerful influence in directing the stream of glucose metabolism into lipogenesis, whereas a relatively low carbohydrate intake tends to minimize the storage of fat."

Forty years ago, none of this was controversial—and the facts have not changed since then. Insulin works to deposit calories as fat and to inhibit the use of that fat for fuel. Dietary carbohydrates are required to allow this fat storage to occur. Since glucose is the primary stimulator of insulin secretion, the more carbohydrates consumed—or the more refined the carbohydrates—the greater the insulin secretion, and thus the greater the accumulation of fat. "Carbohydrate is driving insulin is driving fat," as the Harvard endocrinologist George Cahill recently summed it up.

Regarding the potential dangers of sugar in the diet, it is important to keep in mind that fructose is converted more efficiently into glycerol phosphate than is glucose. This is another reason why fructose stimulates the liver so readily to convert it to triglycerides, and why fructose is considered the most lipogenic carbohydrate. Fructose, however, does not stimulate the pancreas to secrete insulin, so glucose is still needed for that purpose. This suggests that the combination of glucose and fructose—either the 50–50 mixture of table sugar (sucrose) or the 55–45 mixture of high-fructose corn syrup—stimulates fat synthesis and fixes fat in the fat tissue more than does glucose alone, which comes from the digestion of bread and starches.

It is important also to know that the fat cells of adipose tissue are "exquisitely sensitive" to insulin, far more so than other tissues in the body. This means that even low levels of insulin, far below those considered the clinical symptom of hyperinsulinemia (chronically high levels of insulin), will shut down the flow of fatty acids from the fat cells. Elevating insulin even slightly will increase the accumulation of fat in the cells. The longer insulin remains elevated, the longer the fat cells will accumulate fat, and the longer they'll go without releasing it.

Moreover, fat cells remain sensitive to insulin long after muscle cells become resistant to it. Once muscle cells become resistant to the insulin in the bloodstream, as Yalow and Berson explained, the fat cells have to remain sensitive to provide a place to store blood sugar, which would otherwise either accumulate to toxic levels or overflow into the urine and be lost to the body. As insulin levels rise, the storage of fat in the fat cells continues, long after the muscles become resistant to taking up any more glucose. Nonetheless, the pancreas may compensate for this insulin resistance, if it can, by secreting still more insulin. This will further elevate the level of insulin in the circulation and serve to increase further the storage of fat in the fat cells and the synthesis of carbohydrates from fat. It will suppress the release of fat from the fat tissue. Under these conditions—*lipid trapping*, as the geneticist James Neel described it—obesity begins to look preordained. Weights will plateau, as Dennis McGarry suggested in *Science* in 1992, only when the fat tissue becomes insulin-resistant as well, or when the fat deposits enlarge to the point where the forces working to release the fat and burn it for fuel—such as the increased concentration of fatty acids inside the fat cells—once again balance out the effect of the insulin itself.

By the mid-1960s, four facts had been established beyond reasonable doubt: (1) carbohydrates are singularly responsible for prompting insulin secretion; (2) insulin is singularly responsible for inducing fat accumulation; (3) dietary carbohydrates are required for excess fat accumulation; and (4) both Type 2 diabetics and the obese have abnormally elevated levels of circulating insulin and a "greatly exaggerated" insulin response to carbohydrates in the diet, as was first described in 1961 by the Johns Hopkins University endocrinologists David Rabinowitz and Kenneth Zierler.

The obvious implication is that obesity and Type 2 diabetes are two sides of the same physiological coin, two consequences, occasionally concurrent, of the same underlying defects—hyperinsulinemia and insulin resistance. This was precisely what von Noorden had suggested in 1905 with his diabetogenous-obesity hypothesis, even down to the notion that obesity would naturally result when muscle tissue becomes resistant to taking up glucose from the circulation before fat tissue does. Now the science had caught up to the speculation. "We generally accept that obesity predisposes to diabetes; but does not mild diabetes predispose to obesity?" as Yalow and Berson wrote in 1965. "Since insulin is a most lipogenic agent, chronic hyperinsulinism would favor the accumulation of body fat."

When Yalow and Berson measured individual insulin and blood-sugar responses to the consumption of carbohydrates, they reported that even lean, healthy subjects exhibit "great biologic variation" in what they called the "insulin-secretory responses." In other words, we secrete more or less insulin in response to the same amount of carbohydrates, or our insulin is more or less effective at lowering blood sugar or at promoting fat accumulation, or it remains elevated in the circulation for longer or shorter periods of time. And because variations of less than 1 percent in the partitioning of calories either for fuel or for storage as fat could lead to the accumulation of tens of pounds of excess fat over a decade, it would take only infinitesimal variations in these "insulin-secretory responses" to mark the difference between leanness and obesity, and between health and diabetes.

Over the years, prominent diabetologists and endocrinologists—from Yalow and Berson in the 1960s through Dennis McGarry in the 1990s—have speculated on this train of causation from hyperinsulinemia to Type

2 diabetes and obesity. Anything that increases insulin, induces insulin resistance, and induces the pancreas to compensate by secreting still more insulin, will also lead to an excess accumulation of body fat.

One of the more insightful of these analyses was by the geneticist James Neel in 1982, when he "revisited" his thrifty-gene hypothesis and rejected the idea (which has since been embraced so widely by public-health authorities and health writers) that we evolved through periods of feast and famine to hold on to fat.* Neel suggested three scenarios of these insulin-secretory responses that could constitute a predisposition to obesity and/or Type 2 diabetes—each of which, he wrote, would be a physiological "response to the excessive glucose pulses that result from the refined carbohydrates/over-alimentation of many civilized diets." Genetic variations in these responses would determine how long it would be before obesity or diabetes appears, and which of the two appears first. The one important caveat about these three scenarios, Neel added, is that they "should not be thought of as mutually exclusive or as exhausting the possible biochemical and physiological sequences" that might induce obesity and/or diabetes once populations take to eating modern Western diets.

The first of these scenarios was what Neel called a "quick insulin trigger." By this Neel meant that the insulin-secreting cells in the pancreas are hypersensitive to the appearance of glucose in the bloodstream. They secrete too much insulin in response to the rise in blood sugar during a meal; that encourages fat deposition and induces a compensatory insulin resistance in the muscles. The result will be a vicious circle: excessive insulin secretion stimulates insulin resistance, which stimulates yet more insulin secretion. In this scenario, we gain weight until the fat cells eventually become insulin-resistant. When the "overworked" pancreatic cells "lose their capacity to respond" to this insulin resistance, Type 2 diabetes appears.

In Neel's second scenario, there is a tendency to become slightly more insulin-resistant than would normally be the case when confronted with a given amount of insulin in the circulation. So even an appropriate insulin response to the waves of blood sugar that appear during meals will result in insulin resistance, and that in turn requires a ratcheting up of the insulin response. Once again, the result is the vicious cycle.

Neel's third scenario is slightly more complicated, but there's evidence to suggest that this one comes closest to reality. Here an appropriate amount of insulin is secreted in response to the "excessive glucose pulses" of a modern meal, and the response of the muscle cells to the insulin is

* Of Neel's two primary papers on thrifty genes, this is the one that is rarely read or referenced.

also appropriate. The defect is in the relative sensitivity of muscle and fat cells to the insulin. The muscle cells become insulin-resistant in response to the "repeated high levels of insulinemia that result from excessive ingestion of highly refined carbohydrates and/or over-alimentation," but the fat cells fail to compensate. They remain stubbornly sensitive to insulin. So, as Neel explained, the fat tissue accumulates more and more fat, but "mobilization of stored fat would be inhibited." Now the accumulation of fat in the adipose tissue drives the vicious cycle.

This scenario is the most difficult to sort out clinically, because when these investigators measure insulin resistance in humans they invariably do so on a whole-body level, which is all the existing technology allows. Any disparities between the responsiveness of fat and muscle tissue to insulin cannot be measured. This is critical, because for the past thirty-five years the American Diabetes Association has recommended that diabetics eat a diet relatively rich in carbohydrates based on the notion that this makes them more sensitive to insulin, at least temporarily, so the diet appears to ameliorate the diabetes. This effect was initially reported in 1971, by the University of Washington endocrinologists Edwin Bierman and John Brunzell,* who then waged a lengthy and successful campaign to persuade the American Diabetes Association to recommend that diabetics eat more carbohydrates rather than less. If Neel's third scenario is correct, however, a likely explanation for why carbohydrate-rich diets appear to facilitate blood-sugar control after meals is that they increase the insulin sensitivity of the fat cells specifically, while the muscle tissue remains insulin-resistant.

One of the few attempts, if not the only one, to measure the insulin sensitivity of fat cells and muscle cells separately in human subjects was made by the University of Vermont investigator Ethan Sims, in his experimental obesity studies of the late 1960s. Sims and his colleagues surgically removed fat samples from their subjects before, during, and after the periods of forced overeating and weight gain. They reported that high-carbohydrate diets had the unique ability to increase the insulin sensitivity of fat cells, and particularly so in fat cells that were already large and overstuffed. They had no similar effect, however, on the insulin resistance of the muscle tissue.

---

* Brunzell and Bierman fed mildly diabetic patients a diet of 85 percent carbohydrates and no fat, and compared their glucose response with that of patients on a more typical American diet of 45 percent carbohydrates and 40 percent fat. Those on the carbohydrate-rich diet had a slightly lower blood-sugar response, and insulin secretion remained unchanged. Brunzell and Bierman interpreted this to mean that a carbohydrate-enriched diet "increase[s] the sensitivity to insulin of tissue sites of insulin action."

If this observation is correct, it means carbohydrates are uniquely capable of prolonging this lipid-trapping condition by keeping the fat cells sensitive to insulin when they might otherwise become insulin-resistant. This might lower blood-sugar levels temporarily and delay or improve the appearance of diabetes—or "mask" the diabetes, as von Noorden put it—but would do so at the cost of increasing fat accumulation and obesity. Sims's observation suggests that Neel's third scenario for the genesis of obesity and diabetes was astute, and it suggests that a carbohydrate-rich diet might temporarily improve the symptoms of diabetes only by furthering the fattening process. Sims's studies have not been repeated in humans, but they have been reproduced and confirmed in animals. Brunzell says he refuses to believe that Sims got this measurement correct, but he also says that he has never tried to do the measurements himself because they're too difficult. But the question of whether Sims got it right requires a definitive answer. Without one, there's no way to know if the ADA recommendations have been helping diabetics or hurting them, let alone to understand the pathology of obesity and diabetes. The impact on the public health could be immense.

Through the 1970s, physiologists and biochemists worked out the mechanisms by which insulin and other hormones regulate not just the amount of fat we carry, but its distribution throughout the body, independent of how much we might happen to eat or exercise. By the end of the decade, they could explain at both a hormonal and an enzymatic level all the vagaries of what Julius Bauer had called lipophilia, or the "exaggerated tendency of some tissues to store fat."

A critical enzyme in this fat-distribution process is known technically as lipoprotein lipase, LPL, and any cell that uses fatty acids for fuel or stores fatty acids uses LPL to make this possible. When a triglyceride-rich lipoprotein passes by in the circulation, the LPL will grab on, and then break down the triglycerides inside into their component fatty acids. This increases the local concentration of free fatty acids, which flow into the cells—either to be fixed as triglycerides if these cells are fat cells, or oxidized for fuel if they're not. The more LPL activity on a particular cell type, the more fatty acids it will absorb, which is why LPL is known as the "gatekeeper" for fat accumulation.

Insulin, not surprisingly, is the primary regulator of LPL activity, although not the only one. This regulation functions differently, as is the case with all hormones, from tissue to tissue and site to site. In fat tissue, insulin increases LPL activity; in muscle tissue, it decreases activity. As a result, when insulin is secreted, fat is deposited in the fat tissue, and the muscles have to burn glucose for energy. When insulin levels drop, the LPL activity on the fat cells decreases and the LPL activity on the muscle cells increases—the fat cells release fatty acids, and the muscle cells take them up and burn them.

It's the orchestration of LPL activity by insulin and other hormones that accounts for why some areas of the body will accumulate more fat than others, why the distribution of fat is different between men and women, and how these distributions change with age and, in women, with reproductive needs. Women have greater LPL activity in their adipose tissue than men do, for example, and this may be one reason why obesity and overweight are now more common in women than in men. In men, the activity of LPL is higher in the fat tissue of the abdominal region than in the fat tissue below the waist, which would explain why the typical male obesity takes the form of the beer belly. Women have more adipose-tissue LPL activity in the hips and buttocks than in the abdominal region, although after menopause the LPL activity in their abdominal region catches up to that of men.

These various fat deposits are also regulated over time by the changing flux of sex hormones, so LPL can be considered the point at which insulin and sex hormones interact to determine how and when we fatten. The male sex hormone testosterone, for instance, suppresses LPL activity in the abdominal fat, but has little or no effect on the LPL in the fat of the hips and buttocks. Increasing fat accumulation in the abdomen as men age may therefore be a product of both increasing insulin and decreasing testosterone. The female sex hormone progesterone increases the activity of LPL, particularly in the hips and buttocks, but estrogen, another female sex hormone, decreases LPL activity.* It's the decrease in estrogen secretion during menopause—and so the increase in LPL activity—that may explain why women frequently gain weight as they pass through menopause. The effect of decreasing estrogen secretion on LPL activity would also explain why women typically fatten after the removal of the uterus in a hysterectomy. The change in hormonal regulation of LPL also explains how and why fat deposition changes during pregnancy and, after birth, with nursing.

In 1981, M. R. C. Greenwood, who was a student of Jules Hirsch and

---

* This explains why preventing estrogen secretion in female rats—by removing the ovaries—will make them obese, hungry, and sedentary, as we discussed in the previous chapter, whereas replacing the estrogen will make them lean again.

was then at Vassar College, proposed what she called the "gatekeeper hypothesis" of obesity, based on the hormonal regulation of LPL. "Conditions which favor increases in adipose tissue LPL," Greenwood wrote, "result in increased fat accumulation and, when food intake is constant, lead to alterations in body composition." Greenwood proposed the hypothesis based on her studies of the obese strain of rats known as Zucker rats, in which LPL activity in the fat tissue is elevated in the womb—apparently the effect of fetal hyperinsulinemia, though it then persists well into adulthood. As a result, Zucker rats grow monstrously obese. But they will actually lay down more fat, Greenwood reported, if they're kept to a strict diet than they will if they're allowed to eat freely to satisfy their hunger. The less they're allowed to eat, however, the smaller their muscles will be; their brains and kidneys will also be "significantly reduced" in size. "In order to develop this obese body composition in the face of calorie restriction," Greenwood wrote, "several developing organ systems in the obese rats were compromised."

Since Greenwood proposed this LPL gatekeeper hypothesis, researchers have reported that obese humans have increased LPL activity in their fat tissue. They've also reported that LPL activity in fat tissue increases with weight loss on a calorie-restricted diet and it decreases in muscle tissue; both reactions will work to maintain fat in the fat tissue, regardless of any negative energy balance that may be induced by the semi-starvation diet. During exercise, LPL activity increases in muscle tissue, enhancing the absorption of fatty acids into the muscles to be burned as fuel. But when the workout is over, LPL activity in the fat tissue increases. The sensitivity of fat cells to insulin will also be "sufficiently altered," as the University of Colorado physiologist Robert Eckel has described it, so as to restock the fat tissue with whatever fat it might have surrendered.

The open question, as Eckel wrote, is whether the particular hormonal environment that leads us to regain weight once we've lost it—elevated LPL activity on the fat cells and decreased LPL activity in the skeletal muscle—is the same as the one that leads us to grow fat to begin with. If insulin drives obesity, then this is an obvious hypothesis. There is no evidence to refute it, so it must be taken seriously. It has to be noted, too, that carbohydrate-rich meals increase LPL activity in the fat tissue, which would be expected, because they increase insulin secretion as well. Fat-rich meals do not. And so, as Eckel, a recent president of the American Heart Association, has put it, "habitual dietary carbohydrate intake may have a stronger effect on subcutaneous fat storage than does dietary fat intake."

Since none of this research is particularly controversial, it's hard to imagine why obesity researchers would not take seriously the hypothesis that carbohydrates have a unique ability to fatten humans—or, as Thomas Hawkes Tanner put it in *The Practice of Medicine* almost 140 years ago, that "farinaceous and vegetable foods are fattening, and saccharine matters are especially so." Researchers who study carbohydrate metabolism have found this science compelling. In 1991, the Belgian physiologist Henri-Géry Hers, an authority on what are known as glycogen-storage diseases, one of which is named after him, put it this way: "Eating carbohydrates will stimulate insulin secretion and cause obesity. That looks obvious to me. . . ." But this simple chain of cause and effect has nonetheless been rejected out of hand by authority figures in the field of human obesity, who believe that the cause of the condition is manifestly obvious and beyond dispute, that the law of energy conservation dictates that obesity has to be caused by eating too much or moving too little.

George Cahill, a former professor at the Harvard Medical School, is a pedagogical example. Cahill had done some of the earliest research on the regulation of fat-cell metabolism by insulin in the late 1950s, and had coedited the 1965 *Handbook of Physiology* on adipose-tissue metabolism. In 1971, when Cahill gave the Banting Memorial Lecture at the annual meeting of the American Diabetes Association, he described insulin as "the overall fuel control in mammals." "The concentration of circulating insulin," he explained, "serves to coordinate fuel storage and fuel mobilization into and out of the various depots with the needs of the organism, and with the availability or lack of availability of fuel in the environment." When I interviewed Cahill in 2005, he told me it was true that "carbohydrate is driving insulin is driving fat." But Cahill did not consider this chain of cause and effect to be a sufficient reason to speculate that carbohydrates drive obesity. Nor did he consider it a possibility that avoiding carbohydrates might reverse the process. Rather, he believed unconditionally that positive caloric balance was the critical factor. When it came to weight regulation, Cahill repeatedly told me, "a caloric is a calorie is a calorie." He acknowledged that the obese ate no more, on average, than the lean, and this is why he believed that the obese must be fundamentally lazy and this was the proximate cause of their obesity.* There was no reason to test com-

---

* He told me that I could confirm this observation by simply going to an airport and noticing, as he always did, that it was the overweight who took the escalators and the lean who walked up the stairs.

peting hypotheses, Cahill said, because any competing hypothesis would contradict the laws of physics as he understood them.

When clinical investigators tried to unravel the connection between diet, insulin, and obesity in human subjects, as the University of Washington endocrinologist David Kipnis did in the early 1970s, the results were invariably analyzed in light of this same preconception. Kipnis had fed ten "grossly obese" women a series of three- and four-week diets that were either high or low in calories, and high or low in carbohydrates. The fat-rich diets lowered insulin levels, Kipnis reported in *The New England Journal of Medicine* in 1971, and the carbohydrate-rich diets raised them, regardless of how many calories were being consumed. Even when these women were semi-starved on fifteen hundred calories a day, a high-carbohydrate content (72 percent carbohydrates and only 1 percent fat) still increased their insulin levels, even compared with the hyperinsulinemia of these obese women on their normal diets.

One interpretation of these results is that we could remove the carbohydrates from the diet and replace them with fat, and weight would be lost, perhaps without hunger, because insulin levels would drop, even if the total calories consumed did not. Kipnis's results, as the University of Heidelberg clinicians Gotthard Schettler and Guenter Schlierf wrote in 1974, underlined the "necessity of restricting carbohydrates in obesity in order to restore insulin levels to normal, thus hopefully decreasing appetite and fat deposition. . . ."

Kipnis, however, refused to believe that carbohydrates might cause obesity, or that avoiding carbohydrates might ameliorate the problem. When I interviewed him over thirty years later, he described the findings of his research as "very obvious." "You manipulate the amount of carbohydrates you give a human," he said, "you can manipulate his or her basal insulin level." He also said that "insulin causes deposition of fat in fat cells." But when it came to the cause of human obesity or weight gain, Kipnis rejected the relevance of these physiological phenomena. "Most people are obese because they eat more than they need to sustain the energy requirements that they have," he said. "They eat too damn much."

For the past quarter century, Americans have become progressively heavier and more diabetic. By 2004, one in three Americans was considered clinically obese; two in three were overweight. One in ten adult Americans had Type 2 diabetes—one in five over the age of sixty. It is now clear that the roots of this epidemic are evident even in infants and in the birth weights of newborns. Among middle-income families in Massachusetts, for example, as a team of researchers led by Matthew Gillman of Harvard reported last year, the prevalence of excessively fat infants increased dramatically between 1980 and 2001. This increase was most conspicuous among children younger than six months of age.

The probable explanation is that as women of childbearing age get heavier and more of them become diabetic, they pass the metabolic consequences on to their children through what is known technically as the intrauterine environment. The nutrient supply from mother to developing child passes across the placenta in proportion to the nutrient concentration in the mother's bloodstream. If the mother has high blood sugar, then the developing pancreas in the fetus will respond to this stimulus by overproducing insulin-secreting cells. "The baby is not diabetic," explains Boyd Metzger, who studies diabetes and pregnancy at Northwestern University, "but the insulin-producing cells in the pancreas are stimulated to function and grow in size and number by the environment they're in. So they start overfunctioning. That in turn leads to a baby laying down more fat, which is why the baby of a diabetic mother is typified by being a fat baby."

This is also the most likely explanation for why children born to women who gain excessive weight during pregnancy also tend to be fatter. As Laura Riley, medical director of labor and delivery at Massachusetts General Hospital, told the *Boston Globe* in response to the Harvard study, she now tells her patients, "If you overdo it during pregnancy, you're setting yourself up for a bigger baby," and that, in turn, means "you are setting your baby up for potentially a lifetime of weight problems." Gillman and his colleagues described the problem this way: "Our observation of a trend of increasing weight among young infants may portend continued increase in childhood and adult obesity."

But if fatter mothers are more likely to make fatter babies, and fatter babies are more likely to make fatter mothers, which is also a well-documented observation, then this is another vicious cycle. It suggests that, once a generation of adolescents and adults start eating the highly refined carbohydrates and sugars now ubiquitous in our diets, even their children will feel the effect, and perhaps their children's children as well. The extreme instance of this phenomenon today is the Pima Indians, whose incidence of diabetes is among the highest of any population in the world. In 2000, NIH investigators reported that Pima born to mothers who were diabetic have a two- to threefold increased risk themselves of becoming diabetic as adults, and so have a two- to threefold increased risk of passing diabetes on to their own children—of "perpetuating the cycle," as the NIH investigators explained. The "vicious cycle" of the "diabetic intrauterine environment," they wrote, can explain much of the post–World War II increase in Type 2 diabetes among the Pima, and may also "be a factor in the alarming rise of this disease nationally."

The question we now face is whether the same vicious cycle may also be a factor in the alarming rise of obesity nationally, as well as internationally. There's no reason to think that the hormonal and metabolic consequences of high blood sugar—from what James Neel in 1982 called the "excessive glucose pulses that result from the refined carbohydrates/over-alimentation of many civilized diets"—do not pass from mother to child through the intrauterine environment, whether the mother is clinically diabetic or not. If so, the longer the obesity epidemic continues, and the longer we go without unambiguously identifying the causes of obesity, metabolic syndrome, and diabetes, the worse this vicious cycle is likely to get.

Chapter Twenty-three

THE FATTENING CARBOHYDRATE DISAPPEARS

We need the help of the psychosocial scientists in finding better ways of communicating with our patients, in explaining to them that obesity is dangerous, that weight is lost slowly, that carbohydrates make fat and so on.

W.J.H. BUTTERFIELD, later vice-chancellor of the University of Cambridge, introductory remarks to the first Symposium of the Obesity Association of Great Britain, October 1968

It is incredible that in twentieth-century America a conscientious physician should have his hard-won professional reputation placed on the line for daring to suggest that an obesity victim might achieve some relief by cutting out sugars and starches.

ROBERT ATKINS, author of *Dr. Atkins' Diet Revolution*, testifying before Congress, April 12, 1973

T HERE ARE TWO MOMENTS IN THE HISTORY OF George McGovern's Senate Select Committee on Nutrition and Human Needs when the competing paradigms of nutrition and obesity can be captured in the act of shifting—one coming, one going. The first was in April 1973, during a hearing that the committee held on the subject of obesity and fad diets. Appearing that day to testify were Robert Atkins—author of *Dr. Atkins' Diet Revolution*, a book that had already sold almost one million copies in the six months since its publication—and three authorities in nutrition and health, who would testify that Atkins's severely carbohydrate-restricted diet was neither revolutionary, effective, nor safe. The tenor of the hearing was inquisitorial, and a pithy condemnation of Atkins and his diet by the Harvard nutritionist Fred Stare was read into the record by Senator Charles Percy of Illinois (Stare did not attend). "The Atkins diet is nonsense," Stare declared. "Any book that recommends unlimited amounts of meat, butter and eggs, as this does, in my opinion is dangerous. The author who makes the suggestion is guilty of malpractice."

A few weeks later, McGovern's committee hosted hearings on "Sugar in the Diet, Diabetes, and Heart Disease." Testimony came from an international panel of authorities, including Peter Cleave, Aharon Cohen of Hadassah University in Jerusalem, George Campbell of the Durban Diabetes Study Program in South Africa, Peter Bennett of the NIH, and Walter Mertz of the U.S. Department of Agriculture. These investigators discussed the potential dangers of refined carbohydrates in the diet, and John Yudkin testified to the particular dangers of sugar. McGovern and his fellow congressmen found the testimony compelling, although difficult to reconcile with the growing acceptance, their own included, of the notion that it was fatty foods that caused heart disease, and carbohydrates that would prevent it.

Those on the committee saw no connection between the two sets of hearings. They believed Atkins was peddling dietary nonsense, whereas Cleave, Campbell, and the others were promoting reasonable science, albeit a minority viewpoint. The congressmen did not comprehend that both sets of hearings were about the role of refined and easily digestible carbohydrates and the damage they might cause. "We weren't thinking of those two things," said the committee staff director, Kenneth Schlossberg, looking back from a perspective of three decades, "which was not very bright."

Three years later, in July 1976, McGovern's committee returned to the subject of diet and disease in the hearings that would lead, a half year later still, to the publication of *Dietary Goals for the United States*. The first witness was Assistant Secretary of Health Theodore Cooper, who repeatedly emphasized the need for further research to establish reliable knowledge about the diet-disease connection. McGovern and his fellow congressmen, however, wanted to tell the American public something more definitive, so McGovern asked Cooper if he could, at least, agree with the proposition that "overconsumption may be as serious a problem of nutrition as underconsumption."

"Particularly overconsumption of the wrong things," Cooper replied. "Very often in the poor we see people who are plump who might be called obese, and people would then conclude that they do not have a deficiency because they look rotund, healthy in one sense of the word. But it is true that the consumption of high carbohydrate sources with the induction of obesity constitutes a very serious public health problem in the underprivileged and economically disadvantaged. I would agree with that."

This response seems clear enough: the overconsumption of "high carbohydrate sources"—a phrase used to describe carbohydrate-dense starches and refined carbohydrates rather than leafy green vegetables and fruits—was associated with obesity in the poor, and perhaps even the cause. McGovern then asked Cooper to provide a "general rule of thumb" about eating habits that would help prevent disease and lengthen our lives, and Cooper reluctantly agreed to do so.

"What kinds of foods in general should we be consuming less of and what should we be eating more of?" McGovern asked.

"I think what we need to consider doing is to reduce our total fat intake," Cooper replied. "Fat adds a caloric substance—almost twice as much—nine calories per gram—as compared to sugar. I think in order to have an effective reduction in weight and realignment of our composition we have to focus on reducing fat intake."

With that answer, Cooper had contradicted himself, and the conventional wisdom on diet and health in America had shifted. The problem was no longer overconsumption of high-carbohydrate sources, but the overconsumption of fatty foods. And if Cooper realized that reducing our total fat intake meant increasing our consumption of carbohydrates, he neglected to say so.

Between 1973 and the mid-1980s, the notion of the fattening carbohydrate, which had persisted in clinical and popular literature for well over a century, was replaced with the belief that it is dietary fat, with its particularly dense calories, that is responsible for overweight and obesity. The prescription of reducing diets that restricted starches and sugars, and perhaps oils and butter as well, was replaced with diets that targeted fat alone—restricting not just butter and oils but meat, eggs, and dairy products—thereby increasing carbohydrate consumption. Obesity was conceptually transformed from a condition commonly associated with the excessive consumption of carbohydrates and carbohydrate craving to one that would be described by prominent nutritionists as a "carbohydrate-deficiency syndrome," which in turn explained why "an increase in dietary carbohydrate content at the expense of fat is the appropriate dietary part of a therapeutical strategy."

What makes this shift all the more perplexing is that it occurred immediately after the science of fat metabolism evolved to explain why carbohydrates were uniquely fattening, and it followed a six-year period in which carbohydrate-restricted diets achieved unprecedented credibility among clinicians. The latter coincided precisely with the genesis of obesity research as what would be considered a legitimate field of scientific study, a transformation marked by the increasingly frequent appearance of conferences and symposiums dedicated to reporting the latest findings in obesity research, all of which, through 1973, had been dominated by discussions of the peculiar efficacy of carbohydrate-restricted diets.

The first was hosted by the University of California, San Francisco, in December 1967. Among the dozen speakers was the veteran UC Berkeley nutritionist Samuel Lepkovsky, who used exactly the same logic as Alfred Pennington had in the 1950s to argue the biological rationale of carbohydrate restriction. "Positive caloric balance may be a result rather than a cause of the [obese] condition," Lepkovsky said. "It seems desirable in the treatment of obesity to direct efforts toward an increased utilization of fat. This effort can be made by restricting the intake of carbohydrates and increasing the ingestion of fat." The one presentation at the conference that was specifically on the dietary treatment of obesity came from a team of U.S. Navy physicians, who had been prescribing an eight-hundred-to-one-thousand calorie "ketogenic" diet to overweight naval personnel. Their diet was 70 percent fat, 20 percent protein, and 10 percent carbohydrate, and it induced "significant weight loss" in all their patients. "Uniformly and without exception," they added, "patients who underwent dieting found that the satiety value of the ketogenic diet was far superior to that of a mixed or high-carbohydrate diet, even though the food selection was minimal. . . ."

In 1968, the newly founded Obesity Association of Great Britain hosted in London its first symposium on obesity. The presentations were dominated by investigators who believed in the fattening nature of carbohydrates and the efficacy of carbohydrate-restricted diets. These included John Yudkin and his colleague Stephen Szanto; W.J.H. Butterfield, who would later become vice-chancellor of the University of Cambridge; Alan Kekwick and Gaston Pawan of the University of London, who were primarily responsible for reviving the concept of Banting's diet in the U.K.; and Denis Craddock, a general practitioner and author of *Obesity and Its Management,* which would be published in 1969 and was one of at most two or three clinical guides to obesity treatment published in the U.K. in the 1960s or 1970s. As Craddock reported at the conference, he had recently completed a survey of a hundred pregnant patients, sixty of whom had begun to fatten excessively during the early months of their pregnancy. "This weight gain was controlled in most cases"—fifty-seven of the sixty—"simply by restricting carbohydrates in the diet," he said.

The conference had been organized by Alan Howard and his colleague Ian McLean Baird. Howard was a biochemist and pathologist at the University of Cambridge who would later become the founding editor, with George Bray, of the *International Journal of Obesity.* Howard had become interested in carbohydrate restriction because he had been twenty pounds overweight, had unsuccessfully dieted for years, then finally lost the weight and kept it off by avoiding flour, starches, and sweets. At the London conference, Howard reviewed the literature on carbohydrate restriction dating back to Banting and concluded that this was the only effective method to induce and maintain weight loss. "A common feature of all who have written on the subject," he said, is "that the patient's hunger is satisfied whilst on a diet high in carbohydrate of the same caloric value, patients complain of hunger."

After the London meeting, obesity conferences evolved from local to international affairs. The first was in Paris in 1971, hosted by European nutrition and dietetics associations. Here the sole presentation on the dietary treatment of obesity was by a collaboration from the French National Institute on Health and Medical Research (INSERM), which is the local counterpart of the NIH in the United States and the Medical Research Council in the United Kingdom. These INSERM investigators had prescribed diets of twelve to eighteen hundred calories to over a hundred obese patients, in either three or seven meals a day, and with varying amounts of carbohydrates. Weight loss increased, they reported, when the subjects divided their calories among seven meals, which served to moderate the insulin response. Moreover, "lowering the carbohydrate content of the diet increased the weight loss at both meal frequencies."

The next conference was hosted by the NIH in Bethesda, Maryland, in October 1973. Six of the presentations at this meeting discussed the treatment of obesity by methods other than drugs or surgery. Two were on physical activity, and neither reported any significant effect of exercise on body weight. Two addressed the benefits of behavioral modification on weight loss, and neither reported any significant benefit. Of the two remaining presentations, one was by Ernst Drenick of UCLA on prolonged fasting to treat obesity—"our experiences are disappointing," said Drenick—and the other was by Charlotte Young of Cornell on dietary treatments.

As Howard had in London, Young reviewed the hundred-year history of carbohydrate-restricted diets, including the research of Pennington and that of Margaret Ohlson and her own trials in the 1950s. Young then discussed her recent studies, in which she had put obese young men on eighteen-hundred-calorie diets with the protein content fixed at 460 calories (26 percent), but with varying proportions of fat and carbohydrates. Over the course of nine weeks, she reported, "weight loss, fat loss, and percent weight loss as fat appeared to be inversely related to the level of carbohydrate in the diets"—in other words, the fewer carbohydrates and the more fat in the diet, the greater the weight loss and the greater the fat loss. "No adequate explanation could be given for the differences in weight losses," she said. All of the carbohydrate-restricted diets, she said, "gave excellent clinical results as measured by freedom from hunger, allaying of excessive fatigue, satisfactory weight loss, suitability for long term weight reduction and subsequent weight control."

The last of these conferences to be held before the nutritional wisdom began to shift definitively was in London in December 1973, just two months after the NIH meeting. This one was organized by Yudkin, and many of those giving presentations had also attended the NIH conference. Their presentations were similar, but here there was more of a tendency to implicate carbohydrates specifically as the cause of obesity. Lester Salans and Edward Horton, both collaborators of Ethan Sims on his experimental obesity studies, discussed the effect of carbohydrates on hyperinsulinemia and the role of hyperinsulinemia in obesity. "It is clear that in both lean and obese subjects the carbohydrate content of the diet influences . . . insulin and glucose concentrations," Horton reported. He added that it was probably hyperinsulinemia that induced both obesity and insulin resistance. Yudkin then gave the only talk on dietary therapy, entitled "The Low-Carbohydrate Diet," noting that these diets are higher in vitamins and minerals than calorie-restricted diets, simply because the foods restricted—starches and sugars—have few or no vitamins and minerals. The diet will "reduce superfluous adiposity," Yudkin said, "but it will not need to be changed when this has been done. . . . The diet is intended as a new but permanent pattern of eating and not simply as a cure for obesity, to be abandoned when an acceptable loss of weight is achieved." Harry Keen, who was then at Guy's Hospital Medical School and would become one of the most influential diabetologists in the U.K.,* said the critical issue wasn't just obesity, but the chronic diseases that accompanied it. "With the chronically failed case of obesity we are dealing with the wreckage of the situation," he said, so it was necessary to set "new patterns of body weight and body size, if we are going to make a serious attempt to reduce the frequency, for example, of atherosclerosis, of diabetes mellitus and of a number of other conditions." Keen and his colleagues had tested the viability of this goal, he reported, on a group of "ostensibly normal men

---

* From 1990 to 1996, Keen was chairman of the British Diabetic Association. He was also elected honorary president of the International Diabetes Federation in 1991, and was chairman of the WHO Expert Committee on Diabetes in 1980 and 1985.

in whom obesity is represented no more frequently than in the population at large." These men were instructed to restrict their carbohydrate intake to less than five hundred calories a day, but to continue eating protein and fat as desired. The result was an average weight loss of fourteen pounds, impressive because these individuals were not necessarily overweight to begin with. That weight loss had been maintained for almost five years. To those who might be pessimistic about the prevention of obesity and over-weight in the public at large, Keen said, this result should be taken as "a word of reassurance and optimism."

By 1972, *The New York Times Natural Foods Dieting Book* was offering both a low-calorie weight-loss plan, at a thousand calories a day, and a low-carbohydrate method. "You strictly curtail the amount of carbohydrates you eat daily," the book explained. "You eat, instead, foods in which the carbohydrate content is very low or nonexistent. Meat . . . fish, poultry, fats, butter, most cheeses and eggs are equally low in that fattening sub-stance, and these are the foods that form the basis for your diet . . . for without carbohydrates you cannot gain weight!"

Two years later, when the nonprofit organization Consumer Guide published its first edition of *Rating the Diets,* a 380-page compendium of the pros and cons of popular diets, carbohydrate restriction seemed firmly established in the canon. *Rating the Diets,* which obesity authorities would repeatedly recommend as a valuable review of the evidence, concluded that a diet including less than sixty grams of carbohydrates each day had "much to recommend it" and so was "helpful and beneficial" for weight loss. It also quoted a medical textbook to the effect that "the difficult-to-treat obese patient," which effectively means *every* obese patient, "appears to suffer from some defect in dealing with carbohydrate which leads to an unnatural conversion of it to fat and to storage of the fat. Avoidance of too much dietary carbohydrate reduces this tendency." The only caveat with these diets, according to *Rating the Diets,* was that they "pay little attention to the kinds of fats you eat" and so might increase heart-disease risk.

The shift in the nutritional wisdom was now taking place, driven by the contagious effect of Ancel Keys's dietary-fat/heart-disease hypothesis on the closely related field of obesity. Any diet that allowed liberal fat con-sumption was to be considered unhealthy. Clinical investigators working on the problem of human obesity concurred.

Through the 1950s, the carbohydrate-restricted diet had challenged only the positive-caloric-balance hypothesis of obesity. Yudkin had man-aged to reconcile carbohydrate restriction with this conventional wisdom by insisting that low-carbohydrate diets were low-calorie diets in disguise.

By doing so, Yudkin made the diets politically acceptable, although he also directed attention away from the underlying science. In the same 1960 *Lancet* article in which Yudkin proclaimed what he called "the inevitability of calories," he had made the point that if the diet was indeed low in calories, then its fat content would also be comparatively low, reconciling his diet with Keys's dietary-fat hypothesis. This was Yudkin's "no bread, no butter" argument. If carbohydrate calories are restricted, fat calories are, too. Though the *proportion* of fat in the diet increases if carbohydrates are avoided, the absolute quantity of fat may actually decrease. This is why Yudkin insisted that the correct terminology for these diets should be "low-carbohydrate" rather than "high-fat." "It is highly implausible," Yudkin wrote in 1974, "that a given amount of fat that is harmless when energy intake is excessive becomes harmful when this excess is corrected by a reduction in the intake of sugar and starch."

As a result of Yudkin's conciliatory efforts, the only carbohydrate-restricted diets that elicited a backlash from nutritionists were those promoted by clinicians whose interpretation of the science disagreed with Yudkin's. This situation was exacerbated by the fact that it was these physicians, without university affiliations, who adopted the diet quickly and then wrote books for the lay public that sold exceptionally well. Because their claims sounded like quackery—*The High-Calorie Way to Stay Thin Forever*, as *Dr. Atkins' Diet Revolution* was subtitled—they were treated as such, and particularly so after the medical and public-health authorities decided that dietary fat might cause heart disease.

The small contingent of influential nutritionists from Fred Stare's department at Harvard provide an example of how this process of entrenchment evolved. In 1952, when Alfred Pennington lectured at Harvard on the benefits of carbohydrate restriction and Keys was only beginning his crusade against dietary fat, Mark Hegsted had suggested, "Dr. Pennington may be on the right track in the practical treatment of obesity." A decade later, and a year after the American Heart Association had officially sided with Keys, the Brooklyn obstetrician Herman Taller published his best-seller, *Calories Don't Count*, based on Pennington's work and Taller's clinical experiences with the diet. Stare called the book "trash," and Jean Mayer described the high-fat aspect of the diet as "potentially dangerous." Philip White, who received his doctorate in nutrition from Stare's department, then wrote a review of *Calories Don't Count* for *JAMA*, accusing Taller of perpetrating "nutrition nonsense and food quackery." In 1973, in response to the publication of *Dr. Atkins' Diet Revolution*, based on Atkins's clinical experience with overweight patients and another decade of science, White edited a critique of carbohydrate-restricted diets in *JAMA*—the first draft of which was written by Ted Van Itallie, another veteran of Stare's nutrition department—that now dismissed the diets as "bizarre concepts of nutrition and dieting [that] should not be promoted to the public as if they were established scientific principles."

Meanwhile, these nutritionists would readily admit that they didn't know what caused obesity (why some people ate too much and others didn't) and that calorie restriction conspicuously failed to cure it. After nearly twenty years in the field, as Jean Mayer wrote in the introduction to his 1968 monograph, *Overweight,* he was "as aware as any man of the gigantic gaps in our knowledge—and of the likelihood that many of our present concepts may be erroneous." He also noted, in his discussion of hormonal influences on obesity, that insulin "favors fat synthesis" and that someone who oversecretes insulin could "tend to become hungry as a result." But when a physician suggested publicly, as Atkins did, that carbohydrates raised insulin levels, that insulin favors fat synthesis, and that a diet lacking carbohydrates might reverse this process, these nutritionists would denounce it, as Mayer himself did in 1973, as "biochemical mumbojumbo."

With the publication of *Dr. Atkins' Diet Revolution* and its subsequent censure by the American Medical Association, the nature of the professional discussions on carbohydrate-restricted diets turned from their clinical utility to the reasons to avoid them. The actual science suddenly mattered less than ever.

Atkins was a Cornell-trained cardiologist. Between 1959 and 1963, coinciding with the early years of his practice in Manhattan, he gained fifty pounds. He eventually decided to try carbohydrate restriction, he said, "because that's what was being taught at the time." His attempt coincided with the 1963 publication in *JAMA* of a lengthy article by the University of Wisconsin endocrinologist Edgar Gordon, entitled "A New Concept in the Treatment of Obesity." Gordon was one of the few clinicians of that era who studied fat metabolism and then designed a diet based specifically on that science. Gordon's diet, as described in *JAMA,* began with a forty-eight-hour fast—"not to produce a spectacular loss of weight, but rather to break a metabolic pattern of augmented lipogenesis"*—and then allowed protein and fat as desired but limited carbohydrates to minimal fruits, green

---

* Fat synthesis and accumulation.

vegetables, and a half-slice of bread every day. "The total caloric value is quite high in terms of reducing diets," wrote Gordon. Atkins later said his attention was caught by Gordon's observation that his subjects lost weight without ever complaining of hunger.

In his diet, Atkins replaced the two-day initiatory fast with a week or more of complete carbohydrate restriction, under the assumption, as the Atlanta physician Walter Bloom had noted, that the two states were physiologically identical. Atkins said he lost twenty-eight pounds in a month and felt energized in the process. In 1964, while Atkins was personally reaping the benefits of his diet, he was also working part-time as a company physician with AT&T. The junior executives noticed his weight loss, so he told them about the diet. Sixty-five of them eventually tried it, as Atkins told it, and all but one reduced to their desired weight. The sole exception wanted to lose eighty pounds but lost only fifty.

Atkins then started treating obese patients out of his cardiology clinic and developed the diet as he came to prescribe it in his book. He instructed his patients to start off with an initiation period, eating no carbohydrates other than a small green salad twice a day. Once they were losing weight at a suitably rapid rate, they could begin adding small amounts of carbohydrates back into their diet until they reached what he called the critical carbohydrate level, when their weight loss either leveled off or could no longer be maintained. Then they would have to back off again on the carbohydrates to experience further benefit from the diet. He also had them check their urine for ketone bodies—with the same ketosticks used commonly by diabetics—to ensure that they remained in ketosis and were still burning body fat. The reliance on ketosis to initiate and maintain weight loss, and the progressive addition of carbohydrates to the diet, are what Atkins considered his contributions to the clinical science of carbohydrate restriction.* His career as a diet doctor grew slowly until 1966, when the women's fashion magazines began recommending his diet, and his business boomed. After *Vogue* popularized the diet in 1970, Atkins set out to write *Diet Revolution,* which was then advertised as "the famous *Vogue* superdiet explained in full."

The gist of *Dr. Atkins' Diet Revolution* can be distilled down to three assertions. The first is that weight could be lost on his diet without hunger,

---

* The progressive addition of carbohydrates was similar to a common treatment of diabetics in the pre-insulin era: Diabetics would be fasted to lower their blood sugar to healthy levels; then protein and fat calories would be increased gradually, until glucose appeared in their urine. That would be considered the critical calorie level, and the diabetics would never be allowed to eat any more than that.

and perhaps without even restricting calories. Atkins said that his patients regularly lost weight eating three thousand calories a day, and that he had one three-hundred-pounder who reduced significantly while eating five thousand. His only explanation was that obesity is caused by the kind of calories we consume and not the quantity, and so if we avoid carbohydrates our bodies function correctly and shed any excess weight. He attributed the absence of hunger to the copious calories, the ketosis (which is probably not the case), the effect of insulin on blood sugar—all overweight people "produce too much insulin," he wrote, and that lowers blood sugar and makes people hungry—and the secretion of what the British clinicians Alan Kekwick and Gaston Pawan had called fat-mobilizing substance. (Virtually all hormones, with the exception of insulin, will mobilize fat from adipose tissue, but none of them will do so effectively when insulin is elevated.)

Atkins's second claim was that his diet was inherently healthy, much more so than a low-fat diet, because refined carbohydrates and starches, not saturated fat, caused heart disease and diabetes. Atkins later said that Peter Cleave's *Saccharine Disease* had been a revelation to him. In *Diet Revolution* he discussed the research from Yudkin, Margaret Albrink, Robert Stout, and Peter Kuo implicating triglycerides as a more significant risk factor for heart disease than cholesterol. He also claimed, on the basis of his experience with "ten thousand" overweight patients, that cholesterol "usually goes down" on his diet, despite the high saturated-fat content, and that triglycerides invariably decrease.

His third claim was what he called the "cruel hoax" of calorie-restricted diets: "The balanced low-calorie diet has been the medical fashion for so long that to suggest any alternative invites professional excommunication," Atkins wrote. "Yet even most doctors admit (at least privately!) the ineffectiveness of low-calorie diets—balanced or unbalanced." Atkins supported his accusation by invoking Albert Stunkard's 1959 "comprehensive review of the thirty years of medical literature," and offering three reasons why calorie-restricted diets inevitably fail. First, they "don't touch the primary cause of most overweight," which is a "disturbed carbohydrate metabolism." They also fail because they reduce energy expenditure. "Dr. George Bray," he wrote, "has demonstrated that people on low-calorie diets *actually develop lower total body energy requirements* and thus burn fewer calories." (Although Atkins didn't say so, this research had led Bray himself to publish an article entitled "The Myth of Diet in the Management of Obesity.") And, finally, Atkins wrote, "The main reason low-calorie diets fail in the long run is because you go hungry on them. . . . And while you may tolerate hunger for a short time, you can't tolerate hunger all your life."

Had Atkins wanted to avoid professional excommunication, he might have published something other than a polemic couched as a diet book. But he was feeling "resentment," he wrote, "that [he] had been duped so long by misinformation given me in the medical literature." The *Diet Revolution* was not just advocating a way to lose weight, which Atkins credited, in any case, to Banting, Pennington, Kekwick, and Pawan, but overthrowing the current nutritional wisdom entirely. Unlike Irwin Stillman, whose 1967 mega–best-seller *The Doctor's Quick Weight Loss Diet* was also based on carbohydrate restriction, Atkins wanted "a revolution, not just a diet." "Martin Luther King had a dream," Atkins wrote. "I, too, have one. I dream of a world where no one has to diet. A world where the fattening refined carbohydrates have been excluded from the diet." Atkins deliberately portrayed his diet as diametrically opposed to the growing orthodoxy on the nature of a healthy diet. Whereas Keys had insisted that the solution to obesity was to convince fat people that overeating was a sin and overeating fat would kill them, Atkins said his patients lost "thirty, forty, 100 pounds" eating "lobster with butter sauce, steak with Bearnaise sauce . . . *bacon cheeseburgers. . . .*" *"As long as you don't take in carbohydrates,"* Atkins wrote, *"you can eat any amount of this 'fattening' food and it won't put a single ounce of fat on you."*

Diet Revolution may have been, as its publisher claimed, the fastest-selling book in history. Nonetheless, its "chief consequence," as John Yudkin noted in 1974, may have been "to antagonize the medical and nutritional establishment." In fact, Atkins had to antagonize only a very small and select group of men to have a profound and lasting effect on how we think about obesity and weight regulation. In obesity research, particularly in the United States in the 1970s, the established wisdom was determined not by any testing of hypotheses or even establishing of consensus but by the judgment of fewer than a dozen men who dominated the field: Jean Mayer, Fred Stare, Jules Hirsch, George Bray, Theodore Van Itallie, Albert Stunkard, George Cahill, Philip White, and perhaps a few others. (And when these men began to retire from the scene in the 1980s, their younger colleagues—Johanna Dwyer, who received her Ph.D. with Mayer; Francis Xavier Pi-Sunyer, who collaborated with Van Itallie; Kelly Brownell, who worked and studied with Stunkard—assumed the leadership and perpetuated their beliefs.)

When these men came of age in their careers, in the 1950s and early 1960s, obesity research was a new and expanding field of science. It had been reinvented in the United States after World War II, and the National

Institutes of Health was just beginning to provide money for research. These men filled the expanding vacuum. They all came out of the North-eastern academic corridor—Harvard, Yale, Columbia, Rockefeller, the University of Pennsylvania—and they all knew each other. Van Itallie befriended his classmate Stunkard on their first day of medical school at Columbia; he then went to work with Mayer at Harvard and enlisted Stunkard's help to test Mayer's theory of hunger, and so Stunkard got to know Mayer as well. Philip White received his doctorate with Mayer at Harvard and remained in Stare's department until 1956, when he became secretary of the American Medical Association's Councils on Foods and Nutrition and wrote an influential nutrition column for *JAMA*. Van Itallie then became a member of White's council and initiated its 1973 public condemnation of Atkins and all similar carbohydrate-restricted diets. White edited the article. If you weren't in the club, you had little influence. ("The Mississippi River is very deep, or at least it used to be," is how the biochemist and diabetologist Gerald Grodsky of the University of California, San Francisco, described the inability of West Coast investigators to influence medical wisdom in the 1960s and 1970s.)

These individuals became the field's "leading authorities," as the news-papers would call them. They hosted the conferences, edited the textbooks, chaired the committees, and determined research priorities. By the end of the 1970s, they had determined what clinicians and researchers in the field would believe, at least in the United States, and what they still believe overwhelmingly. When McGovern's committee held its post-facto hearings in February 1977 to address the *Dietary Goals for Americans*, only members of this club testified on obesity* (Mayer had been the committee's consultant), and they all embraced the committee's recommendation of a national diet richer in carbohydrates and poorer in fat. Although Van Itallie also testified that he was unaware of any research to support their opinions: "Thus, what I am saying is an assumption rather than a statement of established fact," he acknowledged.

None of these authorities actually specialized in the clinical treatment of obesity except Stunkard, who did so as a psychiatrist treating an eating disorder. Nor were they necessarily the best scientists in the field. Fred Stare and Philip White never studied obesity at all. Cahill's research on fat metabolism and fuel partitioning was seminal, but he didn't see why it should be relevant to human obesity. Stunkard's primary contribution to obesity research through the 1970s was his observation that the obese

* Van Itallie, Stunkard, Bray, Cahill, and Dwyer.

rarely lose weight on diets, and if they do, they don't keep it off. But he never noted, and as a result neither did anyone else, that the only dietary studies he addressed in his seminal analysis were of semi-starvation, so what he had confirmed was that semi-starvation failed, not that all diets did.

Van Itallie and Bray deserve a disproportionate share of the responsibility for effectively removing the concept of the fattening carbohydrate from the nutritional canon, and thus the carbohydrate-restricted diet as well. Virtually everything we believe about what constitutes an effective weight-loss diet can be traced back to the 1970s and the efforts of these two men.

Before Van Itallie decided to write what he called the AMA-sponsored "denunciation" of Atkins in 1973, his only substantive involvement in the science of obesity, as either a researcher or a clinician, was his work with Mayer twenty years earlier. In the intervening years, he had worked on intravenous feeding of hospital patients and dietary influences on cholesterol, among other subjects, but he returned to the subject of weight in 1971, only when one of his post-docs developed an interest in the subject. This led to what Van Itallie considered his primary contribution to obesity research, the development of a feeding machine to study food intake: "You could basically feed yourself by pushing a button," Van Itallie explained. "The machine would deliver a measured quantity of formula diet into your mouth, and then keep a record of how much you took."

Van Itallie felt that *Diet Revolution* was full of what he called "gross inaccuracies," and that there were far too many reasons to believe that the diet could be dangerous to disseminate it so widely. There may have been some personal enmity as well: Van Itallie had been chief of medicine at St. Luke's Hospital in New York when Atkins had served under him as a cardiology resident in the late 1950s. Van Itallie said he didn't work with Atkins closely enough to know him personally, but he did not find him "an appealing personality" nonetheless. Stunkard, talking about all his colleagues in the field, said, "We just despised [Atkins]. We thought he was a jerk, an idiot, who just wants to make money."

The critique that Van Itallie drafted and White edited, which was then published as the official AMA declaration on carbohydrate-restricted diets, was not a balanced assessment of the science, nor was it absent its own gross inaccuracies. It was akin to the diatribes that had been aimed at Banting in the 1860s, Pennington in the 1950s, and Taller, by White himself, in the early 1960s. Atkins, like Banting, Pennington, and Taller, was censured for advocating a diet that was "neither new nor revolutionary." The article accused *Diet Revolution* of lacking "scientific merit," primarily by implying that here was a "way of circumventing the first law of thermodynamics." The diet itself was denounced as "grossly unbalanced," because it "interdicts the 45 percent of calories that is usually consumed as carbohydrates," and so cannot "provide a practicable basis for long-term weight reduction or maintenance, i.e., a lifetime change in eating and exercise habits." That this was the opinion of a nutritionist and a physician, neither of whom had worked clinically with obese patients, was lost in the publication of the critique under the auspices of the AMA itself. Mayer dedicated one of his syndicated newspaper columns to condemning Atkins, based on the AMA critique, which he cited repeatedly as though it were the considered opinion of the entire American Medical Association and not just his former collaborator, edited by his former student. "The American Medical Association," wrote Mayer, has "taken the unusual step of warning the U.S. public against the latest do-it-yourself diet as propounded in 'Dr. Atkins's Diet Revolution.' " The AMA report, Mayer wrote, "explains why the 'diet revolution' cannot work."

After Van Itallie drafted the AMA attack on Atkins, he spent the next decade as the principal arbitrator of the risks and benefits of weight-reduction diets. This coincided with his rise to prominence in the field, after receiving a 1974 NIH grant—"a few hundred thousand dollars," he says, "which was a lot of money at the time"—to start the first federally funded clinical obesity center in the United States, now known as the Theodore Van Itallie Center for Weight Control at St. Luke's Roosevelt Hospital Center. That same year, Van Itallie gave the review presentation on dietary approaches to obesity at the First International Congress on Obesity, although he had yet to do any research personally on the dietary treatment of obesity, or to treat obese patients. In 1975, Van Itallie cowrote (with Pi-Sunyer) the review chapter on obesity and diabetes in the textbook *Diabetes Mellitus,* having now been in the field, in a part-time capacity, for at most four years. He then wrote the chapters on dietary therapy for obesity in the 1978 symposium report *Obesity: Basic Mechanisms and Treatment,* edited by Stunkard; in the 1979 NIH report *Obesity in America,* edited by Bray; in Bray's 1980 textbook, *Obesity: Comparative Methods of Weight Control;* and in Stunkard's 1980 textbook, *Obesity.* In 1983, Van Itallie cochaired the Fourth International Congress on Obesity. In 1984, he coauthored the obesity chapter in the fifth edition of *Present Knowledge in Nutrition,* which had been a standard nutrition reference since its first edition was published thirty years earlier. Because Van Itallie was also engaged as chair of the medical department at Columbia University's Presbyterian College of Physicians and Surgeons, he says, he had no time to do research himself, and relied almost entirely on his collaborators for the few studies he did publish.

Throughout this period, Van Itallie's reviews of dietary therapy for obesity were singularly dedicated to dismissing any evidence that favored the use of carbohydrate-restricted diets. They would invariably begin with the declaration that carbohydrate-restricted diets were just another way to restrict calories, and they would proceed to refute claims made about the diets on the basis that these claims (not to be confused with observations of the diets' efficacy) had not been established beyond reasonable doubt. By the end of these reviews, Van Itallie would promote the continued treatment of obesity by balanced, calorie-restricted diets, while acknowledging that there was "increasing recognition of [their] ineffectiveness."* He would reject any suggestion that carbohydrate-restricted diets should be tried instead, while simultaneously acknowledging that *these* diets were "quite popular and have been followed with varying degrees of success by many dieters."

George Bray's influence in removing the fattening carbohydrate and carbohydrate-restricted diets from the nutritional wisdom was more subtle than Van Itallie's, but may have been ultimately more significant. Bray was a graduate of Harvard Medical School. In the late 1960s, he studied animal models of obesity at UCLA's Harbor General Hospital in Torrance, California. He also collaborated peripherally with Ethan Sims on his experimental obesity studies (Bray had been a medical-school classmate of Sims's colleague Ed Horton) and had notable disagreements with Sims about how this research should be interpreted. In 1973, Bray cochaired the NIH's first obesity conference; he then edited and drafted the subsequent NIH report, *Obesity in Perspective.* In 1977, he chaired the Second International Congress on Obesity and a second NIH conference on obesity. He then edited the NIH report *Obesity in America,* which was published in 1979. Meanwhile, he edited or wrote three of the half-dozen textbooks or clinical manuals on obesity that were published in the United States during the decade—*Treatment and Management of Obesity* (1974), *The Obese Patient* (1976), and *Obesity: Comparative Methods of Weight Control* (1980)—which means effectively all of those not edited or written by Stunkard.†

Bray believed that all diets worked by restricting calories, and since

---

* Van Itallie attributed this ineffectiveness, as was common at the time, to the fact that "a varied diet reduced in energy content remains highly palatable" and so too tempting. "Even the Lord's Prayer does not call for resisting temptation," he would say; "it asks that the supplicant be not led into temptation."

† This does not include several texts specifically on the psychology and behavioral treatment of obesity.

restricting calories eventually failed, nothing else need be discussed. He dismissed as irrelevant the work of those investigators who did actively study the dietary treatment of obesity, like Charlotte Young, who gave the presentation on dietary therapy at the NIH conference on obesity that Bray organized and chaired in 1973. Young specialized in the study of body composition, and she had been studying diets and obesity at Cornell since 1950. In the official NIH report on the conference, *Obesity in Perspective.* Bray treated her discussion of carbohydrate-restricted diets as naïve and of no consequence. In the book he coedited the year after the conference, *Treatment and Management of Obesity,* Young's observations on carbohydrate-restricted diets are described as still requiring further "confirmation before they can be fully accepted. . . . The question of the value of a low carbohydrate diet and its effectiveness in weight loss is still unresolved." In *The Obese Patient,* published three years after the NIH conference, Bray wrote of Young's studies, "The data are suggestive and require careful replication with larger groups of individuals." Yet nowhere in the NIH report on the conference, including a lengthy list of research priorities and "gaps in our current knowledge," did Bray raise the possibility that further research was needed on any dietary therapy for obesity, let alone, as Bray's own textbooks had suggested, the unresolved question of the value of carbohydrate restriction. Bray then proceeded to become the leading proponent of the hypothesis that obesity, like heart disease, was caused primarily by the dense calories of dietary fat, and thus could be cured or prevented by replacing the fat in the diet with carbohydrates.

The dissociation of the science of fat metabolism from any discussions of the cause or treatment of obesity was particularly conspicuous throughout this era and could be considered its legacy. When Bray, Van Itallie, Cahill, and Hirsch gave review talks at these conferences, as they did throughout this period, they would raise the issue of carbohydrate-restricted diets only to refute the claims that such diets offered a metabolic advantage over low-calorie diets. They would omit any mention of research that might explain the reported efficacy of the diets, even when that research was discussed at the same conferences and by investigators they knew personally. In 1977, for instance, Donald Novin, director of the Brain Research Institute at UCLA, discussed what he called the "carbohydrate hypothesis of ingestive behavior" at Bray's Second International Congress on Obesity. Novin suggested that the "widespread popularity of the low carbohydrate diets" could be explained by the effect of carbohydrates on insulin, and then of insulin on fat deposition and thus hunger. Bray, who had worked closely with Novin at UCLA, gave the summary talk at the con-

ference on obesity therapies and omitted mention of Novin's hypothesis.*
When M. R. C. Greenwood discussed the effect of insulin on the enzyme
lipoprotein lipase, LPL, the "gatekeeper" for fat accumulation in cells, at
the Fourth International Congress on Obesity, Hirsch ignored the implications in his review of dietary therapy, even though Greenwood had
received her doctoral degree with Hirsch.

In retrospect, the influential figures in the clinical investigation of
human obesity in the 1970s can be divided into two groups. There were
those who believed carbohydrate-restricted diets were the only efficacious
means of weight control—Denis Craddock, Robert Kemp, John Yudkin,
Alan Howard, and Ian McLean Baird in England, and Bruce Bistrian and
George Blackburn in the U.S.—and wrote books to that effect, or developed variations on these diets with which they could treat patients. These
men invariably struggled to maintain credibility. Then there were those
who refused to accept that carbohydrate restriction offered anything more
than calorie restriction in disguise—Bray, Van Itallie, Cahill, Hirsch, and
their fellow club members. These men rarely if ever treated obese patients
themselves, and they repeatedly suggested that since no diet worked nothing was to be learned by studying diets.

Bray would routinely equate the carbohydrate-restricted diet to every fad
diet that came along—the grapefruit diet, the banana diet, the ice-cream
diet. But when he testified before McGovern's subcommittee in 1977 and
described McGovern's *Dietary Goals* of a carbohydrate-rich diet for the
entire nation as "highly commendable," he also submitted as part of his
testimony a two-hundred-page report by the British Medical Research
Council entitled *Research on Obesity,* apparently ignoring the fact that the
report contradicted his own testimony. Published the same year, it referred
to carbohydrate restriction as the diet "commonly prescribed by general
practitioners" and considered it more effective and certainly more worthy
of discussion than the prescription of diets that depended on restricting
calories. The report also noted that the best weight-reduction results on
record were those reported by Robert Kemp and Denis Craddock, both
British practitioners who prescribed carbohydrate-restricted diets to their
patients and published their results, Kemp in the medical journals and
Craddock in *Obesity and Its Management.*

When a new diet book was published every few years touting yet

---

* According to Novin, when he wrote up his presentation for the conference proceedings Bray
removed the last four pages, all of which were on the link between carbohydrates, insulin, hunger,
and weight gain. "I couldn't believe he would make that kind of arbitrary decision," Novin said.

another physician's variation on carbohydrate restriction, it was treated by Bray and his colleagues as the ultimate evidence that the diet itself didn't work. "If such diets are truly successful," asked Van Itallie in his AMA denunciation of Atkins, "why then, do they fade into obscurity within a relatively short period of time only to be resurrected some years later in slightly different guise and under new sponsorship. Moreover, despite the claims of universal and painless success for such diets, no nationwide decrease in obesity has been reported." Of course, the efficacy of the diet could explain the continued popularity of such books. The diet had survived more or less continuously for over a century and had certainly thrived since the end of World War II. That the medical and nutrition establishments refused to take it seriously, and had even taken to advocating carbohydrate-rich diets instead, could explain the continued high prevalence of obesity.

This nihilistic argument became a mantra. "The evergrowing list of diets are an affirmation of the fact that no diet yet described is by itself a solution to the problem of obesity," Bray said in his 1977 testimony to McGovern's Senate committee. When Hirsch gave the review talk on obesity treatments at the Fourth International Congress on Obesity in 1981, he said: "The proliferation and seemingly endless concern with diets for the treatment of obesity suggests that this search is more motivated by financial rewards for the promoters rather than by an earnest desire to provide healthy and safe diets."

This theme of financial rewards for the promoters of these diets would also be echoed repeatedly. A "common factor of reducing regimens is their commercialism—someone stands to make money from their promotion," wrote George Mann, another veteran of Stare's nutrition department, in *The New England Journal of Medicine* in 1974. This didn't explain those like Pennington, Ohlson, Young, Gordon, or Kekwick and Pawan, who never wrote popular diet books and advocated similar advice to their obese patients, but it was an easy way to dismiss those like Atkins and Taller who did.* They were "instant monetary nutritionists," wrote Stare, who liked to point out that Atkins made over $1 million in one year from *Diet Revolution*, while simultaneously treating five hundred patients weekly in his "very lucrative private medical practice."

But this conflict-of-interest accusation, as we've discussed, often cuts both ways. Stare and his Harvard colleagues played the decisive role in

---

* Indeed, Mayer would divide those who endorsed carbohydrate-restricted diets into those who were sincere and misguided, and those who were simply insincere.

ensuring that anyone who claimed that carbohydrates were uniquely fattening would carry the taint of quackery. When White, Mayer, and Stare publicly condemned Herman Taller's *Calories Don't Count* it was a year after the Harvard nutrition department broke ground on a new $5 million building that was paid for largely through private donations. What Stare called the "lead gift" of $1,026,000 came from the General Foods Corporation, the maker of the very carbohydrate-rich Post cereals, Kool-Aid, and Tang breakfast drink. Over the next decade, Stare became the most public defender of sugar* and additives in modern diets, while his department continued to receive significant funding from the sugar industry; from Oscar Mayer, the maker of hot dogs; from Coca-Cola and the National Soft Drinks Association. Would the resident nutritionists in Stare's department have been more accepting of the efficacy of a diet that restricted refined carbohydrates and sugars if the money had come from another source? If so, would this have effected how other clinical investigators in the field came to interpret the controversy?

The funding of research projects, laboratories, and entire academic centers by the food and pharmaceutical industries is now a fact of life in modern medical research, which is why many journals require that their authors declare potential conflicts of interest. But it raises important questions, just the same. When *Science* dedicated special issues to obesity research in 1998 and again in 2003, James Hill from the University of Colorado was selected both times to write the review article on diet and lifestyle factors that influence weight gain. In those articles Hill argued that passive overeating and sedentary behavior were the causes of obesity, and he recommended reducing fat in the diet. Hill had long been a defender of the role of carbohydrates and particularly sugar in weight regulation. He even wrote an article, paid for by the Sugar Association, promoting the use of sugar in weight-loss diets, under the assumption that a high-carbohydrate diet, even if loaded with sugar, would "reduce the likelihood of overeating, rather than increasing it, as some popular diet theories purport." ("The theory that dietary sugar equals high insulin levels equals excess fat deposits is unproven and makes little biological sense," Hill wrote.) Over the years, as Hill has acknowledged in his conflict-of-interest statements, he has also received consulting fees from Coca-Cola, Kraft Foods, and Mars (makers of Snickers, M&M's and Mars Bars), companies that would stand to suffer significant setbacks if the notion of the fattening

---

* It was not "even remotely true," Stare wrote, "that modern sugar consumption contributes to poor health."

carbohydrate was institutionalized as a fact of science. He has also received over $2 million in what are technically termed "gifts" to his laboratory from Procter & Gamble, the maker of the fat substitute olestra, which has been described in the press as potentially a "dieter's dream." Olestra's only reason for existence is that it will allegedly help us manage our weight by replacing fat in the diet and making it easier for us to consume a low-fat, low-calorie diet. If carbohydrates are the fattening nutrients in human diets rather than fat or all calories, as Atkins suggested, then these diets have no role in weight loss or weight regulation, and olestra's rationale vanishes.

If the study of weight regulation were a legal issue, rather than a medical and scientific one, the support from Procter & Gamble would have been considered reason enough for Hill to recuse himself from any discussions of the dietary treatment of obesity or participation in any dietary trials that might directly influence Procter & Gamble's profitability, and thus perhaps Hill's interests.

In 2002 and 2003, Hill also received over $300,000 a year from the NIH to do a clinical trial testing the Atkins diet against a low-calorie, low-fat diet and, by implication, the justification for olestra as a fat substitute in a weight-reduction diet. And Hill was one of three principal investigators in the follow-up trial of the Atkins diet, for which the NIH provided $5 million. The salient question is whether Hill and the other academics in this pursuit are any less open to having their interpretation of the evidence influenced by financial considerations than Atkins or Taller or any of the other diet-book authors.

"A resolution of the very controversial question of the efficacy of low carbohydrate diets has great practical and theoretical significance," wrote Donald Novin of UCLA in 1978. Because a generation of obesity authorities were determined to dismiss the practical significance of carbohydrate-restricted diets, they dismissed the potential theoretical significance at the same time. Obesity researchers today say they still have no hypothesis of weight regulation that can explain obesity and leanness, let alone account for a century of paradoxical observations. They insist that obesity is inevitably caused by overeating and thus consuming more calories than we expend, but when asked what causes someone to overeat, they have no answer. Yet the research on insulin and fat metabolism offers one, and it has for several decades.

Chapter Twenty-four

THE CARBOHYDRATE HYPOTHESIS, III: HUNGER AND SATIETY

There is only one way to lose weight, and that is to grow accustomed to feeling hungry. This simple fact, known to most people in affluent countries, seems somehow lost on the authors of the diet, weight-loss, and exercise books that find their lucrative way through the drugstore book racks. Two questions, then: Why do they fail to mention it? And why is it so?

Emory University anthropologist Melvin Konner,
The Tangled Wing, 2003

IN 1975, THE DUKE UNIVERSITY PEDIATRICIAN James Sidbury, Jr., described a "rational basis" for the dietary treatment of childhood obesity, one that would neither torment his young patients with hunger nor rely on pharmaceutical means to prevent it. Such a diet, he wrote, would induce weight loss with a "minimum of anguish and struggle." Sidbury had an advantage over other investigators treating obese patients in that he had spent his career studying disorders of carbohydrate metabolism and, indeed, had already earned international renown for his development of a diet, still used today, to treat what are called glycogen storage diseases. The same year Sidbury published his description of a "Program for Weight Reduction in Children," however, he left his clinic at Duke to become director of the National Institute of Child Health and Human Development at NIH. By then, he had written only one short textbook chapter discussing his dietary treatment and one three-page article for an obscure journal called *Connecticut Medicine*. In them, he described an approach to obesity therapy that differed from Robert Atkins's only in the details of the application: Sidbury's diet was very low in both carbohydrates and calories, and Sidbury was writing for medical professionals, not the general public.

He based the design of his diet on several key observations. Fasted children "rarely, if ever, complained of hunger," Sidbury noted, and the "enzymes of lipogenesis"—insulin—rapidly decrease during fasting. Insulin is chronically elevated in obese patients, and the obese children referred to him in his practice typically consumed a diet dominated by carbohydrates—"crackers, potato chips, French fries, cookies, soft drinks, and the like." These foods are digested and absorbed as simple sugars, Sidbury explained, "chiefly glucose, which is the most potent stimulator of insulin release and synthesis." Since insulin will "facilitate lipogenesis" and inhibit the release of fat in the adipose tissue, this in turn created what Sidbury called the "milieu for positive fat balance" in the cells of the adipose tissue. "Thus it was reasoned," Sidbury wrote, "that a low carbohydrate diet would create the conditions vis-à-vis insulin metabolism which would decrease the constant stimulation of the [insulin-secreting] cells of the pancreas. The decreased insulin levels would then permit normal fatty acid mobilization."

The diet that Sidbury eventually used in his clinic and claimed to be uniquely effective contained only 15 percent carbohydrates—"the remaining being apportioned approximately equally between protein and fat"— and from three to seven hundred total calories a day, depending on the child's age. The older the child, the more calories allowed. "Many parents do not believe their child can be satisfied with so little food," Sidbury wrote. "Their attitude changes completely," however, when they see the "obvious change in the amount of food which satisfies the children."*

The phenomena of hunger and satiety have been the running subtext of all our discussions of obesity: the "persistent clamor of hunger" that attends semi-starvation diets; the absence of hunger during fasting and carbohydrate restriction; the question of whether insulin works as a fattening hormone or a hunger hormone when used to treat anorexia. And then, of course, there is the association of hunger, or at least positive caloric balance, with weight gain. If there's one thing the law of energy conservation does indeed tell us, it's that anything that works to increase or decrease our body mass must have compensatory effects on the balance of calories consumed and calories expended. Thus, any viable hypothesis of obesity must also be a hypothesis of hunger and satiety, and perhaps, as Alfred Pennington noted, of energy expenditure as well.

The study of human obesity, however, has included only a few vague conceptions of the physiological underpinnings of hunger and satiety. One common assumption is that when the stomach is empty it contracts, and that signals hunger. By this logic, dietary fat contributes to satiety by prolonging the drainage of nutrients out of the stomach. Another assumption is that hunger is a sensation that exists in the brain, having little or noth-

---

* In treating obese children, Sidbury noted, there is "concern that a low calorie diet will be harmful to growth." On this low-carbohydrate diet, however, the children experienced "continued normal linear growth," even though it was also very low in calories.

ing to do with the immediate metabolic needs of the body itself. Though many obesity researchers will reflexively disagree with this statement, it is essential to the conventional wisdom—that the ability to remain indefinitely on a calorie-restricted diet is a matter of willpower, and the failure to remain on such a diet is a failure of character. Once the pursuit of a therapy for obesity left physiology and biochemistry behind and became a subdiscipline of psychology and psychiatry, and once it was "established" that the only way to lose weight, as Melvin Konner suggested, is to grow accustomed to feeling hungry, the natural focus of virtually all obesity research became, and has remained, the brain.

By the early 1970s, a handful of hypotheses had been proposed to explain how the brain might induce hunger and satiety, and in turn regulate weight by limiting caloric consumption to match expenditure. Two received the most attention and have entered the textbooks as the most likely explanations. Both hypotheses date to the 1950s; neither took into account the evolving research on insulin, insulin resistance, and fat metabolism. Both had conspicuous deficiencies that would be overlooked.

One is Jean Mayer's glucostat hypothesis or, technically, the glucostatic regulation of food intake, and it is invoked to explain the short-term initiation of meals. Receptors in the hypothalamus, said Mayer, metabolize glucose, initiating the sensation of hunger when the available supply of glucose falls, and provoking satiety when it rises. This glucostatic regulation, as Mayer put it, is an "essential component of the mechanism by which the needs of the body make themselves felt in the satiety centers." It couldn't be the only one, however, because it offered no explanation for what Mayer called "the problem of the nature of the very fine adjustment knob . . . the mechanism which will make you regain the weight you lost after an illness, and which makes so difficult the maintenance of weight loss after an arduous weight reduction course."

The second hypothesis, what the Cambridge University physiologist Gordon Kennedy called lipostatic regulation or the lipostat, would evolve in the 1970s into the remarkably durable notion that we are all endowed with a certain set point of body weight or adiposity that we defend against both caloric deprivation and (perhaps less vigorously) caloric surplus. By Kennedy's logic, the lipostat is also located in the hypothalamus and accomplishes its fine-tuning job by monitoring the amount of fat in the body or some by-product of metabolism that is released into the bloodstream in relationship to our adiposity. When this adiposity signal dips below an acceptable level—the set point—the lipostat responds by increasing food intake or decreasing energy expenditure. When the adiposity signal moves above this set point, the lipostat works to suppress food intake and perhaps increase expenditure. According to this hypothesis, the fundamental difference between the lean and the obese is the amount of fat stores that the hypothalamus is set to defend—the set point—not the manner or vigor with which it is defended. Whatever our weight, if we find ourselves in a situation where our current level of body fat is beneath that of our set point, we will fatten easily until we've reached our predetermined level.

This hypothesis is a reformulation based on animal research of what had been considered a fact of life in pre–World War II nutrition textbooks, that "weight loss triggers the dual pressures of increased food intake and decreased caloric expenditure," as Stunkard put it. Nonetheless, obesity authorities have typically considered it unacceptably nihilistic. "It is not appealing from the therapeutic point of view," as Stunkard said, "because it sounds kind of . . . hopeless. If you're fat and your set point is elevated, you're in bad shape." And, of course, if we're fat, or very fat, it is difficult to argue that our set point is not elevated. Moreover, the hypothesis simply failed to explain how the brain manages to monitor our fat stores, and then raise or lower food intake and energy expenditure in response. Saying that we're all endowed with a lipostat that monitors our adiposity and then regulates hunger appropriately is just another way of saying that our weight remains remarkably stable, whether we're lean or obese, and then assigning the cause to a mysterious mechanism in the brain whose function is to achieve this stability.

The more fundamental criticism is that the concept of a set point or a lipostat has little precedent in physiology, whereas the long-term stability of body weight can be explained by a much simpler mechanism that does. Life is dependent on homeostatic systems that exhibit the same relative constancy as body weight, and none of them require a set point, like the temperature setting on a thermostat, to do so. Moreover, it is always possible to create a system that exhibits set-point-like behavior or a settling point, without actually having a set-point mechanism involved. The classic example is the water level in a lake, which might, to the naïve, appear to be regulated from day to day or year to year, but is just the end result of a balance between the flow of water into the lake and the flow out. When Claude Bernard discussed the stability of the *milieu intérieur,* and Walter Cannon the notion of homeostasis, it was this kind of dynamic equilibrium they had in mind, not a central thermostatlike regulator in the brain that would do the job rather than the body itself.

This is where physiological psychologists provided a viable alternative hypothesis to explain both hunger and weight regulation. In effect, they rediscovered the science of how fat metabolism is regulated, but did it from an entirely different perspective, and followed the implications through to the sensations of hunger and satiety. Their hypothesis explained the relative stability of body weight, which has always been one of the outstanding paradoxes in the study of weight regulation, and even why body weight would be expected to move upward with age, or even move upward on average in a population, as the obesity epidemic suggests has been the case lately. And this hypothesis has profound implications, both clinical and theoretical, yet few investigators in the field of human obesity are even aware that it exists.

This is yet another example of how the specialization of modern research can work against scientific progress. In this case, endocrinologists studying the role of hormones in obesity, and physiological psychologists studying eating behavior, worked with the same animal models and did similar experiments, yet they published in different journals, attended different conferences, and thus had little awareness of each other's work and results. Perhaps more important, neither discipline had any influence on the community of physicians, nutritionists, and psychologists concerned with the medical problem of human obesity. When physiological psychologists published articles that were relevant to the clinical treatment of obesity, they would elicit so little attention, said UCLA's Donald Novin, whose research suggested that the insulin response to carbohydrates was a driving force in both hunger and obesity, that it seemed as though they had simply tossed the articles into a "black hole."

The discipline of physiological psychology was founded on Claude Bernard's notion of the stability of the internal environment and Walter Cannon's homeostasis. Its most famous practitioner was the Russian Ivan Pavlov, whose career began in the late nineteenth century. The underlying assumption of this research is that behavior is a fundamental mechanism through which we maintain homeostasis, and in some cases—energy balance in particular—it is the primary mechanism. From the mid-1920s through the 1940s, the central figure in the field was Curt Richter of Johns Hopkins. "In human beings and animals, the effort to maintain a constant internal environment or homeostasis constitutes one of the most universal and powerful of all behavior urges or drives," Richter wrote.

Throughout the first half of the twentieth century, a series of experimental observations, many of them from Richter's laboratory, raised questions about what is meant by the concepts of hunger, thirst, and palatability, and how they might reflect metabolic and physiological needs. For example, rats whose adrenal glands are removed cannot retain salt, and will die within two weeks on their usual diet, from the consequences of salt depletion. If given a supply of salt in their cages, however, or given the choice of drinking salt water or pure water, they will choose either to eat or to drink the salt and, by doing so, keep themselves alive indefinitely. These rats will develop a "taste" for salt that did not exist prior to the removal of their adrenal glands. Rats that have had their parathyroid glands* removed will die within days of tetany, a disorder of calcium deficiency. If given the opportunity, however, they will drink a solution of calcium lactate rather than water—not the case with healthy rats—and will stay alive because of that choice. They will appear to *like* the calcium lactate more than water. And rats rendered diabetic voluntarily choose diets devoid of carbohydrates, consuming only protein and fat. "As a result," Richter said, "they lost their symptoms of diabetes, i.e., their blood sugar fell to its normal level, they gained weight, ate less food and drank only normal amounts of water."

The question most relevant to weight regulation concerns the quantity of food consumed. Is it determined by some minimal caloric requirement, by how the food tastes, or by some other physical factor—like stomach capacity, as is still commonly believed? This was the question addressed in the 1940s by Richter and Edward Adolph of the University of Rochester, when they did the experiments we discussed earlier (see pages 309–10), feeding rats chow that had been diluted with water or clay, or infusing nutrients directly into their stomachs. Their conclusion was that eating behavior is fundamentally driven by calories and the energy requirements of the animal. "Rats will make every effort to maintain their daily caloric intake at a fixed level," Richter wrote. Adolph's statement of this conclusion still constitutes one of the single most important observations in a century of research on hunger and weight regulation: "Food acceptance and the urge to eat in rats are found to have relatively little to do with 'a local condition of the gastro-intestinal canal,' little to do with the 'organs of taste,' and very much to do with quantitative deficiencies of currently metabolized materials"—in other words, the relative presence of usable fuel in the bloodstream.[†]

The physiological hypothesis of weight regulation and hunger that then

---

* Four glands that lie either behind or embedded in the thyroid gland.

† Adolph's studies were also noteworthy because, if humans are anything like rats, they contradict the popular notion that we gain weight by eating energy-dense foods or can lose weight and keep it off by decreasing the density of our diets—by eating soups, for instance, in which the calories are diluted by water, or fiber-rich greens and salads rather than calorie-dense meats—and so learning how to "feel full on fewer calories," as the Penn State nutritionist Barbara Rolls has advocated.

emerged in the mid-1970s evolved directly from the work of the French physiological psychologist Jacques Le Magnen, one of the more remarkable figures in the past century of science. Le Magnen was blind, the result of an attack of encephalitis when he was thirteen years old. He compensated by developing what his colleagues described as a "phenomenal" and "encyclopedic" memory, particularly for the nuances of relevant scientific research. "Jacques Le Magnen knew everything," as his obituary in the journal *Chemical Senses* commented after his death in 2002. He was also "incredibly brilliant," says the University of Cincinnati physiological psychologist Stephen Woods, which seems to be a consensus opinion among those who knew his work. Le Magnen joined the prestigious Collège de France in 1944, and he remained there for forty years, much of it spent working in the office and laboratory that had originally belonged to Claude Bernard. His Laboratory of Sensory and Behavioral Neurophysiology would eventually grow to become perhaps the largest in the world focused on issues related to hunger and weight regulation.

Le Magnen's research on eating behavior began in the early 1950s, when he designed a device to monitor food intake in rats over entire twenty-four-hour cycles. This led him to report that rats ate discrete meals separated in time by discrete intervals. He then set out to establish what factors regulated the size of the meals and the length of the intervals between meals.

Le Magnen's research resulted in two fundamental observations, both confirming Adolph's observation that eating behavior in animals, and thus hunger, is driven by those "quantitative deficiencies of currently metabolized materials."

Le Magnen learned that when rats are allowed to eat whenever they want, the size of the meal determines how long rats will go before they get hungry again. As a new supply of ingested calories is exhausted by the rat's energy expenditure, the animal is motivated to eat again. "All increase or decrease in the two sides of this balance (calories eaten in meals versus metabolic expenditures) will lead to an immediate shortening or lengthening of the meal-to-meal interval," Le Magnen explained. And this is "the major and direct agent of the regulation of food intake."

The second observation was one that is obviously true for humans as well: Rats eat to excess during their waking hours, which means their intake exceeds their expenditure of energy, and so they are hyperphagic while awake, storing fat during this period. While they're sleeping, the rats are in negative energy balance—hypophagic—and they live off the fat accumulated during the waking hours. Weight peaks as the rats are going to sleep and it ebbs as they awake. In humans, this cycle would explain, among other things, why hunger doesn't (or at least shouldn't) wake us from the depths of a night's sleep so we can raid the refrigerator.

While rats are sleeping, they progressively mobilize more and more fatty acids from their adipose tissue and use these fatty acids for fuel. "The restitution of these stored fats and their utilization to cover an important part of the current metabolism reduces the concomitant requirement for an external supply of calories by food intake," Le Magnen wrote. When he used insulin to suppress this mobilization of free fatty acids, the rats ate immediately. Fatty acids released from the adipose tissue, Le Magnen concluded, simply replace or "spare" the available glucose and, by doing so, delay the onset of hunger and the impetus to feed. The liberal availability of these fatty acids in the blood promotes satiety and inhibits hunger.

Another way to phrase this is that anything that induces fatty acids to escape from the fat tissue and then be burned as fuel will promote satiety by providing fuel to the tissues. Anything that induces lipogenesis, or fat synthesis and storage, will promote hunger by removing the available fuel from the circulation. And so hypophagia and hyperphagia, satiety and hunger, Le Magnen wrote, are "indirect and passive consequences" of "the neuroendocrine pattern of fat mobilization or synthesis."

By the mid-1970s, Le Magnen had demonstrated that insulin is the driver of this diurnal cycle of hunger, satiety, and energy balance. At the beginning of waking hours, the insulin response to glucose—the "insulin secretory responsiveness," Le Magnen called it—is enhanced, and it's suppressed during sleep. This pattern is "primarily responsible" for the fat accumulation during the waking hours and the fat mobilization during the sleeping hours. "The hyperinsulin secretion in response to food" during the period when the animals are awake and eating, and the "opposite train" when they are asleep, he explained, produces "a successive fall and elevation" of the level of fatty acids in the blood on a twenty-four-hour cycle—twelve hours during which the fatty acids are depressed and glucose is the primary fuel, and then twelve hours in which they're elevated and fat is the primary fuel. Both hunger, or the urge to eat, and satiety, or the inhibition of eating, are compensatory responses to these insulin-driven cycles of fat storage followed by fat mobilization. Insulin secretion is released in the morning upon waking and drives us to eat, Le Magnen concluded, and it ebbs after the last meal of the day to allow for prolonged sleep without hunger.

This hypothesis of eating behavior did away with set points and lipostats and relied instead on the physiological notion of hunger as a response to the availability of internal fuels and to the hormonal mechanisms of fuel partitioning. Hunger and satiety are manifestations of metabolic needs and physiological conditions at the cellular level, and so they're driven by the body, no matter how much we like to think it's our brains that are in control.

Several variations on this hypothesis were published from the mid-1970s onward by Le Magnen and others. The most comprehensive account was published in 1976 by Edward Stricker at the University of Pittsburgh, and Mark Friedman, then at the University of Massachusetts and now at the Monell Chemical Senses Center in Philadelphia. Their article, "The Physiological Psychology of Hunger: A Physiological Perspective," should be required reading for anyone seriously interested in eating behavior and weight regulation.

The hypothesis is based on three fundamental propositions. The first, as Friedman and Stricker explained, is that the supply of fuel to all body tissues must always remain "adequate for them to function during all physiological conditions and even during prolonged food deprivation." The second proposition is Hans Krebs's revelation from the 1940s that each of the various metabolic fuels—protein, fats, and carbohydrates—is equally capable of supplying energy to meet the demands of the body. The third is that the body has no way of telling the difference between fuels from internal sources—the fat tissue, liver glycogen, muscle protein—and fuels that come from external sources—i.e., whatever we eat that day.

With these propositions in mind, the simplest possible explanation for feeding behavior is that we eat to maintain this flow of energy to cells—to maintain "caloric homeostasis"—rather than maintain body fat stores or some preferred weight. If the cells themselves are receiving sufficient fuel to function, the size of the fat reserves is a secondary concern. As Friedman and Stricker explained, "Hunger appears and disappears according to normally occurring fluctuations in the availability of utilizable metabolic fuels, regardless of which fuels they are and how full the storage reserves." In 1993, the Princeton physiological psychologist Bartley Hoebel described the hypothesis in terms that echoed the origins of the theory in the work of Claude Bernard: "The primitive goal of feeding behavior," Hoebel explained, "is to maintain constancy of the nutrient concentration of the milieu intérieur."

From this perspective, we're not much more complicated than insects, which will seek out food and consume it until their guts are full. External taste receptors signal whether they've come upon something they can benefit from eating; gut receptors signal when sufficient food has been consumed to inhibit the hunger. The role of the brain is to integrate the sensory signals from the gut and the taste receptors and couple them to motor reflexes to initiate eating behavior or inhibit it. In both flies and mosquitoes, if the neural connection between gut and brain is severed, the insect loses its hunger inhibitor and continues to eat until its gut literally ruptures. As Edward Stricker explained in *The New England Journal of Medicine* in 1978, hunger is little more than a disturbing stimulus, like an itch, that "feeding behavior removes or attenuates." Satiety, on the other hand, "is more than the absence of hunger; it is the active suppression of interest in food and of feeding behavior."

The primary difference between humans and insects, by this logic, is that we have two primary fuel tanks (three if we include glycogen stored in the liver, and four if we include protein in the muscles), and they effectively have one. In our case, fuel is stored initially in the gut for the short term, and then in the fat tissue for the medium to longer term. The fat tissue extends the time we can go between meals by hours, days, or more. The fuel supply to the cells is maintained by the filling and emptying of both these energy reserves. "Energy metabolism," Friedman and Stricker wrote, "is maintained by alternating tides of nutrients that sweep in from the intestines or adipose tissue at regular intervals depending on when food consumption occurs." The fat tissue participates actively in metabolism by acting as an energy buffer: it provides storage for nutrients that arrive with the meal but are not immediately necessary for energy, and then it releases them back into the circulation as this absorptive phase is coming to an end. In effect, the fat tissue prevents dramatic shifts in the energy supply, which would otherwise be unavoidable considering the fact that, unlike cattle or sheep, we don't graze continually but, rather, eat episodically in discrete meals.

We can think of eating and satiety as a cycle that begins with the meal and fills the gastrointestinal reserve—the gut. As nutrients are absorbed into the circulation, some are used for fuel immediately, and the rest restock the fat reserves, the glycogen reserves in the liver, and the protein in the muscles. As the gut empties, and this dietary fuel is either stored or oxidized, the fat reserves become the primary source of fuel. As the fat reserves begin to empty and the fuel flow shows signs of faltering, the inhibition of hunger is lifted, we are motivated again to fill the gut, and the cycle begins anew.

This "harmony of tissue metabolisms" is orchestrated by the hypothalamus, via the central nervous system and the endocrine system of hormones. These regulate the filling and emptying of the various storage depots in response to an environment that might require that we suddenly expend more or less energy, or store more or less fat, to accommodate seasonal variations. The hypothalamus does what the brains of insects do: it integrates sensory signals from the body and the rest of the brain, and couples them to motor reflexes that permit or restrain eating behavior. It also adjusts this filling and emptying of the fuel reserves to accommodate the immediate need for fuel and the anticipated need for fuel.

According to this hypothesis, weight stability is nothing more than an equilibrium between the fatty acids flowing into the energy buffer of the fat tissue and the fatty acids flowing out. What the body regulates, as Le Magnen suggested, is the fuel flow to the cells; the amount of body fat we accumulate is a secondary effect of the fuel partitioning that accomplishes this regulation.

The implication of this hypothesis is that both weight gain and hunger will be promoted by factors that work to deposit fatty acids in the fat tissue and inhibit their mobilization—i.e., anything that elevates insulin. Satiety and weight loss will be promoted by factors that increase the release of fatty acids from the fat tissue and direct them to the cells of the tissues and organs to be oxidized—anything that lowers insulin levels. Le Magnen himself demonstrated this in his animal experiments. When he infused insulin into rats, it lengthened the fat-storage phase of their day-night cycle, and it shortened the fat-mobilization-and-oxidation phase accordingly. Their diurnal cycle of energy balance was now out of balance: the rats accumulated more fat during their waking hours than they could mobilize and burn for fuel during their sleeping hours. They no longer balanced their overeating with an equivalent phase of undereating. Not only were their sleep-wake cycles disturbed, but the rats would be hungry during the daytime and continue to eat, when normally they would be living off the fat they had stored at night.* Indeed, when Le Magnen infused insulin into sleeping rats, they immediately woke and began eating, and they continued eating as long as the insulin infusion continued. When during their waking hours he infused adrenaline—a hormone that promotes the mobilization of fatty acids from the fat tissue—they stopped eating.

If this hypothesis holds for humans, it means we gain weight because our insulin remains elevated for longer than nature or evolution intended, and so we fail to balance the inevitable fat deposition with sufficient fat oxidation. Our periods of satiety are shortened, and we are driven to eat more

---

* Rats are nocturnal.

often than we should. If we think of this system in terms of two fuel supplies, the immediate supply in the gut and the reserve in our fat deposits, both releasing fuel into the circulation for use by the tissues, then insulin renders the fat deposits temporarily invisible to the rest of the body by shutting down the flow of fatty acids out of the fat cells, while signaling the cells to continue burning glucose instead. As long as insulin levels remain elevated and the fat cells remain sensitive to the insulin, the use of fat for fuel is suppressed. We store more calories in this fat reserve than we should, and we hold on to these calories even when they're required to supply energy to the cells. We can't use this fat to forestall the return of hunger. "It is not a paradox to say that animals and humans that become obese gain weight because they are no longer able to lose weight," as Le Magnen wrote.

This alternative hypothesis may also tell us something profound about the relationship between nutrition and fertility. That shouldn't be surprising, because reproductive biologists, as we discussed earlier (see pages 373–74), have long considered the availability of food to be the most important environmental factor in fertility and reproduction. By this hypothesis, the critical variable in fertility is not body fat, as is commonly believed, but the immediate availability of metabolic fuels. This was suggested in the late 1980s, when the reproductive biologists George Wade and Jill Schneider described their research on hamsters, which were chosen because of their clockwork four-day estrous cycles. The experiments were remarkably consistent. These animals will go into heat whether they are fat or lean, and they will continue to cycle, as long as they can eat as much food as they want. If both fatty-acid *and* glucose oxidation are inhibited, however, and they're not allowed to increase their food intake in response, their estrous cycles stop. They'll remain infertile, whether they are gaining or losing weight at the time. These animals are responding to the general availability of metabolic fuels. The same observation has been made about pigs, sheep, and cattle. Monkeys will shut down their secretion of the hormone that triggers ovulation if they go twenty-four hours without food, but they'll re-establish secretion immediately upon eating. The more the monkeys are allowed to eat, the more hormone they'll secrete.

If it is true that fertility is determined by the availability of metabolic fuels, as Wade and Schneider explained, then "it would be expected that ovulatory cycles would be inhibited by treatments that direct circulating metabolic fuels away from oxidation and into storage in adipose tissue." This is what insulin does, of course, and, indeed, infusing insulin into animals will shut down their reproductive cycles. In hamsters, insulin infusion "totally blocks" estrous cycles, unless the animals are allowed to increase their normal food intake substantially to compensate. This hypothesis can also explain the infertility associated with obesity in both humans and lab animals. If "an excessive portion of available calories" is locked away in fat tissue, then the animal will act as if it's starving. In such a situation, Wade and Schneider said, "there will be insufficient calories to support both the reproductive and the other physiological processes essential for survival"; reproductive activity shuts down until more food is available to compensate.

This metabolic-fuel hypothesis of fertility has escaped the attention of clinicians. The clear implication is that a woman struggling with infertility or amenorrhea (the suppression of menstruation) will benefit more from a diet that lowers insulin but still provides considerable calories—a low-carbohydrate, high-fat diet—and thus repartitions the fuel consumed so that more is available for oxidation and less is placed in storage.

If this hypothesis of hunger, satiety, and weight regulation is correct, it means that obesity is caused by a hormonal environment—increased insulin secretion or increased sensitivity to insulin—that tilts the balance of fat storage and fat burning. This hypothesis also implies that the only way to lose body fat successfully is to reverse the process; to create a hormonal environment in which fatty acids are mobilized and oxidized in excess of the amount stored. A further implication is that any therapy that succeeds at inducing long-term fat loss—not including toxic substances and disease—has to work through these local regulatory factors on the adipose tissue.

If the principal effect of a drug, for example, is to suppress in the brain the desire to eat, and thus reduce food consumption, then the body will perceive the consequences as caloric deprivation and compensate accordingly. Energy expenditure will be reduced, and weight loss will be temporary at best. On the other hand, any drug that works locally on the fat cells to release fatty acids into the circulation will inhibit hunger because it will be increasing the flow of fuel to the cells. This could also be the case for any treatment that appears to increase metabolism or energy expenditure. A weight-loss drug that works in the brain to increase metabolism will also increase hunger, unless it also works on the fat tissue to mobilize fatty acids that can supply the necessary fuel.

Consider nicotine, for instance, which may be the most successful weight-loss drug in history, despite its otherwise narcotic properties. Cigarette smokers will weigh, on average, six to ten pounds less than nonsmokers. When they quit, they will invariably gain that much, if not more; approximately one in ten gain over thirty pounds. There seems to be nothing smokers can do to avoid this weight gain.

The common belief is that ex-smokers gain weight because they eat more once they quit. They will, but according to studies only in the first two or three weeks. After a month, former smokers will be eating no more than they would have been had they continued to smoke. The excess of calories consumed is not enough to explain the weight gain. Moreover, as Judith Rodin, now president of Rockefeller University, reported in 1987, smokers who quit and then gain weight apparently consume no more calories than those who quit and do not gain weight. (They do eat "significantly more carbohydrates," however, Rodin reported, and particularly more sugar.) Smokers also tend to be less active and exercise less than nonsmokers, so differences in physical activity also fail to explain the weight gain associated with quitting.

The evidence suggests that nicotine induces weight loss by working on fat cells to increase their insulin resistance, while also decreasing the lipoprotein-lipase activity on these cells, both of which serve to inhibit the accumulation of fat and promote its mobilization over storage, as we discussed earlier (see pages 397–99). Nicotine also seems to promote the mobilization of fatty acids directly by stimulating receptors on the membranes of the fat cells that are normally triggered by hormones such as adrenaline. The drug also increases lipoprotein-lipase activity on muscles, and this may explain the steep rise in metabolic rate that occurs immediately after smoking. All of this fits with the observations that smokers use fatty acids for a greater proportion of their daily fuel than nonsmokers, and heavy smokers burn more fatty acids than light smokers. In short, nicotine appears to induce weight loss and fat loss not by suppressing appetite but by freeing up fatty acids from the fat cells and then directing them to the muscle cells, where they're taken up and oxidized, providing the body with some excess energy in the process. When smokers quit, they gain weight because their fat cells respond to the absence of nicotine by significantly increasing lipoprotein-lipase activity. (There's also evidence that the weight-reduction drug fenfluramine—the "fen" half of the popular weight-loss drug phen/fen, which was banned by the FDA in 1997—works in a similar manner, by decreasing lipoprotein-lipase activity in the fat tissue.)

This alternative hypothesis of obesity and its physiological perspective on hunger forces us to rethink virtually all our cherished notions about how weight changes and why. By this hypothesis, any long-term variations in weight, appetite, and energy expenditure—even our inclination to exercise or go for a walk—are likely to be induced at a fundamental level by changes in the regulation of fat metabolism and the partitioning and availability of metabolic fuels in the body. These in turn are driven, first and foremost, by changes in insulin secretion and how our fat and muscle tissue respond to that insulin. In this sense, insulin becomes what researchers who study hibernation and other seasonal weight variations in animals refer to as the adjustable regulator. Increase or decrease the circulating levels of insulin, and weight, hunger, and energy expenditure increase or decrease accordingly. It's insulin that regulates the equilibrium between the forces of fat deposition and the forces of fat mobilization at the adipose tissue.

What's been clear for almost forty years is that the levels of circulating insulin in animals and humans will be proportional to body fat. "The leaner an individual, the lower his basal insulin, and vice versa," as Stephen Woods, now director of the Obesity Research Center at the University of Cincinnati, and his colleague Dan Porte observed in 1976. "This relationship has also been shown to occur in every commonly used model of altered body weight, including . . . genetically obese rodents and overfed humans. In fact, the relationship is sufficiently robust that it exists in the presence of widespread metabolic disorder, such as diabetes mellitus, i.e., obese diabetics have elevated basal insulin levels in proportion to their body weight." Woods and Porte also noted that when they fattened rats to "different proportions of their normal weights," this same relationship between insulin and weight held true. "There are no known major exceptions to this correlation," they concluded. Even the seasonal weight fluctuations in hibernators agree with this correlation; the evidence suggests that annual fluctuations in insulin secretion drive the yearly cycle of weight and eating behavior, although this has never been established with certainty.

This same mechanism might explain the annual patterns of weight fluctuation in humans as well—heavier in the fall and winter and lighter in the spring and summer—that are commonly attributed to increased physical activity supposedly accompanying the joys of spring or driven by the peer pressure and anxiety of the coming of bathing-suit season. When researchers have measured seasonal variations in insulin levels in humans, they have invariably reported that insulin is highest in late fall and early winter—twice as high, according to one 1984 study—and lowest in late spring and early summer. Moreover, as the University of Colorado's Robert Eckel has reported, lipoprotein-lipase activity in fat tissue elevates in late fall and decreases in spring and summer; its activity in skeletal muscle follows an opposite pattern. This would stimulate weight loss in the spring and weight gain in the fall, whether we consciously desire either or not, and would certainly make it easier to lose weight in the spring and gain it in the fall.

One of the most radical implications of this hypothesis is that even such an intractable condition as anorexia nervosa—which, like obesity, is now universally considered a behavioral and psychological disorder—may be caused fundamentally by a physiological defect of fat metabolism and insulin. The behavior of undereating may be a compensatory response to a physiological condition, just as the behavior of overeating can. Any hormonal abnormality that makes it difficult to store calories as fat—the fat cells, for example, becoming prematurely or abnormally resistant to insulin—could conceivably induce a compensatory inhibition of eating behavior and/or an increase in energy expended. What appears to be purely a behavioral phenomenon, the anorexia itself (and perhaps even bulimia nervosa), would be the compensatory response to a physiological problem, the inability to store calories after a meal in the energy buffer of the fat tissue. Correctly identifying cause and effect in these conditions would be difficult, if not impossible, without the understanding that there is an alternative hypothesis to explain the observations.

One final point has to be made about this physiological hypothesis of hunger and weight regulation, and it's almost as counterintuitive as it is important. This is what the hypothesis says about our perception of taste. One seemingly obvious relationship between diet and obesity has always been that the more palatable the food, the more we're likely to overindulge and so grow fat.

In the 1960s and 1970s, obesity researchers referred to this supposed effect of taste on food intake and weight as the palatability hypothesis. But these researchers defined palatability on the basis of how much their experimental animals ate. If their rats or mice ate more of one food than another, the researchers assumed that they did so because they liked it better. The problem is that this concept of palatability "arises mainly from human experience; its existence in animals is an inference," as the physiological psychologist Mark Friedman explained in 1989. In other words, the animals' preference for certain foods could have been explained by other factors.

In fact, our perception of what tastes good depends very much on circumstances. Le Magnen made this observation early in his career, and it's one reason why the subject of his own research evolved from olfactory stimuli to food intake. Le Magnen first noted that our assessment of odor changes with food consumption. The smell of a cinnamon bun baking in the oven will be considerably more enticing when we're hungry than after we've eaten. Our subjective interpretation of taste changes as well. With the possible exception of inordinately expensive meals at fashionable restaurants, the memorable meals of our lives are likely to be those we ate when we were particularly hungry—after a day of hard work or a particularly strenuous workout. "It is often said and not without reason," as Pavlov wrote in the 1890s, "that 'hunger is the best sauce.' "

Le Magnen established that an animal's response to a particular food correlates with how depleted the animal happens to be at the time, with the caloric value of the food, and with how rapidly it fulfills the animal's nutritional requirements. Rats given the choice between caloric sugar solutions and zero-calorie but equally sweet saccharine solutions initially drink similar amounts of both, Le Magnen reported. They both taste good. But the rats will drink more of the sugar solution with each passing day— drinking three times as much on day five as on day one—while rejecting the saccharine solution after three or four days, having apparently concluded, metabolically, that it offers no nutritive value. If the rats drinking the saccharine solution, however, are simultaneously infused with calorie-bearing glucose directly into their stomachs, they will continue to drink the saccharine solution as long as they get the calories along with it. The taste hasn't changed, but their post-absorption metabolic responses have. Foods that supply calories and other nutritional requirements quickly and efficiently will come to be perceived as tasting good, and so we learn to prefer them over others.

This offers up an alternative scenario to the common assumption that we are born with an innate preference for sugar because it would have been evolutionarily beneficial, prompting us to seek out those foods that are the densest source of calories in a world in which calories were supposedly hard to come by. "In evolution," as the Yale psychologist Linda Bartoshuk told the *New York Times* in 1989, "we needed the energy of sweet-tasting, sugary foods, especially during times of scarcity." The research of Le Magnen and others suggests that these preferences have little to do with the presence of famine in our evolutionary history (as discussed on pages 246–47) and everything to do with the absence of these refined carbohydrate foods. We come to prefer these foods, according to the alternative hypothesis, because they induce an exaggerated version of the post-absorption responses to naturally occurring sources of glucose and fructose—either plant foods that are difficult to digest (the kinds of roots, tubers, or fruit eaten by Paleolithic populations) or the protein in meat and the relatively slow conversion of its amino acids into glucose.

Since insulin plays the critical role in our post-absorption responses to particular foods, it's not surprising that insulin may play the critical role in our determination of palatability. A little-discussed observation in obesity research is that insulin is secreted in waves from the pancreas. The first wave begins within seconds of eating a "palatable" food, and well before the glucose actually enters the bloodstream. It lasts for perhaps twenty minutes. After this first wave ebbs, insulin secretion slowly builds back up in a more measured second wave, which lasts for several hours.* The apparent function of the first insulin wave is to prime the body for what's coming. It takes insulin almost ten minutes to have a measurable effect on blood-glucose levels; it takes twice that long to have any significant effect. Meanwhile, glucose is entering the bloodstream from the meal and continuing to stimulate insulin secretion. When blood sugar is at a maximum, the signal to the pancreas to secrete insulin is also highest, but by this time enough insulin has already been secreted to do the necessary job of glucose disposal. "The pancreas has no idea what's going on elsewhere in the body," says University of California, San Francisco, biochemist Gerald Grodsky, who pioneered much of this work. "All it sees is the glucose." The way we apparently evolved to deal with this systems-engineering problem is the flooding of insulin into the circulation immediately upon beginning a meal; this prepares the body in advance to start taking up the glucose as soon as it appears.

Le Magnen described this first wave of insulin as increasing "the metabolic background of hunger." In other words, this wave of insulin shuts down the mobilization of fat from the adipose tissue and stores away blood glucose in preparation for the imminent arrival of more. This leaves the circulation relatively depleted of nutrients. As a result, hunger increases. And this makes the food seem to taste even better. "In man," suggested Le Magnen, "it is reflected by the increased feeling of hunger at the beginning of a meal expressed in the popular adage in French: L'appétit vient en mangeant"—i.e., "the appetite comes while eating." As the meal continues and our appetite is satisfied, the metabolic background of hunger ebbs

---

* Because diabetologists and clinical investigators typically measure insulin in humans or laboratory animals at longer intervals—say, thirty minutes or an hour or two after a meal—they pay little attention to the details of what's happening in between, which means missing this first great wave of insulin secretion.

with the flood of nutrients into the circulation, and so the perceived palatability of the food wanes as well. Palatability, by this logic, is a learned response, conditioned largely by hunger, which in turn is a response to the pattern of insulin secretion and the availability of fatty acids and/or glucose in the circulation.

A related observation that has been a part of scientific study since Pavlov's famous research in the nineteenth century is that the smell, sight, or even thought of food will induce a cascade of physiological reactions. These include the secretion of saliva, gastric juices, and, not surprisingly, insulin. By the 1970s, these cephalic* reflexes had been studied in humans, rats, monkeys, cats, sheep, and rabbits. Le Magnen's student Stylianos Nicolaidis had demonstrated that rats will secrete insulin in response to the mere taste of a sweet substance, and it doesn't matter whether it is sugar or a no-calorie sugar substitute. The perceived taste of sweetness is sufficient to stimulate insulin secretion. Just as Pavlov demonstrated that dogs will salivate at the sound of a bell they have learned to associate with feeding, Stephen Woods and his colleagues demonstrated that rats will secrete insulin when confronted with similar eating-related stimuli. (These researchers arbitrarily chose the smell of mentholatum, a mixture of menthol and petroleum jelly, more commonly used as a topical rub for chest colds.) Humans will do the same. This reflexive release of insulin, Nicolaidis suggested, is "pre-adaptive": it anticipates the effects of a meal or a particular food, and so prepares the body. As Mark Friedman describes it, this cephalic release of insulin also serves to clear the circulation of "essentially anything an animal or a person can use for fuel. Not just blood sugar, but fatty acids, as well. All those nutrients just go away." Hence, the thought of eating makes us hungry, because the insulin secreted in response depletes the bloodstream of the fuel that the peripheral tissues and organs need to survive.

This cephalic secretion of insulin in preparation for the act of eating provides yet another mechanism that may work to induce hunger, weight gain, and obesity in a world of palatable foods, which could mean, of course, simply those foods that induce excessive insulin secretion to handle the unnaturally easy digestibility of their carbohydrates. The idea was suggested in 1977 by the psychologist Terry Powley, who was then at Yale and is now at Purdue University. Powley was discussing the obesity-

---

* "Of or about the head," referring to the fact that these reflexes are not mediated by the peripheral organs themselves—just as the two waves of insulin secretion are an inherent property of the pancreatic cells that secrete insulin—but are stimulated by nerve signals sent directly from the brain.

inducing effect of lesions in the hypothalamus and speculated that the lesions cause the animal to hypersecrete insulin when just thinking about, smelling, or tasting food, and this amplifies its perception of hunger and palatability. The result would be what Powley called a "self-perpetuating situation"—i.e., a vicious cycle. "Rather than secreting quantities of insulin and digestive enzymes appropriate for effective utilization of the ingested material," Powley wrote, "the lesioned animal over-secretes and must then ingest enough calories to balance the hormonal and metabolic adjustments."

Powley did not go so far as to suggest that this same phenomenon was at work in humans, but his then colleague Judith Rodin did. Rodin reported in 1980 that those individuals whose eating behavior is most responsive to the smell or sight of food—a grilling steak, in her experiments— were those who also had the greatest cephalic-phase insulin response. Insulin had to be considered a "major candidate," Rodin suggested, "for an intervening physiological mechanism that might be responsive to environmental stimuli." By 1985, Rodin was speculating that the chronic hyperinsulinemia of the obese would also exacerbate this phenomenon. "A feedback loop is suggested by these findings in which hyperinsulinemia in turn leads to increased consumption, which, unless compensated for, could lead to further weight gain," she wrote. "Because acute hyperinsulinemia can also be produced in some individuals by simply looking at or thinking about food, it, too, can in turn lead to increased consumption and possible weight gain."

The possibility that insulin determines what Le Magnen called the metabolic background of hunger also explains two observations we discussed in the sections on fattening and reducing diets.

The first is the observation by Ethan Sims that he could stuff his convict subjects with as much as ten thousand calories a day of mostly carbohydrate and they would still feel "hunger late in the day," and yet subjects fed eight hundred superfluous calories of fat "developed marked anorexia." On a more familiar level: why is it that most of us can imagine eating a large bag (twenty ounces) of movie popcorn—more than eleven hundred calories if popped in oil,* as it typically is—but not so the equivalent caloric amount of cheese: say, fifteen slices of American cheese, or a cup and a half of melted Brie?

---

* The USDA's standard nutrient database says eleven hundred calories. The Center for Science in the Public Interest puts the number at sixteen hundred.

The simple explanation is that the insulin induced by the carbohydrates serves to deposit both fats and carbohydrates (fatty acids and glucose) as fat in the adipose tissue, and it keeps those calories fixed in the adipose tissue once they get there. As long as we respond to the carbohydrates by secreting more insulin, we continue to remove nutrients from our bloodstream in expectation of the arrival of more, so we remain hungry, or at least absent any feeling of satiation. It's not so much that fat fills us up as that carbohydrates prevent satiety, and so we remain hungry.

The second observation is the carbohydrate craving associated with obesity. Here the metabolic background of hunger is established by chronic hyperinsulinemia rather than the immediate insulin secretion during a carbohydrate-rich meal. In both cases the insulin induces hunger or prevents satiety. In the case of hyperinsulinemia and obesity, however, this happens even between meals, when the cells should be living off a fuel mixture of predominantly fatty acids. Instead, the insulin traps the fat in the fat tissue, and it signals the cells to burn glucose. As far as the body is concerned, the elevated insulin is the indication that we've just eaten—"high levels of insulin herald the 'fed' state," as George Cahill put it—and the signal that carbohydrates are available to be burned. But in this case, they're not. Now the homeostatic system that evolved to maintain blood sugar in a healthy range establishes an internal environment in which the cells are primed to burn glucose for fuel, and only glucose can satisfy that demand, yet there's no expendable glucose in the system. High insulin levels even prevent the liver from releasing the glucose that's stored there as glycogen. As a result, it's glucose that we crave. Even if we eat fat and protein—our cheese slices, for instance—the hyperinsulinemia will work to store these nutrients rather than allow them to be used for fuel.

The practical implication of this situation is critical to how we perceive the dietary treatment of obesity, or simply the maintenance of a healthy weight, in a world of inexpensive, easily digestible carbohydrate-rich foods. Among the more pessimistic arguments wielded against carbohydrate-restricted diets is that all diets fail eventually because the subjects inevitably fall of the diet, just as they do calorie-restricted diets. But this argument is based on the assumption that all diets work by limiting the calories consumed. It also ignores any physiological difference between a craving for carbohydrates and the hunger that results from semi-starvation. The latter is caused by the absence of sufficient calories to satisfy physiological demands. The craving for carbohydrates is more closely akin to an addiction, which is how it was described by the British clinician Robert Kemp in 1963. It is the consequence of hyperinsulinemia, which in turn is caused initially by the presence of carbohydrates in the diet, just as an addiction to nicotine or cocaine or any other addictive substance is caused by the use of these substances. There is nothing inherently natural about such addictions. The hunger that accompanies calorie restriction is an unavoidable physiological condition; the craving for carbohydrates is not.

Sugar (sucrose) is a special case. Just like cocaine, alcohol, nicotine, and other addictive drugs, sugar appears to induce an exaggerated response in that region of the brain known as the reward center—the nucleus accumbens. This suggests that the relatively intense cravings for sugar—a sweet tooth—may be explained by the intensity of the dopamine secretion in the brain when we consume sugar. When the nucleus accumbens "is excessively activated by sweet food or powerful drugs," says Bartley Hoebel of Princeton, "it can lead to abuse and even addiction. When this system is under-active, signs of depression ensue." Rats can be easily addicted to sugar, according to Hoebel, and will demonstrate the physical symptoms of opiate withdrawal when forced to abstain.

Whether the addiction is in the brain or the body or both, the idea that sugar and other easily digestible carbohydrates are addictive also implies that the addiction can be overcome with sufficient time, effort, and motivation, which is not the case with hunger itself (except perhaps in the chronic condition of anorexia). Avoiding carbohydrates will lower insulin levels even in the obese, and so ameliorate the hyperinsulinemia that causes the carbohydrate craving itself. "After a year to eighteen months, the appetite is normalized and the craving for sweets is lost," said James Sidbury, Jr., about the effects on children of his carbohydrate-restricted diet. "This change can often be identified within a specific one to two week period by the individual."

If the more easily digestible carbohydrates are indeed addictive, this changes the terms of all discussions about the efficacy of carbohydrate-restricted diets. That someone might find living without starches, flour, and sugar to be difficult, and that there might be physical symptoms accompanying the withdrawal process, does not speak to the possibility that they might be healthier and thinner for the effort. No one would argue that quitting smoking (or any other addictive drug) is not salutary, even though ex-smokers invariably miss their cigarettes, and many will ultimately return to smoking, the addiction eventually getting the better of them. The same may be true for these carbohydrates.

It also makes us question the admonitions that carbohydrate restriction cannot "generally be used safely," as Theodore Van Itallie wrote in 1979, because it has "potential side effects," including "weakness, apathy, fatigue, nausea, vomiting, dehydration, postural hypotension, and occasional exacerbation of preexisting gout." The important clinical question is whether these are short-term effects of carbohydrate withdrawal, or chronic effects that might offset the benefits of weight loss. The same is true for the occasional elevation of cholesterol that will occur with fat loss—a condition known as *transient hypercholesterolemia*—and that is a consequence of the fact that we store cholesterol along with fat in our fat cells. When fatty acids are mobilized, the cholesterol is released as well, and thus serum levels of cholesterol can spike. The existing evidence suggests that this effect will vanish with successful weight loss, regardless of the saturated-fat content of the diet. Nonetheless, it's often cited as another reason to avoid carbohydrate-restricted diets and to withdraw a patient immediately from the diet should such a thing be observed, under the mistaken impression that this is a chronic effect of a relatively fat-rich diet.

In 1963, when Robert Kemp discussed his clinical experience with carbohydrate-restricted diets and the apparent problem of carbohydrate addiction, he made the point that the necessary step was to establish beyond reasonable doubt whether carbohydrates indeed were the cause of obesity and overweight. By doing so, we could then make informed decisions about the risks and benefits of our cravings. Many former cigarette smokers would likely still be smoking today without the certain knowledge that tobacco causes lung cancer. "At least half of our patients, win or lose, cannot be persuaded that they must permanently alter their eating habits to save their lives," Kemp wrote. "This is undoubtedly a battle for the mind where unfortunately the patient is completely unsettled by the confusion of advice offered from both professional and lay sources." This statement is still true today. Carbohydrate-restricted diets will always be tempting, if for no other reason than their efficacy at inducing weight loss. But to make a permanent change in diet requires the confidence that we will be healthier for doing so. For that, we need the support of physicians, nutritionists, and the public-health authorities, and we need advice that is based on rigorous science, not century-old preconceptions about the penalties of gluttony and sloth.

EPILOGUE

The community of science thus provides for the social validation of scientific work. In this respect, it amplifies that famous opening line of Aristotle's *Metaphysics:* "All men by nature desire to know." Perhaps, but men of science by culture desire to know that what they know is really so.
ROBERT MERTON, *Behavior Patterns of Scientists,* 1968

The first principle is that you must not fool yourself—and you are the easiest person to fool.
RICHARD FEYNMAN, in his Commencement Address at Caltech, 1974

ON FEBRUARY 7, 2003, THE EDITORS OF *Science* published a special issue dedicated to the critical concerns of obesity research. It included four essays written by prominent authorities, all communicating the message of the toxic-environment hypothesis of the obesity epidemic and the belief that obesity is caused by "consuming more food energy than is expended in activity." The one article that offered a potential solution to the national and global problem of burgeoning waistlines—other than the promise of future obesity-fighting drugs—was written by James Hill of the University of Colorado, John Peters of Procter & Gamble, and two colleagues. Hill and Peters introduced the concept of an "energy gap" that could purportedly explain the existence of the obesity epidemic and illuminate a path of action by which it might be halted or reversed. By their calculation, the obesity epidemic represented an energy gap of a hundred calories per person among the American public per day that had been consumed but not expended. To undo the epidemic, Hill and Peters suggested, Americans would have to make either comparable increases in daily energy expenditure—walking one extra mile, perhaps—or decreases in energy consumption, such as "eating 15% less (about three bites) of a typical premium fast-food hamburger." Two years later, when the U.S. Department of Agriculture released the sixth edition of its *Dietary Guidelines for Americans,* it offered similar advice based on the identical logic: "For most adults a reduction of 50 to 100 calories per day may prevent gradual weight gain."

This proposition should evoke a distinct sensation of déjà vu, because it is the precise argument that Carl von Noorden made over a century ago. Hill, Peters, and the USDA authorities, like von Noorden, were treating the regulation of body weight as though it were a purely arithmetical process, in which a small excess of calories consumed, day in and day out, accumulates into pounds of flesh and then tens of pounds, and a small deficit, day in and day out, does the opposite. That this argument is now the cornerstone of the official U.S. government recommendations for obesity prevention made the single caveat in Hill and Peters's *Science* article all that much more remarkable. Speaking of the hundred-calorie energy gap, they said that their "estimate is theoretical and involves several assumptions"—in particular, "Whether increasing energy expenditure or reducing energy intake by 100 kcal/day would prevent weight gain remains to be empirically tested."

The more important point, though, which Hill and Peters did not discuss, was why a century of research had not produced such an empirical test. Two immediate possibilities suggest themselves: Either the accumulated research and observations on weight regulation in humans or animals had never provided sufficient reason to believe that such a proposition should be true, which is a necessary condition for anyone to expend the effort to test it; or, perhaps, nobody cared to test it. In either case, we have to wonder whether the individuals involved in the pursuit of the cure and prevention of human obesity, as Robert Merton would have put it, have the desire to know that what they know is really so.

In the 1890s, Francis Benedict and Wilbur Atwater, pioneers of the science of nutrition in the United States, spent a year in the laboratory testing the assumption that the law of energy conservation applied to humans as well as animals. They did so not because they doubted that it did, but precisely because it seemed so obvious. "No one would question" it, they wrote. "The quantitative demonstration is, however, desirable, and an attested method for such demonstration is of fundamental importance for the study of the general laws of metabolism of both matter and energy."

This is how functioning science works. Outstanding questions are identified or hypotheses proposed; experimental tests are than established either to answer the questions or to refute the hypotheses, regardless of how obviously true they might appear to be. If assertions are made without the empirical evidence to defend them, they are vigorously rebuked. In science, as Merton noted, progress is made only by first establishing whether one's predecessors have erred or "have stopped before tracking down the implications of their results or have passed over in their work what is there to be seen by the fresh eye of another." Each new claim to knowledge, therefore, has to be picked apart and appraised. Its shortcomings have to be established unequivocally before we can know what questions remain to be asked, and so what answers to seek—what we know is really so and what we don't. "This unending exchange of critical judgment," Merton wrote, "of praise and punishment, is developed in science to a degree that makes the monitoring of children's behavior by their parents seem little more than child's play."

The institutionalized vigilance, "this unending exchange of critical judgment," is nowhere to be found in the study of nutrition, chronic disease, and obesity, and it hasn't been for decades. For this reason, it is difficult to use the term "scientist" to describe those individuals who work in these disciplines, and, indeed, I have actively avoided doing so in this book. It's simply debatable, at best, whether what these individuals have practiced for the past fifty years, and whether the culture they have created, as a result, can reasonably be described as science, as most working scientists or philosophers of science would typically characterize it. Individuals in these disciplines think of themselves as scientists; they use the terminology of science in their work, and they certainly borrow the authority of science to communicate their beliefs to the general public, but "the results of their enterprise," as Thomas Kuhn, author of *The Structure of Scientific Revolutions*, might have put it, "do not add up to science as we know it."

Though the reasons for this situation are understandable, they offer scant grounds for optimism. Individuals who pursue research in this confluence of nutrition, obesity, and chronic disease are typically motivated by the desire to conserve our health and prevent disease. This is an admirable goal, and it undeniably requires reliable knowledge to achieve, but it cannot be accomplished by allowing the goal to compromise the means, and this is what has happened. Practical considerations of what is too loosely defined as the "public health" have consistently been allowed to take precedence over the dispassionate, critical evaluation of evidence and the rigorous and meticulous experimentation that are required to establish reliable knowledge. The urge to simplify a complex scientific situation so that physicians can apply it and their patients and the public embrace it has taken precedence over the scientific obligation of presenting the evidence with relentless honesty. The result is an enormous enterprise dedicated in theory to determining the relationship between diet, obesity, and disease, while dedicated in practice to convincing everyone involved, and the lay public, most of all, that the answers are already known and always have been—an enterprise, in other words, that purports to be a science and yet functions like a religion.

The essence of the conflict between science and nutrition is time. Once we decide that science is a better guide to a healthy diet than whatever our parents might have taught us (or our grandparents might have taught our parents), then the sooner we get reliable guidance the better off we are. The existence of uncertainty and competing hypotheses, however, does not change the fact that we all have to eat and we have to feed our children. So what do we do?

There are two common responses to this question, as there will be to the arguments made in this book. One response is to take into account the uncertainties about the health effects of fats and carbohydrates and then suggest that we simply eat in moderation. This in turn implies eating a *balanced* diet in moderation. "Perhaps our most sensible public health recommendation should be moderation in all things, and moderation in *that*," as the University of Michigan professor of public health Marshall Becker suggested back in 1987. But some of us do eat with admirable restraint of the four major food groups and yet are obese or overweight anyway, and presumably have an increased risk of other chronic diseases because of it; some of us are suitably lean, eat balanced diets in moderation, and exercise regularly and yet are insulin-resistant and maybe even diabetic.

The more optimistic response is a compromise position: to take virtually every reasonable hypothesis from the past fifty years that can coexist with the saturated-fat/cholesterol hypothesis of heart disease and fold them all into one seemingly reasonable diet that might do us good and *probably* won't do harm. Thus, the current conception of a healthy diet is one that minimizes salt content and maximizes fiber; has plenty of good fats (monounsaturated and omega-three polyunsaturated fats) and minimal bad fats (saturated fats and trans fats); has plenty of olive oil and fish, and little red meat, butter, lard, and dairy products. When meat is consumed, it's lean, which keeps saturated-fat content down and reduces energy density and thus, supposedly, calories. Dairy is low-fat or no-fat. The diet has plenty of nuts and legumes and good carbohydrates, which are those with copious vitamins, minerals, antioxidants, and fiber (vegetables, fruits, and unrefined grains), but few bad carbohydrates, which are energy-dense and thus contribute to obesity (highly refined carbohydrates and sugars).

It may be true that such a diet is uniquely healthy—but we have no idea if that's really so. The diet has the advantage of being politically correct; it can be recommended without fear of ostracism from the medical community. Whether it is healthier, however, than, say, a meat diet of 70–80 percent fat calories and absent carbohydrates almost entirely, as Stefansson suggested in the 1920s, or any diet of animal products (meat, fish, fowl, eggs, and cheese) and green vegetables but absent entirely starches, sugar, and flour or even sugar alone, is still anybody's guess. And whether such a diet would prevent us from fattening or reverse obesity, or do it better than a mostly meat diet, has also never been tested. If it doesn't, then it's probably not the healthiest diet, because excessive fat accumulation is certainly associated with increased risk of chronic disease.

I have spent much of the last fifteen years reporting and writing about issues of public health, nutrition, and diet. I have spent five years on the research for and writing of this book alone. To a great extent, the conclusions I've reached are as much a product of the age we live in as they are my own skeptical inquiry. Just ten years ago, the research for this book would have taken the better part of a lifetime. It was only with the development of the Internet, of search engines and the comprehensive databases of the Library of Medicine, the Institute for Scientific Information, research libraries, and secondhand-book stores worldwide now accessible online that I was able, with reasonable facility, to locate and procure virtually any written source, whether published a century ago or last week, and to track down and contact clinical investigators and public-health officials, even those long retired.

Throughout this research, I tried to follow the facts wherever they led. In writing the book, I have tried to let the science and the evidence speak for themselves. When I began my research, I had no idea that I would come to believe that obesity is not caused by eating too much, or that exercise is not a means of prevention. Nor did I believe that diseases such as cancer and Alzheimer's could possibly be caused by the consumption of refined carbohydrates and sugars. I had no idea that I would find the quality of the research on nutrition, obesity, and chronic disease to be so inadequate; that so much of the conventional wisdom would be founded on so little substantial evidence; and that, once it was, the researchers and the public-health authorities who funded the research would no longer see any reason to challenge this conventional wisdom and so to test its validity.

As I emerge from this research, though, certain conclusions seem inescapable to me, based on the existing knowledge:

1. Dietary fat, whether saturated or not, is not a cause of obesity, heart disease, or any other chronic disease of civilization.

2. The problem is the carbohydrates in the diet, their effect on insulin secretion, and thus the hormonal regulation of homeostasis—the entire harmonic ensemble of the human body. The more easily digestible and refined the carbohydrates, the greater the effect on our health, weight, and well-being.

3. Sugars—sucrose and high-fructose corn syrup specifically—are particularly harmful, probably because the combination of fructose and glucose simultaneously elevates insulin levels while overloading the liver with carbohydrates.

4. Through their direct effect on insulin and blood sugar, refined carbohydrates, starches, and sugars are the dietary cause of coronary heart disease and diabetes. They are the most likely dietary causes of cancer, Alzheimer's disease, and the other chronic diseases of civilization.

5. Obesity is a disorder of excess fat accumulation, not overeating, and not sedentary behavior.

6. Consuming excess calories does not *cause* us to grow fatter, any more than it causes a child to grow taller. Expending more energy than we consume does not lead to long-term weight loss; it leads to hunger.

7. Fattening and obesity are caused by an imbalance—a disequilibrium—in the hormonal regulation of adipose tissue and fat metabolism. Fat synthesis and storage exceed the mobilization of fat from the adipose tissue and its subsequent oxidation. We become leaner when the hormonal regulation of the fat tissue reverses this balance.

8. Insulin is the primary regulator of fat storage. When insulin levels are elevated—either chronically or after a meal—we accumulate fat in our fat tissue. When insulin levels fall, we release fat from our fat tissue and use it for fuel.

9. By stimulating insulin secretion, carbohydrates make us fat and ultimately cause obesity. The fewer carbohydrates we consume, the leaner we will be.

10. By driving fat accumulation, carbohydrates also increase hunger and decrease the amount of energy we expend in metabolism and physical activity.

In considering these conclusions, one must address the obvious question: can a diet mostly or entirely lacking in carbohydrates possibly be a healthy pattern of eating? For the past half century, our conceptions of the interaction between diet and chronic disease have inevitably focused on the fat content. Any deviation from some ideal low-fat or low-saturated-fat diet has been considered dangerous until long-term, randomized control trials might demonstrate otherwise. Because a diet restricted in carbohydrates is by definition relatively fat-rich, it has therefore been presumed to be unhealthy until proved otherwise. This is why the American Diabetes Association even recommends against the use of carbohydrate-restricted diets for the management of Type 2 diabetes. How do we know they're safe for long-term consumption?

The argument in their defense is the same one that Peter Cleave made forty years ago, when he proposed what he called the saccharine-disease hypothesis. Evolution should be our best guide for what constitutes a healthy diet. It takes time for a population or a species to adapt to any new factor in its environment; the longer we've been eating a particular food as a species, and the closer that food is to its natural state, the less harm it is likely to do. This is an underlying assumption of all public-health recommendations about the nature of a healthy diet. It's what the British epidemiologist Geoffrey Rose meant when he wrote his seminal 1985 essay, "Sick Individuals and Sick Populations," and described the acceptable measures of prevention that could be recommended to the public as those that remove "unnatural factors" and restore " 'biological normality'—that is . . . the conditions to which presumably we are genetically adapted." "Such normalizing measures," Rose said, "may be presumed to be safe, and therefore we should be prepared to advocate them on the basis of a reasonable presumption of benefit."

The fat content of the diets to which we presumably evolved, however, will always remain questionable. If nothing else, whatever constituted the typical Paleolithic hunter-gatherer diet, the type and quantity of fat consumed assuredly changed with season, latitude, and the coming and going of ice ages. This is the problem with recommending that we consume oils in any quantity. Did we evolve to eat olive oil, for example, or linseed oil? And maybe a few thousand years is sufficient time to adapt to a new food but a few hundred is not. If so, then olive oil could conceivably be harmless or even beneficial when consumed in comparatively large quantities by the descendants of Mediterranean populations, who have been consuming it for millennia, but not to Scandinavians or Asians, for whom such an oil is new to the diet. This makes the science even more complicated than it already is, but these are serious considerations that should be taken into account when discussing a healthy diet.

There is no such ambiguity, however, on the subject of carbohydrates. The most dramatic alterations in human diets in the past two million years, unequivocally, are (1) the transition from carbohydrate-poor to carbohydrate-rich diets that came with the invention of agriculture—the addition of grains and easily digestible starches to the diets of hunter-gatherers; (2) the increasing refinement of those carbohydrates over the past few hundred years; and (3) the dramatic increases in fructose consumption that came as the per-capita consumption of sugars—sucrose and now high-fructose corn syrup—increased from less than ten or twenty pounds a year in the mid-eighteenth century to the nearly 150 pounds it is today. Why would a diet that excludes these foods specifically be expected to do anything other than return us to "biological normality"?

It is not the case, despite public-health recommendations to the contrary, that carbohydrates are required in a healthy human diet. Most nutritionists still insist that a diet requires 120 to 130 grams of carbohydrates, because this is the amount of glucose that the brain and central nervous system will metabolize when the diet is carbohydrate-rich. But what the brain uses and what it requires are two different things. Without carbohydrates in the diet, as we discussed earlier (see page 319), the brain and central nervous system will run on ketone bodies, converted from dietary fat and from the fatty acids released by the adipose tissue; on glycerol, also released from the fat tissue with the breakdown of triglycerides into free fatty acids; and on glucose, converted from the protein in the diet. Since a carbohydrate-restricted diet, unrestricted in calories, will, by definition, include considerable fat and protein, there will be no shortage of fuel for the brain. Indeed, this is likely to be the fuel mixture that our brains evolved to use, and our brains seem to run more efficiently on this fuel mixture than they do on glucose alone. (A good discussion of the rationale for a minimal amount of carbohydrates in the diet can be found in the 2002 Institute of Medicine [IOM] report, *Dietary Reference Intakes*. The IOM sets an "estimated average requirement" of a hundred grams of carbohydrates a day for adults, so that the brain can run exclusively on glucose, "without having to rely on a partial replacement of glucose by [ketone bodies]." It then sets the "recommended dietary allowance" at 130 grams to allow margin for error. But the IOM report also acknowledges that the brain will be fine without these carbohydrates, because it runs perfectly well on ketone bodies, glycerol, and the protein-derived glucose.)

Whether a carbohydrate-restricted diet is deficient in essential vitamins and minerals is another issue. As we also discussed (see page 320–26), animal products contain all the amino acids, minerals, and vitamins essential for health, with the only point of controversy being vitamin C. And the evidence suggests that the vitamin C content of meat products is more than sufficient for health, as long as the diet is indeed carbohydrate-restricted, with none of the refined and easily digestible carbohydrates and sugars that would raise blood sugar and insulin levels and so increase our need to obtain vitamin C from the diet. Moreover, though it may indeed be uniquely beneficial to live on meat and only meat, as Vilhjalmur Stefannson argued in the 1920s, carbohydrate-restricted diets, as they have been prescribed ever since, do not restrict leafy green vegetables (what nutritionists in the first half of the twentieth century called 5 percent vegetables) but only starchy vegetables (e.g., potatoes), refined grains and sugars, and thus only those foods that are virtually without any essential nutrients unless they're added back in the processing and so *fortified*, as is the case with white bread. A calorie-restricted diet that cuts all calories by a third, as John Yudkin noted, will also cut essential nutrients by a third. A diet that prohibits sugar, flour, potatoes, and beer, but allows eating to satiety meat, cheese, eggs, and green vegetables will still include the essential nutrients, whether or not it leads to a decrease in calories consumed.

My hope is that this book will change our views of the nature of a healthy diet, as the research for it changed my own; that future discussions of the nature of a healthy diet will begin with the quantity and quality of the carbohydrates contained, rather than the fat. As a challenge to the conventional wisdom on diet, obesity, and chronic disease, however, it presents a dilemma to public-health authorities; to nutritionists and physicians who believe that the advice they have been giving for the past few decades has been correct and based in sound science; and to all of us who simply want to eat healthy but have trouble accepting that everything we have come to believe could be as misguided as I have portrayed it. The resolution to this dilemma is to test the carbohydrate hypothesis rigorously, just as the fat-cholesterol hypothesis of heart disease *should* have been tested forty years ago.

In the past decade, the National Institutes of Health finally began funding randomized-control trials of carbohydrate-restricted diets, as has the Dr. Robert C. Atkins Foundation, but these trials have been designed to test only the hypothesis that such diets can be used safely and effectively as a means to lose weight. The subjects are overweight and obese, and the studies compare weight loss and heart-disease risk factors with the results of low-fat or calorie-restricted diets. These trials are neither planned nor interpreted as tests of the hypothesis that it is the carbohydrates in the diet—"the sugar and starchy elements of food," as *The Lancet* phrased it 140 years ago—that cause fattening and obesity to begin with. Rather, the underlying assumption here, too, is that weight loss is caused inevitably by negative caloric balance—consuming fewer calories then we expend—and the investigators perceive these trials as testing whether carbohydrate restriction allows us to do so with more or less facility than semi-starvation diets that reduce calories directly or reduce fat calories specifically.

A direct test of the carbohydrate hypothesis asks the opposite question: not whether the absence of refined and easily digestible carbohydrates and sugars causes weight loss and is safe, but whether the presence of these carbohydrates causes weight gain and chronic disease. Such a trial would ideally be done with lean, healthy individuals, or with a spectrum of subjects from lean through obese, including those with metabolic syndrome and Type 2 diabetes. They would be randomized into two groups, one of which would consume the sugary and starchy elements of food and one of which would not, and then we would see what happens. We might randomly assign a few thousand individuals to eat the typical American diet of today—including its 140–50 pounds of sugar and high-fructose corn syrup a year, nearly 200 pounds of flour and grain, 130-plus pounds of potatoes, and 27 pounds of corn—and we could assign an equal number to eat a diet of mostly animal products (meat, fish, fowl, eggs, cheese) and leafy green vegetables. Since the latter diet would be relatively high in fat and saturated fat and calorically dense, the conventional wisdom is that it would cause heart disease and, perhaps, obesity and diabetes. So this would test the dietary-fat/cholesterol hypothesis of heart disease, as well as the carbohydrate hypothesis.

Such a trial would not be ideal, because many dietary variables would differ between the two groups—calories and fats among them. The subjects would also know what diet they're consuming, and so the study would not be done blindly (although, ideally, the physicians who treated the subjects and the investigators themselves would be unaware). Nonetheless, it would be a good starting point. Would those eating the carbohydrate-rich diet be more likely to become glucose-intolerant, hyperinsulinemic, and insulin-resistant? Would they be fatter and have a greater incidence of obesity, metabolic syndrome, and Type 2 diabetes? Would they have more heart disease and cancer? Would they die prematurely or live longer? These are the questions we need to answer.

Another question that needs to be addressed urgently regards the health effects of sugar and high-fructose corn syrup alone. Since the 1980s, as we discussed (see page 198), sugar and high-fructose corn syrup have been exonerated as causes of chronic disease on the basis that the evidence was ambiguous. Since then, virtually no studies have been funded; there have been no attempts to clarify the picture. Today I can imagine no research more important to the public health than rigorous, controlled trials of the long-term health effects of sugar and high-fructose corn syrup.

For the past decade, the National Institutes of Health has been funding trials that test whether "lifestyle modification" will prevent diabetes and metabolic syndrome. But these trials are done only in the context of the conventional wisdom on diet, obesity, and disease. In the largest of these trials to date, the $150 million Diabetes Prevention Program, the lifestyle modification included 150 minutes of exercise each week and a low-fat, low-calorie diet. The results confirmed that such a program of diet and exercise will indeed prevent or delay the appearance of diabetes and metabolic syndrome, but they said nothing about what aspect of this lifestyle modification was responsible. Was it the reduction in fat calories or total calories? Was it the exercise? Or was it a change in the type of carbohydrates consumed or a reduction in the total amount of carbohydrates? As we discussed (see page 317), even if the goal of a diet is to reduce calories by preferentially reducing fat, it will inevitably cut back on carbohydrates as well, and usually sugars in particular.*

The NIH is currently spending $200 million on a decade-long trial called Look AHEAD to test the hypothesis that if obese diabetics lose weight they'll be healthier for the effort. This is "the largest, most expensive trial ever funded by NIH for obesity outcome research," says the Baylor University psychologist John Foreyt, who is one of the trial's principal investigators. But once again, the trial tests only the conventional wisdom. The goal of Look AHEAD is to induce five thousand obese diabetics to lose weight by the same lifestyle modification used in the Diabetes Prevention Program: cutting calories and fat calories, and exercising. If these obese diabetics do lose weight, and if they do end up healthier for it, we still won't know whether it was the calories, the fat calories, the exercise, some combination of all three, or maybe just the carbohydrates or the sugar that made the difference. And we won't know whether, if they restricted carbohydrates alone and ate protein and fat to their hearts' content, they would have been healthier still.

Because these trials are planned as a test of only one hypothesis—and a poorly defined hypothesis at that—the research ensures that we won't have the kind of reliable answers that we so desperately need. If the Diabetes Prevention Program had included a test of the carbohydrate hypothesis, the investigators could have compared the effect of a low-fat, low-calorie

---

* After a year, subjects participating in the lifestyle-modification trial had reduced their total food intake, on average, by 450 calories a day. They ate more fruits and vegetables (one to two servings more a day); they decreased their grain consumption by four servings a day and "sweets" by five. The calories from all carbohydrates increased, on average, by over 5 percent, but because of the decrease in total calories, the total amount of carbohydrates consumed decreased.

diet and exercise to the effect of carbohydrate restriction alone, and that would have told us whether it's the carbohydrates or the calories and the sedentary behavior that cause these chronic diseases. If Look AHEAD were to include a test of the carbohydrate hypothesis, we might at least know the answer in another decade. It doesn't, and we won't.

The scientific obligation, as I said in the prologue, is to establish the cause of obesity, diabetes, and the chronic diseases of civilization beyond reasonable doubt. By doing so, we can take the necessary steps to prevent these disorders, rather than trying to cure them or ameliorate them after the fact. If there are competing hypotheses, it does us little good to test one alone. It does little good to continue basing public-health recommendations and dietary advice on association studies (the Framingham Heart Study and the Nurses Health Study are prominent examples) that are incapable of reliably establishing cause and effect. What's needed now are randomized trials that test the carbohydrate hypothesis as well as the conventional wisdom. Such trials would be expensive. Like the Diabetes Prevention Program and Look AHEAD, they'll cost tens or hundreds of millions of dollars. And even if such trials are funded, it might be another decade or two before we have reliable answers. But it's hard to imagine that this controversy will go away if we don't do them, that we won't be arguing about the detrimental role of fats and carbohydrates in the diet twenty years from now. The public will certainly not be served by attempts of interest groups and industry to make this controversy go away. If the tide of obesity and diabetes continues to rise around the world, it's hard to imagine that the cost of such trials, even a dozen or a hundred of them, won't ultimately be trivial compared with the societal cost.

Notes

PROLOGUE: A BRIEF HISTORY OF BANTING

ix *Epigraph.* "Farinaceous . . .": Tanner 1869b:219.

ix ". . . corpulence notoriety": Anon. 1864b. ". . . size or weight": Banting 1864:14.

x "Knowing too that . . .": Harvey 1872:69–70.

x Banting began dieting: Banting 1864:18–19. "I have not felt better . . .": Banting 1869.

x United States, Germany: Banting 1869. "the emperor of the French . . .": Anon. 1864c. "If he is gouty . . .": Quoted in "banting" entry, OED 1989.

x A paper was presented: Anon. 1864f. See also Anon. 1864d; Anon. 1864a. "is tolerably complete . . .": Anon. 1864g. Banting responded: Banting 1869.

xi Banting acknowledged: Banting 1869. Alfred William Moore: Anon. 1864g. John Harvey: Harvey 1864.

xi Brillat-Savarin: Brillat-Savarin 1986 ("fat . . . ," 237–39; ". . . rigid abstinence . . . ," 251).

xi Dancel: Dancel 1864 ("All food . . . ," 59; "The hippopotamus . . . ," 54).

xii "We advise Mr. Banting . . .": Anon. 1864g.

xii "fair trial" and ". . . starchy elements . . .": Anon. 1864e.

xii "To attribute obesity . . .": Mayer 1968:6.

xii Sir William Osler: Osler 1901:439–40. Oertel prescribed a diet: Oertel 1895. See also French 1907:951. Bismarck lost sixty pounds: Schwartz 1986:103–4. Ebstein insisted: Ebstein 1884 ("of meat *every* kind . . . ," 33).

xiii "*Foods to be avoided* . . .": Greene 1951:348.

xiii "The great progress . . .": Bruch 1957:352.

xiv "The overappropriation . . .": French 1907:14. Rony reported: Rony 1940 (". . . marked preference . . . ," 59; "an extremely obese laundress . . . ," 62).

xiv "In Great Britain obesity . . .": Davidson and Passmore 1963:382.

xiv "On the day of the races . . .": Tolstoy 2000:200. "the dearth of proteins . . .": Lampedusa 1988:255.

xv What Dr. Spock taught: Spock 1946:361; Spock 1957:436; Spock 1968:449; Spock 1976:493; Spock 1985:536; Spock and Rothenberg 1992:380. 50 million copies: Pace 1998. "All popular 'slimming regimes' . . .": Davidson and Passmore 1963:389. "The first thing . . .": Brody 1985:18.

xv Brody recommending potatoes, etc.: Brody 1985:18–20. "We need to eat . . .": Brody 1981a:97. ". . . at the height of fashion . . .": Brody 1985:78. "the previous nutritional advice . . .": James 1983:20. *Footnote.* See Barr et al. 1953b; Eppright et al. 1955; Blix 1964; Wilson 1969; McLean, Baird, and Howard 1969; Apfelbaum 1973.

xvi "bizarre concepts . . .": Anon. 1973:1419.

xvi Charlotte Young: C. M. Young 1976 ("The diets developed by Ohlson . . . ," 364; "No adequate explanation . . . ," 365).

xvii "people who cut down . . .": Squires 1985.

xvii "sparingly": USDA 1992.

xvii "There is always an easy solution . . .": Mencken 1982:443.

xvii Less red meat, fewer eggs: Putnam et al. 2002. Fat intake has dropped: USDA Center for Nutrition Policy and Promotion 1998. Fall in cholesterol levels: Gregg et al. 2005.

xviii Ten-year study of heart-disease mortality: Rosamond et al. 1998. See also Rosamond et al. 2001; McGovern et al. 2001. AHA statistics: Thom et al. 2006.

xviii Percentage of smokers has dropped: National Center for Health Statistics 2004.

xviii Incidence of obesity increasing: National Center for Health Statistics 2005:9, 275 (table 73). Diabetes rates: Fox et al. 2006; Cowie et al. 2006.

xix "What we see instead . . .": Interview, William Harlan.

xix Best-selling diet books: Mackarness 1958; Taller 1961; Stillman and Baker 1968; Atkins 1972; Tarnower and Baker 1978; Sears and Lawren 1995; Eades and Eades 1996; Steward et al. 1998; Agatston 2003.

xxi Fixated on cholesterol: This idea came from David Kritchevsky, who, among other accomplishments, authored the first textbook on cholesterol, published in 1958.

xxi Series of expert reports: USDA and USDHEW 1980; USDHHS 1988; NRC 1989; U.K. Department of Health 1994.

xxi "Each science . . .": Whitehead 1980:14–15.

xxv "If science is to progress . . .": Feynman 1967:148.

PART ONE: THE FAT-CHOLESTEROL HYPOTHESIS

3 *Epigraph.* "Men who have excessive faith . . .": Bernard 1957:38.

CHAPTER ONE:
THE EISENHOWER PARADOX

3 *Epigraph.* "In medicine . . .": Bernard 1957:55.

3 The details of Eisenhower's heart attack: Lasby 1997:70–80.

3 White's press conference and Ike's recovery: Ibid.:83–93.

3 Eisenhower's weight, cholesterol, and blood pressure: Ibid.:257–58; interview, George Mann.

4 Ten times a year: Lasby 1997:70. Eisenhower's diet and Snyder's responses: Ibid.:258–59.

4 "He eats nothing . . .": Ibid.

4 "He was fussing . . .": Ibid.:260.

4 Keys made the cover of *Time:* Anon. 1961 (". . . know the facts," 52). First official endorsement: AHA 1961.

5 Eisenhower's half-dozen heart attacks: Lasby 1997:293–323.

5 "great epidemic": White 1971:220.

5 "drastic development . . .": Mayer 1975a:138. Decline in deaths due to eating less fat: See, for instance, Sykowski et al. 1990; Hunink et al. 1997; NCEP 2002:II–26.

6 Osler wrote in 1910: Cited in Cassidy 1946. "If it had been common . . .": White 1971:52. "part and parcel . . ." and ". . . cripples and kills . . .": White 1945:475.

6 Herrick, the ECG, and the early history of cardiology: Liebowitz 1970:146–76. "Medical diagnosis . . ." and ". . . after the publication . . .": Levy 1932.

7 Census numbers: Cooper 1972; Preston et al. 1972. *Fortune* article: Anon. 1950. Cassidy's point: Cassidy 1946.

7 Mitigating against the "epidemic": Levy 1932. See also Tunstall Pedoe 1984.

7 AHA 1957 report: Page et al. 1957 ("great difference . . . ," 165).

8 Between 1949 and 1968: Harper 1996. See also Harper 1983. Proportion of heart-disease deaths dropping: Harper 1996; interviews, Harry Rosenberg, chief of mortality statistics, National Center for Health Statistics, and Thomas Thom, a statistician at the National Heart, Lung, and Blood Institute. WHO committee report: Lozano et al. 2001 (". . . the apparent increase . . . ," 14). About the situation in the United States, see also Woolsey and Moriyama 1948.

8 NHI 1949 allocations: Haseltine 1949. NHI 1960 research budget is from NIH, n.d., NIH Almanac.

9 "a private organization . . .": White 1971:114. 1945 charitable contributions: Anon. 1945. Rome Betts: Moore 1983:57.

9 AHA fund-raising campaign and its success: Anon. 1948a; Anon. 1948b; Davies 1950; Moore 1983:77. "great epidemic . . . ": White 1971:220.

9 Compelling arguments: Mann 1957; Page et al. 1957; Harper 1983. "unobserved publications": Kritchevsky 1992. "They don't fit . . .": Interview, David Kritchevsky.

9 "The present high level . . .": Keys 1953.

10 "The simple fact . . .": Select Committee 1977a:1. CSPI pamphlet: Brewster and Jacobson 1978. "Within this century . . .": Brody 1985:2.

10 Keys's argument: Keys 1953.

10 History of food disappearance statistics: USDA 1953; Call and Sanchez 1967.

11 "Until World War II . . .": Interview, David Call.

11 Historians of dietary habits: See, for instance, Schwartz 1986:46; Cummings 1940:10–24. One French account: Levenstein 1999. USDA 1830s estimate: Appen 1933, cited in Cummings 1940:15. "with plenty of beef-steaks . . .": Trollope 1932.

11 "considered by the general public . . .": Ward 1911.

11 FTC report: FTC 1919 (". . . the amount of meat consumed . . . ," 84).

12 "nationwide propaganda . . .": Stiebeling 1939.

12 *The Jungle:* Sinclair 2003 ("overlooked for days . . . ," 91). Meat sales dropped by half: Young 1981. "The effect was long-lasting . . .": Root and de Rochemont 1995:211.

12 Trends for vegetables, fruits, etc.: USDA 2000.

12 "The preponderance of meat . . .": Clendening 1936:7.

13 Food consumption from the end of World War II: Friend et al. 1979.

13 "medical villain *cholesterol*": Blakeslee and Stamler 1966:28.

13 "biological rust . . .": Ibid. 1966:24.

14 Anitschkow reported: Anitschkow and Chalatow 1913. The problem with rabbits as animal models: See, for instance, Ahrens, Hirsch, et al. 1957; Altshule 1966. ". . . 'cholesterol disease . . .": Leary 1935.

14 Stamler's chicken experiments: Blakeslee and Stamler 1966:36. Naturally occurring atherosclerosis: Altshule 1966; Lindsay and Chaikoff 1963. In baboons: McGill et al. 1960.

14 Exercise lowers it: Goldberg et al. 1984; Heath et al. 1983; Huttunen et al. 1979. Weight gain raises it: Anderson, Lawler, and Keys, 1957. Weight loss lowers it: Milch et al. 1957; Jolliffe et al. 1962. Fluctuate seasonally: Bleiler et al. 1963; Antonis et al. 1965. Change with body position: Tan et al. 1973. Stress will raise cholesterol: Frideman et al. 1958. Male and female hormones: Laskarzewski et al. 1983.

Diuretics: Ames and Hill 1976. Sedatives: Wallace et al. 1980. Alcohol: Fraser et al. 1983. 20 to 30 percent: Kritchevsky 1958:181.

14 "Some works . . .": Gofman and Lindgren 1950.

15 One out of every three women: Stone et al. 1974. Rarely die of heart attacks: Discussed in Ahrens, Hirsch, et al. 1957.

15 Sperry and Landé's research: Landé and Sperry 1936.

15 Heart surgeons and cardiologists: See, for instance, James 1980; interview, Alan Sniderman. Debakey reported: Garrett et al. 1964.

16 ". . . why people get sick . . .": Anon. 1961. The K ration: Sullivan 2004. *Biology of Human Starvation:* Keys, Brozek, et al. 1950.

16 "frank to the point of bluntness . . .": Blackburn n.d. "pretty ruthless" and "Mr. Congeniality": Interview, David Kritchevsky.

16 Keys launched his crusade: Keys, Mickelsen, et al. 1950; Keys et al. 1956. Rittenberg and Schoenheimer: Rittenberg and Schoenheimer 1937. Researchers agreed: See, for instance, Quintao et al. 1971.

17 "a few questions . . .": Hoffman 1979. ". . . members of the Rotary Club . . ." and "a similar picture": Keys 1994.

17 "fatty diet . . .": Keys 1994. "Direct evidence . . .": Keys 1952.

17 Keys's chain of observations: Keys 1994.

18 1950 report from Sweden: Malmros 1950. Similar phenomena: See Keys 1975. Keys concluded: Keys 1994. Skeptics observed: See, for instance, Mann 1957.

18 Keys argued the same proposition: Keys 1953. "remarkable relationship . . .": Keys and Anderson 1955:189.

18 Researchers wouldn't buy it: Yerushalmy and Hilleboe 1957. "magic method . . .": Gould 1996:272.

18 ". . . not very profitable game": Mann 1957.

19 "This causality . . .": Ibid.

19 "uncritically . . ." and ". . . worse than useless": Yerushalmy and Hilleboe 1957.

19 Clinically meaningless: See, for instance, Howell et al. 1997.

19 Keys insisted that all fat: A good example is Keys et al. 1955.

19 Vegetable oil vs. animal fats: Kinsell et al. 1952; Groen et al. 1952.

20 Keys eventually accepted: Anderson, Keys, and Grande 1957.

20 This saturation factor: Kinsell et al. 1958; Ahrens, Insull, et al. 1957. "handicap to clear thinking": Ahrens 1957. Fat content of beef, lard, and chicken fat: USDA n.d.

20 AHA opposed Keys: Page et al. 1957.

20 A new AHA report: AHA 1961. "acceptable compromise" and "some undue pussyfooting": Anon. 1960.

21 *Time* cover story: Anon. 1961.

CHAPTER TWO:
THE INADEQUACY OF LESSER EVIDENCE

22 *Epigraph.* "Another reason . . .": Friedman 1969:77.

22 "unmanageable proportions . . . ": Kaunitz 1977. "totality of data": Stamler et al. 1972:45. "two strikingly polar attitudes . . . ": Blackburn 1975.

22 "It must still be admitted . . .": Dawber 1978. "overwhelming evidence . . .": Dawber 1980:141.

23 "highest level": Sackett 2002.

23 "final scientific proof": Anthony Gotto in Select Committee 1977d:312.

23 *Wall Street Journal* reported: Bishop 1961.

24 "dotting the final i": Anon. 1964b.
24 "The absence of final . . .": Quoted in Blakeslee and Keys 1966:10.
24 Popper's observations: Popper 1979 ("The method of science . . . ," 81; infinite wrong conjectures, 15).
25 "each new research . . .": Keys 1957. Cholesterol and heart disease among Japanese men: Marmot et al. 1975.
25 Navajo Indians: Page et al. 1956. Irish immigrants: Trulson et al. 1964. African nomads: Mann et al. 1964. Swiss Alpine farmers: Gsell and Mayer 1962. Benedictine and Trappist Monks: Groen et al. 1962. Explained away by Keys: Keys 1963; Keys 1975.
25 Mann examined the Masai: Mann et al. 1964. The Samburu had low cholesterol: Shaper 1962. "fully as high . . ." and "It has been estimated . . .": Keys 1963.
26 "feed-back mechanism . . .": Keys 1975.
26 Mann's further research: Mann et al. 1972. Masai moved into Nairobi: Day et al. 1976. "The peculiarities of those primitive nomads . . .": Keys 1975.
26 "The data scarcely warrant . . .": Keys 1975. Roseto study: Stout et al. 1964. "few conclusions . . .": Keys 1966.
26 Framingham risk factors, "reasonably typical": Dawber 1962.
27 Cholesterol and women, "no predictive value": Kannel et al. 1971.
27 Framingham dietary research: Kannel and Gordon 1968 ("promised to be . . . " 2; "cautionary note . . . ," 15); interviews, George Mann and Tavia Gordon.
28 Puerto Rico study: Garcia-Palmieri et al. 1980. Honolulu: Yano et al. 1978; McGee et al. 1984. Chicago: Paul et al. 1963. Tecumseh: Nichols et al. 1976. Evans County: Stulb et al. 1965. Israel: Kahn et al. 1969.
28 "The human understanding . . .": Bacon 1994:58.
29 Western Electric study: Paul et al. 1963.
29 Stamler's return to Western Electric: Shekelle et al. 1981.
30 "The new report . . .": Cohn 1981. "The message of these findings . . .": Brody 1981d. "The Cholesterol Facts": LaRosa et al. 1990. Footnote. Trulson et al. 1964; Kushi et al. 1985.
30 Risks of changing the fats consumed: Ahrens 1979a.
31 Details of Seven Countries Study: Keys 1980. $200,000 support: Anon. 1961.
31 Results were first published: Keys 1970. Mortality rates: Keys 1980:65.
31 Three lessons: Keys 1980:332–35.
32 Finns vs. Cretans: Keys 1970:1–168. Keys's diet books: Keys and Keys 1959; Keys and Keys 1975.
32 1984 report: Keys et al. 1984.
33 "seems to furnish . . .": Pearl 1940:15.
34 "A common feature . . .": Bailar 1980.
35 Hungarian trial: Korányi 1963. British trial: Research Committee 1965.
36 Anti-Coronary Club trial: Christakis, Rinzler, Archer, and Kraus 1966.
36 "Diet Linked . . .": Plumb 1962. "Special Diet . . .": Schmeck 1964. "urged the government . . .": Anon. 1964a.
36 February report: Christakis, Rinzler, Archer, et al. 1966. November report: Ibid.
37 Dayton's Hospital trial: Dayton et al. 1969.
37 Helsinki Study: Miettinen et al. 1972.
38 Proponents of Keys's hypothesis: See, for instance, Steinberg 2005.
38 Minnesota Coronary Survey: Frantz et al. 1989; interview, Ivan Frantz, Jr. Footnote. Frantz et al. 1975.
39 For background on the HRT episode, see Kolata and Petersen 2002. WHI article

on CVD: Manson et al. 2003. On breast cancer: Chlebowski et al. 2003. On stroke: Wassertheil-Smoller et al. 2003. On quality of life: Hays et al. 2003. For more information, see the WHI website at NHLBI (http://www.nhlbi.nih.gov/whi/). *Footnote.* Hulley et al. 1998.

39   "disastrous inadequacy . . .": Sackett 2002.

40   1962 NHI grants: Baker et al. 1963. Ahrens's NIH committee: Review Panel of the NHI 1969 ("The essential reason . . . ," 2). "would be so expensive . . .": Interview, Pete Ahrens.

40   Task Force on Arteriosclerosis: USDHEW 1971 ("formidable" costs, I-21).

40   Two smaller trials: Ibid.: I-22.

CHAPTER THREE:
CREATION OF CONSENSUS

42   *Epigraph.* "In sciences . . .": Bacon 1994:51.

42   Roots of this movement: Levenstein 1993:131–43, 178–94. See also Belasco 1989. "Villagers in Dahomey . . .": Anon. 1962. Famines in the 1960s and early 1970s: See, for instance, Devereux n.d. "hundreds of millions . . .": Ehrlich 1968:11.

42   "enormous appetite . . .": Mayer 1974a:395.

43   "world's most essential commodity": Grant 1974.

43   Argument made by Lappé: Lappé 1971:7–9. "A shopper's decision . . .": Belasco 1989:57.

43   "How do you get people . . .": Cross 1974. AHA recommending meat restriction: See, for instance, Rensberger 1974; Blakeslee 1973.

43   "the battle to feed . . .": Ehrlich 1968:11. Borlaug had created: Easterbrook 1997. "more than any other . . .": Quoted in Hesser 2006:132.

43   AHA revised its recommendations: See, for instance, Brody 1973. "including infants . . .": Inter-Society Commission for Heart Disease Resources 1970.

44   Manufacturers' programs to educate doctors: Levine 1986:40 ("Listen to your heart," 61). Revised version of Stamler's book: Blakeslee and Stamler 1966.

44   Polyunsaturated fats can cause cancer: Pearce and Dayton 1971; Brody 1973.

44   *Dietary Goals:* Select Committee 1977a ("the first comprehensive . . . ," 1).

45   "Premature or not . . .": Brody 1981a:11.

45   McGovern's Committee and its history: Levenstein 1993; interviews, Mark Hegsted, Chris Hitt, Marshall Matz, George McGovern, Nick Motern, Kenneth Schlossberg, and Alan Stone.

45   ". . . totally naïve . . .": Interview, Marshall Matz. McGovern and the Pritikin Center: Interview, George McGovern. *Footnote.* Broad 1979b.

46   Mottern's experience: Interview, Nick Mottern.

46   Mottern, Hegsted, and their perspectives: Interviews, Nick Mottern and Mark Hegsted.

46   Goals number one and two: Select Committee 1977a:31, 37–42, 75.

47   Notably the AHA: Interview, Marshall Matz.

47   "The question to be asked . . .": Select Committee 1977a:3. "all hell . . .": Interview, Mark Hegsted.

47   Levy's testimony: Select Committee 1977b:8–33. "Arguments for lowering . . .": Levy and Ernst 1976.

47   Other prominent investigators: Select Committee 1977b. AMA letter: Select Committee 1977c:670–77.

47 Revised edition of *Dietary Goals:* Select Committee 1977b (avoid overweight, xxxiii; "decrease consumption of animal fat . . . ," xxxix). Pressure from the livestock industry: Interviews, Nick Mottern and George McGovern.

48 Attempts to justify recommendations: Select Committee 1977b ("some witnesses . . . ," "After further review . . . ," xxxiii; "important questions . . . ," and "Does lowering . . . ," xxxvii).

48 "strong, forceful, competent": Burros 1977. "people were getting . . ." and "Tell us . . .": Interview, Carol Foreman.

49 NAS/USDA contract and Leveille's speech: Broad 1979a. Handler and Fredrickson responses: Interview, Carol Foreman.

49 Forman hired Hegsted: Interviews, Mark Hegsted and Carol Foreman.

49 Report by a committee: Task Force Sponsored by the American Society for Clinical Nutrition 1979. "*not* to draw . . . ," "full range . . . ," and "considerable": Ahrens 1979b. ". . . clear majority . . ." and production of *Dietary Guidelines:* Interview, Mark Hegsted. (McGinnis did not respond to repeated requests for interviews.) *Footnote.* Glueck 1979:2642.

50 "Avoid Too Much . . .": USDA and USDHEW 1980.

50 *Toward Healthful Diets:* Food and Nutrition Board NRC 1980. "excoriated in the press": Interview, David Kritchevsky. The first criticisms: Altman 1980; Brody 1980. ". . . in the pocket . . .": Interview, Jane Brody. See also Wade 1980. Details of industry connections: Broad 1980; Handler 1980. Leaked to the press: Risser 1980; interviews, Carol Foreman and James Risser. "embraced a low-fat diet . . .": Baum 1995.

50 House subcommittee hearings: Wade 1980.

51 Handler testified: Handler 1980.

51 Nutritionists in academia: Levenstein 1993 (work closely, 134–35; "unholy alliance," 188).

51 Olson explained: Anon. 1980.

52 "To be a dissenter . . .": Mann 1977. ". . . as big a pusher . . .": Interview, David Kritchevsky.

52 Jacobson's exposé: Rosenthal et al. 1976. "The important question . . .": Stare 1987. Stamler's funding from manufacturers: Blakeslee and Stamler 1966:x.

53 Funded by Frito-Lay: Interview, Mark Hegsted.

53 Honolulu study: Yano et al. 1978. In Framingham and Puerto Rico: Gordon et al. 1981.

53 "reconciling [their] study findings . . .": Gordon et al. 1981.

54 Stamler's Chicago studies: Dyer et al. 1981. Dayton and others: Pearce and Dayton 1971. Swiss Red Cross: Nydegger and Butler 1970. Six ongoing studies: Rose et al. 1974. British, Hungarian, and Czech: Anon. 1978. Study after study: Beaglehole et al. 1980; Kark et al. 1980; Garcia-Palmieri et al. 1981; Miller et al. 1981; Stemmermann et al. 1981; Kozarevic et al. 1981. Framingham Study: Williams et al. 1981. "surprise and chagrin": Kolata 1981. *Footnote.* McGee et al. 1985.

55 Norwegian study: Westlund and Nicolayson 1972. First NHLBI workshop: Feinleib 1981. Second workshop: Feinleib 1982. Levy's comments: Kolata 1981.

55 Third workshop: Feinleib 1983. "the perplexing inconsistencies" and "not preclude . . .": Feinleib 1983.

55 "throw the kitchen sink": Interview, Stephen Hully. Details of MRFIT: MRFIT Research Group 1982.

56 $115 million: Kolata 1982.

56 "... Test Collapses": Bishop 1982. Slightly more deaths: MRFIT Research Group 1982. *Footnote*. Shaten et al. 1997.

56 Details of LRC Trial: LRC Program 1979.

57 Results of the LRC trial: LRC Program 1984a; LRC Program 1984b. "conclusive ...": Quoted in Moore 1989:68 (see endnote 278).

57 "could and should ...": Consensus Conference 1985. "It is now indisputable ...": Anon. 1984a.

57 "unwarranted, unscientific ...": Quoted in Wallis 1984. "unconscionable ...": Quoted in Kolata 1985.

57 Rifkind later explained: Interview, Basil Rifkind.

58 "a massive health campaign": Levy's testimony in Select Committee 1977d:19. "Sorry, It's True ...": Anon 1984a. *Time* follow-up story, including Gotto's quote: Wallis 1984.

58 December consensus conference: Anon. 1984b; Ahrens 1985; Oliver 1985; interviews, Pete Ahrens, David Kritchevsky, Robert Olson, Basil Rifkind, and Daniel Steinberg. "Many people ...": Quoted in Kolata 1985.

59 "were selected to include ...": Oliver 1985. "no doubt ...": Consensus Conference 1985. "you wouldn't have ...": Interview, Daniel Steinberg.

CHAPTER FOUR:
THE GREATER GOOD

60 *Epigraph.* "In reality ...": Arthus 1943:15.

60 NCEP 1987 guidelines: Anon. 1988. "The edict ...": Thompson and Squires 1987. *Nutrition and Health:* USDHHS 1988 (two-thirds of 2.1 million deaths, 4). "exhorts Americans ...": Toufexis 1988. "disproportionate consumption ...": Koop 1988: iii. "The depth ...": Koop 1988:iii–iv. *Diet and Health:* NRC 1989 ("Highest priority," 13).

61 Writers of the surgeon general's report: Interviews, Marion Nestle, managing editor of the report, and Nancy Ernst of NHLBI. Writers of the *Diet and Health* chapters: Interviews, Susma Palmer, program director at the NRC, and Henry Blackburn. *Footnote*. Interview, Henry McGill.

61 Jacobson scolding the authors: Sugarman 1989 (includes Motulsky quote); Burros 1989.

62 Hungarian study: Korányi 1963. British study: Research Committee 1965.

62 "... some indication ...": Ernst and Levy 1984. A new generation: See, for instance, Rovner 1988.

62 Cholesterol and sudden cardiac death: Kannel and Thomas 1982; Dawber 1980 ("The lack of association ...," 131).

63 Stamler's MRFIT reanalysis: Stamler et al. 1986. *Chart.* Martin et al. 1986.

64 Whether we would live longer: Ibid. *Chart.* Ibid.

65 The Harvard study: Taylor et al. 1987.

65 UCSF study: Browner et al. 1991. McGill study: Grover et al. 1994.

66 "They would have liked ...": Interview, Marion Nestle. "I am sensitive ...": Letter from Browner to McGinnis, Feb 14, 1991. I am grateful to Warren Browner for sharing this correspondence with me.

66 "... small or negligible ...": NRC 1989:6.

66 "The mass approach ...": Rose 1981.

67 "People will not ...": Ibid. "The modern British diet ...": Quoted in Le Fanu 1999:307. *Footnote*. Interview, William Taylor.

67 Assumption underpinning mass prevention: Rose 1985 ("would lead us . . . ," 32; "differences between . . . ," 34).

68 Unintended side effects, "unnatural factors," and " 'biological normality' . . .": Rose 1981:1851.

68 ". . . no time for significant . . .": Scrimshaw and Dietz 1995.

69 "nuts, fruits . . ." and "substantial amounts . . .": Blakeslee and Stamler 1966: 41–42.

69 Analysis of hunter-gatherer diets: Eaton and Konner 1985. Low-fat recommendations: NRC 1989:41.

69 "made a mistake": Interview, Boyd Eaton. Revised analysis: Cordain et al. 2000 ("would have contributed . . . ," 690). Paleolithic diets high in protein: Interviews, Loren Cordain, Melvin Konner, John Speth, Craig Stanford. See also Abrams 1987; Harris 1985; Stanford 2001; Stefansson 1946.

70 Histories of the saccharine controversy: Cummings 1986; Merrill 1981. *Footnote.* Interview, Melvin Konner.

70 Motulsky told the *Post:* Quoted in Sugarman 1989.

70 "If the public's diet . . .": Ahrens 1979a.

71 Cited in *Dietary Goals:* Select Committee 1977a:33–34. NAS report: Committee on Diet, Nutrition, and Cancer 1982 ("could be used . . . ," 15). ACS low-fat diet: American Cancer Society 1984.

71 When Japanese women immigrate: See testimonies of Ernst Wynder and Gio Gori in Select Committee 1976:164–208. Adding fat to rat diets: Tannenbaum 1942.

71 Higginson noted: Maugh 1979. "difficult to reconcile": Williams et al. 1981.

72 Critical test from the Nurses Health Study: Willett et al. 1987. "a good study . . .": AP 1987. NCI researchers published: Jones et al. 1987. "perhaps because no one . . .": Marshall 1993b.

72 Results from eight years of Nurses Health Study: Willett et al. 1992. Fourteen years: Holmes et al. 1999.

72 Greenwald had responded: Schatzkin et al. 1989.

73 "indisputable" and "a high-fat, high-calorie . . .": Ibid. "supplemented with" polyunsaturates: Rogers and Longnecker 1988.

73 Kritchevsky published an article: Kritchevsky et al. 1984. Kritchevsky later reported: Klurfeld et al. 1989. Pariza's similar results: Boissoneault et al. 1986. "If you restrict . . .": Interview, Mike Pariza. "overwhelmingly striking . . .": Interview, Demetrius Albanes.

74 Neither "convincing" nor even "probable": World Cancer Research Fund and American Institute for Cancer Research 1997: 252,261–9. "largely null": Interview, Arthur Schatzkin.

74 ACS guidelines: Byers et al. 2002 ("limit consumption . . ."); Kushi et al. 2006 ("there is little . . . ," "major contributors . . . ," "diets high in fat . . . ," and "may have an effect . . .").

74 Details of the WHI: Ritenbaugh et al. 2003.

75 WHI results on breast cancer: Prentice et al. 2006. On heart disease and stroke: Howard, Van Horn, et al. 2006. On colon cancer: Beresford et al. 2006. Elizabeth Nabel stated: NHLBI Communication Office. The accompanying *JAMA* editorials: Buzdar 2006.

76 WHO press release: WHO 2006. Basis of the early controversy: Marshall 1993a. *Footnote.* Howard, Manson, et al. 2006.

76 Bacon would have called: Bacon 1994 ("wishful science," 59; "stuck fast . . . " and ". . . downhill ever since," 84).

77 "... no dishonesty involved ..." and "pathological science": Langmuir 1989. "If you throw money ...": Interview, Wolfgang Panofsky.

77 "Most drugs ...": Interview, Richard Kronmal.

78 Keys, the changing American diet, and the epidemic: Keys 1953. "no basis": Keys 1971.

78 Keys in the 1950s on Japanese men: Keys 1957. The Seven Countries Study: Keys 1980:86; Keys et al. 1994. Japan in the 1990s: Koga et al. 1994 ("... progressive increases ...," "remarkable reduction," and "It is suggested ..."). Average American cholesterol values: National Center for Health Statistics 2006.

78 Keys dismissing misdiagnosis: Keys 1957. "might have been misled ...": Keys et al. 1984.

79 "I've come to think ...": Boffey 1987.

79 French-Italian-Spanish and Australian paradoxes: Powles 2001. Footnote. Guberan 1979.

79 MONICA details and results: Kuulasmaa et al. 2000. "far and away ..." and "whatever the results ...": Interview, Hugh Tunstall-Pedoe.

79 "... classical risk factors ...": Interview, Hugh Tunstall-Pedoe.

80 Jacobs visited Japan: Interview, David Jacobs. Cholesterol, stroke, and Japan: Blackburn and Jacobs 1989.

80 Framingham investigators provided: Anderson et al. 1987.

81 Most striking result: Ibid.

81 The NHLBI workshop: Jacobs et al. 1992. Footnote. Hulley et al. 1992.

81 Rifkind's interpretation: Interview, Basil Rifkind. Cf. Jacobs et al. 1992.

82 "Questions should be pursued ...": Jacobs et al. 1992.

82 Feynman's lectures: Feynman 1967 ("... if your bias ..." and "... absolutely sure ...," 147).

83 Meta-analysis: Mann 1990 provides a good review.

83 Cochrane Collaboration: Taubes 1996; the Cochrane Collaboration Web site (www.cochrane.org).

83 "reduced or modified ...": Hooper et al. 2001.

83 "A major lesson ...": Keys 1975.

84 "The pooled effects suggest ...": Ebrahim et al. 2006.

85 Evidence indeed suggested: Malmros 1950; Schornagel 1953; Vartiainen and Kanerva 1947.

PART TWO: THE CARBOHYDRATE HYPOTHESIS

87 Epigraph. "The world ...": Furnas and Furnas 1937: 62–63.

CHAPTER FIVE:
DISEASES OF CIVILIZATION

89 Epigraph. "The potato ...": Sai 1967.

89 Schweitzer in Lambaréné: Schweitzer 1998:136–39.

89 Appendicitis, "On my arrival in Gabon ...": Schweitzer 1957.

90 Hutton's experience: Hutton n.d. ("The Eskimo ...," 9; "tea, bread ...," 36; "The most striking ...," 35; "living on a 'settler' dietary," 37; "... puny and feeble," 21–22).

90 WHO on "nutrition transition": WHO 2003.

91 Keys on isolated populations: Keys in Blix 1964:54–55. Few likely to live long enough: Keys 1975.

91 "nasty, brutish, and short": Hobbes 1997:100.
91 *Diseases of civilization:* Trowell and Burkitt 1981b.
91 Tanchou's observations: Quoted in Barker 1924:50–51.
92 "natives mingled . . .": Hollander 1923. "dietetic and other . . .": Blair 1923. Fouché reported: Fouché 1923.
92 Hrdlička described: Hrdlička 1908:187–91.
93 Native Americans lived longer: Ibid.:39–41.
93 Levin's survey: Levin 1910.
93 The question of cancer: Hoffman 1915; Williams 1908:12–49, 50–78 (Fiji and Borneo, 42; New York and Philadelphia, 76).
94 *Cancer and Diet:* Hoffman 1937. "at a more or less alarming . . .": Hoffmann 1915:30–33.
94 "Among some 63,000 . . .": Hoffman 1915:151.
94 ". . . no known reasons . . .": Ibid.:147.
94 "It is commonly stated . . .": Brown et al. 1952. See also Fog-Pulson 1949. Canadian physicians: Schaefer et al. 1975. The most comprehensive discussion of cancer in the Inuit is Stefansson 1960a.
95 "In a series of one hundred . . .": Orenstein 1923. "It ran an uninterrupted course . . .": Prentice 1923.
95 "to whom the fleshpot . . .": Anon. 1899. Cancer absent in carnivorous populations: Williams 1908:44–45. "hardly holds good . . .": Levin 1910.
95 "demanding conservation . . .": Hoffman 1937:118.
96 "far-reaching changes": Ibid.
96 Flour, sugar, and appendicitis: Rendle Short 1920. For an intelligent, early discussion of diseases of civilization, see Rabagliati 1897.
96 White flour had its proponents: For a good review of the refining of cereal grains, see Davidson and Passmore 1963:262–82 ("more attractive to the eye," 265; "less liable than . . . ," 267).
97 Sugar consumption skyrocketed: Friend et al. 1979. The English were already eating: Aykroyd 1967:105. Asian nations: Davidson and Passmore 1963:275.
97 Darwin tells: Darwin 1989:291. "acquired a fondness . . .": Murdoch 1892. Primary items of trade: Mountford 1960:14–16. *Footnotes.* In *Across Australia:* Spencer and Gillen 1912:230. "consisted of white flour . . .": Ibid.
97 "The true staff . . .": Quoted in Le Fanu 1987:52. "One great curse . . .": Quoted in Kellock 1985:128. Lane's hypothesis: Lane 1929.
98 Era of nutritional research: For a review of the early vitamin research, see McCollum 1957:201–318. Cancer as a deficiency disease: Barker 1924.
98 ". . . use of vitamin-poor white flour . . .": McCarrison 1961:64. McCarrison's 1921 lecture: McCarrison 1922.
98 McCarrison's research and observations: McCarrison 1961:23–26.
99 Enrichment of white flour in the United States: Levenstein 1993:22. In England: Davidson and Passmore 1963:269–70. "protective foods": McCarrison 1922.

CHAPTER SIX:
DIABETES AND THE CARBOHYDRATE HYPOTHESIS

100 *Epigraphs.* "The consumption of sugar . . .": Allen 1913:146–47. "Sugar and candies . . .": Duncan 1935:59.
100 Hindu physicians: Trowell 1975a.
100 "This ancient belief . . .": Allen 1913:147.

101   The leading authorities: Ibid.:148–49.
101   "in the absence . . .": Ibid.:150.
102   "If he is a poor laborer . . .": Ibid.:152.
102   Diabetes a disease of civilization, "the rich ones . . .": Ibid.:148. *Footnote.* Donnison 1938:23–24.
102   British Medical Association symposium: Charles 1907.
103   Physicians increasingly diagnosed diabetes: Joslin et al. 1935.
103   At Johns Hopkins and Massachusetts General Hospital: Gale 2002. Death rate from diabetes: Emerson and Larimore 1924 ("It is apparent . . .").
104   "synonymous": Interview, Ronald Arky, former president of the ADA. Allen's declining reputation: Bliss 1982:239.
104   Joslin on apple consumption: Emerson and Larimore 1924. Emerson countered: Ibid.
104   "A high percentage . . ." and "must stand in some relation": Joslin 1923:145. A third factor: Joslin 1927.
105   "painstakingly accumulated": White and Joslin 1959:70.
105   Joslin and Himsworth piggybacked: See White and Joslin 1959:70–71; Himsworth 1935; Joslin et al. 1934; Mills 1930. Joslin on insulin and leveling-off of mortality rates: Joslin et al. 1933. *Footnote.* Himsworth 1936.
105   Himsworth on insulin-dependent and non-insulin-dependent diabetes: Himsworth 1936.
105   "a smaller proportion of carbohydrate . . .": Himsworth 1935:142.
106   "striking," "The progressive rise . . . ," and "The diabetic mortality rate . . .": Himsworth 1949a.
106   Himsworth on ". . . coloured races . . .": Himsworth 1935:134–35.
106   Himsworth on Inuit: Ibid. 122–24. Diabetes among Alaskan Eskimos in 1956: Scott and Griffith 1957. Baffin Island study: Heinbecker 1928.
107   "fisherfolk" study: Mitchell 1930. "It would thus appear . . .": Himsworth 1935.
107   In the 1946 and 1959 editions: Joslin et al. 1946:75–76; Joslin et al. 1959:70–71. *Joslin's Diabetes Mellitus:* Marble et al. 1971.
107   "Though the consumption of fat . . .": Himsworth 1949b.
108   Cohen reported: Cohen 1963. "a significantly greater prevalence": Cohen et al. 1961.
108   "The quantity of sugar . . .": Ibid.
108   Prior studied Maoris: Prior et al. 1964.
109   Campbell's research: Campbell's testimony in Select Committee 1973a:208–18.
109   "remarkable difference . . ." and "country cousins": Ibid.
109   Campbell's surveys of Natal population: Campbell 1963; Cleave and Campbell 1966 ("veritable explosion . . . ," 25; numbers in India, 19–24; diabetes among Zulu, 34–35).
110   "a figure in many countries . . ." and "were enormously fat . . .": Campbell in Select Committee 1973a:213.
110   Zulus eating excessive amounts of sugar: Campbell 1963.
110   Campbell's research on sugarcane cutters: Truswell et al. 1971. "diabetes is virtually absent," "huge output . . . ," and ". . . few occupations . . .": Cleave and Campbell 1966:35. Later generations of diabetologists: Interview, Ron Arky.
111   "remarkably constant period . . .": Campbell 1963. See also Cleave and Campbell 1966:46–49.
111   "acute excess": White and Joslin 1959:70. *Footnote.* Joslin et al. 1946:76.
112   "related to sugar": Cleave and Campbell 1966:iv.

112   "His ideas deserved . . .": Quoted in Galton 1976:17.
112   Cleave was an outsider: See Wellcome Library n.d.
112   H. L. Cleave spent the war years: Galton 1976:15; Cleave 1962:68–70.
113   Cleave's intuition: Cleave and Campbell 1966:6–13.
113   Cavities like the canary: Ibid.:11–12.
113   Diabetics prone to heart disease: Joslin 1927; Wahlberg and Thomasson 1968. Dia-
      betes, gallstones, and obesity: Joslin 1927. "The destruction of teeth . . .": Joslin et
      al. 1946:532. See also Shlossman et al. 1990.
114   "The Law of Adaptation . . .": Cleave and Campbell 1966:1. "Whereas cooking . . .":
      Quoted in Galton 1976:8.
114   "eating of a small . . ." and "A person can take down . . .": Cleave 1975:8.
114   Peptic ulcers and lack of protein: Cleave and Campbell 1966:85–88.
114   "Assume that what strains . . .": Ibid.:18.
115   "insufficient appreciation . . ." and "While the consumption . . .": Ibid.:iii.
115   Cleave contested Joslin's belief: Cleave 1956. "what was the opposite . . .": Cleave
      and Campbell 1966:16.
115   Saturated fat increases: Friend et al. 1979.
116   Increase in sugar consumption: Cummings 1940:236. Chart. Cleaveland Campbell
      1966: 16.
117   NAS authors did not differentiate: NRC 1989:273–90.
117   Keys on the 1950s Japanese: Keys, Kimura, et al. 1958. Sugar consumption in
      Japan: Insull et al. 1968. In the United States: Cummings 1940:236. In the U.K.:
      Aykroyd 1967:105.
117   Our understanding of the Mediterranean diet: See Willett et al. 1995. The Seven
      Countries Study on Crete: Kafatos et al. 1997. The Rockefeller study: Allbaugh 1953
      (sugar and flour, 18 and table a.51).
118   Similar studies in China: See, for instance, You et al. 2000; Chen et al. 1990. Doll
      and Armstrong's analysis: Armstrong and Doll 1975. "The degree to which . . .":
      World Cancer Research Fund and American Institute for Cancer Research
      1997:379.
119   "they wouldn't have a sweet tooth . . .": Cleave's testimony in Select Committee
      1973a:248.
119   Yudkin's nutrition department first in Europe: Galton 1976:99. *This Slimming Busi-
      ness:* Yudkin 1958.
119   "remarkable relationship": Keys and Anderson 1955:189. Yudkin took Keys to task:
      Yudkin 1957. Yudkin distanced himself from Cleave: Yudkin testimony in Select
      Committee 1973a:225. Joslin on diabetics and triglycerides: Joslin 1927. Yudkin's
      sugar studies: Akinyanju et al. 1968; Yudkin et al. 1969; Szanto and Yudkin 1969.
120   "Although there is . . .": Masironi 1970.
120   Keys went after Yudkin: Keys 1971. The Seven Countries Study on sugar: Keys
      1980:252–53.
120   Truswell on fat, sugar, and onions: Truswell 1977.
121   "Yudkin was so discredited . . .": Interview, Sheldon Reiser.

CHAPTER SEVEN: FIBER

122   *Epigraph.* "The thing is . . .": Quoted in Sabbagh 2002:130.
122   Background on the hearings: Interview, Kenneth Schlossberg, then staff director of
      McGovern's committee.
122   Committee hearings on sugar, diabetes, and heart disease: Select Committee 1973a.

122 "The only question . . .": Ibid.:256.
123 ". . . die at a very early age": Ibid.:155.
123 "direct relationship": Ibid.:202. ". . . For a modern disease . . .": Ibid.:246.
123 Yudkin and McGovern dialogue: Ibid.:228–29.
124 "If men define . . .": Thomas and Thomas 1929.
125 Burkitt's life story: Galton 1976; Kellock 1985 (ten-thousand-mile trek, 59–65). "one of the world's . . .": Auerbach 1974.
125 Doll told Burkitt about Cleave: Galton 1976:6. "perceptive genius . . .": Burkitt 1979b:12. "What he was saying . . .": Burkitt 1991b. Burkitt tours U.S. hospitals: Burkitt 1970.
125 Burkitt testing Cleave's theory: Burkitt 1991b ("I was able to ask . . ."). Burkitt 1991a ("anecdotal multiplied . . . ," "written off . . . " and "Now, just because . . .").
126 "These 'western' diseases . . .": Burkitt 1971.
126 1920 article: Rendle Short 1920.
126 Appendicitis in Africa and elsewhere: Burkitt 1971 ("very rare in Africa"). See also Burkitt 1969.
126 Burkitt focused on constipation: Kellock 1985 ("normal bowel constituents," 182).
127 "Finished bowel transit tests . . .": Kellock 1985:134. Burkitt and Walker in Africa: Kellock 1985:134–35. "more white bread, sugar . . .": Walker 1962. Walker's *BMJ* article on bowel motility: Walker and Walker 1969.
127 "diets containing the natural amount . . .": Burkitt et al. 1972.
128 "All these diseases . . . ": Ibid.
128 Himsworth on Sherlock Holmes: Galton 1976:21. ". . . simply an integral part . . .": Walker et al. 1978.
129 "three million men . . .": Trowell 1981:4. *Footnote.* Higginson 1997. Trowell's list of African diseases: Trowell 1960:465–66.
129 Trowell's African experiences: Galton 1976 ("ancient Egyptians" and "Hundreds of x-rays . . . ," 63). Trowell first diagnosed heart disease: Trowell 1956. "an amazing spectacle . . .": Trowell 1975c.
130 "Western diseases" and footnote: Trowell and Burkitt 1981a:xiii. Trowell reasoned: Galton 1976:68–69. Fibrous foods harder to chew: Rodale 1974. Trowell accommodated Keys's logic: Trowell 1975b (quotes on 221).
130 "not dismissed completely . . .": Trowell 1975a:38. Burkitt said as much: Burkitt 1991a. ("I recognised that when it came to things like coronary heart disease, excess fat was just as important as diminished fibre. But Cleave would never accept that fat played any role in ill health at all.") Fat and absence of fiber could be blamed: Trowell 1975a:25; Trowell 1975b:221.
130 "major modification": Trowell and Burkitt 1975:343.
131 "Special ethnic groups . . .": Trowell 1975b:221.
131 Trowell's pair of articles in *AJCN*: Trowell 1972a; Trowell 1972b. "regenerative" agriculture: Anon. 1990. "The natural fiber . . .": Rodale 1973.
131 "changes in gastrointestinal behavior": Burkitt et al. 1974. "the tonic for our time": Auerbach 1974. *Reader's Digest* and the reaction: Kellock 1985:166–67.
131 Burkitt's lecture: Kellock 1985:175–85 ("catastrophic drop . . . " and "We eat three times more . . . ," 180–81).
132 Burkitt's disputes over fat or fiber: Kellock 1985:146–47. "furor over fiber" and "A good diet . . .": Mayer and Dwyer 1977.
132 Forty-seven thousand male health professionals: Giovannucci et al. 1994. Eighty-nine thousand nurses: Fuchs et al. 1999. The half-dozen control trials: McKeown-

Eyssen et al. 1994; MacLennan et al. 1995; Alberts et al. 2000; Schatzkin et al. 2000; Bonithon-Kopp et al. 2000; Pfeiffer et al. 2003. Fruits and vegetables: Michels et al. 2000. The WHI results: Beresford et al. 2006; Howard, Van Horn, et al. 2006; Howard, Manson, et al. 2006; Prentice et al. 2006.

133 "Burkitt's hypothesis . . .": Interview, Richard Doll.

133 *NEJM* editorial: Byers 2000.

133 "Observational studies . . . ": Ibid. The American Cancer Society recommendations: Byers and Doyle 2003. The NCI recommendations for preventing colorectal cancer can be found at http://www.nci.nih.gov/cancerinfo/pdq/prevention/colorectal/patient/.

134 "Scientists have known . . .": Boodman 1998.

134 Negative news in *Times:* Stolberg 1999; Kolata 2000b. "If preventing colon cancer . . .": Brody 2000. "Keep the Fiber Bandwagon . . .": Brody 1999a.

134 "Health Advice . . .": Kolata 2000a.

135 "Plenty of Reasons . . .": Burros 2000. "Vindication for the Maligned . . .": Brody 2000. By 2004, Brody advocating fiber: Brody 2004b.

CHAPTER EIGHT:
THE SCIENCE OF THE CARBOHYDRATE HYPOTHESIS

136 *Epigraph.* "Forming hypotheses . . .": Kleiber 1961:273.

136 Tokelau: The primary sources for Tokelau and the Tokelau Island Migration Study are Wessen et al. 1992 (U.S. Exploration Expedition, 37–40); Hunstman and Hooper 1996 (details of TIMS, 1–20; staples of the diet, 286–94); Wessen 2001.

136 TIMS a remarkably complete study: Wessen et al. 1992:18 (99 percent of all known Tokelauns were examined in round one and 82 percent by round three).

137 Dietary changes on Tokelau: Wessen et al. 1992:288–94 (pounds of fish, 30).

137 In the decades that followed: Tuia 2001; Wessen et al. 1992 (modern medical services, 267; cholesterol 306–10; weight changes, 299; *Cenpac Rounder,* 290–91).

137 Migrants to New Zealand: Wessen et al. 1992 ("immediate and extensive changes," 291; "exceptionally high incidence . . ."and "migrants were at higher risk . . . ," 377–78). ". . . big increase in sucrose consumption": Prior et al. 1978.

138 A number of factors: Wessen et al. 1992:383–88 ("substantially higher" and "in fact, obesity . . . ," 384; migrant lifestyle rigorous, 295–96; saturated fat, 292).

138 Difficult to explain simply: Ibid.:384–86 ("that a different set . . . ," 384).

139 Cleave's saccharine disease: Cleave and Campbell 1966 (simplest possible explanation, 6–13).

139 Proposed the name Syndrome X: Reaven 1988. Over the years: Reaven 2005.

140 NHLBI belatedly recognized: NCEP 2002. *Footnote.* The first time: Sugarman 1999. The second time: Lindner 2001. A couple of thousand: LexisNexis search, over seventeen hundred articles between 1977, the beginning of the database for the *Washington Post,* and November 23, 1999, that included the word "cholesterol" in the headline, lead paragraph, or search terms.

141 "What you're faced with . . .": Interview, Scott Grundy.

141 "concept of the nature . . .": Krebs 1971.

142 Bernard observed: Bernard 1957; Bernard 1974 ("harmonious ensemble" and "with such a degree of perfection . . . ," 48). The Cooks' translation of Bernard is occasionally awkward, and so I have used Cannon's translation—"All the vital mechanisms . . . ,"—which is from Cannon 1939:38. "No more pregnant sen-

tence . . .": Quoted in Cannon 1939:38. Cannon also discussed homeostasis in Cannon 1929.

142 "Somehow the unstable stuff . . .": Cannon 1939:23.

142 Homeostasis and body temperature: See, for instance, Greene 1970:23–29.

143 Homeostasis, hypothalamus, and hormonal functions: Wilson et al. 1998. Hormonal effects on *fuel partitioning:* Newsholme and Stuart 1973:329–36.

143 Role of insulin: Catt 1971:106–21.

144 Yudkin observed: Yudkin 1986:116.

144 "The changing dietary patterns . . .": Neel 1999.

145 "We really must learn . . .": Bernard 1957:89. "the wholeness of the organism . . .": Krebs 1971.

145 Hypertension as "insulin-resistant state": McFarlane et al. 2005. Relationship among hypertension, triglycerides, cholesterol, diabetes and heart disease: Rocchini 1998; Hall et al. 2003; Wilson et al. 1998.

146 Textbooks recommend salt reduction: See, for instance, McFarlane et al. 2005. Century-old salt hypothesis: See, for instance, Foster 1922. Salt binge increases blood pressure: Interview, Franklin Epstein. For a more comprehensive treatment of the salt/blood pressure, see Taubes 1998.

146 "inconclusive and contradictory": Stamler 1967:261. "inconsistent . . .": Cooper et al. 1983. "the deadly white powder": Jacobson 1978. Systematic reviews: See, for instance, Graudal et al. 1998; He and MacGregor 2004.

147 Donnison on blood pressure in African natives: Donnison 1938:15–17; Donnison 1929 ("It tends to come down . . ."). Observations confirmed: See, for instance, Page et al. 1974.

147 Hypertension in Kenya and Uganda: Trowell 1981.

148 Donnison on the stress of civilized life: Donnison 1938:43–46. Absence of hypertension as compelling evidence: Intersalt Cooperative Research Group 1988; Taubes 1998; Colburn 1995. (To the question whether salt raises blood pressure, Jeremiah Stamler responds, "Even if you stand on your head, the answer is yes." He then says that the most compelling evidence is that "populations that habitually take in a lot of salt have blood pressures that are higher than other populations that take in less salt.").

148 Shaper's studies: Shaper 1967; Shaper et al. 1969. Prior's studies: Prior et al. 1964; Prior 1971. "the antecedents of cardiovascular disease . . .": Page et al. 1974.

149 Voit on carbohydrates and blood pressure: Rony 1940:154. "With diets predominantly carbohydrate . . .": Benedict et al. 1919:195. New generation: Kekwick and Pawan 1957. Rationalize popularity: See, for instance, Anon. 1973.

149 "remarkable sodium and water . . .": Gordon 1964:1301. Bloom's research: Bloom 1962; Bloom 1967.

149 "low blood pressure resulting . . .": White and Selvey 1974:48. "fluid balance" and "avoid large shifts . . .": Dwyer and Lu 1993:246. Insulin and sodium metabolism: See DeFronzo 1981b for a review. Insulin higher in hypertensives: Welborn et al. 1966.

150 "the major pathogenic defect . . .": Christlieb et al. 1994.

150 Insulin-induced hypertension in textbooks: Randall 1973. Focus on salt hypothesis: kark and Oyama 1973. "One claim . . .": Bray 1978. Low-calorie diets recommended to reduce blood pressure: DeFronzo 1981a. "carbohydrate overeating": Kolanowski 1981.

151 Landsberg's research: Landsberg 1986; Landsberg 2001; Interview, Lewis Landsberg.

CHAPTER NINE:
TRIGLYCERIDES AND THE COMPLICATIONS OF CHOLESTEROL

153 *Epigraph.* "Oversimplification has been . . .": McCollum 1957:37.

153 Gofman's *Science* article: Gofman and Lindgren 1950.

154 Gofman's background: Interview, John Gofman.

155 "At a particular cholesterol level . . .": Gofman and Lindgren 1950.

155 Test carried out by four groups: Cooperative Study of Lipoproteins and Atherosclerosis 1956.

155 Split between Gofman and other investigators: Interviews, John Gofman and Max Lauffer.

155 Four groups published a report: Cooperative Study of Lipoproteins and Atherosclerosis 1956 ("The lipoprotein measurements are so complex . . . ," 724).

156 "While it is true . . .": Gofman et al. 1958:29–30.

156 Carbohydrates elevate VLDL: Gofman 1958.

156 "carbohydrate factor": Ibid.

156 Measurement of total cholesterol: Ibid. ("false and highly . . . ," 281; "generalizations such as . . . ," 273); Gofman et al. 1958 ("Neglect of . . . ," 45).

157 "that the lipemic plasma . . .": Ahrens et al. 1961. "The percent of fat . . .": Joslin 1927.

157 "an exaggerated form . . .": Ahrens et al. 1961.

158 "especially in the areas . . .": Ibid. "We know of no solid . . .": Ahrens et al. 1957.

158 Peters a "contrarian": Interview, Margaret Albrink. Albrink's research: Albrink 1963; Albrink 1962.

159 "Rockefeller Institute Report . . .": Osmundsen 1961. ". . . brought the house down . . .": Interview, Margaret Albrink.

159 Albrink's results confirmed: Kuo 1967; Carlson and Bottiger 1972; Goldstein et al. 1973. *JAMA* published an editorial: Anon. 1967.

160 No consideration to alternative hypothesis: See Bishop 1961; Baker et al. 1963. Seven Countries Study: Keys 1970:I-7.

160 Five-part *NEJM* series: Fredrickson et al. 1967a–e.

161 Four of the five lipoprotein disorders: Fredrickson et al. 1967c:149 (table 2). Warned against low-fat diets: See, for instance, Fredrickson et al. 1967b:219. "sometimes considered synonymous . . .": Fredrickson et al. 1967a:273. "Patients with this syndrome . . .": Lees and Wilson 1971. *Footnote.* Select Committee 1976:37.

161 HDL protection against heart disease proposed: Barr et al. 1951a and b. Confirmed: See, for instance, Nikkila 1953; Gofman et al. 1966; Levy et al. 1966.

162 "negative relation . . .": Gordon 1988. *Footnote.* Gordon 1988; Interview, Tavia Gordon.

162 Controls from all five populations: Castelli et al. 1977. Evidence from Framingham alone: Gordon et al. 1977 ("total cholesterol per se . . . ," 712; "marginal," 710).

162 "striking" revelation and "Of all the lipoproteins . . .": Gordon et al. 1977:707.

163 "fragmentary information . . .": Castelli et al. 1977. See also Hulley et al. 1972.

163 HDL directing attention away from triglycerides: See Hulley et al. 1980.

164 ". . . greeted with a silence . . .": Interview, Tavia Gordon.

164 "the findings re-emphasize . . .": Brody 1977.

165 VA twenty-center trial of gemfibrozil: Rubins et al. 1999.

165 Lowering LDL appears more important: See, for instance, NCEP 2002:II-11.

166 "Whatever the underlying disorder . . .": Kannel et al. 1979.

166 Justifying total-cholesterol measurements: Gordon 1988. "marginal risk factor":

Gordon et al. 1977:710. *"powerful* predictor" and "a *significant* contribution . . .": Kannel et al. 1979. *Footnote*. NCEP 2002:II-1.

166 "lipid profile": See, for instance, Kannel and Castelli 1979. Added little predictive power: Gordon et al. 1977:710 (table VIII).

167 "from a practical point of view . . .": Gordon et al. 1977:712.

167 "In the search . . .": Kannel et al. 1979.

167 Rarely mentioned carbohydrates: See Chait et al. 1993. *Footnote*. "that epidemiological studies have . . .": Chait et al. 1993:3014.

168 Monounsaturated fats: Mattson and Grundy 1985; Grundy 1986. Keys assumed neutrality: Keys et al. 1957. Never been tested: Interview, Scott Grundy. Lyon Diet Heart Trial: Lorgeril et al. 1999. GISSI-Prevenzione: Gruppo Italiano per lo Studio della Sopravvivenza nell'Infarto Miocardico 1999.

168 Stearic acid metabolizes to oleic: Grundy 1994. A good review of the effects of different fats on LDL and HDL cholesterol can be found in Katan et al. 1995.

169 "Everything should be made . . .": Shapiro 2006:231. This quote may be a paraphrase of the following statement: "The supreme goal of all theory is to make the irreducible basic elements as simple and as few as possible without having to surrender the adequate representation of a single datum of experience" (see http://en.wikiquote.org/wiki/Albert_Einstein).

170 "marginal risk factor": Gordon et al. 1977:710. Only a few percentage points higher: Castelli et al. 1977. "If you look in the literature . . .": Ibid.

170 AHA nutrition guidelines: Krauss et al. 1996; Krauss et al. 2000. "30-percent-fat recommendation . . .": Interview, Ronald Krauss.

170 "this conventional notion . . .":Ibid.

171 "blazingly obvious . . . ": Ibid. *Footnote*. Adams and Schumaker 1969; Hammond and Fisher 1971.

171 Krauss's three papers: Shen et al. 1981; Krauss and Burke 1982; Teng et al. 1983. "remarkable heterogeneity . . .": Interview, Ronald Krauss.

172 First report of apo B elevation in heart-disease patients: Sniderman et al. 1980. Disproportionate elevation in apo B: Teng et al. 1983.

172 Small, dense LDL more *atherogenic:* Teng et al. 1983. "little bits of sand": Interview, Allan Sniderman. Role of *oxidized* LDL: Witztum and Steinberg 1991.

172 Pattern A and B and the *atherogenic profile:* Austin et al. 1988. Diabetics have identical pattern: See Chait and Bierman 1994.

173 Diet and the atherogenic profile: See Krauss 2005 for a recent review. "average American diet": Interview, Ronald Krauss. The more saturated fat: Dreon et al. 1998.

173 Renamed *atherogenic dyslipidemia:* See, for instance, Grundy, Hansen, et al. 2004.

174 "Well, I would rather . . .": Interview, Melissa Austin.

174 Best predictor apo B: Walldius et al. 2001. "doesn't tell you anything . . .": Interview, Goran Walldius.

175 LDL clearance and disposal mechanism: Brown and Goldstein 1985.

175 For an overall review of VLDL and LDL metabolism and how to increase LDL cholesterol by increasing VLDL, see Mayes and Botham 2004. See also Berneis and Krauss 2002; DeFronzo 1992.

176 "It's the overproduction of VLDL . . .": Interview, Ernst Schaefer.

176 Krauss's model: Berneis and Krauss 2002.

176 "I am now convinced . . .": Interview, Ronald Krauss.

176 Ahrens on high-carbohydrate diets in undernourished populations: Ahrens et al. 1961.

177 Poverty in the Mediterranean after World War II: See, for instance, Allbaugh 1953.

CHAPTER TEN:
THE ROLE OF INSULIN

178   *Epigraph.* "The suppression of inconvenient . . .": Greene 1953.
178   Vague on "android obesity," etc.: Vague 1956.
179   Gofman on the obesity/heart-disease association: Gofman and Young 1963.
179   Speculation voiced by Joslin: Joslin 1928:103. Man and Peters measured choles-
      terol: Man and Peters 1935. Albrink reported: Albrink et al. 1962. Joslin's similar
      observation: Joslin et al. 1959:275. Albrink confirmed Gofman's observation:
      Albrink and Meigs 1965.
179   "abnormal metabolic patterns": Albrink 1963.
180   "purified carbohydrates": Albrink 1965.
180   Arcane tests before 1960: Interviews, Gerold Grodsky and Roger Unger. "a revolu-
      tion in . . .": Karolinska Institute 1977.
180   Yalow and Berson showed: Yalow and Berson 1960. Obese had elevated insulin lev-
      els: Yalow et al. 1965.
180   *Insulin-resistant:* Berson and Yalow 1965; Berson and Yalow 1970 ("a state . . . ,"
      389).
181   "it is desirable . . .": Berson and Yalow 1970:390.
181   Reaven began his investigations: Reaven et al. 1963.
181   Reaven's two-part hypothesis: Interview, Gerald Reaven.
182   Reaven and Farquhar had reported: Farquhar et al. 1966; Reaven et al. 1967.
182   Working to establish validity of hypothesis: See Reaven and Olefsky 1978. First
      insulin-resistance test: Shen et al. 1970. DeFronzo refined the "gold standard":
      DeFronzo et al. 1979.
182   Reaven's Banting Lecture: Reaven 1988 ("Although this concept . . .").
183   Three Framingham-like studies: Eschwege et al. 1985; Pyorala 1979; Welborn and
      Wearne 1979.
183   "a whole host of . . .": Interview, Ralph DeFronzo. NCEP diagnostic criteria: NCEP
      2002:II-27.
183   Reaven's article on Syndrome X: Reaven and Chen 1996.
184   Silverman on Reaven's results: Quoted in Kolata 1987.
184   *Cognitive dissonance:* Festinger 1957. Kuhn 1970:77–91 ("the awareness . . . ," 81;
      "They will devise . . . ," 78).
184   Krauss and Reaven reported: Reaven et al. 1993. "coequal partner . . .": NCEP
      2002:II-26.
185   Metabolic syndrome officially entered: See NCEP 2002 ("the primary driving
      force . . . ," II-36; "mass elevations . . . ," II-28); Grundy, Hansen, et al. 2004;
      Grundy, Brewer, et al. 2004. Grundy acknowledged: Interview, Scott Grundy.
185   "commonly in persons . . .": NCEP 2002:II-11. *Footnote.* Grundy, Hansen, et al.
      2004 ("very high-carbohydrate . . . ," 553).

CHAPTER ELEVEN:
THE SIGNIFICANCE OF DIABETES

186   *Epigraph.* "Does carbohydrate cause . . .": Joslin 1927.
186   "extraordinarily high incidence": Bradley 1971:446.
186   "numerous and as yet . . .": Ibid.:460.
186   Assumption that saturated fat is the nutritional agent: USDHHS 1988:257–58
      ("The frequent . . . ," 258). The ADA recommendations: ADA 1971.

188   Atherogenic American diet high in fat *and* salt: NCEP 2002:II-18.
188   The *vascular* complications of diabetes: Donnelly et al. 2000.
189   "the effects of insulin . . .": Feener and Dzau 2005:874. "another possibility . . .":
      Johnstone and Nesto 2005:978.
189   First reported in rabbits: Duff and McMillan 1949. In chickens: Katz et al. 1958
      ("one factor . . .").
190   In dogs: Cruz et al. 1961.
190   Stout published studies: Stout 1968; Stout and Vallance-Owen 1969 ("ingestion of
      large quantities . . . " and "The carbohydrate is disposed . . ."); Stout 1969; Stout
      1970; Stout et al. 1975. *Footnote.* "atherogenic hormone": DeFronzo 1997.
191   For a good review of the *oxidative stress* hypothesis, see Giugliano et al. 1996.
191   "conform to a tightly . . .": Bunn and Higgins 1981.
191   For a relatively simple discussion of glycation and AGEs, see Cerami et al. 1987. My
      discussion of AGEs was also based on interviews with John Baynes, Michael
      Brownell, Frank Bunn, Anthony Cerami, Vincent Monnier, Ben Szwergold, and
      Helen Vlassara.
192   Cerami's work on hemoglobin A1c: Koenig et al. 1976. Bunn's work: Gabbay et al.
      1977. See also Bunn et al. 1978.
192   AGEs and the eye: See Stitt 2001. AGEs and other diabetic complications: See
      Singh et al. 2001 for a review.
193   AGEs, collagen, and diabetes as accelerated aging: Monnier et al. 1984.
193   "If you remove the aorta . . .": Interview, Anthony Cerami.
193   Oxidized LDL and heart disease: Steinberg 1997. Oxidized LDL, reactive oxygen
      species and glycation: Bucala et al. 1993. "markedly elevated": Stitt et al. 1997. *Foot-
      note.* "rendering the HDL . . .": Hedrick et al. 2000.
194   "the adverse cardiovascular . . .": Susic et al. 2004.
194   "Current evidence points . . .": Peppa et al. 2003.

CHAPTER TWELVE:
SUGAR

195   *Epigraph.* "M. Delacroix . . .": Brillat-Savarin 1986:104.
196   Reaven initiated study of glycemic index: Crapo et al. 1977 ("traditionally held
      tenet . . ."). Reaven more interested in insulin: Interview, Gerald Reaven. Jenkins and
      Wolever's research: Jenkins et al. 1981. "tremendous": Interview, Thomas Wolever.
197   Vitriolic debate: Interviews, Gerald Reaven ("Ice cream has . . ."), David Jenkins,
      and Thomas Wolever. See also Coulston and Reaven 1997 and the response,
      Wolever 1997.
197   Fructose and the glycemic index: Mayes 1993.
198   "We see no reason . . .": Bantle et al. 1983. Official government position: Glins-
      mann et al. 1986:s65–66. "sucrose or sucrose-containing foods . . .": ADA 2006.
198   "no conclusive evidence . . .": Glinsmann et al. 1986:s15.
198   *Surgeon General's Report:* USDHHS 1988:III. *Diet and Health:* NRC 1989:9.
198   HFCS and climbing sugar consumption: Putnam et al. 2002:8.
199   Sugar and starch consumption over the twentieth century: USDA 2000; Putnam et
      al. 2002. Sugar as dietary nuisance: See, for instance, Mayer 1976.
199   Fructose content of fruit: "Sweetener" entry, *Encyclopædia Brittanica.* Fructose per-
      ceived as healthy: See, for instance, Brody 1983b; Donohue 1988.
200   Metabolism of glucose and fructose: Shafrir 1991. "constitutes a metabolic
      load . . .": Interview, Eleazar Shafrir. *Footnote.* Higgins 1916.

200 "In the 1980s . . .": Interview, Judith Hallfrisch. Sugar raises cholesterol: Swanson et al. 1992. For unbiased reviews of metabolic effects of fructose, see Hollenbeck 1993; IOM 2002:297–303.
200 "pattern of fructose metabolism": Mayes 1993. Fructose causes insulin resistance: Shafrir 1985. Reiser observed in humans: Reiser et al. 1981. *Fructose-induced hypertension:* See Hodges and Rebello 1983; Hwang et al. 1987.
201 "This is really the harmful . . .": Interview, Peter Mayes.
201 Fructose and AGEs: Bunn and Higgins 1981; Dills 1983. Ten times more effective: McPherson et al. 1988; Suárez et al. 1989. More resistant AGEs: Suárez et al. 1995. Increases LDL oxidation: Mowri et al. 2000.
201 The COMA report: U.K. Department of Health 1989:43.
202 "The panel concluded . . .": Ibid.
202 Dedicated an entire issue: *ACJN,* November 1993. "Further studies . . .": Tappy and Jéquier 1993.
202 Institute of Medicine spent twenty pages: IOM 2002:295–324 ("insufficient evidence," 323; no reason to pursue research, 323–24).
203 Half a dozen research projects: NIH CRISP database search, keywords "fructose" and "sucrose."
203 "no conclusive evidence . . .": Glinsmann et al. 1986:s65–66.

CHAPTER THIRTEEN:
DEMENTIA, CANCER, AND AGING

204 *Epigraph.* "The bottom line . . .": Tanzi and Parson 2000:201.
204 NIH funding of Alzheimer's research 1970s and 1980s: NIH CRISP database search, keywords "Alzheimer's" and "dementia."
205 Apo E4 and Alzheimer's: Strittmatter et al. 1993. Alzheimer's researchers blame cholesterol and saturated fat: See, for instance, Mattson 2004. *Footnote.* See, for instance, Marx 2001 ("link between").
205 Japanese Americans vs. Japanese: Graves et al. 1996. African Americans vs. rural Africans: Hendrie et al. 2001.
205 Studies in large populations: Ott et al. 1999 (Rotterdam: "direct or indirect"); Leibson et al. 1997 (Minnesota); Luchsinger et al. 2001 (Manhattan); Arvanitakis et al. 2004 (Midwest); Peila et al. 2002 (Honolulu). Hyperinsulinemia and metabolic syndrome: Kuusisto et al. 1997; Vanhanen et al. 2006.
206 Confusion of Alzheimer's with vascular dementia: See Kalaria 2002; Zekry et al. 2002; Korczyn 2002.
206 Snowdon's nun study: Snowdon 2003.
206 Accumulation of vascular dementia accelerates Alzheimer's: See, for instance, Ravona-Springer et al. 2003.
207 Amyloid precursor protein exists naturally: Interview, Rudolph Tanzi.
207 AGEs in plaques and tangles: Yan et al. 1994; Smith et al. 1994; Vitek et al. 1994. In immature plaques: Sasaki et al. 1998.
207 The AGEs-Alzheimer's hypothesis: Grossman 2003; Obrenovich and Monnier 2004; Moreira et al. 2005.
207 Involvement of insulin: Qiu et al. 1998.
208 Animal experiments: Farris et al. 2003; Miller et al. 2003; Farris et al. 2004. *Footnote.* Kim et al. 2007.
208 Boosting insulin enhances memory: Craft et al. 1996. In 2003, Craft reported: Watson et al. 2003. "We're not saying . . .": Interview, Suzanne Craft.

208 Selkoe and Tanzi on "attendant therapeutic implications . . .": Farris et al. 2004.
209 Higginson's studies of cancer incidence: Reviewed in Higginson 1981; Higginson 1997. "It would seem, therefore . . .": Quoted in Doll and Peto 1981:1197.
209 *At least* 75 to 80 percent: Doll and Peto 1981:1256–60.
209 Role of man-made chemicals minimal, diet maximal: Ibid.:1256 (table 20).
210 "extrinsic" and "environmental factors": See Maugh 1979 and Doll and Peto 1981:1197. "carcinogenic soup": Greenberg 1979. "It appears that . . .": Higginson 1983. Geneva vs. Birmingham, Sweden vs. Japan: Maugh 1979. *Footnote*. Quoted in Maugh 1979.
210 Cold Spring Harbor talks: Hiatt et al. 1977:605–956.
210 "gross aspects . . ." and "ingestion of traces . . .": Doll and Peto 1981:1258. Cancer in Seventh-day Adventists: Phillips 1975.
211 Cancer in Mormons: Lyon and Sorenson 1978; Lyon et al. 1980. "among the biggest . . .": Doll and Armstrong 1981:103. For the next twenty years: See, for instance, Wynder et al. 1983; Carroll and Kritchevsky 1993; U.K. Department of Health 1998.
211 Failed to identify diet-related carcinogens: Interviews, W. Robert Bruce, Richard Doll and Robert Weinberg. Cancer epidemiologists made little attempt: See, for instance, World Cancer Research Fund and American Institute for Cancer Research 1997: 509–19; U.K. Department of Health 1998:189–207.
211 Cleave had suggested: Cleave 1975:28–38. Yudkin on five nations: Yudkin 1986:137. *Diet and Health* on carbohydrates: NRC 1989:282–83.
212 "strikingly similar": Giovanucci 2001.
212 Rous's semi-starvation research: Rous 1914. McCay reported: McCay et al. 1935. Tannenbaum's research: Reviewed in Tannenbaum 1959 ("many types of tumors . . . ," 530; "pathologic changes . . . ," 523).
213 Hormone-dependent factors linked to cancers: Armstrong 1977. Increase in cancer incidence with weight gain: Doll and Peto 1981:1234; World Cancer Research Fund and American Institute for Cancer Research 1997:371–73. Obesity, cancer, and estrogen production: See, for instance, Ballard-Barbash 1999.
213 Warburg's fermentation work: See Warburg 1956. Tumors starved of fuel: Tannenbaum 1959:530. *Footnote*. Tannenbaum 1959:524.
213 Early observations of glucose intolerance in cancer patients: Glicksman et al. 1956. See also Kessler 1971. Unless insulin was added: Temin 1967; Temin 1968. Adrenal and liver-cell cancers: Koontz and Iwahashi 1981. For a review of this research, see Del Giudice et al. 1998. "intensely stimulated . . .": Heuson et al. 1967:359. "exquisitely sensitive . . .": Osborne et al. 1976:4539.
213 Greater number of insulin receptors: Giorgino et al. 1991 ("selective growth . . . ," 452).
214 The Darwinian model of cancer development: Weinberg 2007: 413–24.
214 Ten thousand trillion and "enormous opportunity:" Weinberg 1996:252.
215 Insulin and IGF: For reviews of their roles in cancer development, see Giovannucci 1995; Kaaks 1996; Burroughs et al. 1999; Kaaks and Lukanova 2001; LeRoith and Roberts 2003; Baserga et al. 2003; Pollak et al. 2004. This section was also informed by interviews with Renato Baserga, Edward Giovannucci, Rudolf Kaaks, Derek LeRoith, Bruce Roberts, and Robert Weinberg.
215 "stumbled" upon: Interview, Renato Baserga.
216 "strong inhibition . . .": Baserga 2004.
216 LeRoith's experiments with IGF-deficient mice: Wu et al. 2002; Wu et al. 2003.

217 Cheresh has demonstrated: Brooks et al. 1997; interview, David Cheresh.

217 2003 meeting in London: Interview, Derek LeRoith; Novartis Foundation 2004. Studies linking hyperinsulinemia and IGF to cancer: See Kaaks and Lukanova 2001.

218 "People were thinking . . ." and "When applied simultaneously . . .": Interview, Rudolf Kaaks.

218 "an environment that favored . . .": Pollak et al. 2004.

218 Live 30 to 50 percent longer: See, for instance, Masoro et al. 1982. For a good review of the history of the calorie-restriction science, see Masoro 2003.

218 Two possibilities: Masoro 2003.

219 Harrison's experiments: Harrison et al. 1984 ("Longevities were related"). Whenever these experiments are done: See, for instance, Bertrand et al. 1980.

219 Oxidative stress, antioxidants, and longevity: Tuma 2001; Weinert and Timiras 2003.

219 Characteristics of long-lived organisms: Bartke 2002; Davenport 2003.

220 Genetic studies of yeast: Lin et al. 2000. Worms: Lin et al. 1997. Fruit flies: Clancy et al. 2001. Mice: Holzenberger et al. 2003; Bluher et al. 2003.

220 "When reduced to essentials . . .": Bishop 1989.

220 Longevity mutations regulate dauer state: Kenyon et al. 1993. "The way these worms work . . .": Interview, Cynthia Kenyon.

221 Ruvkun reported: Kimura et al. 1997; interview, Gary Ruvkun. Long-lived fruit-fly mutants: Clancy et al. 2001. See also Kenyon 2001.

221 Gene knockout experiments in mice: Holzenberger et al. 2003. Kahn's research: Bluher et al. 2003; interview, C. Ronald Kahn.

222 "When food becomes limiting . . .": Kenyon 2001:168.

222 Kenyon began a series of experiments: Interview, Cynthia Kenyon.

223 "Could a low-carb . . .": Kenyon's slide from her conference presentation. I'm grateful to Professor Kenyon for providing the slides.

223 Kenyon's restriction of carbohydrate consumption: Interview, Cynthia Kenyon.

223 " . . . attendant therapeutic implications . . .": Farris et al. 2004:1432. "dream of 60 million . . .": Joslin Diabetes Center 2003. Diabetologists take the same tack: See, for instance, LeRoith 2004 ("normalize" and "intensive . . .").

224 NCEP merges both tacks: NCEP 2002 ("atherogenic diet," II-20; "pharmaceutical modification . . . ," II-26).

224 "Weight sits like a spider . . .": Willett 2001:35. "Excess weight . . .": Stamler 1962:57.

PART THREE: OBESITY AND THE REGULATION OF WEIGHT

227 Epigraphs. "How may the medical . . .": Stunkard and McClaren-Hume 1959. "To cultivate the faculty . . .": Tanner 1869a:1.

CHAPTER FOURTEEN:
THE MYTHOLOGY OF OBESITY

229 Epigraph. "A colleague once defined . . .": Cohen 1989:viii.

229 "To have our first idea . . .": Bernard 1957:32–33.

229 "overweight and obesity result . . .": USDHHS 2001:1.

229 "Most studies comparing . . .": NRC 1989:583.

230 Percentage of obese Americans: NCHS 2005:275 (table 73).

230 Proportion consistent throughout society: Ogden et al. 2003. Interview, Katherine Flegal. Children not exempt: NCHS 2005:9, 279 (table 74). *Footnote.* Friedman 2003; interviews, Jeffrey Friedman and Katherine Flegal.

230 "toxic environment . . .": Quoted on Brownell's Yale University faculty information Web page (http://www.yale.edu/psychology/FacInfo/Brownell.html).

231 "Cheeseburgers and french fries . . .": Brownell and Horgen 2004:8. "improved prosperity . . .": Nestle 2003.

231 "risen three-fold" and "As incomes rise . . .": WHO 2004.

232 CDC, "attributable primarily . . .": Wright et al. 2004.

232 USDA on increases in nutrient intake: Gerrior and Bente 2001:table 1. Chart. Wright, et al. 2004. *Footnote.* USDA Center for Nutrition Policy and Promotion 1998.

234 "It appears that efforts . . .": Heini and Weinsier 1997.

234 Lack of CDC evidence on physical activity: Interview, William Dietz, director of CDC Division of Nutrition and Physical Activity. No less active at end of 1990s: CDC 2001. $200 million a year: Lichtenstein 1972. 2005 numbers: International Health, Racquet and Sportsclub Association 2005. *Footnote.* Interview, Mike May, spokesman for the Sporting Goods Manufacturers Association.

234 "exercise explosion . . .": Gilmore 1977. "new fitness revolution . . .": Cohn 1980.

235 Obesity prevalent among the poorest: Bray 1998. NHANES studies confirmed: NCHS 2005:275 (table 73). Stunkard reported: Goldblatt et al. 1965. See also Stunkard 1976a.

235 Obesity blamed on high-fructose corn syrup: Critser 2003:138–40; Bray et al. 2004; Pollan 2006:100–108.

236 "As the typical American diet . . ." and "If the Pima Indians could . . .": NIDDK 1995:19.

236 Russell noted: Russell 1975 ("exhibit a degree . . . ," 66; Fat Louisa, 67).

236 "Especially well-nourished . . .": Hrdlička 1908:156–57.

237 Pima had lived as hunter-gatherers and agriculturalists: Aldritch 1966; Audubon 1906; Bartlett 1965; Castetter and Bell 1942; Cook and Whittemore 1893; Cunning-ham 1996; Curtain 1949; Davis 1962; Dobyns 1978; Dobyns 1989; Eccleston 1950; Ezell 1961; Griffin 1943 ("sprightly," "fine health," and "the greatest abudance . . . ," 34); Harris 1960; Hrdlička 1906; Hrdlička 1908; Jones 1967; Rea 1983; Reid 1858; Russell 1975 ("unidentified worms," 81; also mentioned in Audubon 1906:150); Smith et al. 1994; Spicer 1962; Spier 1978; Webb 1992.

237 "by way of Tucson . . .": Russell 1975:30.

237 "the years of famine": Smith et al. 1994:409. "entirely absorbed . . .": Spicer 1962:148–50.

237 "Certain articles . . .": Russell 1975:66. Hrdlička suggested: Hrdlička 1908:156–57.

238 Women worked as pack animals: Russell 1975:66.

238 "everything obtainable . . .": Hrdlička 1906. "sugar, coffee and canned . . .": Dobyns 1989:61.

238 Sioux diet with government rations: Jackson 1994.

238 According to Kraus: Kraus 1954. Hesse noted: Hesse 1959.

239 Over the next twenty years: Price et al. 1993. "large quantities of refined . . . ," "started to carry . . . ," and "Soda pop . . .": Justice 1994:116–17.

239 "The only question . . ." Select Committee 1973a:256–57. "to some extent a result . . ." and "many of the poorer individuals . . .": Dobyns 1989: 100–101.

239 Obesity in the South Dakoa Sioux: Stene and Roberts 1928. Arizona Apaches: Clifford 1963. North Carolina Cherokees: Stein et al. 1965. Oklahoma tribes: West 1981 ("Men are very fat . . . ," 132).

240 "An overwhelming majority . . .": Meriam et al. 1928. University of Chicago report: Stene and Roberts 1928.

240 ". . . visit to Czechoslovakia . . .": Osancova 1975.

241 Obesity in African Americans in Charleston: Grant and Groom 1959. In Chile: Arteaga 1974. In Trinidad: McCarthy 1966.

241 In Jamaica: Richards and de Casseres 1974.

241 In Rarotonga: Prior 1971. Among Zulus: Slome et al. 1960. Bantu "pensioners": Walker 1964. "Although dietary habits . . .": Adadevoh 1974. For other studies of obesity in impoverished populations, see Reichley et al. 1987; Seftel et al. 1965; Haddock 1969; Johnson 1970; Tulloch 1962:72–75.

242 "It is difficult to explain . . .": Richards and de Casseres 1974.

242 "different only in degree . . .": Gladwell 1998.

243 "Such genes would be advantageous . . .": Diamond 2003. "exquisitely efficient . . ." and "Your genes match nicely . . .": Brownell and Horgen 2003:6. "are trying to find these genes . . .": Gladwell 1998.

243 "thrifty mechanisms to defend . . .": Ravussin 2005.

243 Neel initially proposed: Neel 1962.

244 Sickle-cell anemia and malaria: Rucknagel and Neel 1961.

244 "might have, during a period . . .": Neel 1962.

244 "If the considerable frequency . . .": Ibid.

244 "a relatively recent phenomenon": Neel 1982. Russell's disease list: Russell 1975:268. Hrdlička's: Hrdlička 1908:182–83. Joslin concluded: Joslin 1940. The Bureau of Indian Affairs studies: Cohen 1954; Parks and Waskow 1961.

245 Glucose intolerance in Yanomamo: Spielman et al. 1982. "The high frequency . . .": Neel 1999.

245 "overindulge in . . .": Neel 1982.

245 "The data on which . . .": Neel 1989. "composition of the diet . . .": Neel 1999.

245 "To me, it would be . . .": Hirsch 1978:3.

246 "exquisitely efficient . . .": Brownell and Horgen 2003:5–6. "conditions of unpredictably alternating . . .": Diamond 2003.

246 "paradise for hunting . . .": Eaton et al. 1988:29. "conspicuously well-nourished . . .:" Cohen 1989:96. *Footnote.* Duncan and Burns 1998:89–90,150.

246 The !Kung study: Lee and DeVore 1968. Lee noted: Lee 1968 ("the third year . . . ," "some relatively abundant . . . ," and "have to walk . . . ," 39).

247 "levels of decay . . .": Cohen 1989:86. When hunting failed: Livingstone 2001: 32–33. Resiliency of hunting and gathering: See, for instance, Lee and Devore 1968. Studies of human remains: Cohen 1989:105–42; Cohen 1987:261–85. "the worst mistake . . .": Diamond 1987.

247 "a uniquely low occurrence . . .": Diamond 2003.

248 Average 150-pound man: Wertheimer 1965. "Survival of the species . . .": Cahill and Renold 1965.

248 "facile and unlikely" and "major objection . . .": Rothwell and Stock 1981: 335–36.

248 "When this animal . . .": Zimmet et al. 2001.

248 Schmidt-Nielsen's sand-rat experiments: Schmidt-Nielsen et al. 1964.

249 Monkeys in captivity: Hamilton and Brobeck 1965. Perhaps 60 percent: Bodkin

et al. 1993; Jen et al. 1985. "This is on the kind . . .": Interview, Barbara Hansen. *Footnote.* Strasser 1968.

249 Mammalian species that fatten regularly: Young 1976 is a good review.

250 "opening a can of Crisco . . .": Interview, Irving Zucker.

250 Genetically pre-programmed: Mrosovsky 1976; Mrosovsky 1985 ("It is very hard to prevent . . ."). *Lipectomy* studies: Mauer et al. 2001.

250 Regulation of type and location of fat: See Young 1976.

251 Select *out* rather than *in:* See, for instance, Sims 1976.

CHAPTER FIFTEEN:
HUNGER

252 *Epigraph.* "Khrushchev, too, looks . . .": Liebling 2004:485.

252 Benedict's semi-starvation studies: Benedict et al. 1919 (fourteen to twenty-one hundred calories, 688–89).

252 His subjects lost the weight: Benedict et al. 1919 ("a continuous gnawing . . . ," 360; "almost impossible to keep warm . . . ," 259; reduced energy expenditure, 694–95; blood pressure, 371; pulse rate, 383; anemia, 364–65; concentration, 680; "a decrease in sexual interest . . . ," 640; "the whole picture . . . ," 698).

253 "One general feature . . .": Benedict et al. 1919:683–85.

253 Keys set out to replicate: Keys, Brozek, et al. 1950 ("guinea pigs," 64; "semi-starvation diet," 74).

253 1,570 calories, "The major food items served . . .": Ibid.:74. *Footnote.* Ibid.

254 Keys's conscientious objectors lost on average: Ibid. (body fat lost, 175–76; physiological responses, "As starvation progressed . . . ," 827–28; ". . . wise man . . . ," 290).

254 "behavior and complaints": Ibid.:819- 53, 881–904.

254 Five of the subjects: Ibid. ("character neurosis," 880; "semi-starvation neurosis," 894; ". . . psychosis," 880; "weeping . . . ," 885; "he suffered a sudden . . . ," 887; ". . . drastic cuts . . ." "his neurotic manifestations . . . ," 890; fifth subject, 891).

255 Relaxation of dietary restriction: Ibid.:76–78, 842–53 ("personality deterioration . . . ," 891.)

255 Last weeks of experiment: Ibid. ("the prodigious level . . . " and "though incapable . . . ," 143; weight and body fat, 182).

255 "conventional reducing diets": Van Gaal 1998.

256 "persistent clamor . . .": Keys, Brozek, et al. 1950:835. Dramatic reduction in energy expenditure: See also Grande et al. 1958.

256 "diminishes proportionately . . .": Strang and Evans 1929. Observation reported by Bray: Bray 1970. See also Bray 1969; Brown and Ohlson 1946. "There is no investigator . . .": Garrow 1978:89.

256 Hirsch's experiments: Leibel et al. 1995.

256 "all the physiological . . .": Interview, Jules Hirsch.

257 "of all the damn . . .": Ibid.

257 "eating fewer calories . . .": USDHHS and USDA 2005:13.

257 Stunkard's analysis: Stunkard and McLaren-Hume 1959 ("remarkably ineffective . . ." and "only 12% . . ."). "paradox" and "the widespread assumption . . .": Anon. 1983.

258 "Attempts at weight reduction . . .": Stunkard 1973.

258 The Cochrane Collaboration review: Pirozzo et al. 2002. The USDA analysis:

Kennedy et al. 2001:419 (table 11). Only one study tracked participants for more than a year: Jeffery et al. 1995. The WHI report on weight: Howard, Manson, et al. 2006.

258 "dietary therapy remains . . .": Van Gaal 1998:875–76.

259 "reduction of caloric intake . . .": Maratos-Flier and Flier 2005:541–42.

259 Sixty or ninety minutes: USDHHS and USDA 2005:viii.

260 "while unusually strenuous . . .": Wilder 1933.

260 "He will have to climb . . .": Newburgh 1942:1085.

260 "Vigorous muscle exercise . . .": Rony 1940:55–56. "There has been ample demonstration . . .": Feinstein 1960:365. *Footnote*. Willett and Stampfer 1998:276.

261 Romance of Mayer's background: Mayer 1955; Gershoff 2001. Interview, Albert Stunkard.

261 Mayer extolling virtues of exercise: Mayer and Stare 1953. "debunked . . .": Tolchin 1959. "almost nil": Mayer 1953b.

261 Mayer on high-school girls: Johnson et al. 1956. "The laws of thermodynamics . . .": Mayer 1968:125–26.

262 Girls at summer camp: Bullen et al. 1964. Infants: Rose and Mayer 1968. "The striking phenomenon . . ." and "some individuals . . .": Mayer 1975b:78.

262 The changing-American-diet story: See Brewster and Jacobson 1978.

262 Descriptions of typical meals: Mayer 1968:77–78. "frequently assumed prodigious . . .": Ross 1987:35–36. "fish, poultry, or game . . .": Quoted in ibid. "The 75-cent special . . ." and "were two or more courses . . .": Schwartz 1986:91.

263 "most important factor . . .": Mayer 1973a. "The development of obesity . . .": Mayer 1968:83.

263 "It is a common observation . . .": Rony 1940:80.

263 "For a long period . . .": Mayer 1965.

264 Mayer's observations on exercise and weight control: Mayer 1968:69–84 ("necessarily," 69). "If exercise is decreased . . .": Quoted in Galton 1961. Mayer's rat study: Mayer et al. 1954. West Bengal study: Mayer et al. 1956. *Footnote*. "J. Mayer has since . . .": Mayer and Thomas 1967.

265 As John Garrow noted: Garrow 1978:48–49. "too little": IOM 2002:884.

265 "As much as Dr. Mayer . . .": Glenn 1965.

265 "false idea . . ." and "facts overwhelmingly demonstrate . . .": Mayer 1968:69.

265 Mayer primary author of Health Service report: Brody 1966.

266 "The successful treatment . . .": Anon. 1969:54. "make weight melt . . .": Mayer and Goldberg 1984. "contrary to popular belief . . .": Mayer and Goldberg 1983.

266 "While it is generally agreed . . .": Gilmore 1977. Seventeen thousand Harvard alumni: Paffenbarger et al. 1978. Exercisers smoke less, attend more to diet: See, for instance, Chave et al. 1978.

266 "Diligent exercisers . . .": Cohn 1980. Brody said exercise "the key": Brody 1983a. "essential" to weight-loss program: Beck 1990. "when exercise isn't enough . . .": Stockton 1989.

267 Björntorp reported: Björntorp 1976. NIH conference report: Rodin 1979 ("the importance of exercise . . .", 57). Pi-Sunyer reviewed the evidence: Segal and Pi-Sunyer 1989. Danish investigators: Janssen et al. 1989.

267 Male rats and exercise: Thomas and Miller 1958. Mayer's rats on motorized treadmill: Mayer et al. 1954. The use of *shock grids*, rats retired from exercise programs: Stern and Lowney 1986.

268 Hamsters and gerbils: Sclafani 1981a ("permanent increases").

268 "When surveying the scientific literature . . .": Stern and Lowney 1986.

268 Ninety minutes of moderate activity: Dietary Guidelines Advisory Committee 2005.
268 Finnish review: Fogelholm and Kukkonen-Harjula 2000.
269 "enemies of exercise" and "pseudo-science": Mayer 1969.

CHAPTER SIXTEEN:
PARADOXES

270 *Epigraph.* "The literature on obesity . . .": Bruch 1957:19.
270 "Those who consume more . . .": Burros 2004a. "chronic condition . . .": Stunkard 1973:32.
270 Hirsch's alternative hypotheses: Hirsch 1985.
271 "for the vast majority . . .": USDHHS 2001:1.
271 Fat deposition different in men and women: See Bauer 1941. "The energy conception . . .": Grafe 1933:148.
271 "similarity may be as striking . . .": Bruch 1957:150. "It is genetics . . .": Friedman 2004. "Infants born to diabetic mothers . . .": Select Committee 1976:137–38.
272 Sheldon commented: Sheldon and Stevens 1942 ("It does not take a science . . . ," 2; ". . . emaciated endomorphs," 8).
272 McGovern's committee in 1977: Select Committee 1977c ("I want to be sure . . ." "constantly hear anecdotes . . . ," 222).
272 Sims's studies: Data and observations from these are scattered over numerous publications; the account given here is taken mostly from Sims 1976; Sims et al. 1973; Goldman et al. 1976; Sims et al. 1968.
273 "marked differences . . .": Sims et al. 1973. "lost weight readily . . .": Quoted in Bennett and Gurin 1982:19. "in response to both . . .": Sims 1976:393.
273 Bouchard's twin study: Bouchard et al. 1990. Levine reported: Levine et al. 1999.
273 "Genetic factors . . .": Bouchard et al. 1990.
273 Animal husbandry: See Mayer 1968:45–46.
274 "Up until that moment . . .": Interview, Ingrid Schmidt.
274 Mayer studied obese mice: Mayer 1968:49.
275 Tanner and Chambers: Tanner 1869b:220–21. *Footnote.* Ibid.:222–23.
275 The paradox developed: For a history of this era of nutrition research, see, for instance, Du Bois 1936:93–125; McCollum 1957:115–33. "The amount of information . . .": Atwater 1888.
275 Rubner discovered: Rubner 1982 ("the effect of specific . . . ," 36). For a biography of Rubner, see Chambers 1952.
276 Benedict and Atwater's experiments: Atwater and Benedict 1899.
276 Von Noorden on obesity: See von Noorden 1907a. "His work contains . . .": Bruch 1957:25.
276 Von Noorden on endogenous and exogenous obesity: von Noorden 1907a: 693–700.
276 "The ingestion of a quantity . . .": Ibid.:693.
277 "far the most common . . .": Ibid.:697.
277 "an extremely popular . . .": Rosenberg 1981.
278 Magnus-Levy had reported: Magnus-Levy 1907:261–62. Lean body mass also increases: See, for instance, James et al. 1978. *Footnote.* Von Noorden 1907a: 699–701.
278 *Metabolic rate:* Magnus-Levy 1907:262. "the metabolism of the individual . . .": Stock and Rothwell 1982:39–40.

278 Benedict's studies of basal metabolism: Benedict and Emmes 1915.

279 Bruch began her career: Bruch 1957:5–6.

279 Bruch published: Bruch 1940 ("The terms used . . . ," 747–48); Bruch and Touraine 1940. "Overeating was often vigorously . . .": Bruch 1973:136.

279 "excessive eating . . .": Bruch and Touraine 1940:141. "keep the child . . .": Bruch 1973:136. "giving up [their] . . .": Bruch 1944. *Footnote*. See, for instance, Brownell and Horgen 2004:8.

280 "perverted appetite . . .": Newburgh and Johnston 1930b. "various human weaknesses . . .": Newburgh and Johnston 1930a.

280 In 1942, Newburgh published: Newburgh 1942 ("endocrine disorder," 1058–73; "his gluttonous habits . . . ," "without any effort . . . ," 1094–95).

281 By the end of Newburgh's review: Ibid. ("it might be true . . . ," 1075; "a more realistic . . . ," 1074; "the secretions . . . ," "She does not resist . . . ," 1079).

281 "The work of Newburgh . . ." and "Newburgh answered that": Anon. 1955a.

281 Fundamental flaw: Burros 2004a. Von Noorden's proposition: Von Noorden 1907a:697. "excess calorie consumption . . .": USDHHS 2001.

281 "If eating behavior did not . . .": Bennett 1987. Big eaters: See, for instance, Waterlow 1986.

282 Salient question in obesity research: Bennett 1987. *Footnote*. Ibid.

282 "is energy balance achieved . . .": Ibid. "An obese person . . .": Rony 1940:47–48.

283 "On the few occasions . . .": Durnin and Passmore 1967:132–33.

283 "Talking of a man . . .": Boswell 1992:1086.

284 We could find two such men: See, for instance, Widdowson 1962. Stunkard's 1959 analysis: Stunkard and McClaren-Hume 1959. "Eventually, calorie balance . . .": Keys and Brozek 1953:311.

285 "This procedure alters . . .": Friedman 2004.

285 "Theories that diseases are . . .": Sontag 1990:55. "playable game": Kuhn 1970:90.

285 Von Noorden sought: Von Noorden 1907a:694–97.

286 "As long as we have . . .": Interview, Kelly Brownell. "Our culture's apparent obsession . . .": Hill and Peters 1998.

286 "why then do we not . . .": Quoted in Rony 1940:201. *Footnote*. Starbucks Coffee Company 2006.

287 Stunkard wrote: Stunkard and McClaren-Hume 1959. Mayer also ridiculed: Mayer 1955.

287 "Obesity is not a sin . . .": Mayer 1968:165.

288 "Fat Americans . . .": Spark 1973.

288 "the combination of weak will . . .": Newburgh and Johnston 1930a:212. "It exists in many non-obese . . .": Rony 1940:63.

289 Ravussin's basal-metabolic study: Ravussin et al. 1988. The observation in infants: Roberts et al. 1988. "If obesity was only caused . . .": Ravussin 1993.

289 "I was shocked . . .": Interview, Eric Ravussin.

289 "revolution in thinking . . .": Whalen 1950. Never embraced this conclusion: See Bruch's comments in Anon. 1955a:123–24.

290 "The literature on behavior . . .": Bruch and Touraine 1940:204. "Life situations": Bruch 1940:770.

290 "critical re-evaluation . . .": Bruch 1957:19. "The efficacy of any treatment . . .": Bruch 1940:775. "When I began to work . . .": Bruch 1957:150–51.

291 "adequately established . . .": Bruch 1957:11–12.

291 "primary metabolic . . ." and "Studies of human obesity . . .": Bruch 1973:32.

CHAPTER SEVENTEEN:
CONSERVATION OF ENERGY

292 *Epigraph.* "The complicated mechanism . . .": Rubner 1982:8.
292 Jean Mayer observed: See Mayer 1954:41–43. See also Mayer 1968.
292 "The fact remains . . .": Brody 1999b.
292 "Let me state . . .": Anon. 1955a:111.
293 "a plethora of calories": MacBryde 1951:657.
294 "primary metabolic . . .": Bruch 1973:32.
294 "separate recognized facts . . .": Rony 1940:6.
294 Compared with growing children: Ibid.:47–49.
295 Reverse causation, pregnancy, and weight in animals: Wade and Schneider 1992.
296 "The statement that primary increase . . .": Rony 1940:58–59.
296 Studies of the *pre-obese:* Ravussin et al. 1988; Roberts et al. 1988. See also Ravussin and Swinburn 1992.
297 "We cannot get away . . .": Yudkin 1959.
297 "for most adults . . .": USDHHS and USDA 2005:14. "There is only one trouble . . .": Bruch 1957:25–26.
297 If we consume an average: Miller and Mumford 1966. "It is conceivable . . .": Du Bois 1936:237.
298 "no stranger phenomenon . . .": Ibid.:252.
298 Gordon Kennedy discussed: Kennedy 1961.
298 "multiple metabolic control mechanisms": Cahill and Renold 1965.
298 Intake and expenditure as dependent variables: See, for instance, Lusk 1928:170–74; Grafe 1933:136–46; Du Bois 1936:231–69; Kleiber 1961:266–90.
299 "Changes in . . . hormones . . .": Shetty 1999. "dieting is difficult . . .": Frayn 1996:245.
299 "usual symptoms . . .": Keys, Brozek et al. 1950:884.
300 "spontaneous impulses . . .": Rony 1940:48.
300 "clue to the puzzle" and "react exactly like . . .": Anon. 1955a:124. Child's growth stunted: See, for instance, Ashworth et al. 1968.
300 *Luxuskonsumption:* See Du Bois 1936:262–69 and Grafe 1933:139–46 for pre–World War II reviews. Term first used by Neumann: Neumann 1902. A modern review of Neumann's work can be found in Bennett and Gurin 1982:79–82.
300 "Food in excess . . ." and "well within the capacity . . .": Lyon and Dunlop 1932.
301 "The idea that people . . .": Garrow 1981:53. Pigs fed low-protein diet: Miller and Payne 1962. Survival advantage: See Sims 1976. *Footnote.* Baxter 1976.
301 "thermochemical tangle . . .": Rubner 1982:329. For a modern discussion of diet-induced thermogenesis, see Schutz and Jéquier 1998. Rationale for high-protein diets: See, for instance, Jolliffe 1952:48.
302 As the external environment changes: Rubner 1982:36,329.
302 Rubner argued: See Krebs 1960. Voit believed: see Du Bois 1936:236 ("impulse to . . .").
302 Thrifty and spendthrift metabolisms: Pennington 1953b is perhaps the best post–World War II discussion of this concept.
303 "total self-regulatory functions . . .": Richter 1976:222.
304 "When well nourished . . ." and "a gradual lowering . . .": Du Bois 1936:254–55.
304 "The appetite mechanism . . ." and "pathological changes . . .": Rony 1940:203.
304 "Whatever may be the mechanisms . . .": Lepkovsky 1948:113.

CHAPTER EIGHTEEN: FATTENING DIETS

305 *Epigraph.* "Oversupply of food . . .": French 1907:14.
305 Speke's travels: Speke 1969 ("such an extent . . ." ". . . no mistake . . . ," 172; "another one . . . ," 189–90).
305 John Garrow attempted: Garrow 1978:70. "whenever the prospect . . ." "I learned that . . .": Interview, John Garrow.
306 Fattening with milk: Speke 1969:172, 189–90. Fattening sessions of the Massa: Garine and Koppert 1991. Ritual fattening ceremonies have been documented among primitive populations throughout Africa and the South Pacific, but it's rare that the reports actually document what was eaten. The Massa and sumo were the only two examples I found in which the composition of whose diet is reported in any detail.
306 Nishizawa on the sumo: Nishizawa et al. 1976. *Footnote:* Ibid.
307 Low-fat diets recommended for weight loss: See, for instance, IOM 1995:109–11; NRC 1989:671. Evidence linking dietary fat consumption to obesity: NRC 1989:567. *Footnote.* Ibid.
308 "I could feed them . . .": Interview, George Bray.
308 Fattening rats on fat and carbohydrates: Sclafani 1980. Fattening monkeys: Interview, Barbara Hansen.
308 Sclafani demonstrated: Sclafani 1980; interview, Anthony Sclafani.
308 Fattening with Crisco: Interview, Anthony Sclafani.
309 Seminal experiments by Adolph: Adolph 1947.
309 Thermodynamic evidence: See von Noorden 1907b:62–64. Flatt's calculations: Flatt 1978. Sims and Danforth explained: Sims and Danforth 1987.
310 Sims's overfeeding studies: Horton et al. 1974 ("obvious question," 233). See also Sims et al. 1968; Sims et al. 1973; Goldman et al. 1976; Sims 1976.
310 Sims and Danforth believed: See Sims and Danforth 1974. "Simply stated . . ." and "In view of these . . .": Danforth 1985:1137. See also Sims and Danforth 1987.
311 "plates of pork chops . . .": Interview, Edward Horton. "The bottom line . . .": Interview, Elliot Danforth.
311 "difficult assignment . . .": Sims and Danforth 1974. "hunger late in the day . . .": Goldman et al. 1976:176.
311 "marked anorexia": Goldman et al. 1976:166.

CHAPTER NINETEEN: REDUCING DIETS

313 *Epigraph.* "Concentrated carbohydrates . . .": Reader et al. 1952.
313 "fad diet": AHA 2005: front jacket flap. "In the instruction . . .": Newburgh 1942:1087.
313 Stanford diet: Cutting 1943. Harvard diet: Williams et al. 1948. Chicago diet: Steiner 1950 ("general rules"). Cornell diet: Reader et al. 1952.
314 Keeping the body in nitrogen equilibrium: See, for instance, Preble 1915.
314 "the inclusion of . . .": Campbell 1936. "All forms of bread . . .": Gardiner-Hill 1925.
315 Lean meat meant any meat: See, for instance, Steiner 1950.
315 Evans's weight-maintenance diet: Evans 1947:582.
315 Evans's very low-calorie diet: Strang et al. 1930 ("composed of . . ."). "The secret of the success . . .": Wilder 1933. "No concession . . .": Evans 1953.
316 "The next question to decide . . .": Croftan 1906.

316 Dunlop believed: Dunlop and Murray-Lyon 1931.

316 Observation echoed: Anderson 1935; Bruch 1944:361–64; Rony 1940:59–62 ("It is easier . . . ," 62); Williams et al. 1948. Common rationale: See, for instance, Evans and Strang 1931; Lyon and Dunlop 1931. *Footnote.* Sidbury and Schwartz 1975.

317 Evans restricted carbohydrates almost entirely: Strang et al. 1930. Newburgh concluded: Newburgh 1942:1094–95. *Footnote.* Strang et al. 1930.

317 "tendency to retain water . . .": Lyon and Dunlop 1932:337. "Changes in body-weight . . .": Benedict and Carpenter 1910:110–12.

318 First meaningful report: Hanssen 1936. Results from University Clinic: Moller 1931.

318 ". . . hungry most of the time . . .": Evans 1953:132. "relatively poor . . .": Hanssen 1936.

319 Nutritionists will insist: See, for instance, Mayer 1974b.

319 Carbohydrates not the only source of glucose: See, for instance, Harper 1971:249. Ketone-body levels in diabetes, etc.: Van Itallie and Nufert 2003.

319 Modified by Harvey and Niemeyer: Harvey 1872. By Oertel: Oertel 1895. By Ebstein: Ebstein 1884 ("the fat of ham . . . ," 33).

320 Stefannson and the Inuit: See, for instance, Stefansson 1936; Stefansson 1946; Stefansson 1960b.

320 "with perhaps 30 percent fish . . .": Stefansson 1946:22. "not proper human food": Stefansson 1960b:33.

320 Inuit paid little attention: Jenness 1959 (". . . added nothing . . . ," 110; "no fruit . . . ," 191). See also Freuchen 1961:9–11, 142. "we lived upon . . .": Dana 1946:251–52.

320 *Newer Knowledge:* McCollum 1922. "A safe rule of thumb . . .": Pellet 1987:164.

321 Deficiency diseases: McCollum 1957 ("water gruel sweetened . . . ," 252–54; Beriberi in the Japanese navy, 188–89; Voegtlin's experiment, 303). See also Carpenter 1986; Carpenter 2000. Guinea pigs given scurvy: Bannerjee 1945.

321 "protective foods": See, for instance, McCarrison 1922.

321 Animal foods contain all: Harris 1985:35–36. See also Abrams 1987:231; Davidson and Passmore 1963:192–252. *Footnote.* Harris 1985:35–36.

322 "a dietery deficiency . . .": Tso 1997:32.

322 Stefansson argued: Stefansson 1936.

323 Research by Louis Newburgh: Newburgh 1923; Newburgh et al. 1930. Anon 1930a; Moulton 1930; Newburgh 1931b. "easier to believe . . .": Stefansson 1946:68.

323 In the winter: Anon. 1928.

323 For three weeks: Stefansson 1946:60–89; Stefansson 1936. "In every individual specimen . . .": McClellan and Du Bois 1930.

324 Inuit and raw meat: See Mowat 1978:96. Calories and nutrients consumed: Lieb 1929.

324 "The only dramatic part . . .": Du Bois 1946:xii. "Both men were in . . .": McClellan and Du Bois 1930. The other eight articles are Lieb 1929; Lieb and Tolstoi 1929; McClellan, Rupp, et al. 1930; McClellan et al. 1931; Tolstoi 1929a; Tolstoi 1929b; Torrey 1930; Torrey and Montu 1931.

324 "Mr. Stefansson makes . . .": Garside 1946. Du Bois's introduction: Du Bois 1946 ("a great many dire . . . ," xii; "Quite evidently . . . ," x).

325 B vitamins depleted from the body: See Carpenter 2000:213–18. ". . . an increased need . . .": Select Committee 1973b:43 44. Vitamin C in Type 2 diabetes: Will and Byers 1996 ("biologically plausible . . ."). Metabolic syndrome and vitamin C: Ford

et al. 2003. See also Bode 1997. My discussion of vitamin deficiencies is also based on interviews with Betti Jane Burri, Tim Byers, Kenneth Carpenter, John Cunningham, and Theodore Van Itallie.

325 Vitamin C similar to glucose: Will and Byers 1996; Basu and Schlorah 1982:121. Glucose and vitamin C compete: Cunningham 1988 ("globally inhibited"); Cunningham 1998. "marked fall": Cox et al. 1974. *Footnote.* Carpenter 1986: 200–204.

CHAPTER TWENTY:
UNCONVENTIONAL DIETS

327 *Epigraphs.* "Here was a treatment . . .": Pennington 1954. "Does it help people . . .": Brody 2002.
327 Liebling's three kinds of writers: Liebling 1975:317.
328 Bruch's fine-boned girl: Bruch 1957:372–73. Kuhn on process of discovery: Kuhn 1970:52–53.
328 Donaldson's history: From his obituary in the *New York Times*, Anon. 1966; and his memoirs, Donaldson 1962 ("fat cardiacs," 32; "the fattest meat . . . ," 34; "hotel portion," 35).
329 Over four decades: Mackarness 1975:63–65. Patients who didn't lose weight: Donaldson 1962 ("bread addiction," 67; "Remember that grapefruit . . . ," 66; "You are out of your mind . . . ," 103).
329 Pennington heard of Donaldson: Pennington 1952.
329 Gehrmann and Du Pont's industrial-medicine division: Kehoe 1960. Gehrmann was prompted: Woody 1950. *Footnote.* Quoted in Pennington 1951b.
330 "We had urged . . .": Quoted in Woody 1950.
330 "Notable was a lack . . .": Pennington 1949. "In a few cases . . .": Pennington 1953c. *Footnote.* Pennington 1952 ("even an apple").
330 *Holiday* magazine: Woody 1950. Harvard symposium: Barr et al. 1953 (Hegsted's comments, 137).
331 Pennington in *NEJM*: Pennington 1953c. *JAMA* took the position: Anon. 1952. *Lancet:* Anon. 1953.
331 "a surprisingly large . . .": Hamlyn 1953. "Pennington's idea . . .": Greene 1953. "Pennington has hardly proved . . .": Anon. 1954.
331 Thorpe at AMA meeting: Thorpe 1957.
332 *JAMA* still insisted: Anon. 1958 ("adequate in all . . . ," "the most reasonable . . . ," "least harmful"). *Footnote.* See White 1962; Anon. 1973.
332 "The edibility . . . ,": Ohlson et al. 1955:173. *Footnote.* Ibid.
333 Ohlson initially tested: Cederquist et al. 1952 ("subjects reported lack . . .").
333 "Without exception . . .": Ibid.
333 Over a ten-year period: Ohlson et al. 1955 ("dry, uninteresting . . . ," "sufficient to provide . . . ," "uniformly," "did not appear . . . ," "they also appeared . . . ," 185).
334 "can only mean that replenishment . . .": Ibid.:177.
334 Young tested diet on women: Young 1952 ("unanimous in saying . . . ," "despite an unusually," "reported that their skins . . . ," "No excessive fatigue . . ."); Young et al. 1953. Male students: Young et al. 1957 ("in every case").
335 Bruch noted: Bruch 1957:353,371–76.
335 Leith reported: Leith 1961.
336 "To be honest . . .": Interview, Per Björntorp. "There's no point . . .": Interview, George Bray.

494 NOTES

336 "excessive fatigue . . .": Cederquist et al. 1952.
336 "The absence of complaints . . .": Wilder 1933 Bistrian and Blackburn reported: Palgi et al. 1985. "People loved it": Interview, George Blackburn. For confirmation of the absence of hunger, see Wadden et al. 1985. *Footnote.* Dietz 1989.
337 1000 calories: Rabast et al. 1978; Rabast et al. 1979. 1,200 calories: Krehl et al. 1967. 1,320 calories: Gordon et al. 1963. 2,200 calories: Palmgren and Sjövall 1957. More than 2,700 calories: Milch et al. 1957. "encouraged to eat . . .": LaRosa et al. 1980.
337 Comparisons with low-calorie diets: Hanssen 1936; Palmgren and Sjövall 1957; Silverstone and Lockead 1963; Peña et al. 1979 ("eat as much . . .").
337 Kemp's three papers: Kemp 1963 ("One is that other . . . ," "a working hypothesis . . . ," "possible for the first time . . ."); Kemp 1966; Kemp 1972.
338 Beginning in 1956: Kemp 1972.
338 Atkins diet: Atkins 1972. The five adult studies: Brehm et al. 2003; Foster et al. 2003; Samaha et al. 2003; Yancy et al. 2004; Gardner et al. 2007 [the Stanford study]. The adolescent study: Sondike et al. 2003.

In two of these studies—Foster et al. 2003, and Gardner et al. 2007—the investigators also included a comparison of weight maintenance at the end of a year. In the former, those randomized to the Atkins diet maintained a greater weight loss than those assigned to the low-calorie, low-fat diet (4.4 ± 6.7 vs. 2.5 ± 6.3 percent of initial body weight), but the difference was not "statistically significant." In the latter, those randomized to the Atkins diet maintained, on average, a 10.4-pound weight loss, compared with 5.7 pounds for a "lifestyle" program that included both a low-fat, low-calorie diet and exercise. The difference between these numbers and the larger numbers that Kemp reported could be due to the fact that these modern studies were more rigorous in following up on patients and measuring weight. It could be because the modern studies provided no counseling after the first two months of the trial, whereas Kemp categorized patients as having "defaulted" if they did not continue to appear at his monthly counseling sessions. Perhaps for this reason, the carbohydrate restriction after the first couple of months in the modern trials was at best modest. In Gardner et al. 2007—see p. 973 (table 2)—the subjects randomized to the Atkins diet consumed, on average, 30 percent of their calories from carbohydrates at six months, and 35 percent at twelve. This problem was particularly significant in the one trial I omitted: Dansinger et al. 2005:46 (table 2). Here the subjects randomized to the Atkins diet consumed an average of 137 grams of carbohydrates per day (32 percent of calories) after only two months of the trial, and 190 grams (over 40 percent) at six months and a year—the equivalent of four to five large baked potatoes every day—effectively identical to subjects randomized to low-fat, low-calorie diets. Thus, the similarity in weight loss on the different diets in this particular trial may simply reflect the similarity in carbohydrate consumption.
338 "first published synthesis . . .": Bravata et al. 2003.
339 "Calories are all alike . . .": Quoted in Berland 1983:7. "the effect of specific . . .": Rubner 1982:36.
339 Bistrian and Blackburn instructed: Palgi et al. 1985. "thousands of patients . . .": Interview, Bruce Bistrian.
340 Paradox relating to hunger: Leith 1961 ("nagging discomfort"); Peña et al. 1979; Hanssen 1936; Krehl et al. 1967 ("more than amply satisfied"). "Isn't the proof . . .": Interview, Bruce Bistrian.
341 Sims's overfeeding experiments. Goldman et al. 1976:167.
341 Bloom's articles on starvation therapy: Azar and Bloom 1963 ("At a cellular level . . ."); Bloom and Azar 1963.

341 "little hunger": Bloom 1958. "In total starvation . . .": Keys, Brozek, et al. 1950:829. "The most astonishing aspect . . .": Drenick et al. 1964. "The gratifying weight loss . . .": Anon. 1964c.

342 "mighty stimulant . . .": Pennington 1954. "Many individuals spontaneously . . .": Dwyer 1985:185.

342 "The fact remains . . .": Anon. 1973.

343 "If you put a restaurant-size . . .": Mayer 1975a:30–31. Fallback for the NIH recommendation: Ernst and Levy 1984:733–34; interview, William Harlan.

343 "Yudkin showed that . . .": Interview, George Bray. Two papers: Yudkin and Carey 1960 [six subjects]; Stock and Yudkin 1970 [eleven subjects].

343 Yudkin explained: Yudkin 1958 ("The irrefutable, unarguable . . . ," 59; "much of the extra fat today . . . ," 141; fat calories will come down, 149).

343 Experimental evidence: Yudkin and Carey 1960.

344 Subjects losing weight consuming considerable calories: Milch et al. 1957; Werner 1955; Rilliet 1954 ("numerous and encouraging").

345 "all-you-can-eat diet[s] . . .": Brody 1981c. "The best definition": Keys, Brozek et al. 1950:32.

345 "persistent clamor of hunger": Keys, Brozek, et al. 1950:835. "nonappetizing nature": Cahill 1975:58–59.

346 "token" amounts: Keys, Brozek, et al. 1950:74.

346 "their appetite-depressing . . .": Spark 1973. "Substances called ketones . . .": Brody 1996.

346 The existing research refutes ketone hypothesis: Drenick et al. 1964 ("It is not clear . . ." ". . . did not reappear"); Sidbury and Schwartz 1975. See also Kinsell 1969.

347 "these foods digest . . .": Brody 2002. Even those investigators: Werner 1955; Kinsell et al. 1964 ("There is a good reason . . .").

347 Yudkin had struggled: Yudkin and Carey 1960 ("for reasons . . ." "It would seem . . .").

347 "It is better . . .": Bernard 1957:37. "inevitability": Yudkin and Carey 1960. "Claims that weight loss . . .": White and Selvey 1974:48.

348 Physicians who took Pennington seriously: Thorpe 1957; Taller 1961.

348 Pennington set out: Pennington 1954 ("voluminous . . . ," "meager . . . ," "These tended . . ."). See also Pennington 1951b.

349 Something Benedict suggested: Benedict 1925:57. And Du Bois believed: Du Bois 1936:254–55. "index of calorie nutrition . . .": Pennington 1953b.

349 "static phase": Rony 1940:47. "His caloric intake . . .": Pennington 1952.

349 Diet-induced decrease: Benedict et al. 1919:694–95; Strang and Evans 1929; Brown and Ohlson 1946. Lusk suggested: Lusk 1928:173. "their tissues are not . . .": Pennington 1952.

349 A conundrum: Pennington 1953d. Stetten reported: Salcedo and Stetten 1943.

350 Applying the same law of energy conservation: Pennington 1952.

350 "the size of the adipose deposits . . .": Ibid.

350 "provides for a more effective . . .": Ibid.

350 "It dawned on me . . ." and "like clockwork": Pennington 1954.

350 Defect explains sedentary behavior: Pennington 1951a.

351 Pennington explained this wasn't the case: Ibid.

351 Consider the kind: The example of Keys's conscientious objectors is my own, based on Pennington 1951a.

352 Maintain weight at seventeen hundred calories: Keys 1949.

352 "What happens when . . ." and "The first noticeable effect . . .": Pennington 1951a.

352 "A more rational form...": Pennington 1953d. directs "measures primarily toward...": Pennington 1951a.

352 Healthy equilibrium reestablished: Pennington 1953b ("Mobilization of increased..."); Adolph 1947; Richter 1976;

353 If fat can be mobilized: Pennington 1953c ("sufficient effectiveness," "no calorie restriction...," "Weight would be lost...," "The result would...").

353 Energy expenditure would increase: Pennington 1953a. Du Bois's observation: McClellan et al. 1931.

354 Four thousand calories a day: Evans in Newburgh 1931a; Werner 1955. Might eat three thousand calories: Pennington 1953b.

354 Pyruvic acid: Pennington 1955. His contemporaries dismiss him: see Yudkin 1959.

CHAPTER TWENTY-ONE:
THE CARBOHYDRATE HYPOTHESIS, I: FAT METABOLISM

355 *Epigraph.* "Looking at obesity...": Bruch 1957:147–48.

355 Astwood discovered: Anon. 1976 ("a brilliant series..."); Cassidy 1976 ("a record perhaps...").

355 "The Heritage of Corpulence:" Astwood 1962.

357 "regulation of ingestive behaviors...": Greenwood 1985:20.

358 "The vast majority...": Ibid.

358 "Something has happened...": Interview, George Cahill.

359 Bergmann's *lipophilia* hypothesis: Bergmann and Stroebe 1927 ("It seems just as illogical....," 593–94). I'm grateful to Richard Frank and Haidi Kuhn Segal for the translation. See also Bauer 1941; Rony 1940:159–75.

360 "A second operation...": Bauer 1941.

361 Case reported in 1913: Rony 1940 ("Adiposity of the lower body," 170–71).

361 "noted Vienna authority...": Anon. 1930b. Bauer's expertise: Anon. 1979. See also Bauer 1945. My primary source for Bauer's observations on obesity is Bauer 1941. "The genes responsible...": Bauer 1940.

362 "A local factor must exist...": Bauer 1941:975.

362 "Like a malignant tumor...": Quoted in ibid.:978.

363 "obese boys in whom...": Ibid.:980.

363 Grafe's textbook: Grafe 1933. "more or less fully accepted": Rony 1940:173–74. "... this conception deserves...": Wilder and Wilbur 1938:310–11.

364 1955 German textbook chapter: Bahner 1955:1023–26. References from German literature: Rony 1940; Rynearson and Gastineau 1949. *Footnote.* Interview, Theodore Van Itallie.

364 Bauer's articles in English: Silver and Bauer 1931; Bauer 1940; Bauer 1941. Newburgh's seminal paper: Newburgh 1942.

365 "indubitable" and "is also probably present...": Cahill 1978.

365 "significantly more weight": Lee and Schaffer 1934. For a similar experiment, see Marx et al. 1942.

366 "These mice will make fat...": Mayer 1968:48. Benedict reported this: discussed in Alonso and Maren 1955, which reported confirmation of the observation in a different strain of mice.

366 Greenwood's Zucker rat studies: Greenwood et al. 1981.

367 Hypothalamic tumor in 1840: Brobeck 1946. Nicolaidis recounted: Interview, Stylianos Nicolaidis.

367 Hypothalamic research in its early years: See Brobeck et al. 1943; Magoun and Fisher 1980. Hetherington and Ranson resolved controversy: Hetherington and Ranson 1939.

367 Brobeck's research: Brobeck et al. 1943 ("the laws of thermodynamics . . . ," 836).

368 Ranson interpreted: Hetherington and Ranson 1942 ("the tremendously de-creased . . ."). "related to the feeding habits": Brobeck et al. 1943:842. *Footnote.* Ibid.

368 Ranson argued: Hetherington and Ranson 1942:615.

369 "concertmaster . . .": Anon. 1940.

369 Ranson studied fluid balance and diabetes insipidus: Fisher et al. 1938:1–2.

369 Hypothalamic lesions cause diabetes insipidus: Ibid.

370 "classic type of experimental obesities": Mayer 1953a. Teitelbaum's experience: Teit-elbaum 1955; interview, Philip Teitelbaum ("Of course they overate . . .").

371 Lesioning the lateral hypothalamus: Anand and Brobeck 1951. Ransom's lab had reported: Magoun and Fisher 1980.

371 Hetherington did research for U.S. Air Force: Interview, John Brobeck. Later edi-tions of Ranson's textbook: See Ranson and Clark 1964:311.

371 Hypothalamus as regulator of eating *behavior:* See, for instance, Sutin 1976; Schachter and Rodin 1974:75–83. Psychologists would "discard": Sclafani 1981b: 409.

372 Brooks reported: Brooks 1946.

372 Brooks could only do so: Brooks 1946; Brooks and Lambert 1946 ("severe and per-manent . . . ," 700; "followed by an augmentation . . . ," 707).

372 Studying hibernators: See Mrosovsky 1976.

372 Dietary models of obesity: Sclafani 1987 (high-sugar diets); Oscai et al. 1984 (high-fat). See also Wade 1982. Regaining weight after fasting: Levitsky et al. 1976. "It doesn't matter . . .": Interview, Irving Faust. Transgenic animals: See, for instance, Bluher et al. 2003; Cohen et al. 2002.

373 Removed ovaries from rats: For an excellent review of this work and the entire field of weight regulation and reproduction in mammals, see Wade and Schneider 1992. It was my interview with George Wade that opened my eyes to the reverse-causality hypothesis of weight gain.

373 "revelation": Interview, George Wade. "The animals overeat and get fat . . .": Inter-view, Tim Bartness.

374 "Hard living . . . retards . . .": Darwin 2004:56. "Fertility is linked . . . ," "partition-ing and utilization . . . ," and "reciprocal, redundant . . .": Wade and Schneider 1992:235–36.

375 Newburgh still promoting his hypothesis: Newburgh 1948. "an excuse for avoid-ance . . .": Rynearson and Gastineau 1949:42.

375 "being ruthless in self-criticism . . .": Krebs 1967.

CHAPTER TWENTY-TWO:
THE CARBOHYDRATE HYPOTHESIS, II: INSULIN

376 *Epigraphs.* "Every woman knows . . .": Passmore and Swindells 1963:331. "The fact that insulin increases . . .": Haist and Best 1966:1350.

376 Newburgh rejected "endocrine abnormality": Newburgh, 1929 lecture, in New-burgh and Johnston 1930a. "theories that attributed obesity . . .": Anon. 1955b.

377 Mayer pointed out: Mayer 1968:67–68.

377 Von Noorden suggested: Von Noorden, 1907c:61–62.

378 As early as 1923: Rony 1940:228.

378 Falta argued in pre-insulin era: Falta 1923:583–84. ("A functionally intact pancreas is necessary for fattening," is more commonly translated as "For fattening, therefore, is necessary a functionally intact pancreas.") Falta argued after insulin discovered: Rony 1940:289.

379 Clinicians in Europe using insulin: Grafe 1933:75–76. "rich in carbohydrates . . .": Rony 1940:289–90. Insulin for depression and schizophrenia: See Rinkel and Himwich 1959. "all the patients gained weight": Nasar 1998:293. "drastic increase": Butscher 2003: 122. *The Bell Jar:* Plath 1996 (twenty pounds, 237; "fatter and fatter," 192).

379 Insulin therapy for diabetics: Jacobson et al. 1994:444; Carlson and Campbell 1993. Rosenzweig portrayed: Rosenzweig 1994:483–84. *Footnote.* Ibid.

380 "an excellent fattening substance": Grafe 1933:75–76. Newburgh insisted: Newburgh 1942:1082–83. See also Conn 1944.

380 Rony reviewed: Rony 1940:115.

381 Clinical investigators would state: Rynearson and Gastineau 1949:34–35. See also Jolliffe 1963:15.

381 McGarry on Minkowski: McGarry 1992.

382 "garbage can": Interview, Bernard Jeanrenaud.

382 "Until recently . . .": Bruch 1957:148. "amazing how little . . .": Bruch 1973:6.

382 "the time-honored assumption . . .": Bruch 1957:148. The first phase: See Wertheimer and Shapiro 1948:452–53 ("no marked quantity . . ." "abundant").

383 Schoenheimer's life and work: Clarke 1941. With David Rittenberg: See Schoenheimer 1961.

383 Their discoveries: Schoenheimer 1961 ("indistinguishable . . . ," 56). Wertheimer's seminal review: Wertheimer and Shapiro 1948 ("Mobilization and deposition," "The 'classical theory . . . ,' " "the lowering of the fat . . . ," 454). *Footnote.* Renold and Cahill 1965a:1–3.

384 "a factor acting directly . . .": Wertheimer and Shapiro 1948:454.

384 Krebs cycle: Krebs 1981 ("the main energy source . . ." "All three major . . . ," 114).

385 "The high degree of metabolic . . .": Bruch 1957:155–56.

385 Path of events to obesity: Ibid. ("the big question . . . ," 156; "Since it is now . . . ," 158).

385 "the fat streams . . .": Magnus-Levy 1907:164. "small component . . .": Benedict 1915. Nutritionists insisted: Cahill and Owen 1968; Newsholme and Start 1973: 212–13.

386 The 1956 papers: Dole 1956; Gordon and Cherkes 1956 ("relation to the need" "the anticipated need"); Laurell 1956.

386 APS *Handbook:* Renold and Cahill 1965. 50 to 70 percent: Fritz 1961. "Adipose tissue is no longer . . .": Renold et al. 1965. An excellent review of the regulation of fat metabolism and adipose tissue is Newsholme and Start 1973:195–246.

387 "This lipogenesis is regulated . . .": Wertheimer 1965:6. *Footnote.* Hollifield and Parson 1965.

387 Second critical point: See Newsholme and Start 1973:197–98.

388 "a ceaseless stream . . .": Brodie et al. 1965:584.

388 half the triglycerides not used for fuel: Reshef et al. 2003. "The storage of triglyceride fat . . .": Gordon 1969:329–30.

388 *Glycerol phosphate:* For a review of the role of this molecule, the triglyceride/fatty-

acid cycle, and the glucose/fatty acid cycle, see Newsholme and Start 1973:214–34. "so that they are unable . . .": Gordon 1970:242.

389 Randle cycle: Randle et al. 1963 is Randle's seminal paper on the glucose/fatty acid cycle.

390 "even in trace amounts . . .": Wertheimer and Shafrir 1960:483.

390 "the principal regulator . . ." and "only the negative . . .": Berson and Yalow 1965:561.

390 Effects of other hormones suppressed by insulin: Gordon 1970; Fritz 1961. Anything that increases insulin: Berson and Yalow 1965. The list of hormones that promote fat mobilization and accumulation is from Steinberg and Vaughn 1965.

391 Insulin secretion in VMH-lesioned animals: Han et al. 1965; Han 1968; Frohman et al. 1969; Han and Frohman 1970. "off the scale": Powley 1977. Severing the vagus nerve: Hustvedt and Lovo 1972. Hypersecretion of insulin: Assimacopoulos-Jeannet and Jeanrenaud 1976. Footnote. Bray 1984.

392 "overwhelming": Woods and Porte 1976:275.

392 "relative unavailability . . .": Gordon 1964:1295.

392 Amount of glycerol phosphate available: Margolis and Vaughan 1962; Renold and Cahill 1965b.

392 "It may be stated categorically . . .": Gordon et al. 1963.

393 "Carbohydrate is driving insulin . . .": Interview, George Cahill.

393 Fructose converted more efficiently: Havel 2005:135–36.

393 "exquisitely sensitive": See, for instance, Cahill and Owen 1968:112. See also Cahill et al. 1959; Wertheimer and Shafrir 1960; Zierler and Rabinowitz 1964. Even low levels of insulin: See Bray 1976a:121.

393 Fat cells remain sensitive: See Berson and Yalow 1965; Neel 1982; McGarry 1992.

394 "greatly exaggerated" insulin response: Rabinowitz and Zierler 1962 and 1961.

394 Diabetogenous-obesity hypothesis: Von Noorden, 1907c:61–62. "We generally accept": Berson and Yalow 1965:554.

394 "great biologic variation" and "insulin-secretory responses": Ibid.:555.

394 Diabetologists and endocrinologists have speculated: Berson and Yalow 1965; McGarry 1992.

395 Neel's three scenarios: Neel 1982.

396 Investigators measure on whole-body level: Interview, Eric Ravussin. ADA rationale for carbohydrate-rich diet: See, for instance, Franz et al. 2003. Reported by Bierman and Brunzell: Brunzell et al. 1971.

396 Sims's obesity studies: Bray 1972; Sims et al. 1973; Salans et al. 1974.

397 "mask" the diabetes: Von Noorden 1907c:61. Reproduced in animals: Maegawa et al. 1986. Brunzell refuses: Interview, John Brunzell.

397 "exaggerated tendency . . .": Silver and Bauer 1931.

397 Lipoprotein lipase: For a review of how LPL regulates use of fatty acids, see, for instance, Newsholme and Leech 1983:246–99. A more recent review can be found in Merkel et al. 2002.

398 Orchestration of LPL activity: Arner et al. 1981; Smith 1985; Rebuffé-Scrive 1987; Arner and Eckel 1998.

398 LPL is where insulin and sex hormones interact: Smith 1985; Björntorp 1985. Testosterone and LPL: Rebuffé-Scrive 1987. Progesterone: Greenwood et al. 1987. Estrogen: Rebuffé-Scrive et al. 1986. Changing fat deposition with pregnancy: Lithell 1987; Greenwood et al. 1987.

399 Greenwood's "gatekeeper hypothesis" and Zucker-rat studies: Greenwood et al. 1981.

399   LPL gatekeeper hypothesis, researchers reported: Kern et al. 1990; Eckel 2003; Arner and Eckel 1998 ("sufficiently altered"). During exercise: Kiens et al. 1989; Hardman and Herd 1998.

399   The open question, "habitual dietary carbohydrate . . .": Yost et al. 1998.

400   "farinaceous and vegetable . . .": Tanner 1869b:217. "Eating carbohydrates will stimulate . . .": discussion period in Gracey et al. 1991:194.

400   Cahill gave Banting Memorial Lecture: Cahill 1971 ("overall fuel control . . ." "The concentration of circulating . . . ," 785). "carbohydrate is driving insulin . . . ," "a calorie is a calorie . . . ," and the obese as fundamentally lazy: Interviews, George Cahill.

401   Kipnis fed ten "grossly obese" women: Grey and Kipnis 1971.

401   "necessity of restricting carbohydrates": Schettler and Schlief 1974:394–95.

401   Kipnis described his findings: Interview, David Kipnis.

401   Americans have become progressively heavier: Ogden et al. 2006. And more diabetic: CDC 2005. Gillman reported: Kim et al. 2006. On heavier infants and newborns, see also Schack-Nielsen et al. 2006 (Denmark); Surkan et al. 2004 (Sweden).

402   "The baby is not diabetic . . .": Interview, Boyd Metzger.

402   "If you overdo it . . .": Quoted in Goldberg 2006. "Our observation of a trend . . .": Kim et al. 2006.

402   Fatter babies more likely: See, for instance, Guo et al. 2002. "perpetuating the cycle . . .": Dabelea et al. 2000.

403   "excessive glucose pulses": Neel 1982.

CHAPTER TWENTY-THREE:
THE FATTENING CARBOHYDRATE DISAPPEARS

404   Epigraphs. "We need the help . . .": Butterfield 1969:8. "It is incredible . . .": Atkins 1973:8.

404   McGovern hearing on Atkins diet: Select Committee 1973b ("The Atkins diet is nonsense . . . ," 17).

404   McGovern hearings on "Sugar in the Diet . . .": Select Committee 1973a.

405   "We weren't thinking . . .": Interview, Kenneth Schlossberg.

405   McGovern's 1976 hearings on diet and disease: Select Committee 1976 ("overconsumption may be as serious . . . ," 9).

405   "Particularly overconsumption of the wrong things . . .": Ibid.:10.

406   "general rule of thumb": Ibid.:19–20.

406   "I think what we need . . .": Ibid.

406   "carbohydrate-deficiency syndrome . . .": Astrup et al. 1994. See also Golay and Bobbioni 1997.

407   Proceedings of the UCSF conference: Wilson 1969. "Positive caloric balance . . .": Lepkovsky 1969:95. Navy "ketogenic" diet study: Piscatelli et al. 1969 ("significant weight loss," 185; "Uniformly and without . . . ," 188).

407   Proceedings of Obesity Association of Great Britain conference: McLean Baird and Howard 1969. "This weight gain was controlled . . .": Craddock in discussion period, in McLean Baird and Howard 1989:124.

407   Howard became interested in carbohydrate restriction: Interview, Alan Howard. "A common feature of all . . .": Howard 1969:104.

408   Proceedings of the Paris conference: Apfelbaum 1973. The INSERM presentation: Debry et al. 1973 ("lowering the carbohydrate").

408 Proceedings of the NIH conference: Bray ed. 1976a. Presentations on physical activity: Lutwak and Coulson 1976; Björntorp 1976. On behavioral modification: Stuart 1976; Stunkard 1976c. On fasting: Drenick 1976 ("our experiences . . . ," 358). Young on diet: Young 1976.

408 Young's presentation: Young 1976 ("weight loss, fat loss . . . ," 365; "No adequate . . . ," 364).

409 Proceedings of the 1973 London conference: Burland et al. 1974. Salans talk: Salans et al. 1974. Horton's presentation: Horton et al. 1974 ("It is clear that . . . ," 225). Horton added that it was probably hyperinsulinemia: Discussion period, in Burland et al. 1974:249. Yudkin gave talk: Yudkin 1974 ("reduce superfluous adiposity . . . ," 276).

409 Harry Keen said: Discussion period, in Burland et al. 1974:361.

410 "You strictly curtail . . .": Tarr 1972:13.

410 Rating the Diets: Berland 1974 ("much to recommend it" "helpful . . . ," 222; "the difficult-to-treat . . . ," 220; "pay little attention . . . ," 347). Obesity authorities would recommend: Bray 1978:254; Dwyer 1985:185.

411 Yudkin's "no bread, no butter" argument: Yudkin and Carey 1960 ("the inevitability of calories"); Yudkin 1972c. "It is highly implausible . . .": Yudkin 1974:274.

411 The High-Calorie Way: Atkins 1972.

411 "Dr. Pennington may be . . .": Barr et al. 1953:137. "trash," "potentially dangerous": Quoted in Yuncker 1962. "nutrition nonsense . . .": White 1962. Written by Van Itallie: Interview, Theodore Van Itallie. "bizarre concepts of nutrition . . .": Anon. 1973.

412 Mayer wrote: Mayer 1968 ("as aware as . . . ,"; "favors fat . . . ," 67; "tend to become . . . ," 203). "biochemical mumbojumbo": Mayer 1973b.

412 "because that's what was being taught . . .": Interview, Robert Atkins. See also Atkins 1972:21–24. Gordon's JAMA article: Gordon et al. 1963 ("not to produce . . ." "The total caloric . . . ," 55). Atkins said attention caught: Interview, Robert Atkins.

413 Bloom had noted: Bloom and Azar 1963; Azar and Bloom 1963. Atkins lost twenty-eight pounds, AT&T experiment: Atkins 1972:26–27.

413 Atkins in Vogue: Pierson 1970. Footnote. Bliss 1976:35.

414 "produce too much insulin": Atkins 1972:32.

414 "ten thousand . . .": Ibid.:2–3. Cleave as inspiration: Interview, Robert Atkins.

414 Third claim: Ibid. ("cruel hoax . . . ," 95; "the balanced low-calorie diet . . . ," 84–5). Bray published: Bray 1969; Bray 1970 ("The Myth of Diet").

415 Atkins's polemic: Atkins 1972 ("resentment . . . ," 26; "a revolution . . . ," 6; "Martin Luther King . . . ," 294; "lobster with butter . . . ," 3; "As long as you . . . ," 15). Stillman's mega–best-seller: Stillman and Baker 1967.

415 Fastest-selling book: Select Committee 1973b:iv. "chief consequence . . .": Yudkin 1974:273–74.

416 Background on Van Itallie, Stunkard, and Mayer: Stunkard 1976b:20; interviews, Albert Stunkard and Theodore Van Itallie. Van Itallie and White: Interview, Theodore Van Itallie. "The Mississippi River . . .": Interview, Gerold Grodsky.

416 McGovern's committee hearings on obesity: Select Committee 1977e ("Thus, what I am saying," 205–6).

417 "denunciation": Interview, Theodore Van Itallie. See also his testimony in Select Committee 1977e:44–64.

417 "gross inaccuracies . . .": Interview, Theodore Van Itallie. "We just despised . . .": Interview, Albert Stunkard.

417 Van Itallie and White's critique: Anon. 1973. Mayer's column: Mayer 1973b.
418 "a few hundred thousand . . .": Interview, Theodore Van Itallie. Van Itallie's writ-
ings: Van Itallie et al. 1976; Pi-Sunyer and Van Itallie 1975; Van Itallie 1978; Van
Itallie 1979; Van Itallie 1980a; Van Itallie 1980b. Fourth International Congress:
Hirsch and Van Itallie 1985. In *Present Knowledge in Nutrition:* Vaselli et al. 1984. No
time to do research: Interview, Theodore Van Itallie.
419 Van Itallie on dietary therapy: Van Itallie 1978 ("increasing recognition . . . ," 610);
Van Itallie 1979; Van Itallie 1980b:250–51. *Footnote.* Van Itallie 1980b:250–51).
419 Bray's disagreements with Sims: See Sims and Danforth 1974. Bray's conference
résumé in the 1970s: Bray 1975; Bray 1976a; Bray ed. 1978; Bray 1979. Textbooks:
Bray and Bethune 1974; Bray 1976b; Bray 1980.
419 Bray believed: Interview, George Bray. Bray's treatment of Young in *Obesity in Perspective:* Bray 1975:43. "confirmation before they . . .": Gwinup 1974:98. "The data
are suggestive . . .": Bray 1976b:312–13. The report on the NIH conference: Bray
1975; Bray 1976a (research priorities and "gaps in our current knowledge," 1–6).
Leading proponent: See, for instance, Bray and Popkin 1998; Bray and Popkin
1999.
420 Novin's 1977 presentation: Novin 1978 ("widespread popularity . . ."). Bray omitted
mention: Bray 1978. Greenwood's "gatekeeper" presentation: Greenwood 1985.
Hirsch ignores implications: Hirsch 1985. *Footnote:* Interview, Donald Novin.
421 Bray would routinely equate: See, for instance, Brody 1981c; Select Committee
1977c:106, 207. "highly commendable": Select Committee 1977e:206. The MRC
report: James 1977 ("commonly prescribed . . . ," 171).
422 "If such diets are truly . . .": Quoted in Anon. 1973.
422 "The evergrowing list of diets . . .": Select Committee 1977e:101. "The proliferation . . .": Hirsch 1985:195.
422 "common factor of reducing . . .": Mann 1974. "instant money nutritionists": Stare
1987:xxx. "very lucrative . . .": Whelan and Stare 1983:26. *Footnote.* Mayer 1968:160.
423 The Harvard nutrition department's $5 million building, and the "lead gift": Stare
1987:xv–xvi. Stare as the defender of sugar and additives: Rosenthal et al. 1976.
Footnote. Whelan and Stare 1983:194 ("not even remotely").
423 Hill's articles in *Science:* Hill and Peters 1998; Hill et al. 2003. "reduce the likelihood . . ." and "The theory that . . .": Hill n.d. Hill's conflict-of-interest statements:
See, for instance, Foster et al. 2003:2089. Received $2 million in "gifts" from Proc-
ter & Gamble: Information gathered via a Freedom of Information Act request to
the University of Colorado in Sept. 2003. "dieter's dream": Potts 1987.
424 Hill received $300,000 from NIH: NIH Extramural Awards by State and Foreign
Site: http://grants1.nih.gov/grants/award/state/state.htm. $5 million: e-mail from
Marguerite Klein, NIH program officer for the Atkins diet trial.
424 "A resolution . . .": Novin 1978:31.

CHAPTER TWENTY-FOUR:
THE CARBOHYDRATE HYPOTHESIS, III: HUNGER AND SATIETY

425 *Epigraph.* "There is only one way . . .": Konner 2003:376.
425 Sidbury described: Sidbury and Schwartz 1975 ("rational basis" "minimum of
anguish . . . ," 66). His three-page article: Schwartz and Sidbury 1974. Sidbury
noted: Sidbury and Schwartz 1975:66–67. Sidbury's diet: Ibid.:67–69. *Footnote.*
Ibid.
426 Dietary fat prolongs drainage of nutrients: Davidson and Passmore 1963:97.

427 Mayer's glucostat hypothesis: Mayer 1968:20–24.
427 lipostatic regulation: Kennedy 1961. Set point: See Keesey 1980.
428 "weight loss triggers . . .": Stunkard 1980:9. "It is not appealing . . .": Quoted in Rovner 1986. Mysterious mechanism: Davis and Wirtshafter 1978.
428 The more fundamental criticism: This idea, and that of the settling point, are in ibid.
429 "black hole": Interview, Donald Novin.
429 "In human beings and animals . . .": Richter 1976:224.
429 Experimental observations on hunger, thirst, and palatability: Ibid. ("As a result . . . ," 198).
430 "Rats will make every . . .": Ibid.:210. "Food acceptance and the urge . . .": Adolph 1947:122. *Footnote.* Rolls and Barnett 2000.
431 "phenomenal . . .": Bellisle et al. 2003. "incredibly brilliant": Interview, Stephen Woods. Le Magnen's life and career: Le Magnen 2001.
431 Rats ate discrete meals: See Le Magnen 1976. See also Le Magnen 1971, his seminal paper on the physiological psychology of hunger.
431 Two fundamental observations: Le Magnen 1971:213–19. "quantitative deficiencies . . .": Adolph 1947:122.
431 "All increase or decrease . . .": Le Magnen 1971:220.
431 While they're sleeping: Ibid.
432 "The restitution . . .": Ibid.:238. "spare": Ibid.:243.
432 "indirect and passive consequences . . .": Le Magnen 1981:315.
432 Insulin is the driver: Le Magnen 1976:99–100.
433 Several variations of the hypothesis: Le Magnen 1984; Toates and Booth 1974; Friedman and Stricker 1976.
433 Three propositions: Friedman and Stricker 1976 ("adequate for them . . . ," 413).
433 The simplest possible explanation: Ibid. ("Hunger appears," 424). "The primitive goal": Hoebel and Hernandez 1993:43.
433 We're not much more complicated than insects: Hoebel and Hernandez 1993; Lepkovsky 1973. "feeding behavior removes . . .": Stricker 1978.
434 "Energy metabolism": Friedman and Stricker 1976:413.
434 "harmony of tissue metabolisms": Ibid.:413.
435 What the body regulates: Le Magnen 1984.
435 Le Magnen demonstrated this: Le Magnen 1981.
436 "It is not a paradox . . .": Le Magnen 1984:517.
436 Food availability most important to fertility: Bronson 1988:88. Body fat, as commonly believed: Frisch and McArthur 1974. Availability of metabolic fuels: Schneider and Wade 1989 and Wade and Schneider 1992.
436 Wade and Schneider explained: Wade and Schneider 1992:246–47.
437 Nicotine as weight-loss drug: Filozof et al. 2004.
438 Excess calories not enough to explain weight gain: Perkins 1993. Rodin reported: Rodin 1987. Physical activity: Perkins 1993.
438 Nicotine and LPL: Chajek-Shaul et al. 1990; Perkins 1993; Sztalryd et al. 1996; Carney and Goldberg 1984. Fenfluramine and LPL: Deshaies et al. 1994.
439 The adjustable regulator: See Mrosovsky 1985:45–46.
439 Woods and Porte observed: Woods and Porte 1976:274. Seasonal weight variations in hibernators: Le Magnen 1988: Florant et al. 1985.
439 Seasonal variations in insulin levels: Fahlen et al. 1971; Behall et al. 1984. In LPL activity: Donahoo et al. 2000. Spring and fall weight changes: See, for instance, Andersson and Rossner 1992.

440 "arises mainly from human . . .": Ramirez et al. 1989.
441 Le Magnen first noted: See Le Magnen 2001. "It is often said . . .": Pavlov 1955:109.
441 Rats given the choice: Le Magnen 1978. For similar experiments done by Anthony Sclafani, see Sclafani and Nissenbaum 1988.
441 "In evolution . . .": Quoted in Goleman 1989.
442 First wave of insulin secretion: Simpson et al. 1968. "The pancreas has no idea . . .": Interview, Gerold Grodsky.
442 "the metabolic background . . .": Le Magnen 1985:59. "In man . . .": Le Magnen 1978.
443 Cephalic reflexes: Pavlov 1955:245–70. Nicolaidis demonstrated: Nicolaidis 1969 ("pre-adaptive"); interview, Stylianos Nicolaidis. Woods and his colleagues: Woods et al. 1977. "essentially anything . . .": Interview, Mark Friedman.
443 Idea suggested by Powley: Powley 1977 ("self-perpetuating . . ." "Rather than secreting . . . ," 102).
444 Rodin reported: Rodin 1980 ("major candidate . . . ," 232). "A feedback loop . . .": Rodin 1985:14.
444 "hunger late in the day": Sims and Danforth 1974. "developed marked anorexia": Goldman et al. 1976:166. Calories in popcorn and cheese: USDA n.d. Footnote. Hurley and Liebman 2003.
445 "high levels of insulin . . .": Cahill 1971:785.
445 Subjects inevitably fall off the diet: See, for instance, Anon. 1973; Select Committee 1977b:9. Described by Kemp: Kemp 1963.
446 Nucleus accumbens: Hoebel et al. 1999 ("is excessively activated . . . ," 559).
446 Avoiding carbohydrates lowers insulin: Grey and Kipnis 1971. "After a year . . .": Sidbury and Schwartz 1975:71–72.
446 "generally be used safely . . .": Van Itallie 1979. Transient hypercholesterolemia: Phinney et al. 2003. Interview, Stephen Phinney. Cited as another reason: Burros 2004b; interview, Edward Ahrens.
447 Kemp discussed: Kemp 1963 ("At least half of our patients . . .").

EPILOGUE

449 Epigraphs. "The community of science . . .": Merton 1973:339. "The first principle . . .": Feynman 1985:343.
449 The four essays: Friedman 2003; Hill et al. 2003 ("energy gap," "eating 15% less"); Nestle 2003 ("consuming more food energy"); Pi-Sunyer 2003. "For most adults . . .": USDHHS and USDA 2005:14.
450 Von Noorden's argument: Von Noorden 1907a:693–700. "estimate is theoretical . . .": Hill et al. 2003.
450 "No one would question . . .": Atwater and Benedict 1899.
451 "have stopped before . . ." and "This unending exchange . . .": Merton 1973:339.
451 "the results of their enterprise . . .": Kuhn 1970:163.
452 "Perhaps our most sensible . . .": Becker 1987.
455 ADA recommends against carbohydrate restriction: Bantle et al. 2006.
455 "Sick Individuals and Sick Populations": Rose 1985.
456 Ten or twenty pounds a year in the mid-eighteenth century: Cummings 1940:236 (U.S.); Aykroyd 1967:105 (U.K.). 150 pounds: Putnam et al. 2002:8 (U.S.).
456 Brains run more efficiently on this fuel mixture: Cahill and Veech 2003. IOM report: IOM 2002:275–80;285–90 ("without having to rely . . . ," 288).

457 Yudkin noted: Yudkin 1974.

457 NIH-funded tests of carbohydrate-restricted diets: Brehm et al. 2003; Foster et al. 2003; Samaha et al. 2003; Sondike et al. 2003; Yancy et al. 2004; Gardner et al. 2007. "the sugar and starchy elements . . .": Anon. 1864c.

458 Typical American diet of today: USDA Economic Research Service 2005.

459 Results of the DPP: Orchard et al. 2005; Knowler et al. 2002.
     Footnote. Mayer-Davis et al. 2004.

459 "the largest, most expensive . . .": Interview, John Foreyt. The goal and methodology of Look AHEAD: Look AHEAD Protocol Review Committee 2006.

Bibliography

Abrams, H. L., Jr. 1987. "The Preference for Animal Protein and Fat: A Cross-Cultural Survey." In Harris and Ross, eds., 1987, 207–23.

Adadevoh, B. K. 1974. "Obesity in the African: Socio-Medical and Therapeutic Considerations." In Burland, Samuel, and Yudkin, eds., 1974, 60–73.

Adams, G. H., and V. N. Schumaker. 1969. "Polydispersity of Human Low-Density Lipoproteins." *Annals of the New York Academy of Sciences*. Nov. 7; 164(1):130–46.

Adolph, E. F. 1947. "Urges to Eat and Drink in Rats." *American Journal of Physiology*. 151:110–25.

Agatston, A. 2003. *The South Beach Diet*. Emmaus, Pa.: Rodale Press.

Ahrens, E. H., Jr. 1985. "The Diet-Heart Question in 1985: Has It Really Been Settled?" *Lancet*. May 11; 325 (8437):1085–87.

———. 1979a. "Dietary Fats and Coronary Heart Disease: Unfinished Business." *Lancet*. Dec. 22/29; 314(8156–57):1345–48.

———. 1979b. "Introduction." *American Journal of Clinical Nutrition* 32 (suppl.):2627–31.

———. 1957. "Nutritional Factors and Serum Lipid Levels." *American Journal of Medicine*. Dec.; 23(6):928–52.

Ahrens, E. H., Jr., J. Hirsch, W. Insull, Jr., T. T. Tsaltas, R. Blomstrand, and M. L. Peterson. 1957. "Dietary Control of Serum Lipids in Relation to Atherosclerosis." *JAMA*. Aug. 24; 164 (17):1905–11.

Ahrens, E. H., Jr., J. Hirsch, K. Oette, J. W. Farquhar, and Y. Stein. 1961. "Carbohydrate-Induced and Fat-Induced Lipemia." *Transactions of the Medical Society of London*. 74:134–46.

Ahrens, E. H., Jr., W. Insull, Jr., R. Blomstrand, J. Hirsch, T. T. Tsaltas, and M. L. Peterson. 1957. "The Influence of Dietary Fats on Serum-Lipid Levels in Man." *Lancet*. May 11; 272 (6976):943–53.

Akinyanju, P. A., R. U. Qureshi, A. J. Salter, and J. Yudkin. 1968. "Effect of an 'Atherogenic' Diet Containing Starch or Sucrose on the Blood Lipids of Young Men." *Nature*. June 8; 218 (5145):975–77.

Alberts, D. S., M. E. Martinez, D. J. Roe, et al. 2000. "Lack of Effect of a High-Fiber Cereal Supplement on the Recurrence of Colorectal Adenomas: Phoenix Colon Cancer Prevention Physicians' Network." *New England Journal of Medicine*. April 20; 342 (16):1156–62.

Albrink, M. J. 1965. "Diet and Cardiovascular Disease." *Journal of the American Dietetic Association*. Jan.; 46:26–29.

———. 1963. "The Significance of Serum Triglycerides." *Journal of the American Dietetic Association*. Jan.; 42:29–31.

———. 1962. "Triglycerides, Lipoproteins, and Coronary Artery Disease." *Archives of Internal Medicine*. 109(3):345–59.

Albrink, M. J., P. H. Lavietes, E. B. Man, and J. R. Paul. 1962. "Relationship Between Serum Lipids and the Vascular Complications of Diabetes from 1931 to 1961." *Transactions of the Association of American Physicians.* 75:235–41.

Albrink, M. J., and J. W. Meigs. 1965. "The Relationship Between Serum Triglycerides and Skinfold Thickness in Obese Subjects." *Annals of the New York Academy of Sciences.* Oct. 8; 131 (1):673–83.

Aldritch, L. D. 1966. *A Journal of the Overland Route to California.* U.S.A.: Readex Microprint.

Allbaugh, L. G. 1953. *Crete: A Case Study of an Underdeveloped Area.* Princeton, N.J.: Princeton University Press.

Allen, F. M. 1913. *Studies Concerning Glycosuria and Diabetes.* Cambridge, Mass.: Harvard University Press.

Alonso, L. G., and T. H. Maren. 1955. "Effect of Food Restriction on Body Composition of Hereditary Obese Mice." *American Journal of Physiology.* 183(2):284–90.

Altman, L. K. 1980. "Report About Cholesterol Draws Agreement and Dissent." *New York Times.* May 20; A16.

Altshule, M. D. 1966. "The Uselessness of Diet in the Treatment of Atherosclerosis." In Ingelfinger, Relman, and Finland, eds., 1966, 69–78.

American Cancer Society. 1984. "Nutrition and Cancer, Cause and Prevention: An American Cancer Society Special Report." *CA: A Cancer Journal for Clinicians.* March–April; 34(2):121–26.

American Diabetes Association (ADA). 2006. "Nutrition Recommendations and Interventions for Diabetes—2006: A Position Statement of the American Diabetes Association." *Diabetes Care.* Sept.; 29(9):2140–57.

———. 1971. "Principles of Nutrition and Dietary Recommendations for Patients with Diabetes Mellitus: 1971." *Diabetes.* Sept.; 20(9):633–34.

American Heart Association [AHA]. 2005. *The No-Fad Diet: A Personal Plan for Healthy Weight Loss.* New York: Clarkson Potter.

———. 1961. "Dietary Fat and Its Relation to Heart Attacks and Strokes: Report by the Central Committee for Medical and Community Program of the American Heart Association." *JAMA.* Feb. 4; 175:389–91.

Ames, R. P., and P. Hill. 1976. "Elevation of Serum-Lipid Levels During Diuretic Therapy of Hypertension." *American Journal of Medicine.* 61:748–57.

Anand, B. K., and J. R. Brobeck. 1951. "Hypothalamic Control of Food Intake in Rats and Cats." *Yale Journal of Biological Medicine.* Nov.; 24(2):123–40.

Anderson, J. H. 1935. "The Treatment of Obesity." *Lancet.* Sept. 14; 226(5846):604–8.

Anderson, J. T., A. Keys, and F. Grande. 1957. "The Effects of Different Food Fats on Serum Cholesterol Concentration in Man." *Journal of Nutrition.* July 10; 62(3):421–24.

Anderson, J. T., A. Lawler, and A. Keys. 1957. "Weight Gain from Simple Overeating. 2. Serum Lipids and Blood Volume." *Journal of Clinical Investigation.* 36(1):81–88.

Anderson, K. M., W. P. Castelli, and D. Levy. 1987. "Cholesterol and Mortality: 30 Years of Follow-Up from the Framingham Study." *JAMA.* April 24; 257(16):2176–80.

Andersson, I., and S. Rossner. 1992. "The Christmas Factor in Obesity Therapy." *International Journal of Obesity and Related Metabolic Disorders.* Dec.; 16(12):1013–15.

Anitschkow, N., and S. Chalatow. 1913. "Über experimentelle Cholester-Insteatose und ihre Bedeutung für die Entstehung einiger pathologischer Prozesse." *Centralblatt für Allgemeine Pathologie und Pathologische Anatomie.* 24:1–9.

Anon. 1990. "Publisher Robert Rodale." *Chicago Tribune.* Sept. 21; 11.

———. 1988. "Report of the National Cholesterol Education Program Expert Panel on

Detection, Evaluation, and Treatment of High Blood Cholesterol in Adults: The Expert Panel." *Archives of Internal Medicine.* Jan.; 148(1):36–69.

———. 1984a. "Sorry, It's True. Cholesterol Really Is a Killer." *Time.* Jan 23; 30.

———. 1984b. "Lowering Blood Cholesterol to Prevent Heart Disease." NIH Consensus Development Conference December 10–12, 1984. Program and Abstracts. National Heart, Lung, and Blood Institute and the NIH Office of Medical Applications of Research.

———. 1983. "This Week's Citation Classic—Stunkard A. & McLaren-Hume M. 'The Results of Treatment for Obesity: A Review of the Literature and Report of a Series.' " *Current Content.* Nov. 23; 67(21):24.

———. 1980. "The Nutritionist Who Prepared the Pro-Cholesterol Report Defends It Against Critics." *People.* June 16; 13:58–64.

———. 1979. "On Julius Bauer." *Lancet.* June 23; 313(8130):1359–60.

———. 1978. "A Co-operative Trial in the Primary Prevention of Ischaemic Heart Disease Using Clofibrate: Report from the Committee of Principal Investigators." *British Heart Journal.* Oct.; 40(10):1069–118.

———. 1976. "Edwin Bennett Astwood." *New England Journal of Medicine.* April 8; 294(15):840–41.

———. 1973. "A Critique of Low-Carbohydrate Ketogenic Weight Reduction Regimens: A Review of 'Dr. Atkins' Diet Revolution.' " *JAMA.* June 4; 224(10):1415–19.

———. 1969. *White House Conference on Food, Nutrition, and Health: Final Report.* Washington, D.C.: U.S. Government Printing Office.

———. 1967. "Coronary Heart Disease and Carbohydrate Metabolism." *JAMA.* Sept. 25; 201(13):164–65.

———. 1966. "Blake Donaldson, Internist, Author." *New York Times.* Feb. 21; 39.

———. 1964a. "Prudent Diet." *Newsweek.* Oct. 19; 19:92.

———. 1964b. "Heart Association Stirs Up a Controversy by Urging Public to Alter Intake of Fats." *Wall Street Journal.* June 10; 6.

———. 1964c. "Starvation and Obesity." *JAMA.* Jan. 11; 187:184.

———. 1962. "Villagers in Dahomey Crawl to Town to Seek Food." *New York Times.* May 20; 13.

———. 1961. "The Fat of the Land." *Time.* Jan. 13; 67(3):48–52.

———. 1960. "Fat in the Fire." *Time.* Dec. 26; 76(27):33.

———. 1958. "Diet for Weight Reduction." *JAMA.* March 29; 166(13):1660.

———. 1955a. "Combined Staff Clinic: Obesity." *American Journal of Medicine.* July; 19(1):111–25.

———. 1955b. "Obesity." *JAMA.* March 26; 157(13):1126.

———. 1954. "Metabolism in Obesity." *Lancet.* Jan. 16; 263(6803):144–45.

———. 1953. "Annotations: Obesity." *Lancet.* July 18; 262(6777):126.

———. 1952. "Freak Diets." *JAMA.* Feb. 16; 148(7):590.

———. 1950. "Why Executives Drop Dead." *Fortune.* June; 41:88–91, 149–56.

———. 1948a. "National Heart Week." *American Heart Journal.* 35:528.

———. 1948b. "Reports of Local Heart Association Activities." *American Heart Journal.* 36:158–59.

———. 1945. "Pooling of Funds Urged in Health Report." *JAMA.* Dec. 8; 129:1037.

———. 1940. "Concertmaster." *Time.* Jan 1. Online at http://www.time.com/time/archive/preview/0,10987,763313,00.html.

———. 1930a. "All-Meat Eskimo Diet Is Declared Harmful by University of Michigan Doctors After Test." *New York Times.* March 17; 15.

————. 1930b. "Vienna Specialist Blames 'Mass Suggestion' for Parrot Fever Scare, Which He Holds Baseless." *New York Times*. Jan. 16; 3.

————. 1928. "Stefansson Testing Meat Diet Theory." *New York Times*. Feb. 28; 8.

————. 1899. "The Month." *Practitioner*. 62:369. Cited in R. N. Proctor, *Cancer Wars* (New York: Basic Books, 1995).

————. 1865. "A Banting Mania." *British Medical Journal*. July 29; 97.

————. 1864a. "Dr. Lankester on the Banting System." *British Medical Journal*. Nov. 12; 565.

————. 1864b. "Popular Medical Works." *British Medical Journal*. Nov. 5; 531.

————. 1864c. "Bantingism." *British Medical Journal*. Oct 22; 481.

————. 1864d. "Bantingism." *British Medical Journal*. Oct 22; 469–70.

————. 1864e. "Bantingism." *Lancet*. Oct. 1; 84(2144):387–88.

————. 1864f. "Papers." *British Medical Journal*. Aug. 13; 203.

————. 1864g. "Bantingism." *Lancet*. May 7; 83(2123):520.

Antonis, A., I. Bersohn, R. Plotkin, D. I. Easty, and H. E. Lewis. 1965. "The Influence of Seasonal Variation, Diet, and Physical Activity on Serum Lipids in Young Men in Antarctica." *American Journal of Clinical Nutrition*. May; 16:428–35.

Apfelbaum, M., ed. 1973. *Régulation de L'Equilibre Energétique chez l'Homme*. [*Energy Balance in Man*.] Paris: Masson.

Appen, B. 1933. "The World Situation in Cattle and Beef." *Foreign Crops and Markets*. Bureau of Agricultural Economics. U.S. Department of Agriculture. Jan. 23; 26:88–90.

Armstrong, B. K. 1977. "The Role of Diet in Human Carcinogenesis with Special Reference to Endometrial Cancer." In Hiatt, Watson, and Winsten, eds., 1977, 557–65.

Armstrong, B. K., and R. Doll. 1975. "Environmental Factors and Cancer Incidence and Mortality in Different Countries with Special Reference to Dietary Practices." *International Journal of Cancer*. 15:617–31.

Arner, P., and R. H. Eckel. 1998. "Adipose Tissue as a Storage Organ." In Bray, Bouchard, and James, eds., 1998, 379–96.

Arner, P., P. Engfeldt, and H. Lithell. 1981. "Site Differences in the Basal Metabolism of Subcutaneous Fat in Obese Women." *Journal of Clinical Endocrinology & Metabolism*. Nov.; 53(5):948–52.

Arteaga, A. 1974. "The Nutritional Status of Latin American Adults." In *Nutrition and Agricultural Development*, ed. N. S. Scrimshaw and B. Moises (New York: Plenum Press), 67–76.

Arthus, M. 1943. *Philosophy of Scientific Investigation*. Trans. H. E. Sigerist. Baltimore: Johns Hopkins Press.

Arvanitakis, Z., R. S. Wilson, J. L. Bienias, D. A. Evans, and D. A. Bennett. 2004. "Diabetes Mellitus and Risk of Alzheimer Disease and Decline in Cognitive Function." *Archives of Neurology*. May; 61(5):661–66.

Ashworth, A., R. Bell, W. P. James, and J. C. Waterlow. 1968. "Calorie Requirements of Children Recovering from Protein-Calorie Malnutrition." *Lancet*. Sept. 14; 292(7568): 600–603.

Assimacopoulos-Jeannet, F., and B. Jeanrenaud. 1976. "The Hormonal and Metabolic Basis of Experimental Obesity." *Clinics in Endocrinology & Metabolism*. July; 5(2):337–65.

Associated Press. 1987. "Link of Fat to Breast Cancer Disputed." *New York Times*. Dec 31; 6.

Astrup, A., B. Buemann, N. J. Christensen, and S. Toubro. 1994. "Failure to Increase Lipid Oxidation in Response to Increasing Dietary Fat Content in Formerly Obese Women." *American Journal of Physiology*. April; 266 (4, pt. 1):E592–99.

Astwood, E. B. 1962. "The Heritage of Corpulence." *Endocrinology.* Aug.; 71:337–41.

Atkins, R. C. 1973. Prepared Statement of Dr. Robert C. Atkins. In Select Committee on Nutrition and Human Needs of the United States Senate 1973b, 4–8.

———. 1972. *Dr. Atkins' Diet Revolution: The High Calorie Way to Stay Thin Forever.* New York: David McKay.

Atwater, W. O. 1888. "What We Should Eat." *Century Illustrated Magazine.* June; 36(2):257.

Atwater, W. O., and F. G. Benedict. 1899. *Experiments on the Metabolism of Matter and Energy in the Human Body.* Bulletin No. 69, U.S. Department of Agriculture. Washington, D.C.: U.S. Government Printing Office.

Audubon, J. W. 1906. *Audubon's Western Journal: 1849–1850.* Cleveland: Arthur H. Clark.

Auerbach, S. 1974. "Roughing It—Tonic for Our Time." *Washington Post.* Aug. 19; B1.

Austin, M. A., J. L. Breslow, C. H. Hennekens, J. E. Buring, W. C. Willett, and R. M. Krauss. 1988. "Low-Density Lipoprotein Subclass Patterns and Risk of Myocardial Infarction." *JAMA.* Oct. 7; 260(13):1917–21.

Aykroyd, W. R. 1967. *The Story of Sugar.* Chicago: Quadrangle Books.

Azar, G. J., and W. L. Bloom. 1963. "Similarities of Carbohydrate Deficiency and Fasting. II. Ketones, Nonesterified Fatty Acids and Nitrogen Excretion." *Archives of Internal Medicine.* Sept.; 112:338–4.

Bacon, F. 1994. *Novum Organum.* Ed. and trans. P. Urbach and J. Gibson. Peru, Ill.: Carus Publishing Company. [Originally published 1620.]

Bahner, F. 1955. "Fettsucht und Magersucht." In *Innersekretorische Krankheiten Fettsucht Magersucht,* vol. VII, no. 1, of *Handbuch der Inneren Medizin,* 4th edition, ed. F. Bahner, H. W. Bansi, G. Fanconi, A. Jores, W. Zimmerman (Berlin: Springer-Verlag), 978–1163.

Bailar, J. C., 3rd. 1980. "Cause and Effect in Epidemiology: What Do We Know About Hypertriglyceridemia?" *New England Journal of Medicine.* June 19; 302(25):1417–18.

Baker, B. M., I. D. Frantz, Jr., A. Keys, et al. 1963. "The National Diet-Heart Study: An Initial Report." *JAMA.* July 13; 185:105–6.

Ballard-Barbash, R. 1999. "Energy Balance, Anthropometry, and Cancer." In *Nutritional Oncology,* ed. D. Heber, G. L. Blackburn, and V. L. Go (San Diego: Academic Press), 137–52.

Banerjee, S. 1945. "Relation of Scurvy to the Adrenalin Content of the Adrenal Glands of Guinea Pigs." *Journal of Biological Chemistry.* 159:327–31.

Banting, W. 1869. *Letter on Corpulence, Addressed to the Public.* 4th edition. London: Harrison. Republished New York: Cosimo Publishing, 2005. Online at http://www.lowcarb.ca/corpulence/index.html.

———. 1864. *Letter on Corpulence, Addressed to the Public.* 3rd edition. London: Harrison.

Bantle, J. P., D. C. Laine, G. W. Castle, J. W. Thomas, B. J. Hoogwerf, and F. C. Goetz. 1983. "Postprandial Glucose and Insulin Responses to Meals Containing Different Carbohydrates in Normal and Diabetic Subjects." *New England Journal of Medicine.* July 7; 309(1):7–12.

Bantle, J. P., J. Wylie-Rosett, A. L. Albright, et al. 2006. "Nutrition Recommendations and Interventions for Diabetes—2006: A Position Statement of the American Diabetes Association." *Diabetes Care.* Sept.; 29(9):2140–45.

Barker, J. E. 1924. *Cancer: How It Is Caused, How It Can Be Prevented.* New York: E. P. Dutton.

Barr, D. P., J. R. Brobeck, H. W. Brosin, et al. 1953. *Overeating, Overweight and Obesity.* Nutrition Symposium Series No. 6. New York: National Vitamin Foundation.

Barr, D. P., E. M. Russ, and H. A. Eder. 1951a. "Protein-Lipid Relationships in Human Plasma. I. In Normal Individuals." *American Journal of Medicine.* Oct.; 11(4):468–79.

————. 1951b. "Protein-Lipid Relationships in Human Plasma. II. In Atherosclerosis and Related Conditions." *American Journal of Medicine*. Oct.; 11(4):480–93.

Bartke, A. 2002. "Insulin-Like Growth Factor 1 and Mammalian Aging." *Science of Aging Knowledge Environment*. April 24; 2002(16):vp4. Online at http://sageke.sciencemag .org/cgi/content/full/sageke;2002/16/vp4.

Bartlett, J. R. 1965. *Personal Narrative of Explorations and Incidents in Texas, New Mexico, California, Sonora and Chihuahua*. Vol. 2. Chicago: Rio Grande Press.

Baserga, R. 2004. "Targeting the IGF-1 Receptor: From Rags to Riches." *European Journal of Cancer*. Sept.; 40(14):2013–15.

Baserga, R., F. Peruzzi, and K. Reiss. 2003. "The IGF-1 Receptor in Cancer Biology." *International Journal of Cancer*. Dec. 20; 107(6):873–77.

Basu, T. K., and C. J. Schlorah. 1982. *Vitamin C in Health and Disease*. Westport, Conn.: Avi Publishing.

Bauer, J. 1945. *Constitution and Disease: Applied Constitutional Pathology*. New York: Grune & Stratton.

————. 1941. "Obesity: Its Pathogenesis, Etiology, and Treatment." *Archives of Internal Medicine*. May; 67(5):968–94.

————. 1940. "Observations on Obese Children." *Archives of Pediatrics*. 57:631–40.

Baum, G. 1995. "Do They Really Enjoy Being Dinner Party Don't-Invite-'Ems?" *Los Angeles Times Magazine*. Aug 6; 14.

Baxter, K. L. 1976. "Energy Utilization and Obesity in Domesticated Animals." In Bray, ed., 1976b, 127–35.

Beaglehole, R., M. A. Foulkes, I. A. Prior, and E. F. Eyles. 1980. "Cholesterol and Mortality in New Zealand Maoris." *British Medical Journal*. Feb. 2; 280(6210):285–87.

Beck, M. 1990. "The Losing Formula." *Newsweek*. April 30; 52.

Becker, M. H. 1987. "The Cholesterol Saga: Whither Health Promotion?" *Annals of Internal Medicine*. April; 106(4):623–26.

Behall, K. M., D. J. Scholfield, J. G. Hallfrisch, J. L. Kelsay, and S. Reiser. 1984. "Seasonal Variation in Plasma Glucose and Hormone Levels in Adult Men and Women." *American Journal of Clinical Nutrition*. Dec.; 40(6 suppl.):1352–56.

Belasco, W. J. 1989. *Appetite for Change: How the Counterculture Took On the Food Industry, 1966–1988*. New York: Pantheon Books.

Bellisle, F., P. Laffort, and E. Köster. 2003. "Jacques Le Magnen (1916–2002)." *Chemical Senses*. 28:85–86.

Benedict, F. G. 1925. "The Measurement and Significance of Metabolism." In *Lectures on Nutrition* (Philadelphia: W. B. Saunders), 17–58.

————. 1915. *A Study of Prolonged Fasting*. Publication No. 203. Washington, D.C.: Carnegie Institution of Washington.

Benedict, F. G., and T. M. Carpenter. 1910. *The Metabolism and Energy Transformations of Healthy Man During Rest*. Washington, D.C.: Carnegie Institution of Washington.

Benedict, F. G., and L. E. Emmes. 1915. "A Comparison of the Basal Metabolism of Normal Men and Women." *Journal of Biology and Chemistry*. 20(3):253–62.

Benedict, F. G., W. R. Miles, P. Roth, and H. M. Smith. 1919. *Human Vitality and Efficiency Under Prolonged Restricted Diet*. Washington, D.C.: Carnegie Institution of Washington.

Bennett, W. 1987. "Dietary Treatments of Obesity." *Annals of the New York Academy of Science*. June 15; 499:250–63.

Bennett, W., and J. Gurin. 1982. *The Dieter's Dilemma: Eating Less and Weighing More*. New York: Basic Books.

Beresford, S. A., K. C. Johnson, C. Ritenbaugh, et al. 2006. "Low-Fat Dietary Pattern and

Risk of Colorectal Cancer: The Women's Health Initiative Randomized Controlled Dietary Modification Trial." *JAMA*. Feb. 8; 295(6):643–54.

Bergmann, G. von, and F. Stroebe. 1927. "Die Fettsucht." In *Handbuch der Biochemie des Menschen un der Tiere*, ed. C. Oppenheimer (Jena, Germany: Verlag von Gustav Fischer), 562–98.

Berland, T. 1983. *Rating the Diets*. Skokie, Ill.: Consumer Guide.

———. 1974. *Rating the Diets*. Skokie, Ill.: Consumer Guide.

Bernard, C. 1974. *Phenomena of Life Common to Animals and Vegetables*. Trans. R. P. Cook and M. A. Cook. Dundee: R. P. and M. A. Cook. [Originally published 1878.]

———. 1957. *An Introduction to the Study of Experimental Medicine*. Trans. H. C. Green. New York: Dover Publications. [Originally published 1865.]

Berneis, K. K., and R. M. Krauss. 2002. "Metabolic Origins and Clinical Significance of LDL Heterogeneity." *Journal of Lipid Research*. Sept.; 43(9):1363–79.

Berry, E. M., S. H. Blondheim, H. E. Eliahou, and E. Shafrir, eds. 1987. *Recent Advances in Obesity Research: V*. London: John Libbey.

Berson, S. A., and R. S. Yalow. 1970. "Insulin 'Antagonists' and Insulin Resistance." In *Diabetes Mellitus: Theory and Practice*, ed. M. Ellenberg and H. Rifkin (New York: McGraw-Hill), 388–423.

———. 1965. "Some Current Controversies in Diabetes Research." *Diabetes*. Sept.; 14(9):549–72.

Bertrand, H. A., F. T. Lynd, E. J. Masoro, and B. P. Yu. 1980. "Changes in Adipose Tissue Mass and Cellularity Through Adult Life of Rats Fed ad Libitum or a Life-Prolonging Restricted Diet." *Journal of Gerontology*. Nov.; 35(6):827–35.

Bishop, J. E. 1982. "Heart Attacks: A Test Collapses." *Wall Street Journal*. Oct. 6; 32.

———. 1961. "Helping the Heart: Major Research Effort Started to See If Diet Can Prevent Attacks." *Wall Street Journal*. Oct. 27; 1.

Bishop, M. J. 1989. "Retroviruses and Oncogenes II." Online at http://nobelprize.org/ nobel_prizes/medicine/laureates/1989/bishop-lecture.html.

Björntorp, P. 1985. "Adipose Tissue in Obesity." In Hirsch and Van Itallie, eds., 1985, 163–70.

———. 1976. "Effects of Physical Conditioning in Obesity." In Bray, ed., 1976b, 397–406.

Björntorp, P., M. Cairella, and A. N. Howard, eds. 1981. *Recent Advances in Obesity Research: III*. London: John Libbey.

Blackburn, H. n.d. "Ancel Keys: An Appreciation." Online at http://mbbnet.umn.edu/ firsts/blackburn_h.html.

———. 1975. "Contrasting Professional Views on Atherosclerosis and Coronary Disease." *New England Journal of Medicine*. Jan. 9; 292(2):105–7.

Blackburn, H., and D. R. Jacobs, Jr. 1989. "The Ongoing Natural Experiment of Cardiovascular Diseases in Japan." *Circulation*. March; 79(3):718–20.

Blair, M. C. 1923. "Freedom of Negro Races from Cancer." *British Medical Journal*. July 21; 130–31.

Blakeslee, A. 1973. "A (Possible) Silver Lining in Food Prices: Better Nutrition." *Washington Post, Times Herald*. Aug. 23; F14.

Blakeslee, A., and J. Stamler. 1966. *Your Heart Has Nine Lives: Nine Steps to Heart Health*. New York: Pocket Books.

Bleiler, R. E., E. S. Yearick, S. S. Schnur, I. L. Singson, and M. A. Ohlson. 1963. "Seasonal Variation of Cholesterol in Serum of Men and Women." *American Journal of Clinical Nutrition*. Jan.; 12:12–16.

Bliss, M. 1982. *The Discovery of Insulin*. Toronto: McClelland & Stewart.

Blix, G., ed. 1964. *Symposia of the Swedish Nutrition Foundation II: Occurrence, Causes, and Prevention of Overnutrition.* Uppsala: Almqvist & Wiksells.

Bloom, W. L. 1967. "Carbohydrates and Water Balance." *American Journal of Clinical Nutrition.* Feb.; 20(2):157–62.

———. 1962. "Inhibition of Salt Excretion by Carbohydrate. *Archives of Internal Medicine.* Jan.; 109:26–32.

———. 1958. "Fasting as an Introduction to the Treatment of Obesity." *Metabolism.* May; 8(3):214–20.

Bloom, W. L., and G. J. Azar. 1963. "Similarities of Carbohydrate Deficiency and Fasting. I. Weight Loss, Electrolyte Excretion, and Fatigue." *Archives of Internal Medicine.* Sept.; 112:333–37.

Bluher, M., B. B. Kahn, and C. R. Kahn. 2003. "Extended Longevity in Mice Lacking the Insulin Receptor in Adipose Tissue." *Science.* Jan. 24; 299(5606):572–74.

Blundell, J. E., and R. J. Stubbs. 1998. "Diet Composition and the Control of Food Intake." In Bray, Bouchard, and James, eds., 1998, 243–72.

Bode, A. M. 1997. "Metabolism of Vitamin C in Health and Disease." *Advances in Pharmacology.* 38:21–47.

Bodkin, N. L., J. S. Hannah, H. K. Ortmeyer, and B. C. Hansen. 1993. "Central Obesity in Rhesus Monkeys: Association with Hyperinsulinemia, Insulin Resistance and Hypertriglyceridemia?" *International Journal of Obesity and Related Metabolic Disorders.* Jan.; 17(1):53–61.

Boffey, P. M. 1987. "Cholesterol: Debate Flares over Wisdom in Widespread Reductions." *New York Times.* July 14; C1.

Boissonneault, G. A., C. E. Elson, and M. W. Pariza. 1986. "Net Energy Effects of Dietary Fat on Chemically Induced Mammary Carcinogenesis in F344 Rats." *Journal of the National Cancer Institute.* Feb.; 76(2):335–38.

Bonithon-Kopp, C., O. Kronborg, A. Giacosa, U. Rath, and J. Faivre. 2000. "Calcium and Fibre Supplementation in Prevention of Colorectal Adenoma Recurrence: A Randomised Intervention Trial. European Cancer Prevention Organisation Study Group." *Lancet.* Oct. 14; 356(9238):1300–1306.

Boodman, S. G. 1998. "The First Line of Defense: Those Old Standbys, Diet and Exercise, Are Key Weapons in the Fight Against Cancer." *Washington Post.* Feb. 10; Z25.

Boswell, J. 1992. *The Life of Samuel Johnson.* New York: Alfred A. Knopf. [Originally published 1791.]

Bouchard, C., A. Tremblay, J. P. Despres, et al. 1990. "The Response to Long-Term Overfeeding in Identical Twins." *New England Journal of Medicine.* May 24; 322(21):1477–82.

Bradley, R. F. 1971. "Cardiovascular Disease." In Marble et al., eds., 1971, 417–77.

Bravata, D. M., L. Sanders, J. Huang, et al. 2003. "Efficacy and Safety of Low-Carbohydrate Diets: A Systematic Review." *JAMA.* April 9; 289(14):1837–50.

Bray, G. A. 1998. "Classification and Evaluation of the Overweight Patient." In Bray, Bouchard, and James, eds., 1998, 831–54.

———. 1984. "Syndromes of Hypothalamic Obesity in Man." *Pediatric Annals.* July; 13(7):525–36.

———, ed. 1980. *Obesity: Comparative Methods of Weight Control.* Westport, Conn.: Technomic Publishing.

———, ed. 1979. *Obesity in America.* DHEW Publication No. (NIH) 79-359. Washington, D.C.: U.S. Government Printing Office.

———. 1978. "To Treat or Not to Treat—That Is the Question?" In Bray, ed., 1978, 248–65.

————, ed. 1978. *Recent Advances in Obesity Research: II.* London: Newman Publishing.

————, ed. 1976a. *The Obese Patient.* Philadelphia: W. B. Saunders.

————, ed. 1976b. *Obesity in Perspective.* DHEW Publication No. (NIH) 76-852. Washington, D.C.: U.S. Government Printing Office.

————, ed. 1975. *Obesity in Perspective.* DHEW Publication No. (NIH) 75-708. Washington, D.C.: U.S. Government Printing Office.

————. 1972. "New Developments in Diabetes, Obesity, and Insulin Resistance." *California Medicine.* Oct.; 119(4):22–26.

————. 1970. "The Myth of Diet in the Management of Obesity." *American Journal of Clinical Nutrition.* Sept.; 23(9):1141–48.

————. 1969. "Effect of Caloric Restriction on Energy Expenditure in Obese Patients." *Lancet.* Aug. 23; 2(7617):397–98.

Bray, G. A., and J. E. Bethune, eds. 1974. *Treatment and Management of Obesity.* New York: Harper & Row.

Bray, G. A., C. Bouchard, and W. P. James, eds. 1998. *Handbook of Obesity.* New York: Marcel Dekker.

Bray, G. A., S. J. Nielsen, and B. M. Popkin. 2004. "Consumption of High-Fructose Corn Syrup in Beverages May Play a Role in the Epidemic of Obesity." *American Journal of Clinical Nutrition.* April; 79(4):537–43.

Bray, G. A., and B. M. Popkin. 1999. "Dietary Fat Affects Obesity Rate." *American Journal of Clinical Nutrition.* Oct.; 70(4):572–73.

————. 1998. "Dietary Fat Intake Does Affect Obesity!" *American Journal of Clinical Nutrition.* Dec.; 68(6):1157–73.

Brehm, B. J., R. J. Seeley, S. R. Daniels, and D. A. D'Alessio. 2003. "A Randomized Trial Comparing a Very Low Carbohydrate Diet and a Calorie-Restricted Low Fat Diet on Body Weight and Cardiovascular Risk Factors in Healthy Women." *Journal of Clinical Endocrinology and Metabolism.* April; 88(4):1617–23.

Brewster, L., and M. F. Jacobson. 1978. *The Changing American Diet.* Washington, D.C.: Center for Science in the Public Interest.

Brillat-Savarin, J. A. 1986. *The Physiology of Taste.* Trans. M. F. Fisher. San Francisco: North Point Press. [Originally published 1825]

Broad, W. J. 1980 "Academy Says Curb on Cholesterol Not Needed." *Science.* June 20; 208 (4450):1354–5.

————. 1979a. "NIH Deals Gingerly with Diet-Disease Link." *Science.* June 15; 204(4398):1175–78.

————. 1979b. "Jump in Funding Feeds Research on Nutrition." *Science.* June 8; 204(4397):1060–64.

Brobeck, J. R. 1993. "Remembrance of Experiments Almost Forgotten." *Appetite.* Dec.; 21(3):225–31.

————. 1946. "Mechanisms of the Development of Obesity in Animals with Hypothalamic Lesions." *Physiological Reviews.* 26(4):541–59.

Brobeck, J. R., J. Tepperman, and C. N. Long. 1943. "Experimental Hypothalamic Hyperphagia in the Albino Rat." *Yale Journal of Biology and Medicine.* 15:831–53.

Brodie, B. R., R. P. Maickel, and D. N. Stern. 1965. "Autonomic Nervous System and Adipose Tissue." In Renold and Cahill, eds., 1965, 583–600.

Brody, J. E. 2004a. "Sane Weight Loss in a Carb-Obsessed World: High Fiber and Low Fat." *New York Times.* May 23; F7.

————. 2004b. "For Unrefined Healthfulness: Whole Grains." *New York Times.* March 4; F5.

————. 2002. "High-Fat Diet: Count Calories and Think Twice." *New York Times*. Sept. 10; F6.

————. 2000. "Vindication for the Maligned Fiber Diet." *New York Times*. May 23; F8.

————. 1999a. "Keep the Fiber Bandwagon Rolling, for Heart and Health." *New York Times*. July 20; F6.

————. 1999b. "Doubts Fail to Deter 'The Diet Revolution.' " *New York Times*. May 25; F7.

————. 1996. "Personal Health." *New York Times*. Dec. 25; C6.

————. 1985. *Jane Brody's Good Food Book: Living the High-Carbohydrate Way*. New York: W. W. Norton.

————. 1983a. "To Lose Weight, More Exercise Is the Key." *New York Times*. Aug. 3; C1.

————. 1983b. "Personal Health." *New York Times*. July 27; 6.

————. 1981a. *Jane Brody's Nutrition Book*. New York: W. W. Norton.

————. 1981b. "Nutritional Factors: What Does a Diet Actually Do?" *New York Times*. Feb. 24; C1.

————. 1981c. "The Science of Dieting: A Fight Against Mind and Metabolism." *New York Times*. Feb. 24; C1.

————. 1981d. "Long-Term Study Links Cholesterol to Hazard of Early Coronary Death." *New York Times*. Jan. 8:A1.

————. 1980. "Experts Assail Report Declaring Curb on Cholesterol Isn't Needed." *New York Times*. June 1; A1.

————. 1977. "Chemical Carriers of Cholesterol Put Light on Heart-Attack Puzzle." *New York Times*. Jan. 18; 13.

————. 1973. "Heart Association Strengthens Its Advice: Cut Down on Fats." *New York Times*. June 28; 54.

————. 1966. "Obesity Called Rising Health Hazard." *New York Times*. July 16; 1.

Bronson, F. H. 1988. *Mammalian Reproductive Biology*. Chicago: University of Chicago Press.

Brooks, C. M. 1946. "The Relative Importance of Changes in Activity in the Development of Experimentally Produced Obesity in the Rat." *American Journal of Physiology*. Dec.; 147:708–16.

Brooks, C. M., and E. F. Lambert. 1946. "A Study of the Effect of Limitation of Food Intake and the Method of Feeding on the Rate of Weight Gain During Hypothalamic Obesity in the Albino Rat." *American Journal of Physiology*. Dec.; 147:695–707.

Brooks, P. C., R. L. Klemke, S. Schon, J. M. Lewis, M. A. Schwartz, and D. A. Cheresh. 1997. "Insulin-Like Growth Factor Receptor Cooperates with Integrin $\alpha v \beta 5$ to Promote Tumor Cell Dissemination in Vivo." *Journal of Clinical Investigation*. March; 99(6):1390–98.

Brown, E. G., and M. A. Ohlson. 1946. "Weight Reduction of Obese Women of College Age. I. Clinical Results and Basal Metabolism." *Journal of the American Dietetic Association*. 22:849–57.

Brown, G. M., L. B. Cronk, and T. J. Boag. 1952. "The Occurrence of Cancer in an Eskimo." *Cancer*. Jan.; 5(1):142–43.

Brown, M. S., and J. L. Goldstein. 1985. "A Receptor-Mediated Pathway for Cholesterol Homeostasis." Online at http://nobelprize.org/nobel_prizes/medicine/laureates/1985/brown-lecture.html.

Brownell, K. D., and J. P. Foreyt, eds. 1986. *Handbook of Eating Disorders*. New York: Basic Books.

Brownell, K. D., and K. B. Horgen. 2004. *Food Fight: The Inside Story of the Food Industry, America's Obesity Crisis, and What We Can Do About It*. New York: McGraw-Hill.

Browner, W. S., J. Westenhouse, and J. A. Tice. 1991. "What If Americans Ate Less Fat? A Quantitative Estimate of the Effect on Mortality." *JAMA*. June 26; 265(24):3285–91.

Bruch, H. 1973. *Eating Disorders: Obesity, Anorexia Nervosa, and the Person Within*. New York: Basic Books.

———. 1957. *The Importance of Overweight*. New York: W. W. Norton.

———. 1944. "Dietary Treatment of Obesity in Childhood." *Journal of the American Dietetic Association*. 20:361–64.

———. 1940. "Obesity in Childhood III. Physiologic and Psychologic Aspects of the Food Intake of Obese Children." *American Journal of Diseases of Children*. 59:739–81.

Bruch, H., and G. Touraine. 1940. "Obesity in Childhood V. The Family Frame of Obese Children." *Psychosomatic Medicine*. April; 2(2):141–206.

Brunzell, J. D., R. L. Lerner, W. R. Hazzard, D. Porte, Jr., and E. L. Bierman. 1971. "Improved Glucose Tolerance with High Carbohydrate Feeding in Mild Diabetes." *New England Journal of Medicine*. March 11; 284(10):521–24.

Bucala, R., Z. Makita, T. Koschinsky, A. Cerami, and H. Vlassara. 1993. "Lipid Advanced Glycosylation: Pathway for Lipid Oxidation in Vivo." *Proceedings of the National Academy of Sciences*. July 15; 90(14): 6434–38.

Bullen, B. A., R. B. Reed, and J. Mayer. 1964. "Physical Activity of Obese and Nonobese Adolescent Girls Appraised by Motion Picture Sampling." *American Journal of Clinical Nutrition*. April; 14:211–23.

Bunn, H. F., K. H. Gabbay, and P. M. Gallop. 1978. "The Glycosylation of Hemoglobin: Relevance to Diabetes Mellitus." *Science*. April 7; 200(4337):21–27.

Bunn, H. F., and P. J. Higgins. 1981. "Reaction of Monosaccharides with Proteins: Possible Evolutionary Significance." *Science*. July 10; 213(4504):222–24.

Burkitt, D. P. 1991a. In interview with Max Blythe, Gloucestershire, 29 October 1991, Interview IV. Royal College of Physicians and Oxford Brookes Medical Sciences Video Archive (MSVA 64).

———. 1991b. In interview with Max Blythe, Gloucestershire, 26 February 1991, Interview III. The Royal College of Physicians and Oxford Brookes Medical Sciences Video Archive (MSVA 64).

———. 1979a. *Don't Forget Fibre in Your Diet*. London: Martin Dunitz Ltd.

———. 1979b. *Eat Right—to Keep Healthy and Enjoy Life More*. New York: Arco Publishing.

———. 1971. "Some Neglected Leads to Cancer Causation." *Journal of the National Cancer Institute*. Nov.; 47(5):913–19.

———. 1970. "Relationship as a Clue to Causation." *Lancet*. Dec. 12; 296(7685):1237–40.

———. 1969. "Related Disease—Related Cause?" *Lancet*. Dec. 6; 294(7632):1229–31.

Burkitt, D. P., and H. C. Trowell, eds. 1975. *Refined Carbohydrate Foods and Disease: Some Implications of Dietary Fibre*. New York: Academic Press.

Burkitt, D. P., A. R. Walker, and N. S. Painter. 1974. "Dietary Fiber and Disease." *JAMA*. Aug. 19; 229(8):1068–74.

———. 1972. "Effect of Dietary Fibre on Stools and the Transit-Times, and Its Role in the Causation of Disease." *Lancet*. Dec. 30; 300(7792):1408–12.

Burland, W. L., P. D. Samuel, and J. Yudkin, eds. 1974. *Obesity*. New York: Churchill Livingstone.

Burros, M. 2004a. "Read Any Good Nutrition Labels Lately?" *New York Times*. Dec. 1; F1.

———. 2004b. "Dieter Sues Atkins Estate and Company." *New York Times*. May 27; A4.

———. 2000. "Plenty of Reasons to Say, 'Please Pass the Fiber.' " *New York Times*. April 26; F5.

————. 1989. "Diet and Health: Old Lesson in New Detail." *New York Times*. March 2; A1.

————. 1977. "Carol Foreman: Moving Inside USDA." *Washington Post*. March 31; 81.

Burroughs, K. D., S. E. Dunn, J. C. Barrett, and J. A. Taylor. 1999. "Insulin-Like Growth Factor—I: A Key Regulator of Human Cancer Risk?" *Journal of the National Cancer Institute*. April 7; 91(7):579–81.

Butscher, E. B. 2003. *Sylvia Plath: Method and Madness: A Biography*. Tucson, Ariz.: Schaffner Press.

Butterfield, W. J. 1969. Introduction. In McLean Baird and Howard, eds., 1969, 3–9.

Buzdar, A. U. 2006. "Dietary Modification and Risk of Breast Cancer." *JAMA*. Feb. 8; 295(6):691–92.

Byers, T. 2000. "Diet, Colorectal Adenomas, and Colorectal Cancer." *New England Journal of Medicine*. April 20; 342(16):1206–7.

Byers, T., and C. Doyle. 2003. "Diet, Physical Activity and Cancer . . . What's the Connection?" Online at American Cancer Society Web site: http: //www.cancer.org/docroot/PED/content/PED_3_1x_Link_Between_Lifestyle_and_CancerMarch03.asp.

Byers, T., M. Nestle, A. McTiernan, et al. (American Cancer Society 2002 Nutrition and Physical Activity Guidelines Advisory Committee). 2002. "American Cancer Society Guidelines on Nutrition and Physical Activity for Cancer Prevention: Reducing the Risk of Cancer with Healthy Food Choices and Physical Activity." *CA: A Cancer Journal for Clinicians*. March–April; 52(2):92–119.

Cahill, G. F., Jr. 1978. "Obesity and Diabetes." In Bray, ed., 1978, 101–10.

————. 1975. "Weight Reduction Diets." In Bray, ed., 1975, 58–59.

————. 1971. "The Banting Memorial Lecture: Physiology of Insulin in Man." *Diabetes*. Dec.; 20(12):785–99.

Cahill, G. F., Jr., B. Jeanrenaud, B. Leboeuf, and A. E. Renold. 1959. "Effects of Insulin on Adipose Tissue." *Annals of the New York Academy of Sciences*. Sept. 25; 82: 4303–11.

Cahill, G. F., Jr., and O. E. Owen. 1968. "Some Observations on Carbohydrate Metabolism in Man." In Dickens, Randle, and Whelan, eds., 1968, vol. 1, 497–522.

Cahill, G. F., Jr., and A. E. Renold. 1965. "Regulation of Adipose Tissue Metabolism Within the Intact Organism." In Renold and Cahill, eds., 1965, 681–84.

Cahill, G. F., Jr., and R. L. Veech. 2003. "Ketoacids? Good Medicine?" *Transactions of the American Clinical Climatological Association*. 114: 149–61.

Call, D. L., and A. M. Sanchez. 1967. "Trends in Fat Disappearance in the United States, 1909–65." *Journal of Nutrition*. Oct.; 93 (2 suppl.):1–28.

Campbell, G. D. 1963. "Diabetes in Asians and Africans in and Around Durban." *South African Medical Journal*. Nov. 30; 37:1995–208.

Campbell, W. R. 1936. "Obesity and Its Treatment." *Canadian Medical Association Journal*. 34: 41–48.

Cannon, W. B. 1939. *The Wisdom of the Body*. New York: W. W. Norton.

————. 1929. "Organization for Physiological Homeostasis." *Physiological Reviews*. 9(3):399–431.

Carlson, L. A., and L. E. Bottiger. 1972. "Ischaemic Heart-Disease in Relation to Fasting Values of Plasma Triglycerides and Cholesterol: Stockholm Prospective Study." *Lancet*. April 22; 299(7756):865–68.

Carlson, M. G., and J. P. Campbell. 1993. "Intensive Insulin Therapy and Weight Gain in IDDM." *Diabetes*. Dec.; 42(12):1700–1707.

Carney, R. N., and A. P. Goldberg. 1984. "Weight Gain After Cessation of Cigarette Smoking: A Possible Role for Adipose-Tissue Lipoprotein Lipase." *New England Journal of Medicine*. March 8; 310(10):614–16.

Carpenter, K. J. 2000. *Beriberi, White Rice, and Vitamin B: A Disease, a Cause and a Cure.* Berkeley: University of California Press.

―――. 1986. *The History of Scurvy and Vitamin C.* New York: Cambridge University Press.

Carroll, K. K., and D. Kritchevsky, eds. 1993. *Nutrition and Disease Update: Cancer.* Champaign, Ill.: AOCS Press.

Cassidy, C. E. 1976. "Commemorative Tribute: Edwin B. Astwood." *Endocrinology.* Nov.; 99(3):1155–60.

Cassidy, M. 1946. "Coronary Disease: The Harveian Oration of 1946." *Lancet.* Oct. 26; 248(6426):587–90.

Castelli, W. P., J. T. Doyle, T. Gordon, et al. 1977. "HDL Cholesterol and Other Lipids in Coronary Heart Disease: The Cooperative Lipoprotein Phenotyping Study." *Circulation.* May; 55(5):767–72.

Castetter, E. F., and W. H. Bell. 1942. *Pima and Papago Indian Agriculture.* Albuquerque: University of New Mexico Press.

Catt, K. J. 1971. *An ABC of Endocrinology.* London: Lancet.

Cederquist, D. C., W. D. Brewer, A. N. Wagoner, D. Dunsing, and M. A. Ohlson. 1952. "Weight Reduction on Low-Fat and Low-Carbohydrate Diets." *Journal of the American Dietetic Association.* Feb.; 28(2):113–16.

Centers for Disease Control and Prevention. 2005. *National Diabetes Fact Sheet: General Information and National Estimates on Diabetes in the United States, 2005.* Atlanta, Ga.: U.S. Department of Health and Human Services, Centers for Disease Control and Prevention.

―――. 2001. "Physical Activity Trends—United States, 1990–1998." *Morbidity and Mortality Weekly Reports.* March 9; 50(9):166–69. Online at http://www.cdc.gov/mmwr/preview/mmwrhtml/mm5009a3.htm.

Cerami, A., H. Vlassara, and M. Brownlee. 1987. "Glucose and Aging." *Scientific American.* May; 256(5):90–96.

Chait, A., and E. L. Bierman. 1994. "Pathogenesis of Macrovascular Disease in Diabetes." In Kahn and Weir, eds., 1994, 648–64.

Chait, A., J. D. Brunzell, M. A. Denke, et al. 1993. "Rationale of the Diet-Heart Statement of the American Heart Association: Report of the Nutrition Committee." *Circulation.* Dec.; 88(6):3008–29.

Chajek-Shaul, T., E. M. Berry, E. Ziv, et al. 1990. "Smoking Depresses Adipose Lipoprotein Lipase Response to Oral Glucose." *European Journal of Clinical Investigation.* June; 20(3):299–304.

Chambers, W. H. 1952. "Max Rubner." *Journal of Nutrition.* 48: 1–12.

Chan, J. M., M. J. Stampfer, E. Giovannucci, et al. 1998. "Plasma Insulin-Like Growth Factor-I and Prostate Cancer Risk: A Prospective Study." *Science.* Jan. 23; 279(5350): 563–66.

Charles, R. H. 1907. "Discussion on Diabetes in the Tropics." *British Medical Journal.* Oct. 19; 1051–64.

Chave, S. P., J. N. Morris, S. Moss, and A. M. Semmence. 1978. "Vigorous Exercise in Leisure Time and the Death Rate: A Study of Male Civil Servants." *Journal of Epidemiology and Community Health.* Dec.; 32(4):239–43.

Chen, J., T. C. Campbell, J. Li, and R. Peto. 1990. *Diet, Lifestyle, and Mortality in China: A Study of Characteristics of 65 Chinese Counties.* Joint publication: Oxford University Press; Ithaca, N.Y.: Cornell University Press; and Beijing: People's Medical Publishing House.

Chlebowski, R. T., S. L. Hendrix, R. D. Langer, et al. (WHI Investigators). 2003. "Influence of Estrogen Plus Progestin on Breast Cancer and Mammography in Healthy Postmenopausal Women: The Women's Health Initiative Randomized Trial." *JAMA.* June 25; 289(24):3243–53.

Christakis, G., S. H. Rinzler, M. Archer, and A. Kraus. 1966. "Effect of the Anti-Coronary Club Program on Coronary Heart Disease. Risk-Factor Status." *JAMA.* Nov. 7; 198(6):597–604.

Christakis, G., S. H. Rinzler, M. Archer, et al. 1966. "The Anti-Coronary Club: A Dietary Approach to the Prevention of Coronary Heart Disease—a Seven-Year Report." *American Journal of Public Health and the Nation's Health.* Feb.; 56(2):299–314.

Christlieb, A. R., A. S. Krolweski, and J. H. Warram. 1994. "Hypertension." In Kahn and Weir, eds., 1994, 817–35.

Cioffi, L. A., W. P. James, and T. B. Van Itallie, eds. 1981. *The Body Weight Regulatory System: Normal and Disturbed Mechanisms.* New York: Raven Press.

Clancy, D. J., D. Gems, L. G. Harshman, et al. 2001. "Extension of Life-Span by Loss of CHICO, a Drosophila Insulin Receptor Substrate Protein." *Science.* April 6; 292(5514):104–6.

Clarke, H. T. 1941. "Rudolf Schoenheimer." *Science.* Dec. 12; 94(2450):553–54.

Cleave, T. L. 1975. *The Saccharine Disease: The Master Disease of Our Time.* New Canaan, Conn.: Keats Publishing.

———. 1962. *Peptic Ulcer.* Bristol: John Wright & Sons.

———. 1956. "The Neglect of Natural Principles in Current Medical Practice." *Journal of the Royal Naval Medical Service.* 42(2):54–83.

Cleave, T. L., and G. D. Campbell. 1966. *Diabetes, Coronary Thrombosis, and the Saccharine Disease.* Bristol: John Wright & Sons.

Clendening, L. 1936. *The Balanced Diet.* New York: D. Appleton–Century.

Clifford, N. J. 1963. "Coronary Heart Disease and Hypertension in the White Mountain Apache Tribe." *Circulation.* Nov.; 28: 926–31.

Cohen, A. M. 1963. "Effect of Environmental Changes on Prevalence of Diabetes and of Atherosclerosis in Various Ethnic Groups in Israel." In *The Genetics of Migrant and Isolate Populations,* ed. E. Goldschmidt. New York: Williams & Wilkins, 127–30.

Cohen, A. M., S. Bavly, and R. Poznanski. 1961. "Change of Diet of Yemenite Jews in Relation to Diabetes and Ischaemic Heart-Disease." *Lancet.* Dec. 23; 278(7217): 1399–1401.

Cohen, B. M. 1954. "Diabetes Mellitus Among Indians of the American Southwest: Its Prevalence and Clinical Characteristics in a Hospitalized Population." *Annals of Internal Medicine.* March; 40(3):588–99.

Cohen, M. N. 1989. *Health and the Rise of Civilization.* New Haven, Conn.: Yale University Press.

———. 1987. "The Significance of Long-Term Changes in Human Diet and Food Economy." In Harris and Ross, eds., 1987, 261–85.

Cohen, P., M. Miyazaki, N. D. Socci, et al. 2002. "Role for Stearoyl-CoA Desaturase-1 in Leptin-Mediated Weight Loss." *Science.* July 12; 297(5579):240–43.

Cohn, V. 1981. "Linking of Heart Disease to High-Cholesterol Diet Reinforced by New Data." *Washington Post.* Jan. 8; A14.

———. 1980. "A Passion to Keep Fit: 100 Million Americans Exercising." *Washington Post.* Aug. 31; A1.

Colburn, D. 1995. "Should You Shake Salt?" *Washington Post.* Sept. 5; Z08.

Collip, P. J., ed. 1975. *Childhood Obesity.* Acton, Mass.: Publishing Sciences Group.

Committee on Diet, Nutrition, and Cancer, Assembly of Life Sciences, National Research Council [NRC]. 1982. *Diet, Nutrition, and Cancer.* Washington, D.C.: National Academy Press.

Conn, J. E. 1944. "Obesity II: Etiological Aspects." *Physiological Reviews.* 24(1):31–45.

Consensus Conference. 1985. "Lowering Blood Cholesterol to Prevent Heart Disease." *JAMA.* April 12; 253(14):2080–86.

Cook, C. H., and I. T. Whittemore. 1893. *Among the Pimas, or, The Mission to the Pima and Maricopa Indians.* Albany, N.Y.: Ladies' Union Mission School Association.

Cooper, R., K. Liu, M. Trevisan, W. Miller, and J. Stamler. 1983. "Urinary Sodium Excretion and Blood Pressure in Children: Absence of a Reproducible Association." *Hypertension.* Jan.–Feb.; 5(1):135–39.

Cooper, T. 1972. "Arteriosclerosis, Policy, Polity, and Parity." *Circulation.* Feb.; 45(2):433–40.

Cooperative Study of Lipoproteins and Atherosclerosis. 1956. "Evaluation of Serum Lipoprotein and Cholesterol Measurements as Predictors of Clinical Complications of Atherosclerosis: Report of a Cooperative Study of Lipoproteins and Atherosclerosis." *Circulation.* Oct.; 14(4, pt. 2):691–742.

Cordain, L., J. B. Miller, S. B. Eaton, N. Mann, S. H. Holt, and J. D. Speth. 2000. "Plant-Animal Subsistence Ratios and Macronutrient Energy Estimations in Worldwide Hunter-Gatherer Diets." *American Journal of Clinical Nutrition.* March; 71(3):682–92.

Coulston, A. M., and G. M. Reaven. 1997. "Much Ado About (Almost) Nothing." *Diabetes Care.* March; 20(3):241–43.

Cowie, C. C., K. F. Rust, D. D. Byrd-Holt, et al. 2006. "Prevalence of Diabetes and Impaired Fasting Glucose in Adults in the U.S. Population: National Health and Nutrition Examination Survey, 1999–2002." *Diabetes Care.* June; 29(6):1263–68.

Cox, B. D., M. J. Whichelow, W. J. Butterfield, and P. Nicholas. 1974. "Peripheral Vitamin C Metabolism in Diabetics and Non-Diabetics: Effect of Intra-Arterial Insulin." *Clinical Science & Molecular Medicine.* July; 47(1):63–72.

Craft, S., J. Newcomer, S. Kanne, et al. 1996. "Memory Improvement Following Induced Hyperinsulinemia in Alzheimer's Disease." *Neurobiology of Aging.* Jan.–Feb.; 17(1):123–30.

Crandall, R. W., and L. B. Lave, eds. 1981. *The Scientific Basis of Health and Safety Regulation: Studies in the Regulation of Economic Activity.* Washington, D.C.: Brookings Institution.

Crapo, P. A., G. Reaven, and J. Olefsky. 1977. "Postprandial Plasma-Glucose and -Insulin Responses to Different Complex Carbohydrates." *Diabetes.* 26(12):1178–83.

Critser, G. 2003. *Fat Land: How Americans Became the Fattest People in the World.* New York: Houghton Mifflin.

Croftan. A. C. 1906. "The Dietetics of Obesity." *JAMA.* July–Dec.; 47: 820–23.

Cross, J. 1974. "The Politics of Food." *Nation.* Aug. 17; 114–16.

Cruz, A. B., Jr., D. S. Amatuzio, F. Grande, and L. J. Hay. 1961. "Effect of Intra-Arterial Insulin on Tissue Cholesterol and Fatty Acids in Alloxan-Diabetic Dogs." *Circulation Research.* Jan.; 9(1):39–43.

Cummings, L. C. 1986. "The Political Reality of Artificial Sweeteners." In Sapolsky, ed., 1986, 116–40.

Cummings, R. O. 1940. *The American and His Food: A History of Food Habits in the United States.* Chicago: University of Chicago Press.

Cunningham, B. 1996. "W. H. 'Bold' Emory's Notes of a Military Reconnaissance: A Survey of Arizona's Gila River, 1846." *Smoke Signal.* Fall; 66:106–15.

Cunningham, J. J. 1998. "The Glucose/Insulin System and Vitamin C: Implications in Insulin-Dependent Diabetes Mellitus." *Journal of the American College of Nutrition.* April; 17(20):105–8.

————. 1988. "Altered Vitamin C Transport in Diabetes Mellitus." *Medical Hypotheses.* Aug.; 26(4):263–65.

Curtain, L. S. 1949. *By the Prophet of the Earth: Ethnobotany of the Pima.* Tucson: University of Arizona Press.

Cutting, W. C. 1943. "The Treatment of Obesity." *Journal of Clinical Endocrinology.* Feb.; 3(2):85–88.

Dabelea, D., W. C. Knowler, and D. J. Pettitt. 2000. "Effect of Diabetes in Pregnancy on Offspring: Follow-Up Research in the Pima Indians." *Journal of Maternal and Fetal Medicine.* Jan.–Feb.; 9(1):83–88.

Dana, R. H., Jr. 1946. *Two Years Before the Mast.* Garden City, N.Y.: Doubleday. [Originally published 1840.]

Dancel, J. F. 1864. *Obesity, or Excessive Corpulence: The Various Causes and the Rational Means of Cure.* Trans. M. Barrett. Toronto: W. C. Chewett.

Danforth, E., Jr. 1985. "Diet and Obesity." *American Journal of Clinical Nutrition.* May; 41(5 suppl.):1132–45.

Dansinger, M. L., J. A. Gleason, J. L. Griffith, H. P. Selker, and E. J. Schaefer. 2005. "Comparison of the Atkins, Ornish, Weight Watchers, and Zone Diets for Weight Loss and Heart Disease Risk Reduction: A Randomized Trial." *JAMA.* Jan. 5; 293(1):43–53.

Darwin, C. 2004. *The Variation of Animals and Plants Under Domestication.* Vol. 2. Whitefish, Mont.: Kessinger Publishing. [Originally published 1868.]

————. 1989. *The Voyage of the Beagle.* London: Penguin Books. [Originally published 1839.]

Davenport, R. J. 2003. "Power to the People." *Science of Aging Knowledge Environment.* Dec. 17; 2003(50): ns8. Online at http://sageke.sciencemag.org/cgi/content/full/2003/50/ns8.

Davidson, S., and R. Passmore. 1963. *Human Nutrition and Dietetics.* 2nd edition. Edinburgh: E. & S. Livingstone.

Davies, L. E. 1950. "$4,000,000 Is Raised in Heart Campaign." *New York Times.* June 25; 37.

Davis, G. P., Jr. 1962. N. B. Carmony and D. F. Brown, eds. *Man and Wildlife in Arizona: The American Exploration Period, 1824–1865.* Scottsdale, Ariz.: Arizona Game and Fish Dept.

Davis, J. D., and D. Wirtshafter. 1978. "Set Points or Settling Points for Body Weight? A Reply to Mrosovsky and Powley." *Behavioral Biology.* Nov.; 24(3):405–11.

Dawber, T. R. 1980. *The Framingham Study: The Epidemiology of Atherosclerotic Disease.* Cambridge, Mass.: Harvard University Press.

————. 1978. "Annual Discourse—Unproved Hypotheses." *New England Journal of Medicine.* Aug. 31; 299(9):452–58.

————. 1962. "The Epidemiology of Coronary Heart Disease—The Framingham Enquiry." In "Symposium on Arteriosclerosis," *Proceedings of the Royal Society of Medicine.* April; 55:265–71.

Day, J., M. Carruthers, A. Bailey, and D. Robinson. 1976. "Anthropometric, Physiological, and Biochemical Differences Between Urban and Rural Maasai." *Atherosclerosis.* 23(2):357–61.

Dayton, S. D., M. L. Pearce, S. Hashimoto, W. J. Dixon, and U. Tomiyasu. 1969. "A Controlled Clinical Trial of a Diet High in Unsaturated Fat in Preventing Complications of Atherosclerosis." *Circulation.* July; 40(1):II-1-62.

Debry, G., R. Rohr, R. Azouaou, L. Vassilitch, and G. Mottaz. 1973. "Influence du Multi-fractionnement et de la Composition en Lipides et en Glucides d'une Ration Hypocalorique sur la Réduction des Obèses." In Apfelbaum, ed., 1973, 305–10.

DeFronzo, R. A. 1997. "Insulin Resistance: A Multifaceted Syndrome Responsible for NIDDM, Obesity, Hypertension, Dyslipidaemia, and Atherosclerosis." *Netherlands Journal of Medicine.* May; 50(5):191–97.

———. 1992. "Insulin Resistance, Hyperinsulinemia, and Coronary Artery Disease: A Complex Metabolic Web." *Journal of Cardiovascular Pharmacology.* 20 (11 suppl.): s1–16.

———. 1981a. "Insulin and Renal Sodium Handling: Clinical Implications." In Björntorp, Cairella, and Howard, eds., 1981, 32–41.

———. 1981b. "The Effect of Insulin on Renal Sodium Metabolism: A Review with Clinical Implications." *Diabetologia.* Sept.; 21(3):165–71.

DeFronzo, R. A., J. D. Tobin, and R. Andres. 1979. "Glucose Clamp Technique: A Method for Quantifying Insulin Secretion and Resistance." *American Journal of Physiology.* Sept.; 237(3):E214–23.

Del Giudice, M. E., I. G. Fantus, S. Ezzat, G. McKeown-Eyssen, D. Page, and P. J. Goodwin. 1998. "Insulin and Related Factors in Premenopausal Breast Cancer Risk." *Breast Cancer Research and Treatment.* Jan.; 47(2):111–20.

Deshaies, Y., A. Dagnault, A. Boivin, and D. Richard. 1994. "Tissue- and Gender-Specific Modulation of Lipoprotein Lipase in Intact and Gonadectomised Rats Treated with Di-Fenfluramine." *International Journal of Obesity and Related Metabolic Disorders.* June; 18(6):405–11.

Devereux, S. n.d. "Famine in the Twentieth Century." IDS Working Paper 105. Online at http://www.ntd.co.uk/idsbookshop/details.asp?id=541.

Diamond, J. 2003. "The Double Puzzle of Diabetes." *Nature.* June 5; 423(6940): 599–602.

———. 1987. "The Worst Mistake in the History of the Human Race." *Discover.* May; 64–66.

Dickens, F., P. J. Randle, and W. J. Whelan, eds. 1968. *Carbohydrate Metabolism and Its Disorders.* 2 vols. New York: Academic Press.

Dietary Guidelines Advisory Committee. 2005. "2005 Dietary Guidelines Advisory Committee Report." U.S. Department of Health and Human Services and U.S. Department of Agriculture. Online at http://www.health.gov/DietaryGuidelines/dga2005/report/.

Dills, W. L., Jr. 1983. "Protein Fructosylation: Fructose and the Maillard Reaction." *American Journal of Clinical Nutrition.* Nov.; 58(5 suppl.):779S–787S.

Dobyns, H. F. 1989. *The Pima-Maricopa.* New York: Chelsea House.

———. 1978. "Who Killed the Gila." *Journal of Arizona History.* 19(1):17–30.

Dole, V. P. 1956. "A Relation Between Non-Esterified Fatty Acids in Plasma and the Metabolism of Glucose." *Journal of Clinical Investigation.* Feb.; 35(2):150–54.

Doll, R., and B. Armstrong. 1981. "Cancer." In Trowell and Burkitt, eds., 1981, 93–112.

Doll, R., and R. Peto. 1981. "The Causes of Cancer: Quantitative Estimates of Avoidable Risks of Cancer in the United States Today." *Journal of the National Cancer Institute.* June; 66(6):1191–308.

Donahoo, W. T., D. R. Jensen, T. Y. Shepard, and R. H. Eckel. 2000. "Seasonal Variation in Lipoprotein Lipase and Plasma Lipids in Physically Active, Normal Weight Humans." *Journal of Clinical Endocrinology & Metabolism.* Sept.; 85(9):3065–68.

Donaldson, B. F. 1962. *Strong Medicine.* Garden City, N.Y.: Doubleday.

Donnelly, R., A. M. Emslie-Smith, I. D. Gardner, and A. D. Morris. 2000. "ABC of Arterial and Venous Disease: Vascular Complications of Diabetes." *British Journal of Medicine.* April 15; 320(7241):1062–66.

Donnison, C. P. 1938. *Civilization and Disease.* Baltimore: William Wood.

———. 1929. "Blood Pressure in the African Native." *Lancet.* Jan. 5; 213(5499):6–7.

Donohue, P. G. 1988. "Advice Softened on Table Sugar." *St. Petersburg Times.* Sept. 13; 2D.

Dousset, J. C., J. B. Gutierres, and N. Dousset. 1986. "Hypercholesterolaemia After Administration of Nicotine Chewing Gum." *Lancet.* Dec. 13; 328(8520):1393–94.

Drenick, E. J. 1976. "Weight Reduction by Prolonged Fasting." In Bray, ed., 1976b, 341–61.

Drenick, E. J., M. E. Swendseid, W. H. Blahd, and S. G. Tuttle. 1964. "Prolonged Starvation as Treatment for Severe Obesity." *JAMA.* Jan. 11; 187:100–105.

Dreon, D. M., H. A. Fernstrom, H. Campos, P. Blanche, P. T. Williams, and R. M. Krauss. 1998. "Change in Dietary Saturated Fat Intake Is Correlated with Change in Mass of Large Low-Density-Lipoprotein Particles in Men." *American Journal of Clinical Nutrition.* May; 67(5):828–36.

Du Bois, E. F. 1946. Introduction: "The Physiological Side." In Stefansson 1946, ix–xiii.

———. 1936. *Basal Metabolism in Health and Disease.* 2nd edition. Philadelphia: Lea & Febiger.

Duff, G. L., and G. C. McMillan. 1949. "Effect of Alloxan Diabetes on Experimental Cholesterol Atherosclerosis in the Rabbit." *Journal of Experimental Medicine.* 89(6):611–30.

Duncan, D., and K. Burns. 1998. *Lewis & Clark: The Journey of the Corps of Discovery.* London: Pimlico.

Duncan, G. G. 1935. *Diabetes Mellitus and Obesity.* Philadelphia: Lea & Febiger.

Dunlop, D. M., and R. M. Murray-Lyon. 1931. "A Study of 523 Cases of Obesity." *Edinburgh Medical Journal.* Oct.; 38:561–77.

Durnin, J. V., and R. Passmore. 1967. *Energy, Work and Leisure.* London: Heinemann Educational Books.

Dwyer, J. 1985. "Classifying Current Popular and Fad Diets." In Hirsch and Van Itallie, eds., 1985, 179–91.

Dwyer, J., and D. Lu. 1993. "Popular Diets for Weight Loss: From Nutritionally Hazardous to Healthful." In Stunkard and Wadden, eds., 1993, 231–52.

Dyer, A. R., J. Stamler, O. Paul, et al. 1981. "Serum Cholesterol and Risk of Death from Cancer and Other Causes in Three Chicago Epidemiological Studies." *Journal of Chronic Diseases.* 34(6):249–60.

Eades, M. R., and M. D. Eades. 1996. *Protein Power.* New York: Bantam Books.

Easterbrook, G. 1997. "Forgotten Benefactor of Humanity." *Atlantic Monthly.* Jan.; 279(1):75–82.

Eaton, S. B., and M. Konner. 1985. "Paleolithic Nutrition: A Consideration of Its Nature and Current Implications." *New England Journal of Medicine.* Jan. 31; 312(5):283–89.

Eaton, S. B., M. Shostak, and M. Konner. 1988. *The Paleolithic Prescription.* New York: Harper & Row.

Ebrahim, S., A. Beswick, M. Burke, G. Davey Smith. 2006. "Multiple Risk Factor Interventions for Primary Prevention of Coronary Heart Disease." *Cochrane Database of Systematic Reviews.* Oct. 18; (4):CD001561.

Ebstein, W. 1884. *Corpulence and Its Treatment, on Physiological Principles.* Trans. E. W. Hoeber. New York: Brentano Bros.

Eccleston, R. 1950. *Overland to California on the Southwestern Trail: Diary of Robert Eccleston.* Berkeley: University of California Press

Eckel, R. H. 2003. "Obesity: A Disease or a Physiological Adaptation for Survival." In Eckel, ed., 2003, 3–30.

Eckel, R. H., ed. 2003. *Obesity: Mechanisms and Clinical Management*. New York: Lippincott Williams & Wilkins.

Efstratiadis, A. 1998. "Genetics of Mouse Growth." *International Journal of Developmental Biology*. 42(7):955–76.

Ehrlich, P. R. 1968. *The Population Bomb*. New York: Ballantine Books.

Emerson, H., and L. D. Larimore. 1924. "Diabetes Mellitus—A Contribution to Its Epidemiology Based Chiefly on Mortality Statistics." *Archives of Internal Medicine*. Nov.; 34(5):585–630.

Encyclopædia Britannica. "Sweetener" entry. Retrieved May 20, 2004, from Encyclopædia Britannica Premium Service. Online at http://www.britannica.com/eb/article?eu=72467.

Eppright, E. S., P. Swanson, and C. A. Iverson, eds. 1955. *Weight Control: A Collection of Papers Presented at the Weight Control Colloquium*. Ames: Iowa State College Press.

Erlichman, J., A. L. Kerbey, and W. P. James. 2002. "Physical Activity and Its Impact on Health Outcomes. Paper 2, Prevention of Unhealthy Weight Gain and Obesity by Physical Activity: An Analysis of the Evidence." *Obesity Review*. Nov.; 3(4):273–87.

Ernst, N. D., and R. I. Levy. 1984. "Diet and Cardiovascular Disease." In Olson et al. eds., 1984, 724–39.

Eschwege, E., J. L. Richard, N. Thibult, et al. 1985. "Coronary Heart Disease Mortality in Relation with Diabetes, Blood Glucose and Plasma Insulin Levels: The Paris Prospective Study, Ten Years Later." *Hormone and Metabolic Research Supplement Series*. 15:41–46.

Evans, F. A. 1953. "A Practical Regimen for the Cure of Obesity." In Barr et al., eds., 1953, 130–35.

———. 1947. "Obesity." In *Diseases of Metabolism*, 2nd edition, ed. G. G. Duncan (Philadelphia: W. B. Saunders), 524–94.

Evans, F. A., and J. M. Strang. 1931. "The Treatment of Obesity with Low Caloric Diets." *JAMA*. July–Dec.; 97:1063–69.

Ezell, P. H. 1961. "The Hispanic Acculturation of the Gila River Pimas." American Anthropological Association Memoir 90. Oct.; 63(5 pt. 2).

Fahlen, M., A. Oden, P. Björntorp, and G. Tibblin. 1971. "Seasonal Influence on Insulin Secretion in Man." *Clinical Science*. Nov.; 41(5):453–58.

Falta, W. 1923. *Endocrine Diseases, Including Their Diagnosis and Treatment*. 3rd edition. Trans. and ed. M. K. Meyers. Philadelphia: P. Blakiston's Son.

Farquhar, J. W., A. Frank, R. C. Gross, and G. M. Reaven. 1966. "Glucose, Insulin, and Triglyceride Responses to High and Low Carbohydrate Diets in Man." *Journal of Clinical Investigation*. Oct.; 45(10):1648–56.

Farris, W., S. Mansourian, Y. Chang, et al. 2003. "Insulin-Degrading Enzyme Regulates the Levels of Insulin, Amyloid Beta-Protein, and the Beta-Amyloid Precursor Protein Intracellular Domain in Vivo." *Proceedings of the National Academy of Sciences*. April 1; 100(7):4162–67.

Farris, W., S. Mansourian, M. A. Leissring, et al. 2004. "Partial Loss-of-Function Mutations in Insulin-Degrading Enzyme That Induce Diabetes Also Impair Degradation of Amyloid Beta-Protein." *American Journal of Pathology*. April; 164(4):1425–34.

Federal Trade Commission. 1919. *Report of the Federal Trade Commission on the Meat-Packing Industry*. Summary and pt. I. June 24. Washington D.C.: U.S. Government Printing Office.

Feener, E. P., and V. J. Dzau. 2005. "Pathogenesis of Cardiovascular Disease in Diabetes." In Kahn et al., eds., 2005, 867–84.

Feinleib, M. 1983. "Review of the Epidemiological Evidence for a Possible Relationship

Between Hypocholesterolemia and Cancer." *Cancer Research.* May; 43(5 suppl.): 2503s–7s.

————. 1982. "Summary of a Workshop on Cholesterol and Non-Cardiovascular Disease Mortality." *Preventive Medicine.* May; 11(3):360–67.

————. 1981. "On a Possible Inverse Relationship Between Serum Cholesterol and Cancer Mortality." *American Journal of Epidemiology.* July; 114(1):5–10.

Feinstein, A. R. 1960. "The Treatment of Obesity: An Analysis of Methods, Results, and Factors Which Influence Success." *Journal of Chronic Diseases.* April; 11:349–93.

Festinger, L. 1957. *A Theory of Cognitive Dissonance.* Stanford, Calif.: Stanford University Press.

Feynman, R. P. 1985. *Surely You're Joking, Mr. Feynman (Adventures of a Curious Character).* New York: W. W. Norton.

————. 1967. *The Character of Physical Law.* Cambridge, Mass.: MIT Press.

Filozof, C., M. C. Fernandez Pinilla, and A. Fernandez-Cruz. 2004. "Smoking Cessation and Weight Gain." *Obesity Reviews.* May; 5(2):95–103.

Fisher, C., W. R. Ingram, and S. W. Ranson. 1938. *Diabetes Insipidus and the Neuro-Hormonal Control of Water Balance: A Contribution to the Structure and Function of the Hypothalammico-Hypophyseal System.* Ann Arbor, Mich.: Edwards Brothers.

Flatt, J. P. 1978. *The Composition of Energy Expenditure and Its Influence on Energy Balance.* In Bray, ed., 1978, 211–28.

Florant, G. L., A. K. Lawrence, K. Williams, and W. A. Bauman. 1985. "Seasonal Changes in Pancreatic B-Cell Function in Euthermic Yellow-Bellied Marmots." *American Journal of Physiology.* Aug.; 249(2, pt. 2):R159–65.

Fogelholm, M., and K. Kukkonen-Harjula. 2000. "Does Physical Activity Prevent Weight Gain—A Systematic Review." *Obesity Reviews.* Oct.; 1(2):95–111.

Fog-Pulson, M. 1949. "Pheochromocytoma in Greenlanders of Pure Eskimo Type." *Nordisk Medicine.* 41:416.

Food and Nutrition Board, National Research Council [NRC]. 1980. *Toward Healthful Diets.* Pamphlet. Washington, D.C.: National Academy of Sciences.

Ford, E. S., A. II. Mokdad, W. H. Giles, and D. W. Brown. 2003. "The Metabolic Syndrome and Antioxidant Concentrations: Findings from the Third National Health and Nutrition Examination Survey." *Diabetes.* Sept.; 52(9):2346–52.

Foster, G. D., H. R. Wyatt, J. O. Hill, et al. 2003. "A Randomized Trial of a Low-Carbohydrate Diet for Obesity." *New England Journal of Medicine.* 22; 348(21): 2082–90.

Foster, N. B. 1922. "Treatment of Hypertension." *JAMA.* 79:1089–94.

Fouché, F. P. 1923. "Freedom of Negro Races from Cancer." *British Medical Journal.* June 30:1116.

Fox, C. S., M. J. Pencina, J. B. Meigs, R. S. Vasan, Y. S. Levitzky, and R. B. D'Agostino, Sr. 2006. "Trends in the Incidence of Type 2 Diabetes Mellitus from the 1970s to the 1990s: The Framingham Heart Study." *Circulation.* June 27; 113(25):2914–18.

Frantz, I. D. Jr., E. A. Dawson, P. L. Ashman, et al. 1989. "Test of effect of lipid lowering by diet on cardiovascular risk: The Minnesota Coronary Survey." *Arteriosclerosis.* Jan.–Feb.; 9(1):129–35.

Frantz, I. D., Jr., E. A. Dawson, K. Kuba, E. R. Brewer, L. C. Gatewood, and G. E. Bartsch. 1975. "The Minnesota Coronary Survey: Effect of Diet on Cardiovascular Events and Deaths." *Circulation.* Oct.; 51 and 52 (2 suppl.):II-4.

Franz, M. J., J. P. Bantle, C. A. Beebe, et al. 2003. "American Diabetes Association: Evidence-Based Nutrition Principles and Recommendations for the Treatment and Prevention of Diabetes and Related Complications." *Diabetes Care.* Jan.; 26 (suppl. 1):S51–61.

Fraser, G. E., J. T. Anderson, N. Foster, R. Goldberg, D. Jacobs, and H. Blackburn. 1983. "The Effect of Alcohol on Serum High Density Lipoprotein (HDL): A Controlled Experiment." *Atherosclerosis.* March; 46(3):275–86.

Frayn, K. N. 1996. *Metabolic Regulation: A Human Perspective.* London: Portland Press.

Fredrickson, D. S., R. I. Levy, and R. S. Lees. 1967a. "Fat Transport in Lipoproteins—An Integrated Approach to Mechanisms and Disorders." *New England Journal of Medicine.* Feb. 2; 276(5):273–81.

———. 1967b. "Fat Transport in Lipoproteins—An Integrated Approach to Mechanisms and Disorders." *New England Journal of Medicine.* Jan. 26; 276(4):215–25.

———. 1967c. "Fat Transport in Lipoproteins—An Integrated Approach to Mechanisms and Disorders." *New England Journal of Medicine.* Jan. 19; 276(3):148–56.

———. 1967d. "Fat Transport in Lipoproteins—An Integrated Approach to Mechanisms and Disorders." *New England Journal of Medicine.* Jan. 12; 276(2):94–103.

———. 1967e. "Fat Transport in Lipoproteins—An Integrated Approach to Mechanisms and Disorders." *New England Journal of Medicine.* Jan. 5; 276(1):34–42.

French, J. M. 1907. *A Text-Book of the Practice of Medicine, for Students and Practitioners.* 3rd, revised edition. New York: William Wood.

Freuchen, P. 1961. *Peter Freuchen's Book of the Eskimos.* Cleveland: World Publishing Company.

Frideman, M., R. H. Rosenman, and V. Carroll. 1958. "Changes in the Serum Cholesterol and Blood Clotting Time in Men Subjected to Cyclic Variation of Occupational Stress." *Circulation.* May; 17(5):852–61.

Friedman, J. M. 2004. "Modern Science Versus the Stigma of Obesity." *Nature Medicine.* June; 10(6):563–69.

———. 2003. "A War on Obesity, Not the Obese." *Science.* Feb. 7; 299(5608):856–58.

Friedman, M. 1969. *Pathogenesis of Coronary Artery Disease.* New York: McGraw-Hill.

Friedman, M. I., and E. M. Stricker. 1976. "The Physiological Psychology of Hunger: A Physiological Perspective." *Psychological Review.* Nov.; 83(6):409–31.

Friend, B., L. Page, and R. Marston. 1979. "Food Consumption Patterns, U.S.A.: 1909–13 to 1976." In *Nutrition, Lipids, and Coronary Heart Disease: A Global View,* ed. R. I. Levy, B. M. Rifkind, B. H. Dennis, and N. Ernst. New York: Raven Press, 489–522.

Frisch, R. E., and J. W. McArthur. 1974. "Menstrual Cycles: Fatness as a Determinant of Minimum Weight for Height Necessary for Their Maintenance or Onset." *Science.* Sept. 13; 185(4155):949–51.

Fritz, I. B. 1961. "Factors Influencing the Rates of Long-Chain Fatty Acid Oxidation and Synthesis in Mammalian Systems." *Physiological Reviews.* Jan.; 41:52–129.

Frohman, L. A., L. L. Bernardis, J. D. Schnatz, and L. Burek. 1969. "Plasma Insulin and Triglyceride Levels After Hypothalamic Lesions in Weanling Rats." *American Journal of Physiology.* June; 216(6):1496–501.

Fuchs, C. S., E. L. Giovannucci, G. A. Colditz, et al. 1999. "Dietary Fiber and the Risk of Colorectal Cancer and Adenoma in Women." *New England Journal of Medicine.* Jan. 21; 340(3):169–76.

Furnas, C. C., and S. M. Furnas. 1937. *Man, Bread, and Destiny: The Story of Man and His Food.* New York: New Home Library.

Gabbay, K. H., K. Hasty, J. L. Breslow, R. C. Ellison, H. F. Bunn, and P. M. Gallop. 1977. "Glycosylated Hemoglobins and Long-Term Blood Glucose Control in Diabetes Mellitus." *Journal of Clinical Endocrinology & Metabolism.* May; 44(5):859–64.

Gale, E. A. 2002. "The Rise of Childhood Type I Diabetes in the 20th Century." *Diabetes.* Dec.; 51(12):3353–61.

Galton, L. 1976. *The Truth About Fiber in Your Food.* New York: Crown.

———. 1961. "Why We Are Overly Larded." *New York Times.* Jan. 15; SM37.

Garcia-Palmieri, M. R., P. D. Sorlie, R. Costas, Jr., and R. J. Havlik. 1981. "An Apparent Inverse Relationship Between Serum Cholesterol and Cancer Mortality in Puerto Rico." *American Journal of Epidemiology.* July; 114(1):29–40.

Garcia-Palmieri, M. R., P. Sorlie, J. Tillotson, R. Costas, Jr., E. Cordero, and M. Rodriguez. 1980. "Relationship of Dietary Intake to Subsequent Coronary Heart Disease Incidence: The Puerto Rico Heart Health Program." *American Journal of Clinical Nutrition.* Aug.; 33(8):1818–27.

Gardiner-Hill, H. 1925. "The Treatment of Obesity." *Lancet.* Nov. 14; 206(5333):1034–35.

Gardner, C. D., A. Kiazand, S. Alhassan, et al. 2007. "Comparison of the Atkins, Zone, Ornish, and LEARN Diets for Change in Weight and Related Risk Factors Among Overweight Premenopausal Women: The A TO Z Weight Loss Study—A Randomized Trial." *JAMA.* March 7; 297(9):969–77.

Garine, I. de, and G. J. Koppert. 1991. "Guru-Fattening Sessions Among the Massa." *Ecology of Food and Nutrition.* 25:1–28.

Garine, I. de, and N. J. Pollock, eds. 1985. *Social Aspects of Obesity.* Amsterdam: Overseas Publishers Association.

Garrett, H. E., E. C. Horning, B. G. Creech, and M. Debakey. 1964. "Serum Cholesterol Values in Patients Treated Surgically for Atherosclerosis." *JAMA.* Aug. 31; 189:655–59.

Garrow, J. S. 1981. *Treat Obesity Seriously.* New York: Churchill Livingstone.

———. 1978. *Energy Balance and Obesity in Man.* New York: Elsevier/North-Holland Biomedical Press.

Garside, E. B. 1946. "Carnivorous Menu." *New York Times.* Oct 27; 173.

Gerrior, S., and L. Bente. 2001. *Nutrient Content of the U.S. Food Supply, 1909–1997.* U.S. Department of Agriculture Center for Nutrition Policy and Promotion Home Economics Research Report No. 54.

Gershoff, S. N. 2001. "Jean Mayer 1920–1993." *Journal of Nutrition.* June; 131(6):1651–54.

Gilmore, C. P. 1977. "Taking Exercise to Heart." *New York Times.* March 27; 211.

Giorgino, F., A. Belfiore, G. Milazzo, et al. 1991. "Overexpression of Insulin Receptors in Fibroblast and Ovary Cells Induces a Ligand-Mediated Transformed Phenotype." *Molecular Endocrinology.* March; 5(3):452–59.

Giovannucci, E. 2001. "Insulin, Insulin-Like Growth Factors, and Colon Cancer: A Review of the Evidence." *Journal of Nutrition.* Nov.; 131(11 suppl.):3109S–20S.

———. 1995. "Insulin and Colon Cancer." *Cancer Causes and Control.* March; 6(2):164–79.

Giovannucci, E., E. B. Rimm, M. J. Stampfer, G. A. Colditz, A. Ascherio, and W. C. Willett. 1994. "Intake of Fat, Meat, and Fiber in Relation to Risk of Colon Cancer in Men." *Cancer Research.* May 1; 54(9):2390–97.

Giugliano, D., A. Ceriello, and G. Paolisso. 1996. "Oxidative Stress and Diabetic Vascular Complications." *Diabetes Care.* March; 19(3):257–67.

Gladwell, M. 1998. "The Pima Paradox." *New Yorker.* Feb. 2. Online at http://www.gladwell.com/1998/1998_02_02_a_pima.htm.

Glenn, M. 1965. "Diet or Exercise?" *New York Times.* May 23; SM34.

Glicksman, A. S., W. P. Myers, and R. W. Rawson. 1956. "Diabetes Mellitus and Carbohydrate Metabolism in Patients with Cancer." *Medical Clinics of North America.* 40(3):887–900.

Glinsmann, W. H., H. Irausquin, and Y. K. Park. 1986. "Report from FDA's Sugars Task

Force, 1986: Evaluation of Health Aspects of Sugars Contained in Carbohydrate Sweeteners." *Journal of Nutrition.* Nov.; 116(11 suppl.):s1–s216.

Glueck, C. J. 1979. "Appraisal of Dietary Fat as a Causative Factor in Atherogenesis." *American Journal of Clinical Nutrition.* Dec.; 32(12 suppl.):2637–43.

Gofman, J. W. 1958. "Diet in the Prevention and Treatment of Myocardial Infarction." *American Journal of Cardiology.* Feb.; 1(2):271–83.

Gofman, J. W., and F. P. Lindgren. 1950. "The Role of Lipids and Lipoproteins in Atherosclerosis." *Science.* Feb. 17; 111(2877):166–86.

Gofman, J. W., A. V. Nichols, and E. V. Dobbin. 1958. *Dietary Prevention and Treatment of Heart Disease.* New York: Putnam.

Gofman, J. W., and W. Young. 1963. "The Filtration Concept of Atherosclerosis and Serum Lipids in the Diagnosis of Atherosclerosis." In Sandler and Bourne, eds., 1963, 197–230.

Gofman, J. W., W. Young, and R. Tandy. 1966. "Ischemic Heart Disease, Atherosclerosis, and Longevity." *Circulation.* Oct.; 34(4):679–97.

Golay, A., and E. Bobbioni. 1997. "The Role of Dietary Fat in Obesity." *International Journal of Obesity and Related Metabolic Disorders.* June; 21(3 suppl.):S2–11.

Goldberg, C. 2006. "Weight Risk on the Rise for Infants." *Boston Globe.* Aug. 10; A1.

Goldberg, L., D. L. Elliot, R. W. Schutz, and F. E. Kloster. 1984. "Changes in Lipid and Lipoprotein Levels After Weight Training." *JAMA.* July 27; 252(4):504–6.

Goldblatt, P. B., M. E. Moore, and A. J. Stunkard. 1965. "Social Factors in Obesity." *JAMA.* June 21; 192:1039–44.

Goldman, R. F., M. F. Haisman, G. Bynum, E. S. Horton, and E. A. Sims. 1976. "Experimental Obesity in Man: Metabolic Rate in Relation to Dietary Intake." In Bray, ed., 1976b, 165–86.

Goldstein, J. L., W. R. Hazzard, H. G. Schrott, E. L. Bierman, and A. G. Motulsky. 1973. "Hyperlipidemia in Coronary Heart Disease. I. Lipid Levels in 500 Survivors of Myocardial Infarction." *Journal of Clinical Investigation.* July; 52(7):1533–43.

Goleman, D. 1989. "What's for Dinner? Psychologists Explore Quirks and Cravings." *New York Times.* July 11; C1.

Goodhart, R. S., and M. E. Shils, eds. 1973. *Modern Nutrition in Health and Disease: Dietotherapy.* 5th edition. Philadelphia: Lea & Febiger.

Gordon, E. S. 1970. "Metabolic Aspects of Obesity." *Advances in Metabolic Disorders.* 4:229–96.

———. 1969. "The Metabolic Importance of Obesity." In *Symposium on Foods: Carbohydrates and Their Roles,* ed. H. W. Schultz. Westport, Conn.: Avi Publishing, 322–46.

———. 1964. "New Concepts of the Biochemistry and Physiology of Obesity." *Medical Clinics of North America.* Sept.; 18(5):1285–1306.

Gordon, E. S., M. Goldberg, and G. J. Chosy. 1963. "A New Concept in the Treatment of Obesity." *JAMA.* Oct. 5; 186:50–60.

Gordon, R. S., Jr., and A. Cherkes. 1956. "Unesterified Fatty Acid in Human Blood Plasma." *Journal of Clinical Investigation.* Feb.; 35(2):206–12.

Gordon, T. 1988. "The Diet-Heart Idea: Outline of a History." *American Journal of Epidemiology.* Feb.; 127(2):220–25.

Gordon, T., W. P. Castelli, M. C. Hjortland, W. B. Kannel, and T. R. Dawber. 1977. "High Density Lipoprotein as a Protective Factor Against Coronary Heart Disease: The Framingham Study." *American Journal of Medicine.* May; 62(5):707–14.

Gordon, T., A. Kagan, M. Garcia-Palmieri, et al. 1981. "Diet and Its Relation to Coronary Heart Disease and Death in Three Populations." *Circulation.* March; 63(3):500–515.

Gould, S. J. 1996. *The Mismeasure of Man.* New York: W. W. Norton.

Gracey, M., N. Kretchmer, and E. Rossi, eds. 1991. *Sugars in Nutrition*. New York: Raven Press.

Grafe, E. 1933. *Metabolic Diseases and Their Treatment*. Trans. M. G. Boise. Philadelphia: Lea & Febiger.

Grande, F., J. T. Anderson, and A. Keys. 1958. "Changes of Basal Metabolic Rate in Man in Semistarvation and Refeeding." *Journal of Applied Physiology*. March; 12(2):230–38.

Grant, F. W., and D. Groom. 1959. "A Dietary Study Among a Group of Southern Negroes." *Journal of the American Dietetic Association*. Sept.; 35:910–18.

Grant, J. P. 1974. "Foreword." In L. Brown and E. Eckholm, *By Bread Alone*. New York: Praeger, ix–xi.

Graudal, N. A., A. M. Galloe, and P. Garred. 1998. "Effects of Sodium Restriction on Blood Pressure, Renin, Aldosterone, Catecholamines, Cholesterols, and Triglyceride: A Meta-Analysis." *JAMA*. May 6; 279(17):1383–91.

Graves, A. B., E. B. Larson, S. D. Edland, et al. 1996. "Prevalence of Dementia and Its Subtypes in the Japanese American Population of King County, Washington State: The Kame Project." *American Journal of Epidemiology*. Oct. 15; 144(8):760–71.

Greenberg, D. S. 1979. "Cancer: The Difference Life Style Makes." *Washington Post*. July 17; A17.

Greene, R. 1970. *Human Hormones*. New York: McGraw-Hill.

———. 1953. "Obesity." *Lancet*. Aug. 1; 262(6770):253.

———, ed. 1951. *The Practice of Endocrinology*. Philadelphia: J. B. Lippincott.

Greenwood, M. R. 1985. "Normal and Abnormal Growth and Maintenance of Adipose Tissue." In Hirsch and Van Itallie, eds., 1985, 20–25.

Greenwood, M. R., M. Cleary, L. Steingrimsdottir, and J. R. Vaselli. 1981. "Adipose Tissue Metabolism and Genetic Obesity." In Björntorp, Cairella, and Howard, eds., 1981, 75–79.

Greenwood, M. R., R. Savard, D. B. West, and R. Kava. 1987. "Energy Metabolism and Nutrient 'Gating' in Pregnancy and Lactation." In Berry et al., eds., 1987, 266–71.

Gregg, E. W., Y. J. Cheng, B. L. Cadwell, et al. 2005. "Secular Trends in Cardiovascular Disease Risk Factors According to Body Mass Index in U.S. Adults." *JAMA*. April 20; 293(15):1868–74.

Grey, N., and D. M. Kipnis. 1971. "Effect of Diet Composition on the Hyperinsulinemia of Obesity." *New England Journal of Medicine*. Oct. 7; 285(15):827–31.

Griffin, J. S. 1943. *A Doctor Comes to California: The Diary of John S. Griffin, Assistant Surgeon with Kearny's Dragoons, 1846–1847*. San Francisco: California Historical Society.

Grobstein, C. 1981. "Saccharin: A Scientist's View." In Crandall and Lave, eds., 1981, 117–30.

Groen, J., B. K. Tjiong, C. E. Kamminga, and A. F. Willebrands. 1952. "The Influence of Nutrition, Individuality, and Some Other Factors, Including Various Forms of Stress, on the Serum Cholesterol: An Experiment of Nine Months Duration in 60 Normal Human Volunteers." *Voeding*. 13:556–87.

Groen, J. J., B. K. Tjiong, M. Koster, A. F. Willebrands, G. Verdonck, and M. Pierloot. 1962. "The Influence of Nutrition and Ways of Life on Blood Cholesterol and the Prevalence of Hypertension and Coronary Heart Disease Among Trappist and Benedictine Monks." *American Journal of Clinical Nutrition*. June; 10:456–70.

Grossman, H. 2003. "Does Diabetes Protect or Provoke Alzheimer's Disease? Insights into the Pathobiology and Future Treatment of Alzheimer's Disease." *CNS Spectrum*. Nov.; 8(11):815–23.

Grover, S. A., K. Gray-Donald, L. Joseph, M. Abrahamowicz, and L. Coupal. 1994.

"Life Expectancy Following Dietary Modification or Smoking Cessation: Estimating the Benefits of a Prudent Lifestyle." *Archives of Internal Medicine.* Aug. 8; 154(15): 1697–704.

Grundy, S. M. 1994. "Influence of Stearic Acid on Cholesterol Metabolism Relative to Other Long-Chain Fatty Acids." *American Journal of Clinical Nutrition.* Dec.; 60(6 suppl.):986S–90S.

———. 1986. "Cholesterol and Coronary Heart Disease, a New Era." *JAMA.* Nov. 28; 256(20):2849–58.

Grundy, S. M., H. B. Brewer, Jr., J. I. Cleeman, S. C. Smith, Jr., and C. Lenfant. 2004. "Definition of Metabolic Syndrome—Report of the National Heart, Lung, and Blood Institute/American Heart Association Conference on Scientific Issues Related to Definition." *Circulation.* Jan. 27; 109(3):433–38.

Grundy, S. M., B. Hansen, S. C. Smith, Jr., J. I. Cleeman, and R. A. Kahn. 2004. "Clinical Management of Metabolic Syndrome—Report of the American Heart Association/ National Heart, Lung, and Blood Institute/American Diabetes Association Conference on Scientific Issues Related to Management." *Circulation.* Feb. 3; 109(4):551–56.

Gruppo Italiano per lo Studio della Sopravvivenza nell'Infarto Miocardico. 1999. "Dietary Supplementation with n-3 Polyunsaturated Fatty Acids and Vitamin E After Myocardial Infarction: Results of the GISSI-Prevenzione Trial." *Lancet.* Aug. 7; 354(9177): 447–55.

Gsell, D., and J. Mayer. 1962. "Low Blood Cholesterol Associated with High Calorie, High Saturated Fat Intakes in a Swiss Alpine Village Population." *American Journal of Clinical Nutrition.* June; 10:471–79.

Guberan, E. 1979. "Surprising Decline of Cardiovascular Mortality in Switzerland: 1951–1976." *Journal of Epidemiology and Community Health.* June; 33(2):114–20.

Guery, P., and M.-C. Secretin. 1991. "Sugars and Non-Nutritive Sweeteners." In Gracey, Kretchmer, and Rossi, eds., 1991, 33–54.

Guo, S. S., W. Wu, W. C. Chumlea, and A. F. Roche. 2002. "Predicting Overweight and Obesity in Adulthood from Body Mass Index Values in Childhood and Adolescence." *American Journal of Clinical Nutrition.* Sept.; 76(3):653–58.

Gwinup, G. 1974. "Effects of Diet and Exercise in the Treatment of Obesity." In Bray and Bethune, eds., 1974, 93–102.

Haddock, D. R. 1969. "Obesity in Medical Out-Patients in Accra." *Ghana Medical Journal.* Dec.: 251–54.

Haist, R. E., and C. H. Best. 1966. "Carbohydrate Metabolism and Insulin." In *The Physiological Basis of Medical Practice,* 8th edition, ed. C. H. Best and N. M. Taylor. Baltimore: Williams & Wilkins, 1329–67.

Hall, J. E., D. W. Jones, J. Henegar, T. M. Dwyer, and J. J. Kuo. 2003. "Obesity, Hypertension, and Renal Disease." In Eckel, ed., 2003, 273–300.

Hamilton, C. L., and J. R. Brobeck. 1965. "Control of Food Intake in Normal and Obese Monkeys." *Annals of the New York Academy of Sciences.* Oct. 8; 131(1):583–92.

Hamlyn, E. C. 1953. "Obesity." *Lancet.* July 25; 262(6778):203.

Hammond, M. G., and W. R. Fisher. 1971. "The Characterization of a Discrete Series of Low Density Lipoproteins in the Disease, Hyper-Pre-Beta-Lipoproteinemia: Implications Relating to the Structure of Plasma Lipoproteins." *Journal of Biological Chemistry.* Sept. 10; 246(17):5454–65.

Han, P. W. 1968. "Energy Metabolism of Tube-Fed Hypophysectomized Rats Bearing Hypothalamic Lesions." *American Journal of Physiology.* Dec.; 215(6): 1343–50.

Han, P. W., and L. W. Frohman. 1970. "Hyperinsulinemia in Tube-Fed Hypophysectomized Rats Bearing Hypothalamic Lesions." *American Journal of Physiology.* Dec.; 219(6): 1632–36.

Han, P. W., C. H. Lin, K. C. Chu, H. Y. Mu, and A. C. Liu. 1965. "Hypothalamic Obesity in Weanling Rats." *American Journal of Physiology.* Sept.; 209(3):627–31.

Handler, P. 1980. "On Bias: Does Where You Stand Really Depend on Where You Sit?" In *Issues and Current Studies.* Washington, D.C.: National Research Council, 1–15.

Hanssen, P. 1936. "Treatment of Obesity by a Diet Relatively Poor in Carbohydrates." *Acta Medica Scandinavica.* 88:97–106.

Hardman, A. E., and S. L. Herd. 1998. "Exercise and Postprandial Lipid Metabolism." *Proceedings of the Nutrition Society.* Feb.; 57(1):63–72.

Harper, A. E. 1996. "Dietary Guidelines in Perspective." *Journal of Nutrition.* April; 126(4 suppl.):1042S–48S.

———. 1983. "Coronary Heart Disease—An Epidemic Related to Diet?" *American Journal of Clinical Nutrition.* April; 37(4):669–81.

Harper, A. H. 1971. *Review of Physiological Chemistry.* 13th edition. Los Altos, Calif.: Lange Medical Publications.

Harris, B. B. 1960. *The Gila Trail: The Texas Argonauts and the California Gold Rush.* Ed. R. H. Dillon. Norman: University of Oklahoma Press.

Harris, M. 1985. *Good to Eat: Riddles of Food and Culture.* New York: Simon & Schuster.

Harris, M., and E. B. Ross, eds. 1987. *Food and Evolution: Toward a Theory of Human Food Habits.* Philadelphia: Temple University Press.

Harrison, D. E., J. R. Archer, and C. M. Astle. 1984. "Effects of Food Restriction on Aging: Separation of Food Intake and Adiposity." *Proceedings of the National Academy of Sciences.* March; 81(6):1835–38.

Harvey, J. 1864. "Bantingism." *Lancet.* May 14; 83(2124):571.

Harvey, W. 1872. *On Corpulence in Relation to Disease: With Some Remarks on Diet.* London: Henry Renshaw.

Haseltine, N. S. 1949. "Public Health Service Awards $8,614,737 to Press Nation's Fight Against Heart Disease." *Washington Post.* Sept. 13; B3.

Havel, P. J. 2005. "Dietary Fructose: Implications for Dysregulation of Energy Homeostasis and Lipid/Carbohydrate Metabolism." *Nutrition Review.* May; 63(5):133–57.

Hays, J., J. K. Ockene, R. L. Brunner, et al. [WHI Investigators.] 2003. "Effects of Estrogen Plus Progestin on Health-Related Quality of Life." *New England Journal of Medicine.* May 8; 348(19):1839–54.

He, F. J., and G. A. MacGregor. 2004. "Effect of Longer-Term Modest Salt Reduction on Blood Pressure." *Cochrane Database of Systematic Reviews.* No. 3:CD004937.

Heath, G. W., A. A. Ehsani, J. M. Hagberg, J. M. Hinderliter, and A. P. Goldberg. 1983. "Exercise Training Improves Lipoprotein Lipid Profiles in Patients with Coronary Artery Disease." *American Heart Journal.* June; 105(6):889–95.

Hedrick, C. C., S. R. Thorpe, M. X. Fu, et al. 2000. "Glycation Impairs High Density Lipoprotein Function." *Diabetologia.* March; 43(3):312–20.

Heinbecker, P. 1928. "Studies on the Metabolism of Eskimos." *Journal of Biological Sciences.* Dec.; 80(2):461–75.

Heini, A. F., and R. L. Weinsier. 1997. "Divergent Trends in Obesity and Fat Intake Patterns: The American Paradox." *American Journal of Medicine.* March; 102(3):259–64.

Hendrie, H. C., A. Ogunniyi, K. S. Hall, et al. 2001. "Incidence of Dementia and Alzheimer Disease in 2 Communities: Yoruba Residing in Ibadan, Nigeria, and African Americans Residing in Indianapolis, Indiana." *JAMA.* Feb. 14; 285(6):739–47.

Hesse, F. G. 1959. "A Dietary Study of the Pima Indian." *American Journal of Clinical Nutrition.* Sept.–Oct.; 7:532–37.

Hesser, L. 2006. *The Man Who Fed the World: Nobel Peace Prize Laureate Norman Borlaug and His Battle to End World Hunger.* Dallas: Durban House.

Hetherington, A. W., and S. W. Ranson. 1942. "The Spontaneous Activity and Food Intake of Rats with Hypothalamic Lesions." *American Journal of Physiology.* June; 136(4):609–17.

———. 1940. "Hypothalamic Lesions and Adiposity in the Rat." *Anatomical Record.* 78:149–72.

———. 1939. "Experimental Hypothalamico-Hypophyseal Obesity in the Rat." *Proceedings of the Society for Experimental Biology and Medicine.* June; 41(2):465–66.

Heuson, J. C., A. Coune, and R. Heimann. 1967. "Cell Proliferation Induced by Insulin in Organ Culture of Rat Mammary Carcinoma." *Experimental Cell Research.* Feb.; 45(2):351–60.

Hiatt, H. H., J. D. Watson, and J. A. Winsten, eds. 1977. *Origins of Human Cancer.* Bk. B, *Mechanisms of Carcinogenesis.* Cold Spring Harbor, N.Y.: Cold Spring Harbor Laboratory.

Higgins, H. L. 1916. "The Rapidity with Which Alcohol and Some Sugars May Serve as Nutriment." *American Journal of Physiology.* Aug.; 41(2):258–265.

Higginson, J. 1997. "From Geographical Pathology to Environmental Carcinogenesis: A Historical Reminiscence." *Cancer Letters.* 117:133–42.

———. 1983. "Developing Concepts on Environmental Cancer: The Role of Geographical Pathology." *Environmental Mutagenesis.* 5(6):929–40.

———. 1981. "Rethinking the Environmental Causation of Human Cancer." *Food and Cosmetics Toxicology.* Oct.; 19(5):539–48.

Hill, D. J., and R. D. Milner. "Insulin as a Growth Factor." *Pediatric Research.* Sept.; 19(9):879–86.

Hill, J. O. n.d. "The Role of Carbohydrates in Weight Management." Accessed Feb. 20, 2003, from the Sugar Organization Web site: www.sugar.org/science/carbohydrates .html.

Hill, J. O., and J. C. Peters. 1998. "Environmental Contributions to the Obesity Epidemic." *Science.* May 29; 280(5368):1371–74.

Hill, J. O., H. R. Wyatt, G. W. Reed, and J. C. Peters. 2003. "Obesity and the Environment: Where Do We Go from Here?" *Science.* Feb. 7; 299(5608):853–55.

Himsworth, H. P. 1949a. "Diet in the Aetiology of Human Diabetes." *Proceedings of the Royal Society of Medicine.* 42(5): 323–26.

———. 1949b. "The Syndrome of Diabetes Mellitus and Its Causes." *Lancet.* March 19; 253(6551):465–73.

———. 1936. "Diabetes Mellitus: Its Differentiation into Insulin-Sensitive and Insulin-Insensitive Types." *Lancet.* Jan. 18; 227(5864):127–30.

———. 1935. "Diet and the Incidence of Diabetes Mellitus." *Clinical Science.* Sept.; 2(1):117–48.

Hirsch, J. 1985. "Dietary Treatment." In Hirsch and Van Itallie, eds., 1985, 192–95.

———. 1978. "Obesity: A Perspective." In Bray, ed., 1978, 1–5.

Hirsch, J., and T. B. Van Itallie, eds. 1985. *Recent Advances in Obesity Research: IV.* London: John Libbey.

Hobbes, T. 1997. *Leviathan.* New York: Simon & Schuster. [Originally published 1651.]

Hodges, R. E., and T. Rebello. 1983. "Carbohydrates and Blood Pressure." *Annals of Internal Medicine.* May; 98(5, pt. 2):838–41.

Hoebel, B., and L. Hernandez. 1993. "Basal Neural Mechanisms of Feeding and Weight Regulation." In Stunkard and Wadden, eds., 1993, 43–62.

Hoebel, B. G., P. V. Rada, G. P. Mark, and F. Pothos. 1999. "Neural Systems for Reinforcement and Inhibition of Behavior: Relevance to Eating, Addiction, and Depression." In *Well-Being: Foundations of Hedonic Psychology*, ed. D. Kahneman, E. Diener, and N. Schwara. New York: Russell Sage Foundation, 558–72.

Hoffman, F. L. 1937. *Cancer and Diet*. Baltimore: Williams & Wilkins.

———. 1915. *The Mortality from Cancer Throughout the World*. Newark, N.J.: Prudential Press.

Hoffman, W. 1979. "Meet Monsieur Cholesterol." *University of Minnesota Update*. Winter. Online at http://www.mbbnet.umn.edu/hoff/hoff_ak.html.

Hollander, B. 1923. "Freedom of Negro Races from Cancer." *British Medical Journal*. July 7:46.

Hollenbeck, C. 1993. "Dietary Fructose Effects on Lipoprotein Metabolism and Risk for Coronary Disease." *American Journal of Clinical Nutrition*. Nov.; 58(5 suppl.): 800s–809s.

Hollifield, G., and W. Parson. 1965. "Effect of Feeding on Fatty Acid Synthesis." In Renold and Cahill, eds., 1965, 393–98.

Holmes, M. D., D. J. Hunter, G. A. Colditz, et al. 1999. "Association of Dietary Intake of Fat and Fatty Acids with Risk of Breast Cancer." *JAMA*. March 10; 281(10):914–20.

Holzenberger, M., J. Dupont, B. Ducos, et al. 2003. "IGF-1 Receptor Regulates Lifespan and Resistance to Oxidative Stress in Mice." *Nature*. Jan. 9; 421(6919): 182–87.

Hooper, L., C. D. Summerbell, J. P. Higgins, et al. 2001. "Reduced or Modified Dietary Fat for Preventing Cardiovascular Disease." *Cochrane Database of Systematic Reviews*. No. 3:CD002137.

Horton, E. S., E. Danforth, Jr., and E. A. Sims. 1974. "Endocrine and Metabolic Alterations Associated with Overfeeding and Obesity in Man." In Burland, Samuel, and Yudkin, eds., 1974, 229–51.

Howard, A. N., ed. 1976. *Recent Advances in Obesity Research: I*. Westport, Conn.: Technomic Publishing.

———. 1969. "Dietary Treatment of Obesity." In McLean Baird and Howard, eds., 1969, 96–111.

Howard, B. V., J. E. Manson, M. L. Stefanick, et al. 2006. "Low-Fat Dietary Pattern and Weight Change over 7 Years: The Women's Health Initiative Dietary Modification Trial." *JAMA*. Jan. 4; 295(1):39–49.

Howard, B. V., L. Van Horn, J. Hsia, et al. 2006. "Low-Fat Dietary Pattern and Risk of Cardiovascular Disease: The Women's Health Initiative Randomized Controlled Dietary Modification Trial." *JAMA*. Feb. 8; 295(6):655–66.

Howell, W. H., D. J. McNamara, M. A. Tosca, B. T. Smith, and J. A. Gaines. 1997. "Plasma Lipid and Lipoprotein Responses to Dietary Fat and Cholesterol: A Meta-Analysis." *American Journal of Clinical Nutrition*. June; 65(6):1747–64.

Hrdlička, A. 1908. *Physiological and Medical Observations Among the Indians of Southwestern United States and Northern Mexico*. Washington D.C.: U.S. Government Printing Office.

———. 1906. "Notes on the Pima of Arizona." *American Anthropologist*. Jan.–March; 8(1):39–46.

Hulley, S. B., D. Grady, T. Bush, et al. 1998. "Randomized Trial of Estrogen Plus Progestin for Secondary Prevention of Coronary Heart Disease in Postmenopausal Women: Heart and Estrogen/Progestin Replacement Study (HERS) Research Group." *JAMA*. Aug. 19; 280(7):605–13.

Hulley, S. B., R. H. Rosenman, R. D. Bawol, and R. J. Brand. 1980. "Epidemiology as a

Guide to Clinical Decisions: The Association Between Triglyceride and Coronary Heart Disease." *New England Journal of Medicine.* June 19; 302(25):1383–89.

Hulley, S. B., J. M. Walsh, and T. B. Newman. 1992. "Health Policy on Blood Cholesterol: Time to Change Directions." *Circulation.* Sept.; 86(3):1026–29.

Hulley, S. B., W. S. Wilson, M. I. Burrows, and M. Z. Nichaman. 1972. "Lipid and Lipoprotein Responses of Hypertriglyceridaemic Outpatients to a Low-Carbohydrate Modification of the A.H.A. Fat-Controlled Diet." *Lancet.* Sept. 16; 300(7777):551–55.

Hunink, M. G., L. Goldman, A. N. Tosteson, et al. 1997. "The Recent Decline in Mortality from Coronary Heart Disease, 1980–1990: The Effect of Secular Trends in Risk Factors and Treatment." *JAMA.* Feb. 19; 277(7):535–42.

Huntsman, J., and A. Hooper. 1996. *Tokalau: A Historical Ethnography.* Auckland: Auckland University Press.

Hurley, J., and B. Liebman. 2003. "The Top Pops." *Nutrition Action Healthletter.* March; 13–15.

Hustvedt, B. A., and A. Lovo. 1972. "Correlation Between Hyperinsulinemia and Hyperphagia in Rats with Ventromedial Hypothalamic Lesions." *Acta Physiologica Scandinavica.* Jan.; 84(1):29–33.

Hutton, S. K. n.d. *Health Conditions and Disease Incidence Among the Eskimos of Labrador.* London: Wessex Press.

Huttunen, J. K., E. Lansimies, E. Voutilainen, et al. 1979. "Effect of Moderate Physical Exercise on Serum Lipoproteins: A Controlled Clinical Trial with Special Reference to Serum High-Density Lipoproteins." *Circulation.* Dec.; 60(6):1220–29.

Hwang, I. S., B. B. Hoffman, and G. M. Reaven. 1987. "Fructose-Induced Insulin Resistance and Hypertension in Rats." *Hypertension.* Nov.; 10(5):512–16.

Ingelfinger, F. J., A. S. Relman, and M. Finland, eds. 1966. *Controversy in Internal Medicine.* Philadelphia: W. B. Saunders.

Institute of Medicine [IOM] of the National Academies. 2002. *Dietary Reference Intakes: Energy, Carbohydrate, Fiber, Fat, Fatty Acids, Cholesterol, Protein, and Amino Acids.* Washington, D.C.: National Academies Press.

———. 1995. *Weighing the Options: Criteria for Evaluating Weight-Management Programs.* Washington, D.C.: National Academy Press.

Insull, W., Jr., T. Oiso, and K. Tsuchiya. 1968. "Diet and Nutritional Status of Japanese." *American Journal of Clinical Nutrition.* July; 21(7):753–77.

International Health, Racquet and Sportsclub Association. 2005. "The Scope of the US Health Club Industry." Online at http://cms.ihrsa.org/IHRSA/viewPage.cfm?pageId=804.

Intersalt Cooperative Research Group. 1988. "Intersalt, an International Study of Electrolyte Excretion and Blood Pressure: Results for 24 Hour Urinary Sodium and Potassium Excretion. *British Medical Journal.* July 30; 297(6644):319–28.

Inter-Society Commission for Heart Disease Resources. 1970. "Report of Inter-Society Commission for Heart Disease Resources: Prevention of Cardiovascular Disease— Primary Prevention of the Atherosclerotic Diseases." *Circulation.* Dec.; 42(6):A55–95.

Jackson, Y. M. 1994. "Diet, Culture, and Diabetes." In Joe and Young, eds., 1994, 381–406.

Jacobs, D., H. Blackburn, M. Higgins, et al. 1992. "Report of the Conference on Low Blood Cholesterol: Mortality Associations." *Circulation.* Sept.; 86(3):1046–60.

Jacobson, A. M., S. T. Hauser, B. J. Anderson, and W. Polonsky. 1994. "Psychosocial Aspects of Diabetes." In Kahn and Weir, eds., 1994, 431–50.

Jacobson, M. 1978. "The Deadly White Powder." *Mother Jones.* July; 12–20.

James, T. N. 1980. "Presidential Address, AHA 53rd Scientific Sessions, Miami Beach,

Florida, November 1980: Sure Cures, Quick Fixes, and Easy Answers—A Cautionary Tale About Coronary Disease." *Circulation*. May; 63(5):1199A–202A.

James, W. P. 1983. "A Discussion Paper on Proposals for Nutritional Guidelines for Health Education in Britain." National Advisory Committee on Nutrition Education (NACNE). The Health Education Council.

James, W. P., ed. 1977. *Research on Obesity: A Report of the DHSS/MRC Group*. London: Her Majesty's Stationery Office.

James, W. P., H. L. Davies, J. Bailes, and M. J. Dauncey. 1978. "Elevated Metabolic Rates in Obesity." *Lancet*. May 27; 311(8074):1122–25.

Janssen, G. M., C. J. Graef, and W. H. Saris. 1989. "Food Intake and Body Composition in Novice Athletes During a Training Period to Run a Marathon." *International Journal of Sports Medicine*. May; 10(1 suppl.):S17–21.

Jeanrenaud, B. 1979. "Insulin and Obesity." *Diabetologia*. Sept.; 17(3):133–38.

Jeffery, R. W., W. L. Hellerstedt, S. A. French, and J. E. Baxter. 1995. "A Randomized Trial of Counseling for Fat Restriction Versus Calorie Restriction in the Treatment of Obesity." *International Journal of Obesity and Related Metabolic Disorders*. Feb.; 19(2):132–37.

Jen, K. L., B. C. Hansen, and B. L. Metzger. 1985. "Adiposity, Anthropometric Measures, and Plasma Insulin Levels of Rhesus Monkeys." *International Journal of Obesity*. 9(3):213–24.

Jenkins, D. J., T. M. Wolever, R. H. Taylor, et al. 1981. "Glycemic Index of Foods: A Physiological Basis for Carbohydrate Exchange." *American Journal of Clinical Nutrition*. March; 34(3):362–66.

Jenness, D. 1959. *The People of the Twilight*. Chicago: University of Chicago Press.

Joe, J. R., and R. S. Young, eds. 1994. *Diabetes as a Disease of Civilization: The Impact of Culture Change on Indigenous Peoples*. New York: Mouton de Gruyter.

Johnson, M. L., B. S. Burke, and J. Mayer. 1956. "Relative Importance of Inactivity and Overeating in the Energy Balance of Obese High School Girls." *American Journal of Clinical Nutrition*. Jan.–Feb.; 4(1):37–44.

Johnson, T. O. 1970. "Prevalence of Overweight and Obesity Among Adult Subjects of an Urban African Population Sample." *British Journal of Preventive and Social Medicine*. 24:105–9.

Johnstone, M. T., and R. Nesto. 2005. "Diabetes Mellitus and Heart Disease." In Kahn et al., eds., 2005, 975–98.

Jolliffe, N. 1963. *Reduce and Stay Reduced on . . . the Prudent Diet*. New York: Simon & Schuster.

———. 1952. *Reduce and Stay Reduced*. New York: Simon & Schuster.

Jolliffe, N., S. H. Rinzler, M. Archer, et al. 1962. "Effect of a Prudent Reducing Diet on the Serum Cholesterol of Over-Weight Middle-Aged Men." *American Journal of Clinical Nutrition*. March; 10:200–211.

Jones, B. M. 1967. *Health-Seekers in the Southwest, 1817–1900*. Norman: University of Oklahoma Press.

Jones, D. Y., A. Schatzkin, S. B. Green, et al. 1987. "Dietary Fat and Breast Cancer in the National Health and Nutrition Examination Survey I. Epidemiologic Follow-Up Study." *Journal of the National Cancer Institute*. Sept.; 79(3):465–71.

Joslin Diabetes Center. 2003. Press Release: "Study Shows It May Someday Be Possible to Stay Slim, Avoid Type 2 Diabetes—and Live Longer—While Eating What You Want." Jan. 23.

Joslin, E. P. 1940. "The Universality of Diabetes: A Survey of Diabetes Morbidity in Arizona." *JAMA*. Oct.–Dec.; 115:2033–38.

———. 1928. *The Treatment of Diabetes Mellitus.* 4th edition. Philadelphia: Lea & Febiger.

———. 1927. "Arteriosclerosis and Diabetes." *Annals of Clinical Medicine.* 5:1061–79.

———. 1923. *The Treatment of Diabetes Mellitus.* 3rd edition. New York: Lea & Febiger.

Joslin, E. P., L. I. Dublin, and H. H. Marks. 1935. "Studies in Diabetes Mellitus. III. Interpretation of the Variations in Diabetes Incidence." *American Journal of the Medical Sciences.* 189:163–91.

———. 1934. "Studies in Diabetes Mellitus. II. Its Incidence and the Factors Underlying Its Variations." *American Journal of the Medical Sciences.* April; 187(4):433–57.

———. 1933. "Studies in Diabetes Mellitus. I. Characteristics and Trends of Diabetes Mortality Throughout the World." *American Journal of the Medical Sciences.* Dec.; 180(6):753–73.

Joslin, E. P., H. F. Root, P. White, and A. Marble, eds. 1959. *The Treatment of Diabetes Mellitus.* 10th edition. Philadelphia: Lea & Febiger.

Joslin, E. P., H. F. Root, P. White, A. Marble, and C. C. Bailey, eds. 1946. *The Treatment of Diabetes Mellitus.* 8th edition. Philadelphia: Lea & Febiger.

Justice, J. W. 1994. "The History of Diabetes Mellitus in the Desert People." In Joe and Young, eds., 1994, 69–127.

Kaaks, R. 1996. "Nutrition, Hormones, and Breast Cancer: Is Insulin the Missing Link?" *Cancer Causes and Control.* Nov.; 7(6):605–25.

Kaaks, R., and A. Lukanova. 2001. "Energy Balance and Cancer: The Role of Insulin and Insulin-Like Growth Factor-I." *Proceedings of the Nutrition Society.* Feb.; 60(1):91–106.

Kafatos A., A. Diacatou, G. Voukiklaris, et al. 1997. "Heart Disease Risk-Factor Status and Dietary Changes in the Cretan Population over the Past 30 Years: The Seven Countries Study." *American Journal of Clinical Nutrition.* June; 65(6):1882–86.

Kahn, C. R., and G. C. Weir, eds. 1994. *Joslin's Diabetes Mellitus.* 13th edition. Media, Pa.: Lippincott Williams & Wilkins.

Kahn, C. R., G. C. Weir, G. L. King, A. M. Jacobson, A. C. Moses, and R. J. Smith, eds. 2005. *Joslin's Diabetes Mellitus.* 14th edition. New York: Lippincott Williams & Wilkins.

Kahn, H. A., J. H. Medalie, H. N. Neufeld, E. Riss, M. Balogh, and J. J. Groen. 1969. "Serum Cholesterol: Its Distribution and Association with Dietary and Other Variables in a Survey of 10,000 Men." *Israeli Journal of Medical Sciences.* Nov.–Dec.; 5(6):1117–27.

Kalaria, R. 2002. "Similarities Between Alzheimer's Disease and Vascular Dementia." *Journal of the Neurological Sciences.* Nov. 15; 203–4:29–34.

Kannel, W. B., and W. P. Castelli. 1979. "Is the Serum Total Cholesterol an Anachronism?" *Lancet.* Nov. 3; 314(8149):950–51.

Kannel, W. B., W. P. Castelli, and T. Gordon. 1979. "Cholesterol in the Prediction of Atherosclerotic Disease: New Perspectives Based on the Framingham Study." *Annals of Internal Medicine.* Jan.; 90(1):85–91.

Kannel, W. B., W. P. Castelli, T. Gordon, and P. M. McNamara. 1971. "Serum Cholesterol, Lipoproteins, and the Risk of Coronary Heart Disease: The Framingham Study." *Annals of Internal Medicine.* Jan.; 74(1):1–12.

Kannel, W. B., and T. Gordon. 1968. *The Framingham Diet Study: Diet and Regulation of Serum Cholesterol.* Sect. 24 of *The Framingham Study: An Epidemiological Investigation of Cardiovascular Disease.* Bethesda, Md.: U.S. Department of Health, Education, and Welfare, Public Health Service, and National Institutes of Health.

Kannel, W. B., and H. E. Thomas, Jr. 1982. "Sudden Coronary Death—The Framingham Study." *Annals of the New York Academy of Sciences.* March; 382:3–21.

Kark, J. D., A. H. Smith, and C. G. Hames. 1980. "The Relationship of Serum Cholesterol to the Incidence of Cancer in Evans County, Georgia." *Journal of Chronic Diseases.* 33(5):311–32.

Kark, R. M., and J. H. Oyama. 1973. "Nutrition and Cardiovascular-Renal Diseases." In Goodhart and Shils, eds., 1973, 852–94.

Karolinska Institute. 1977. Press Release: The 1977 Nobel Prize in Physiology or Medicine. Online at http://nobelprize.org/nobel_prizes/medicine/laureates/1977/press.html.

Katan, M. B., P. L. Zock, and R. P. Mensink. 1995. "Dietary Oils, Serum Lipoproteins, and Coronary Heart Disease." *American Journal of Clinical Nutrition.* June; 61(6 suppl.):1368S–73S.

Katz, L. N., J. Stamler, and R. Pick. 1958. *Nutrition and Atherosclerosis.* Philadelphia: Lea & Febiger.

Kaunitz, H. 1977. "Importance of Lipids in Arteriosclerosis: An Outdated Theory." In Select Committee on Nutrition and Human Needs of the United States Senate 1977c, 42–54.

Keesey, R. 1980. "A Set-Point Analysis of the Regulation of Body Weight." In Stunkard, ed., 1980, 144–65.

Kehoe, R. A. 1960. "In Memoriam—George Howard Gehrmann, 1890–1959." *Archives of Environmental Health.* Sept.; 1:177–80.

Kekwick, A., and G. L. Pawan. 1957. "Metabolic Study in Human Obesity with Isocaloric Diets High in Fat, Protein, or Carbohydrate." *Metabolism Clinical and Experimental.* 6(5):447–60.

Kellock, B. 1985. *The Fibre Man: The Life-Story of Dr Denis Burkitt.* Herts, England: Lion Publishing.

Kemp, R. 1972. "The Over-All Picture of Obesity." *Practitioner.* Nov.; 209(253):654–60.

———. 1966. "Obesity as a Disease." *Practitioner.* March; 196(173):404–9.

———. 1963. "Carbohydrate Addiction." *Practitioner.* March; 190:358–64.

Kennedy, E. T., S. A. Bowman, J. T. Spence, M. Freedman, and J. King. 2001. "Popular Diets: Correlation to Health, Nutrition, and Obesity." *Journal of the American Dietetic Association.* April; 101(4):411–20.

Kennedy, G. C. 1961. "The Central Nervous Regulation of Calorie Balance." *Proceedings of the Nutrition Society.* 20:58–64.

Kenyon, C. 2001. "A Conserved Regulatory System for Aging." *Cell.* April 20; 105(2):165–68.

Kenyon, C., J. Chang, E. Gensch, A. Rudner, and R. Tabtiang. 1993. "A C. Elegans Mutant That Lives Twice As Long As Wild Type." *Nature.* Dec. 2; 366(6454):461–64.

Kern, P. A., J. M. Ong, B. Saffari, and J. Carty. 1990. "The Effects of Weight Loss on the Activity and Expression of Adipose-Tissue Lipoprotein Lipase in Very Obese Humans." *New England Journal of Medicine.* April 12; 322(15):1053–59.

Kessler, I. I. 1971. "Cancer and Diabetes Mellitus: A Review of the Literature." *Journal of Chronic Diseases.* Jan.; 23(8):579–600.

Keys A. 1994. "The Inception and Pilot Surveys." In *The Seven Countries Study: A Scientific Adventure in Cardiovascular Disease Epidemiology,* ed. D. Kromhout, A. Menotti, and H. Blackburn. Utrecht: Brouwer, 15–26.

———. 1980. *Seven Countries: A Multivariate Analysis of Death and Coronary Heart Disease.* Cambridge, Mass.: Harvard University Press.

———. 1975. "Coronary Heart Disease—The Global Picture." *Atherosclerosis.* Sept.–Oct.; 22(2):149–92.

———. 1971. "Sucrose in the Diet and Coronary Heart Disease." *Atherosclerosis.* Sept.–Oct.; 14(2):193–202.

————, ed. 1970. "Coronary Heart Disease in Seven Countries." *Circulation.* April 4; 41(4 suppl.):I-1-211.

————. 1966. "Arteriosclerotic Heart Disease in Roseto, Pennsylvania." *JAMA.* Jan. 10; 195(2):93–95.

————. 1963. "The Role of the Diet in Human Atherosclerosis and Its Complications." In Sandler and Bourne, eds., 1963, 263–301.

————. 1957. "Diet and the Epidemiology of Coronary Heart Disease." *JAMA.* Aug. 24; 164(17):1912–19.

————. 1953. "Atherosclerosis: A Problem in Newer Public Health." *Journal of Mount Sinai Hospital, New York.* July–Aug.; 20(2):118–39.

————. 1952. "Human Atherosclerosis and the Diet." *Circulation.* Jan.; 5(1):115–18.

————. 1949. "The Calorie Requirement of Adult Man." *Nutrition Abstracts and Reviews.* July; 19:1–10.

Keys, A., and J. T. Anderson. 1955. "The Relationship of the Diet to the Development of Atherosclerosis in Man." In *Symposium on Atherosclerosis*, pub. no. 338. Washington, D.C.: National Research Council, National Academy of Sciences, 181–97.

Keys, A., J. T. Anderson, F. Fidanza, M. H. Keys, and B. Swahn. 1955. "Effects of Diet on Blood Lipids in Man, Particularly Cholesterol and Lipoproteins." *Clinical Chemistry.* Feb.; 1(1):34–52.

Keys, A., J. T. Anderson, and F. Grande. 1957. "Prediction of Serum-Cholesterol Responses of Man to Changes in Fats in the Diet." *Lancet.* Nov. 16; 273(7003):959–66.

Keys, A., J. T. Anderson, O. Mickelsen, S. F. Adelson, and R. Fidanza. 1956. "Diet and Serum Cholesterol in Man: Lack of Effect of Dietary Cholesterol." *Journal of Nutrition.* May 10; 59(1):39–56.

Keys, A., and J. Brozek. 1953. "Body Fat in Adult Man." *Physiological Reviews.* July; 33(3):245–325.

Keys, A., J. Brozek, A. Henschel, O. Mickelsen, and H. L. Taylor. 1950. *The Biology of Human Starvation.* 2 vols. Minneapolis: University of Minnesota Press.

Keys, A., and M. Keys. 1975. *How to Eat Well and Stay Well: The Mediterranean Way.* Garden City, N.Y.: Doubleday.

————. 1959. *Eat Well and Stay Well.* Garden City, N.Y.: Doubleday.

Keys, A., N. Kimura, A. Kusukawa, B. Bronte-Stewart, N. Larsen, and M. H. Keys. 1958. "Lessons from Serum Cholesterol Studies in Japan, Hawaii and Los Angeles." *Annals of Internal Medicine.* Jan.; 48(1):83–94.

Keys, A., A. Menotti, C. Aravanis, et al. 1984. "The Seven Countries Study: 2,289 Deaths in 15 Years." *Preventive Medicine.* March; 13(2):141–54.

Keys, A., O. Mickelsen, E. O. Miller, and C. B. Chapman. 1950. "The Relation in Man Between Cholesterol Levels in the Diet and in the Blood." *Science.* July 21; 112(2899):79–81.

Kiens, B., H. Lithell, K. J. Mikines, and R. E. Richter. 1989. "Effects of Insulin and Exercise on Muscle Lipoprotein Lipase Activity in Man and Its Relation to Insulin Action." *Journal of Clinical Investigation.* Oct.; 84(4):1124–29.

Kim, J., K. E. Peterson, K. S. Scanlon, et al. 2006. "Trends in Overweight from 1980 Through 2001 Among Preschool-Aged Children Enrolled in a Health Maintenance Organization." *Obesity.* July; 14(7):1107–12.

Kim, M., L. B. Hersh, M. A. Leissring, et al. 2007. "Decreased Catalytic Activity of the Insulin Degrading Enzyme in Chromosome 10–Linked Alzheimer's Disease Families." *Journal of Biological Chemistry.* March 16; 282(11):7825–32.

Kimura, K. D., H. A. Tissenbaum, Y. Liu, and G. Ruvkun. 1997. "Daf-2, an Insulin Recep-

tor–Like Gene That Regulates Longevity and Diapause in Caenorhabditis Elegans."
Science. Aug. 15; 277(5328):942–46.

Kinsell, L. W. 1969. "Dietary Composition—Weight Loss: Calories Do Count." In Wilson, ed., 1969, 177–84.

Kinsell, L. W., R. W. Friskey, G. D. Michaels, and S. Splitter. 1958. "Essential Fatty Acids, Lipid Metabolism, and Atherosclerosis." *Lancet.* Feb. 15; 271(7016):334–39.

Kinsell, L. W., B. Gunning, G. D. Michaels, J. Richardson, S. E. Cox, and C. Lemon. 1964. "Calories Do Count." *Metabolism.* March; 13:195–204.

Kinsell, L. W., J. Partridge, L. Boling, S. Margen, and G. Michaels. 1952. "Dietary Modification of Serum Cholesterol and Phospholipid Levels." *Journal of Clinical Endocrinology & Metabolism.* July; 12(7):909–13.

Kleiber, M. 1961. *The Fire of Life: An Introduction to Animal Energetics.* New York: John Wiley.

Klurfeld, D. M., C. B. Welch, L. M. Lloyd, and D. Kritchevsky. 1989. "Inhibition of DMBA-Induced Mammary Tumorigenesis by Caloric Restriction in Rats Fed High-Fat Diets." *International Journal of Cancer.* May 15; 43(5):922–25.

Knowler, W. C., E. Barrett-Connor, S. E. Fowler, et al. 2002. [Diabetes Prevention Program Research Group.] "Reduction in the Incidence of Type 2 Diabetes with Lifestyle Intervention or Metformin." *New England Journal of Medicine.* Feb. 7; 346(6):393–403.

Koenig, R. J., C. M. Peterson, R. L. Jones, C. Saudek, M. Lehrman, and A. Cerami. 1976. "Correlation of Glucose Regulation and Hemoglobin A1c in Diabetes Mellitus." *New England Journal of Medicine.* Aug. 19; 295(8):417–20.

Koga, Y., R. Hashimoto, H. Adachi, M. Tsuruta, H. Tashiro, and H. Toshima. 1994. "Recent Trends in Cardiovascular Disease and Risk Factors in the Seven Countries Study: Japan." In *Lessons for Science from the Seven Countries Study,* ed. H. Toshima, Y. Koga, and H. Blackburn. Tokyo: Springer-Verlag, 63–74.

Kolanowski, J. 1981. "Evaluation of the Possible Role of Insulin and Glucagons in Changes of Sodium Balance During Fasting." In Björntorp, Cairella, and Howard, eds., 1981, 25–31.

Kolata, G. 2000a. "Health Advice: A Matter of Cause, Effect and Confusion." *New York Times.* April 25; F1.

———. 2000b. "2 Fiber Studies Find No Benefit for the Colon." *New York Times.* April 20; A1.

———. 1987. "High-Carb Diets Questioned." *Science.* Jan. 9; 235(4785):164.

———. 1985. "Heart Panel's Conclusions Questioned." *Science.* Jan. 4; 227(4682):40–41.

———. 1982. "Heart Study Produces a Surprise Result." *Science.* Oct. 1; 218(4567):31–32.

———. 1981. "Data Sought on Low Cholesterol and Cancer." *Science.* March 27; 211(4489):1410–11.

Kolata, G., and M. Petersen. 2002. "Hormone Replacement Study a Shock to the Medical System." *New York Times.* July 10; 1.

Konner, M. 2003. *The Tangled Wing: Biological Constraints on the Human Spirit.* 2nd edition. New York: Henry Holt.

Koontz, J. W., and M. Iwahashi. 1981. "Insulin as a Potent, Specific Growth Factor in a Rat Hepatoma Cell Line." *Science.* Feb. 27; 211(4485):947–49.

Koop, C. E. 1988. "Message from the Surgeon General." In U.S. Department of Health and Human Services, 1988.

Korányi, A. 1963. "Prophylaxis and Treatment of the Coronary Syndrome." *Therapeutica Hungarica.* 11:17.

Korczyn, A. D. 2002. "Mixed Dementia—The Most Common Cause of Dementia." *Annals of the New York Academy of Sciences.* Nov.; 977:129–34.

Kozarevic, D., D. McGee, N. Vojvodic, et al. 1981. "Serum Cholesterol and Mortality: The Yugoslavia Cardiovascular Disease Study." *American Journal of Epidemiology.* July; 114(1):21–28.

Kraus, B. R. 1954. *Indian Health in Arizona: A Study of Health Conditions Among Central and Southern Arizona Indians.* Tucson: University of Arizona Press.

Krauss, R. M. 2005. "Dietary and Genetic Probes of Atherogenic Dyslipidemia." *Arteriosclerosis, Thrombosis, and Vascular Biology.* Nov.; 25(11):2265–72.

Krauss, R. M., and D. J. Burke. 1982. "Identification of Multiple Subclasses of Plasma Low Density Lipoproteins in Normal Humans." *Journal of Lipid Research.* Jan.; 23(1):97–104.

Krauss, R. M., R. J. Deckelbaum, N. Ernst, et al. 1996. "Dietary Guidelines for Healthy American Adults: A Statement for Health Professionals from the Nutrition Committee, American Heart Association." *Circulation.* Oct. 1; 94(7):1795–800.

Krauss, R. M., R. H. Eckel, B. Howard, et al. 2000. "AHA Dietary Guidelines, Revision 2000: A Statement for Healthcare Professionals from the Nutrition Committee of the American Heart Association." *Stroke.* Nov.; 31(11):2751–66.

Krebs, H. A. 1981. *Reminiscences and Reflections.* Oxford: Clarendon Press.

———. 1971. "How the Whole Becomes More Than the Sum of the Parts." *Perspectives in Biology and Medicine.* Spring; 14(3):448–57.

———. 1967. "The Making of a Scientist." *Nature.* Sept. 30; 215(5109):1441–45.

———. 1960. "The Cause of the Specific Dynamic Action of Food-Stuffs." *Arzneimittelforschung.* May; 10:369–73.

Krehl, W. A., A. Lopez, E. I. Good, and R. E. Hodges. 1967. "Some Metabolic Changes Induced by Low Carbohydrate Diets." *American Journal of Clinical Nutrition.* Feb.; 20(2):139–48.

Kritchevsky, D. 1992. "Unobserved Publications." In *Human Nutrition: A Continuing Debate,* ed. M. Eastwood, C. Edwards, and D. Parry. (London: Chapman and Hall.)

———. 1958. *Cholesterol.* New York: John Wiley.

Kritchevsky, D., M. M. Weber, and D. M. Klurfeld. 1984. "Dietary Fat Versus Caloric Content in Initiation and Promotion of 7,12-Dimethylbenz(a)anthracene–Induced Mammary Tumorigenesis in Rats." *Cancer Research.* Aug.; 44(8):3174–77.

Kuhn, T. S. 1970. *The Structure of Scientific Revolutions.* 2nd edition. Chicago: University of Chicago Press.

Kuo, P. T. 1967. "Hyperglyceridemia in Coronary Artery Disease and Its Management." *JAMA.* July 10; 201:87–94.

Kushi, L. H., T. Byers, C. Doyle, et al. 2006. [American Cancer Society 2006 Nutrition and Physical Activity Guidelines Advisory Committee.] "American Cancer Society Guidelines on Nutrition and Physical Activity for Cancer Prevention: Reducing the Risk of Cancer with Healthy Food Choices and Physical Activity." *CA: A Cancer Journal for Clinicians.* Sept.–Oct.; 56(5):254–81.

Kushi, L. H., R. A. Lew, F. J. Stare, et al. 1985. "Diet and 20-Year Mortality from Coronary Heart Disease: The Ireland-Boston Diet-Heart Study." *New England Journal of Medicine.* March 28; 312(13):811–18.

Kuulasmaa, K., H. Tunstall-Pedoe, A. Dobson, et al. 2000. "Estimation of Contribution of Changes in Classic Risk Factors to Trends in Coronary-Event Rates Across the WHO MONICA Project Populations." *Lancet.* Feb. 26; 355(9205):675–87.

Kuusisto, J., K. Koivisto, L. Mykkanen, et al. 1997. "Association Between Features of the Insulin Resistance Syndrome and Alzheimer's Disease Independently of Apolipoprotein E4 Phenotype: Cross Sectional Population Based Study." *British Medical Journal.* Oct. 25; 315(7115):1045–49.

Lampedusa, G. di 1988. *The Leopard.* Trans. A. Colquhoun. New York: Random House. [Originally published 1958.]

Landé, K. E., and W. M. Sperry. 1936. "Human Atherosclerosis in Relation to the Cholesterol Content of the Blood." *Archives of Pathology.* 22:301–312.

Landsberg, L. 2001. "Insulin-Mediated Sympathetic Stimulation: Role in the Pathogenesis of Obesity-Related Hypertension (or, How Insulin Affects Blood Pressure, and Why)." *Journal of Hypertension.* March; 19(3, pt. 2):523–28.

———. 1986. "Diet, Obesity and Hypertension: An Hypothesis Involving Insulin, the Sympathetic Nervous System, and Adaptive Thermogenesis." *Quarterly Journal of Medicine.* New series. Dec.; 61(236):1081–90.

Lane, W. A. 1929. *The Prevention of the Diseases Peculiar to Civilization.* London: Faber & Faber.

Langmuir, I. 1989. "Pathological Science." *Physics Today.* Oct.; 42(10):36–48.

Lappé, F. M. 1971. *Diet for a Small Planet.* New York: Ballantine Books.

LaRosa, J. C., A. Gordon, R. Muesing, and D. R. Rosing. 1980. "Effects of High-Protein, Low-Carbohydrate Dieting on Plasma Lipoproteins and Body Weight." *Journal of the American Dietetic Association.* Sept.; 77(3):264–70.

LaRosa, J. C., D. Hunninghake, D. Bush, et al. 1990. "The Cholesterol Facts: A Summary of the Evidence Relating Dietary Fats, Serum Cholesterol, and Coronary Heart Disease—A Joint Statement by the American Heart Association and the National Heart, Lung, and Blood Institute, the Task Force on Cholesterol Issues, American Heart Association." *Circulation.* May; 81(5):1721–33.

Lasby, C. G., 1997. *Eisenhower's Heart Attack: How Ike Beat Heart Disease and Held on to the Presidency.* Lawrence: University Press of Kansas.

Laskarzewski, P. M., J. A. Morrison, J. Gutai, T. Orchard, P. R. Khoury, and C. J. Glueck. 1983. "High and Low Density Lipoprotein Cholesterols in Adolescent Boys: Relationships with Endogenous Testosterone, Estradiol, and Quetelet Index." *Metabolism.* March; 32(3):262–71.

Laurell, S. 1956. "Plasma Free Fatty Acids in Diabetic Acidosis and Starvation." *Scandinavian Journal of Clinical and Laboratory Investigation.* 8(1):81–82.

Leary, T. 1935. "Atherosclerosis: The Important Form of Arteriosclerosis, a Metabolic Disease." *JAMA.* Aug. 17; 105(7):475–81.

Lee, M. O., and N. K. Schaffer. 1934. "Anterior Pituitary Growth Hormone and the Composition of Growth." *Journal of Nutrition.* March; 7(3):337–63.

Lee, R. B. 1968. "What Hunters Do for a Living, or, How to Make Out on Scarce Resources." In Lee and DeVore, eds., 1968, 30–48.

Lee, R. B., and I. DeVore. 1968. "Problems in the Study of Hunters and Gatherers." In Lee and DeVore, eds., 1968, 3–12.

Lee, R. B., and I. DeVore, eds. 1968. *Man the Hunter.* New York: Aldine Publishing.

Lees, R. S., and D. E. Wilson. 1971. "The Treatment of Hyperlipidemia." *New England Journal of Medicine.* Jan. 28; 284(4):186–95.

Le Fanu, J. 1999. *The Rise and Fall of Modern Medicine.* New York: Carroll & Graf.

———. 1987. *Eat Your Heart Out: The Fallacy of the Healthy Diet.* London: Macmillan.

Leibel, R. L., M. Rosenbaum, and J. Hirsch. 1995. "Changes in Energy Expenditure Resulting from Altered Body Weight." *New England Journal of Medicine.* March 9; 332(10):621–28.

Leibson, C. L., W. A. Rocca, V. A. Hanson, et al. 1997. "Risk of Dementia Among Persons with Diabetes Mellitus: A Population-Based Cohort Study." *American Journal of Epidemiology.* Feb. 15; 145(4):301–8.

Leith, W. 1961. "Experiences with the Pennington Diet in the Management of Obesity." *Canadian Medical Association Journal.* June 24; 84:1411–14.

Le Magnen, J. 2001. "My Scientific Life: 40 Years at the Collège de France." *Neuroscience & Biobehavioral Reviews.* July; 25(5):375–94.

———. 1988. "Lipogenesis, Lipolysis and Feeding Rhythms." *Annals of Endocrinology (Paris).* 49(2):98–104.

———. 1985. *Hunger.* Cambridge: Cambridge University Press.

———. 1984. "Is Regulation of Body Weight Elucidated?" *Neuroscience & Biobehavioral Reviews.* Winter; 8(4):515–22.

———. 1981. "Neuroendocrine Bases for the Liporegulatory Mechanism." In Cioffi, James, and Van Itallie, eds., 1981, 315–22.

———. 1978. "Metabolically Driven and Learned Feeding Responses in Man." In Bray, ed., 1978, 45–53.

———. 1976. "Interactions of Glucostatic and Lipostatic Mechanisms in the Regulatory Control of Feeding." In Novin et al., eds., 1976, 89–101.

———. 1971. "Advances in Studies on the Physiological Control and Regulation of Food Intake." In *Progress in Physiological Psychology,* ed. E. Stellar and J. M. Sprague. New York: Academic Press, 203–61.

Lepkovsky, S. 1973. "Newer Concepts in the Regulation of Food Intake." *American Journal of Clinical Nutrition.* March; 26(3):271–84.

———. 1969. "Fundamental Problems of Appetite Control." In Wilson, ed., 1969, 91–98.

———. 1948. "The Physiological Basis of Voluntary Food Intake." In *Advances in Food Research,* vol. 1, eds. E. M. Mrak and G. F. Steward. New York: Academic Press, 105–48.

LeRoith, D. 2004. "Blast from the Past—Insulin Does It Again!" *Journal of Clinical Endocrinology & Metabolism.* July; 89(7):3103–4.

LeRoith, D., R. Baserga, L. Helman, and C. T. Roberts, Jr. 1995. "Insulin-Like Growth Factors and Cancer." *Annals of Internal Medicine.* Jan. 1; 122(1):54–59.

LeRoith, D., and C. T. Roberts, Jr. 2003. "The Insulin-Like Growth Factor System and Cancer." *Cancer Letters.* June 10; 195(2):127–37.

Levenstein, H. 1999. "The Perils of Abundance: Food, Health, and Morality in American History." In *A Culinary History of Food,* ed. J.-L. Flandrin, M. Montanari, and A. Sonnenfeld. New York: Columbia University Press, 516–29.

———. 1993. *Paradox of Plenty: A Social History of Eating in Modern America.* New York: Oxford University Press.

Levin, I. 1910. "Cancer Among the North American Indians and Its Bearing Upon the Ethnological Distribution of Disease." *Zeitschrift für Krebsforschung.* Oct.; 9(3):422–35.

Levine, J. A., N. L. Eberhardt, and M. D. Jensen. 1999. "Role of Nonexercise Activity Thermogenesis in Resistance to Fat Gain in Humans." *Science.* Jan. 8; 283(5399):212–14.

Levine, J. M. 1986. "Hearts and Minds: The Politics of Diet and Heart Disease." In Sapolsky, ed., 1986, 40–79.

Levitsky, D. A., I. Faust, and M. Glassman. 1976. "The Ingestion of Food and the Recovery of Body Weight Following Fasting in the Naive Rat." *Physiology & Behavior.* Oct.; 17(4):575–80.

Levy, R. I., and N. Ernst. 1976. "Diet, Hyperlipidemia, and Atherosclerosis." In Goodhart and Shils, eds., 1973, 895–918.

Levy, R. I., R. S. Lees, and D. S. Fredrickson. 1966. "The Nature of Pre Beta (Very Low Density) Lipoproteins." *Journal of Clinical Investigation.* Jan.; 45(1):63–77.

Levy, R. L. 1932. "Discussion on Arterial Thrombosis." *Transactions of the Association of American Physicians.* 47:77.

Lichtenstein, G. 1972. "Fraud Complaints on Health-Spas Rise." *New York Times.* Dec. 26; 69.

Lieb, C. W. 1929. "The Effects on Human Beings of a Twelve Months' Exclusive Meat Diet—Based on Intensive Studies on Two Arctic Explorers Living Under Average Conditions in a New York Climate." *JAMA.* July–Dec.; 96:20–22.

Lieb, C. W., and E. Tolstoi. 1929. "Effect of an Exclusive Meat Diet on Chemical Constituents of the Blood." *Proceedings of the Society for Experimental Biology and Medicine.* Jan.; 26(4):324–25.

Liebling, A. J. 2004. "The Earl of Louisiana." In *Just Enough Liebling: Classic Work by the Legendary New Yorker Writer,* ed. D. Remnick. New York: North Point Press, 471–519.

———. 1975. *The Press.* 2nd edition. New York: Ballantine Books.

Liebowitz, J. O. 1970. *The History of Coronary Heart Disease.* Berkeley: University of California Press.

Lin, K., J. B. Dorman, A. Rodan, and C. Kenyon. 1997. "Daf-16: An HNF-3/Forkhead Family Member That Can Function to Double the Life-Span of Caenorhabditis Elegans." *Science.* Nov. 14; 278(5341):1319–22.

Lin, S. J., P. A. Defossez, and L. Guarente. 2000. "Requirement of NAD and SIR2 for Life-Span Extension by Calorie Restriction in Saccharomyces Cerevisiae." *Science.* Sept. 22; 289(5487):2126–28.

Lindner, L. 2001. "When Good Carbs Turn Bad: Fine-Tuning the Carbohydrate-to-Fat Ratio May Reduce Heart Risks for Those with 'Syndrome X.' " *Washington Post.* June 19; T09.

Lindsay, S., and I. L. Chaikoff. 1963. "Naturally Occurring Arteriosclerosis in Animals: A Comparison with Experimentally Induced Lesions." In Sandler and Bourne, eds., 1963, 350–438.

Lipid Research Clinics [LRC] Program. 1984a. "The Lipid Research Clinics Coronary Primary Prevention Trial Results. I. Reduction in Incidence of Coronary Heart Disease." *JAMA.* Jan. 20; 251(3):351–64.

———. 1984b. "The Lipid Research Clinics Coronary Primary Prevention Trial Results. II. The Relationship of Reduction in Incidence of Coronary Heart Disease to Cholesterol Lowering." *JAMA.* Jan. 20; 251(3):365–74.

———. 1979. "The Coronary Primary Prevention Trial: Design and Implementation." *Journal of Chronic Diseases.* 32(9–10):609–31.

Lithell, H. 1987. "Site Differences in Lipoprotein Lipase Activity." In Berry et al., eds., 1987, 77–81.

Livingstone, D. 2001. *Missionary Travels in South Africa.* Vol. 1. Santa Barbara, Calif.: Narrative Press.

Look AHEAD Protocol Review Committee. 2006. "Protocol Action for Health in Diabetes, Look AHEAD Clinical Trial, November 4, 2006." 5th revision. Online, downloaded Oct. 17, 2006, at https://lookahead.phs.wfubmc.edu/index.cfm?ev=studyProtocol.

Lorgeril, M. de, P. Salen, J. L. Martin, I. Monjaud, J. Delaye, and N. Mamelle. 1999. "Mediterranean Diet, Traditional Risk Factors, and the Rate of Cardiovascular Complications After Myocardial Infarction: Final Report of the Lyon Diet Heart Study." *Circulation.* Feb. 16; 99(6):779–85.

Lozano, R., C. J. Murray, A. D. Lopez, and T. Sato. 2001. "Miscoding and Misclassification of Ischaemic Heart Disease Mortality." World Health Organization. Online at http://www.who.int/entity/healthinfo/paper12.pdf.

Luchsinger, J. A., M. X. Tang, Y. Stern, S. Shea, and R. Mayeux. 2001. "Diabetes Mellitus and Risk of Alzheimer's Disease and Dementia with Stroke in a Multiethnic Cohort." *American Journal of Epidemiology*. Oct. 1; 154(7):635–41.

Lusk, G. 1928. *The Elements of the Science of Nutrition*. Philadelphia: W. B. Saunders.

Lutwak, L., and A. Coulson. 1976. "Activity and Obesity." In Bray, ed., 1976a, 393–96.

Lyon, D. M., and D. M. Dunlop. 1932, "A Comparison of the Effects of Diet and of Thyroid Extract." *Quarterly Journal of Medicine*. 1:331–52.

Lyon, J. L., J. W. Gardner, and D. W. West. 1980. "Cancer Incidence in Mormons and Non-Mormons in Utah During 1967–75." *Journal of the National Cancer Institute*. Nov.; 65(5):1055–61.

Lyon, J. L., and A. W. Sorenson. 1978. "Colon Cancer in a Low-Risk Population." *American Journal of Clinical Nutrition*. Oct.; 31(10 suppl.):S227–30.

MacBryde, C. M. 1951. "Obesity." In *Textbook of Medicine*, ed. R. L. Cecil and R. F. Loeb. Philadelphia: W. B. Saunders, 647–53.

Mackarness, R. 1975. *Eat Fat and Grow Slim*. Revised edition. Glasgow: Fontana/Collins.

———. 1958. *Eat Fat and Grow Slim*. London: Harvill Press.

MacLennan, R., F. Macrae, C. Bain, et al. 1995. "Randomized Trial of Intake of Fat, Fiber, and Beta Carotene to Prevent Colorectal Adenomas: The Australian Polyp Prevention Trial." *Journal of the National Cancer Institute*. Dec. 6; 87(23):1760–66.

Maegawa, H., M. Kobayashi, O. Ishibashi, Y. Takata, and Y. Shigeta. 1986. "Effect of Diet Change on Insulin Action: Difference Between Muscles and Adipocytes." *American Journal of Physiology*. Nov.; 251(5, pt. 1):E616–23.

Magnus-Levy, A. 1907. *The Physiology of Metabolism*. Vol. 1 of *Metabolism and Practical Medicine*, ed. C. von Noorden. Chicago: W. T. Keener.

Magoun, H. W., and C. Fisher. 1980. "Walter R. Ingram at Ranson's Institute of Neurology, 1930–1936." *Perspectives in Biology and Medicine*. Autumn; 24(1):31–56.

Malmros, H. 1950. "The Relation of Nutrition to Health: A Statistical Study of the Effect of the War-Time on Arteriosclerosis, Cardiosclerosis, Tuberculosis, and Diabetes." *Acta Medica Scandinavica*. 246(suppl.):137–53.

Man, E. B., and J. P. Peters. 1935. "Serum Lipoids in Diabetes." *Journal of Clinical Investigation*. 14(5):579–94.

Mann, C. C. 1990. "Meta-Analysis in the Breech." *Science*. Aug. 3; 249(4968):476–80.

Mann, G. V. 1977. "Diet-Heart: End of an Era." *New England Journal of Medicine*. Sept. 22; 297(12):644–50.

———. 1974. "The Influence of Obesity on Health (Second of Two Parts)." *New England Journal of Medicine*. Aug. 1; 291(5):226–32.

———. 1957. "Epidemiology of Coronary Heart Disease." *American Journal of Medicine*. Sept.; 23(3):463–80.

Mann, G. V., R. D. Shaffer, R. S. Anderson, and H. H. Sandstead. 1964. "Cardiovascular Disease in the Masai." *Journal of Atherosclerosis Research*. July–Aug.; 4:289–312.

Mann, G. V., A. Spoerry, M. Gray, and D. Jarashow. 1972. "Atherosclerosis in the Masai." *American Journal of Epidemiology*. Jan.; 95(1):26–37.

Manson, J. E., J. Hsia, K. C. Johnson, et al. [Women's Health Initiative Investigators.] 2003. "Estrogen Plus Progestin and the Risk of Coronary Heart Disease." *New England Journal of Medicine*. Aug. 7; 349(6):523–34.

Maratos-Flier, E., and J. S. Flier. 2005. "Obesity." In Kahn et al., eds., 2005, 533–45.

Marble, A., P. White, R. F. Bradley, and L. P. Krall, eds. 1971. *Joslin's Diabetes Mellitus*. 11th edition. Philadelphia: Lea & Febiger.

Margolis, S., and M. Vaughan. 1962. "α-Glycerophosphate Synthesis and Breakdown in Homogenates of Adipose Tissue." *Journal of Biological Chemistry*. Jan.; 237:44–48.

Marmot, M. G., S. L. Syme, A. Kagan, H. Kato, J. B. Cohen, and J. Belsky. 1975. "Epidemiologic Studies of Coronary Heart Disease and Stroke in Japanese Men Living in Japan, Hawaii, and California: Prevalence of Coronary and Hypertensive Heart Disease and Associated Risk Factors." *American Journal of Epidemiology*. Dec.; 102(6): 514–25.

Marshall, E. 1993a. "Big Science Enters the Clinic." *Science*. May 7; 260(5109):744–47.

———. 1993b. "Epidemiology: Search for a Killer—Focus Shifts from Fat to Hormones." *Science*. Jan. 29; 259(5095):618–21.

Martin, M. J., S. B. Hulley, W. S. Browner, L. H. Kuller, and D. Wentworth. 1986. "Serum Cholesterol, Blood Pressure, and Mortality: Implications from a Cohort of 361,662 Men." *Lancet*. Oct. 25; 328(8513):933–36.

Marx J. 2001. "Alzheimer's Disease: Bad for the Heart, Bad for the Mind?" *Science*. Oct. 19; 294(5542):508–9.

Marx, W., M. E. Simpson, W. O. Reinthardt, and H. M. Evans. 1942. "Response to Growth Hormone of Hypophysectomized Rats When Restricted to Food Intake of Controls." *American Journal of Physiology*. Feb.; 135(3):614–18.

Masironi, R. 1970. "Dietary Factors and Coronary Heart Disease." *Bulletin of the World Health Organization*. 42(1):103–14.

Masoro, E. J. 2003. "Subfield History: Caloric Restriction, Slowing Aging, and Extending Life." *Science of Aging Knowledge Environment*. Feb. 26; 2003(8):RE2. Online at http://sageke.sciencemag.org/cgi/content/full/sageke;2003/8/re2.

Masoro, E. J., B. P. Yu, and H. A. Bertrand. 1982. "Action of Food Restriction in Delaying the Aging Process." *Proceedings of the National Academy of Sciences*. July; 79(13): 4239–41.

Mattson, F. H., and S. M. Grundy. 1985. "Comparison of Effects of Dietary Saturated, Monounsaturated, and Polyunsaturated Fatty Acids on Plasma Lipids and Lipoproteins in Man." *Journal of Lipid Research*. Feb.; 26(2):194–202.

Mattson, M. P. 2004. "Pathways Towards and Away from Alzheimer's Disease." *Nature*. Aug. 5; 430(7000):631–39.

Mauer, M. M., R. B. Harris, and T. J. Bartness. 2001. "The Regulation of Total Body Fat: Lessons Learned from Lipectomy Studies." *Neuroscience & Biobehavioral Reviews*. Jan.; 25(1):15–28.

Maugh, T. H. 1979. "Cancer and Environment: Higginson Speaks Out." *Science*. Sept. 28; 205(4413):1363–66.

Mayer, J. 1976. "The Bitter Truth About Sugar." *New York Times Magazine*. June 20:26–34.

———. 1975a. *A Diet for Living*. New York: David McKay.

———. 1975b. "Obesity During Childhood." In Winnick, ed., 1975, 73–80.

———. 1974a. "By Bread Alone." *New York Times Book Review*. Dec. 15:19.

———. 1974b. "Banning Carbohydrates: Mistaken, Possibly Dangerous." *Washington Post*. Feb. 28; F4.

———, ed. 1973. *U.S. Nutrition Policies in the Seventies*. San Francisco: W. H. Freeman.

———. 1973a. "Heart Disease: Plans for Action." In Mayer, ed., 1973, 44–52.

———. 1973b. "Diet Revolution: Basically Old Hat." *Washington Post*. April 14; F1.

———. 1969. "Obesity: 'Disease of Civilization.' " *Washington Post*. Nov. 30:33.

———. 1968. *Overweight: Causes, Cost, and Control*. Englewood Cliffs, N.J.: Prentice-Hall.

———. 1965. "The Best Diet Is Exercise." *New York Times Magazine*. April 25; 34, 40, 42, 49, 50, 52, 54–5.

———. 1955. "Appetite and Obesity." *Atlantic*. Sept.; 58–62.

———. 1954. "Multiple Causative Factors in Obesity." In *Fat Metabolism*, ed. V. A. Najjar. Baltimore: Johns Hopkins University Press, 22–43.

————. 1953a. "Traumatic and Environmental Factors in the Etiology of Obesity." *Physiological Reviews*. Oct.; 33(4):472–508.

————. 1953b. "Decreased Activity and Energy Balance in the Hereditary Obesity-Diabetes Syndrome of Mice." *Science*. May 8; 117(3045):504–5.

Mayer, J., and J. Dwyer. 1977. "Nutrition." *Washington Post*. May 12; 93.

Mayer, J., and J. Goldberg. 1984. "Nutrition." *Washington Post*. July 1; H10.

————. 1983. "Nutrition." *Washington Post*. Dec. 11; H14.

Mayer, J., N. B. Marshall, J. J. Vitale, J. H. Christensen, M. B. Mashayekhi, and F. J. Stare. 1954. "Exercise, Food Intake and Body Weight in Normal Rats and Genetically Obese Adult Mice." *American Journal of Physiology*. June; 177(3):544–48.

Mayer, J., P. Roy, and K. P. Mitra. 1956. "Relation Between Caloric Intake, Body Weight, and Physical Work: Studies in an Industrial Male Population in West Bengal." *American Journal of Clinical Nutrition*. March–April; 4(2):169–75.

Mayer, J., and F. J. Stare. 1953. "Exercise and Weight Control." *Journal of the American Dietetic Association*. April; 29(4):340–43.

Mayer, J., and D. W. Thomas. 1967. "Regulation of Food Intake and Obesity." *Science*. April 21; 156(773):328–37.

Mayer-Davis, E. J., K. C. Sparks, K. Hirst, et al. [Diabetes Prevention Program Research Group.] 2004. "Dietary Intake in the Diabetes Prevention Program Cohort: Baseline and 1-Year Post Randomization." *Annals of Epidemiology*. Nov.; 14(10):763–72.

Mayes, P. A. 1993. "Intermediary Metabolism of Fructose." *American Journal of Clinical Nutrition*. Nov.; 58(5 suppl.):754S–65S.

Mayes, P. A., and K. M. Botham. 2004. "Lipid Transport & Storage." In *Harper's Illustrated Biochemistry*, 26th edition, ed. R. K. Murray, D. K. Granner, P. A. Mayes, and V. W. Rodwell (New York: Lange Medical Books/McGraw-Hill.) 205–18.

McCarrison, R. 1961. *Nutrition and Health*. London: Faber & Faber.

————. 1922. "An Address on Faulty Food in Relation to Gastro-Intestinal Disorder." *Lancet*. Feb. 4; 199(5136):207–12.

McCarthy, C. 1966. "Dietary and Activity Patterns of Obese Women in Trinidad." *Journal of the American Dietetic Association*. Jan.; 48:33–37.

McCay, C. M., M. F. Crowell, and L. A. Maynard. 1935. "The Effect of Retarded Growth upon the Length of Life Span and upon the Ultimate Body Size." *Journal of Nutrition*. July; 10(1):63–79.

McClellan, W. S., and E. F. Du Bois. 1930. "Clinical Calorimetry. XLV. Prolonged Meat Diets with a Study of Kidney Function and Ketosis." *Journal of Biological Chemistry*. July; 87(3):651–68.

McClellan, W. S., V. R. Rupp, and V. Toscani. 1930. "Clinical Calorimetry. XLVI. Prolonged Meat Diets with a Study of the Metabolism of Nitrogen, Calcium, and Phosphorus." *Journal of Biological Chemistry*. July; 87(3):669–80.

McClellan, W. S., H. J. Spencer, and E. A. Falk. 1931. "Clinical Calorimetry. XLVII. Prolonged Meat Diets with a Study of the Respiratory Metabolism." *Journal of Biological Chemistry*. Oct.; 93(2):419–34.

McCollum, E. V. 1957. *A History of Nutrition: The Sequence of Ideas in Nutrition Investigations*. Cambridge, Mass.: Riverside Press.

————. 1922. *The Newer Knowledge of Nutrition: The Use of Food for the Preservation of Vitality and Health*. New York: Macmillan.

McFarlane, S. I., J. Castro, D. Kirpichnikov, and J. R. Sowers. 2005. "Hypertension in Diabetes Mellitus." In Kahn et al., eds., 2005, 968–74.

McGarry, D. J. 1992. "What If Minkowski Had Been Ageusic? An Alternative Angle on Diabetes." *Science*. Oct. 30; 258(5083):766–70.

McGee, D., D. Reed, G. Stemmerman, G. Rhoads, K. Yano, and M. Feinleib. 1985. "The Relationship of Dietary Fat and Cholesterol to Mortality in 10 Years: The Honolulu Heart Program." *International Journal of Epidemiology.* March; 14(1):97–105.

McGee, D. L., D. M. Reed, K. Yano, A. Kagan, and J. Tillotson. 1984. "Ten-Year Incidence of Coronary Heart Disease in the Honolulu Heart Program: Relationship to Nutrient Intake." *American Journal of Epidemiology.* May; 119(5):667–76.

McGill, H. C., Jr., J. P. Strong, R. L. Holman, and N. T. Werthessen. 1960. "Arterial Lesions in the Kenya Baboon." *Circulation Research.* 8:670–79.

McGovern, P. G., D. R. Jacobs, Jr., E. Shahar, et al. 2001. "Trends in Acute Coronary Heart Disease Mortality, Morbidity, and Medical Care from 1985 through 1997: the Minnesota Heart Survey." *Circulation.* July 3; 104(1):19–24.

McKeown-Eyssen, G. E., E. Bright-See, W. R. Bruce, et al. 1994. "A Randomized Trial of a Low Fat High Fibre Diet in the Recurrence of Colorectal Polyps: Toronto Polyp Prevention Group." *Journal of Clinical Epidemiology.* May; 47(5):525–36.

McLean Baird, I., and A. N. Howard, eds. 1969. *Obesity: Medical and Scientific Aspects. Proceedings of the First Symposium of The Obesity Association of Great Britain Held in London, October 1968.* London: E. & S. Livingstone.

McPherson, J. D., B. H. Shilton, and D. J. Walton. 1988. "Role of Fructose in Glycation and Cross-Linking of Proteins." *Biochemistry.* March 22; 27(6):1901–7.

Mencken, H. L. 1982. "The Divine Afflatus." In *A Mencken Chrestomathy.* New York: Vintage Books. [Originally published 1949.]

Meriam, L., R. A. Brown, R. H. Cloud, et al. 1928. "The Problem of Indian Administration." Baltimore: Johns Hopkins University Press. Online at http://www.alaskool.org/native_ed/research_reports/IndianAdmin/Indian_Admin_Problms.html.

Merkel, M., R. H. Eckel, and J. J. Goldberg. 2002. "Lipoprotein Lipase: Genetics, Lipid Uptake, and Regulation." *Journal of Lipid Research.* Dec.; 43(12):1997–2006.

Merrill, R. A. 1981. "Saccharin: A Regulator's View." In Crandall and Lave, eds., 1981, 153–72.

Merton, R. K. 1973. *The Sociology of Science: Theoretical and Empirical Investigations.* Chicago: University of Chicago Press.

Michels, K. B., E. Giovannucci, K. J. Joshipura, et al. 2000. "Prospective Study of Fruit and Vegetable Consumption and Incidence of Colon and Rectal Cancers." *Journal of the National Cancer Institute.* Nov. 1; 92(21):1740–52.

Miettinen M., O. Turpeinen, M. J. Karvonen, R. Elosuo, and E. Paavilainen. 1972. "Effect of Cholesterol-Lowering Diet on Mortality from Coronary Heart-Disease and Other Causes: A Twelve-Year Clinical Trial in Men and Women." *Lancet.* Oct. 21; 300(7782):835–38.

Milch, L. J., W. J. Walker, and N. Weiner. 1957. "Differential Effect of Dietary Fat and Weight Reduction on Serum Levels of Beta-Lipoproteins." *Circulation.* Jan.; 15(1):31–34.

Miller, B. C., E. A. Eckman, K. Sambamurti, et al. 2003. "Amyloid-Beta Peptide Levels in Brain Are Inversely Correlated with Insulysin Activity Levels in Vivo." *Proceedings of the National Academy of Sciences.* May 13; 100(10):6221–26.

Miller, D. S., and P. Mumford. 1966. "Obesity: Physical Activity and Nutrition." *Proceedings of the Nutrition Society.* 25(2):100–107.

Miller, D. S., and P. R. Payne. 1962. "Weight Maintenance and Food Intake." *Journal of Nutrition.* 78(3):255–62.

Miller, S. R., P. I. Tartter, A. E. Papatestas, G. Slater, and A. H. Aufses, Jr. 1981. "Serum Cholesterol and Human Colon Cancer." *Journal of the National Cancer Institute.* Aug.; 67(2):297–300.

Mills, C. A. 1930. "Diabetes Mellitus: Sugar Consumption in Its Etiology." *Archives of Internal Medicine.* 46:582–84.

Mintz, S. W. 1986. *Sweetness and Power: The Place of Sugar in Modern History.* New York: Penguin Books.

Mitchell, H. S. 1930. "Nutrition Survey in Labrador and Northern Newfoundland." *Journal of the American Dietetic Association.* 6:29–35.

Moller, E. 1931. "Results of Exclusively Dietary Treatment in 46 Cases of Obesity." *Acta Medica Scandinavica.* 74(4):341–52.

Monnier, V. M., R. R. Kohn, and A. Cerami. 1984. "Accelerated Age-Related Browning of Human Collagen in Diabetes Mellitus." *Proceedings of the National Academy of Sciences.* Jan.; 81(2):583–87.

Moore, T. J. 1989. *Heart Failure.* New York: Simon & Schuster.

Moore, W. W. 1983. *Fighting for Life: The Story of the American Heart Association, 1911–1975.* Dallas: American Heart Association.

Moreira, P. I., M. A. Smith, X. Zhu, A. Nunomura, R. J. Castellani, and G. Perry. 2005. "Oxidative Stress and Neurodegeneration." *Annals of the New York Academy of Sciences.* June; 1043:545–52.

Moulton, C. R. 1930. "Is a Meat Diet a Menace?" *JAMA.* July–Dec.; 95:1762.

Mountford, C. P., ed. 1960. *Anthropology and Nutrition,* vol. 2 of *Records of the American-Australian Scientific Expedition to Arnhem Land.* Melbourne: Melbourne University Press.

Mowat, F. 1978. *People of the Deer.* New York: Jove/HBJ Books.

Mowri, H. O., B. Frei, and J. F. Keaney, Jr. 2000. "Glucose Enhancement of LDL Oxidation Is Strictly Metal Ion Dependent." *Free Radical Biology & Medicine.* Nov. 1; 29(9):814–24.

Mrosovsky, N. 1985. "Cyclical Obesity in Hibernators: The Search for the Adjustable Regulator." In Hirsch and Van Itallie, eds., 1985, 45–56.

———. 1976. "Lipid Programmes and Life Strategies in Hibernators." *American Zoologist.* 16:685–97.

Multiple Risk Factor Intervention Trial [MRFIT] Research Group. 1982. "Multiple Risk Factor Intervention Trial: Risk Factor Changes and Mortality Results." *JAMA.* Sept. 24; 248(12):1465–77.

Murdoch, J. 1892. *Ethnological Results of the Point Barrow Expedition.* Ninth Annual Report of the Bureau of Ethnology. Washington, D.C.: U.S. Government Printing Office.

Nasar, S. 1998. *A Beautiful Mind.* New York: Simon & Schuster.

National Cancer Institute Web site. Colorectal Cancer (PDQ®): Prevention. Online at http://www.nci.nih.gov/cancerinfo/pdg/prevention/colorectal/patient/.

National Center for Health Statistics [NCHS]. 2006. *Health, United States, 2006, with Chartbook on the Health of Americans.* Washington, D.C.: U.S. Government Printing Office.

———. 2005. *Health, United States, 2005, with Chartbook on Trends in the Health of Americans.* Washington, D.C.: U.S. Government Printing Office.

———. 2004. "FASTATS A to Z (2004)." Online at http://www.cdc.gov/nchs/fastats/.

National Cholesterol Education Program [NCEP]. 2002. "Third Report of the National Cholesterol Education Program (NCEP) Expert Panel on Detection, Evaluation, and Treatment of High Blood Cholesterol in Adults: (Adult Treatment Panel III) Final Report." *Circulation.* Dec. 17; 106(25):3143–421.

National Heart, Lung, and Blood Institute [NHLBI] Communication Office. 2006. Press release: "News from the Women's Health Initiative: Reducing Total Fat Intake May

Have Small Effect on Risk of Breast Cancer, No Effect on Risk of Colorectal Cancer, Heart Disease, or Stroke." Feb. 6.

National Institute of Diabetes and Digestive and Kidney Diseases [NIDDK]. 1995. *The Pima Indians: Pathfinders for Health*. NIH Publication No. 95-3821.

National Institutes of Health. n.d. "The NIH Almanac—Appropriations." Online at http://www.nih.gov/about/almanac/appropriations/index.htm

National Research Council [NRC], Committee on Diet and Health, Food and Nutrition Board, Commission on Life Sciences. 1989. *Diet and Health: Implications for Reducing Chronic Disease Risk*. Washington, D.C.: National Academy Press.

Neel, J. V. 1999. "The 'Thrifty Genotype' in 1998." *Nutrition Reviews*. May; 57(5, pt. 2):S2–9.

———. 1994. *Physician to the Gene Pool: Genetic Lessons and Other Stories*. New York: John Wiley.

———. 1989. "Update to 'The Study of Natural Selection in Primitive and Civilized Human Populations.' " *Human Biology*. Oct.–Dec.; 61(5–6):811–23.

———. 1982. "The Thrifty Genotype Revisited." In *The Genetics of Diabetes Mellitus*, ed. J. Köbberling and R. Tattersall. New York: Academic Press, 283–93.

———. 1962. "Diabetes Mellitus: A 'Thrifty' Genotype Rendered Detrimental by 'Progress'?" *American Journal of Human Genetics*. Dec.; 14:353–62.

Nestle, M. 2003. "The Ironic Politics of Obesity." *Science*. Feb. 7; 269(5608):781.

Neumann, R. O. 1902. "Experimentelle Beiträge zur Lehre von dem Täglichen Nahrungsbedarf des Menschen unter besonderer Berücksichtigung der Notwendigen Eiweissmenge." *Archiv für Hygiene*. 45:1–87.

Newburgh, L. H. 1948. "Energy Metabolism in Obese Patients." *Bulletin of the New York Academy of Medicine*. 24(4):227–38.

———. 1944. "Obesity I. Energy Metabolism." *Physiological Reviews*. 24:18–31.

———. 1942. "Obesity." *Archives of Internal Medicine*. Dec.; 70:1033–96.

———. 1931a. "The Cause of Obesity." *JAMA*. Dec. 5; 97(23):1659–63.

———. 1931b. "Is a Meat Diet a Menace?" *JAMA*. Jan.–June; 96:289–90.

———. 1923. "The Etiology of Nephritis." *Medicine*. Feb.; 2(1):77–104.

Newburgh, L. H., M. Falcon-Lesses, and M. W. Johnston. 1930. "The Nephropathic Effect in Man of a Diet High in Beef Muscle and Liver." *American Journal of Medical Sciences*. 179:305–10.

Newburgh, L. H., and M. W. Johnston. 1930a. "Endogenous Obesity—A Misconception." *Annals of Internal Medicine*. Feb.; 8(3):815–25.

———. 1930b. "The Nature of Obesity." *Journal of Clinical Investigation*. Feb.; 8(2): 197–213.

Newsholme, E. A., and A. R. Leech. 1983. *Biochemistry for the Medical Sciences*. New York: John Wiley.

Newsholme, E. A., and C. Start. 1973. *Regulation in Metabolism*. New York: John Wiley.

Nichols, A. B., C. Ravenscroft, D. E. Lamphiear, and L. D. Ostrander, Jr. 1976. "Independence of Serum Lipid Levels and Dietary Habits: The Tecumseh Study." *JAMA*. Oct. 25; 236(17):1948–53.

Nicolaidis, S. 1969. "Early Systematic Responses to Orogastric Stimulation in the Regulation of Food and Water Balance: Functional and Electrophysiological Data." *Annals of the New York Academy of Sciences*. May 15; 157(2):1176–20.

Nikkila, E. 1953. *Studies on the Lipid-Protein Relationships in Normal and Pathological Sera and the Effect of Heparin on Serum Lipoproteins*. Helsinki: Mercatorin Kirjapaino.

Nishizawa, T., I. Akaoka, Y. Nishida, Y. Kawaguchi, and E. Hayashi. 1976. "Some Factors

Related to Obesity in the Japanese Sumo Wrestler." *American Journal of Clinical Nutrition*. Oct.; 29(10):1167–74.

Noorden, C. von. 1907a. "Obesity." Trans. D. Spence. In von Noorden and Hall, eds., 1907, 693–715.

Noorden, C. von. 1907b. "Overfeeding." Trans. R. W. Marsden. In *Metabolism: Physiology and Pathology with Its Application to Practical Medicine*, ed. C. von Noorden and I. W. Hall (Chicago: W. T. Keener & Co.), 62–89.

Noorden, C. von. 1907c. *Diabetes Mellitus*. Vol. 7 of *Clinical Treatises on the Pathology and Therapy of Disorders of Metabolism and Nutrition*. New York: E. B. Treat.

Noorden, C. von, and I. W. Hall, eds. 1907. *The Pathology of Metabolism*. Vol. 3 of *Metabolism and Practical Medicine*. Chicago: W. T. Keener & Co.

Novartis Foundation. 2004. *Biology of IGF-1: Its Interaction with Insulin in Health and Malignant States*. Chichester, U.K.: John Wiley.

Novin, D. 1978. "Some Expected and Unexpected Effects of Glucose on Food Intake." In Bray, ed., 1978, 27–32.

Novin, D., W. Wyrwicka, and G. A. Bray, eds. 1976. *Hunger: Basic Mechanisms and Clinical Implications*. New York: Raven Press.

Nydegger, U. R., and R. E. Butler. 1970. "Serum Lipoprotein Levels in Patients with Cancer." *Cancer Research*. Aug.; 32:1756–60.

Obrenovich, M. E., and V. M. Monnier. 2004. "Glycation Stimulates Amyloid Formation." *Science of Aging Knowledge Environment*. Jan. 4; No. 2:pe3. Online at http://sageke.sciencemag.org/cgi/content/full/2004/2/pe3.

Oertel, M. J. 1895. "Obesity." In *Nutritive Disorders*. Vol. 2 of *Twentieth Century Practice*, ed. T. L. Stedman. New York: William Wood, 625–728.

Ogden, C. L., M. D. Carroll, L. R. Curtin, M. A. McDowell, C. J. Tabak, and K. M. Flegal. 2006. "Prevalence of Overweight and Obesity in the United States, 1999–2004." *JAMA*. April 5; 295(13):1549–55.

Ogden, C. L., M. D. Carroll, and K. M. Flegal. 2003. "Epidemiologic Trends in Overweight and Obesity." *Endocrinology and Metabolism Clinics of North America*. Dec.; 32(4):741–60.

Ohlson, M. A., W. D. Brewer, D. Kereluk, A. Wagoner, and D. C. Cederquist. 1955. "Weight Control Through Nutritionally Adequate Diets." In Eppright et al., eds., 1955, 170–87.

Oliver, M. F. 1985. "Consensus or Nonsensus Conferences on Coronary Heart Disease." *Lancet*. May 11; 325(8437):1087–89.

Olson, R. E., H. P. Broquist, C. O. Chichester, W. J. Darby, A. C. Kobye, Jr., and R. M. Stalvey, eds. 1984. *Present Knowledge in Nutrition*. 5th edition. Washington, D.C.: Nutrition Foundation.

Orchard, T. J., M. Temprosa, R. Goldberg, et al. [Diabetes Prevention Program Research Group.] 2005. "The Effect of Metformin and Intensive Lifestyle Intervention on the Metabolic Syndrome: The Diabetes Prevention Program Randomized Trial." *Annals of Internal Medicine*. April 19; 142(8):611–19.

Orenstein, A. J. 1923. "Freedom of Negro Races from Cancer." *British Medical Journal*. Aug. 25; 342.

Osancova, K. 1976. "Trends of Dietary Intake and Prevalence of Obesity in Czechoslovakia." In Howard, ed., 1976, 42–50.

Osborne, C. K., G. Bolan, M. E. Monaco, and M. E. Lippman. 1976. "Hormone Responsive Human Breast Cancer in Long-Term Tissue Culture: Effect of Insulin." *Proceedings of the National Academy of Sciences*. Dec.; 73(12):4536–40.

Oscai, L. B., M. M. Brown, and W. C. Miller. 1984. "Effect of Dietary Fat on Food Intake, Growth, and Body Composition in Rats." *Growth*. Winter; 48(4):415–24.

Osler, W. 1901. *The Principles and Practice of Medicine*. New York: D. Appleton.

Osmundsen, J. A. 1961. "New Views Given on Fats in Diet: Foods Rich in Starches and Sugars Appear to Raise Level of Triglycerides." *New York Times*. May 4; 39.

Ott, A., R. P. Stolk, F. van Harskamp, H. A. Pols, A. Hofman, and M. M. Breteler. 1999. "Diabetes Mellitus and the Risk of Dementia: The Rotterdam Study." *Neurology*. Dec. 10; 53(9):1937–42.

Oxford English Dictionary. 1989. OED Online. 2nd edition. Oxford University Press.

Pace, E. 1998. "Benjamin Spock, World's Pediatrician, Dies at 94." *New York Times*. Mar. 17: A1.

Paffenbarger, R. S., Jr., A. L. Wing, and R. T. Hyde. 1978. "Physical Activity as an Index of Heart Attack Risk in College Alumni." *American Journal of Epidemiology*. Sept.; 108(3):161–75.

Page, I. H., L. A. Lewis, and J. Gilbert. 1956. "Plasma Lipids and Proteins and Their Relationship to Coronary Disease Among Navajo Indians." *Circulation*. May; 13(5):675–79.

Page, I. H., F. J. Stare, A. C. Corcoran, H. Pollack, and C. F. Wilkinson, Jr. 1957. "Atherosclerosis and the Fat Content of the Diet." *Circulation*. Aug.; 16(2):163–78.

Page, L. B., A. Damon, and R. C. Moellering, Jr. 1974. "Antecedents of Cardiovascular Disease in Six Solomon Islands Societies." *Circulation*. June; 49(6):1132–46.

Palgi, A., J. L. Read, I. Greenberg, M. A. Hoefer, R. R. Bistrian, and G. L. Blackburn. 1985. "Multidisciplinary Treatment of Obesity with a Protein-Sparing Modified Fast: Results in 668 Outpatients." *American Journal of Public Health*. Oct.; 75(10):1190–94.

Palmgren, B., and B. Sjövall. 1957. "Studier Rörande Fetma: IV. Forsook Med Pennington-Diet." *Nordisk Medicin*. 28(3):457–58.

Parra-Covarrubias, A., I. Rivera-Rodriguez, and A. Almaraz-Ugalde. 1971. "Cephalic Phase of Insulin Secretion in Obese Adolescents." *Diabetes*. Dec.; 20(12):800–802.

Parks, J. H., and E. Waskow. 1961. "Diabetes Among the Pima Indians of Arizona." *Arizona Medicine*. April; 18(4):99–106.

Passmore, R., and Y. E. Swindells. 1963. "Observations on the Respiratory Quotients and Weight Gain of Man After Eating Large Quantities of Carbohydrate." *British Journal of Nutrition*. 17:331–39.

Paul, O., M. H. Lepper, W. H. Whelan, et al. 1963. "A Longitudinal Study of Coronary Heart Disease." *Circulation*. July; 28:20–31.

Pavlov, I. P. 1955. *Selected Works*. Trans. S. Belsky. Moscow: Foreign Language Publishing House.

Pearce, M. L., and S. Dayton. 1971. "Incidence of Cancer in Men on a Diet High in Polyunsaturated Fat." *Lancet*. March 6; 297(7697):464–67.

Pearl, R. 1940. *Introduction to Medical Biometry and Statistics*. Philadelphia: W. B. Saunders.

Peila, R., B. L. Rodriguez, and L. F. Launer. 2002. "Type 2 Diabetes, APOE Gene, and the Risk for Dementia and Related Pathologies—The Honolulu-Asia Aging Study." *Diabetes*. April; 51(4):1256–62.

Pellet, P. L. 1987. "Problems and Pitfalls in the Assessment of Human Nutritional Status." In Harris and Ross, eds., 1987, 163–79.

Peña, L., M. Peña, J. Gonzalez, and A. Claro. 1979. "A Comparative Study of Two Diets in the Treatment of Primary Exogenous Obesity in Children." *Acta Paediatrica Academiae Scientiarum Hungericae*. 20(1):99–103.

Pennington, A. W. 1955. "Pyruvic Acid Metabolism in Obesity." *American Journal of Digestive Diseases*. Feb.; 22(2):33–37.

————. 1954. "Treatment of Obesity: Developments of the Past 150 Years." *American Journal of Digestive Diseases*. March; 21(3):65–69.

————. 1953a. "Obesity: Overnutrition or Disease of Metabolism?" *American Journal of Digestive Diseases*. Sept.; 20(9):268–74.

————. 1953b. "Treatment of Obesity with Calorie Unrestricted Diets." *American Journal of Clinical Nutrition*. July–Aug.; 1(5):343–48.

————. 1953c. "A Reorientation on Obesity." *New England Journal of Medicine*. June 4; 248(23):959–64.

————. 1953d. "An Alternate Approach to Obesity." *American Journal of Clinical Nutrition*. Jan.; 1(2):100–106.

————. 1952. "Obesity." *Medical Times*. July; 80(7):389–98.

————. 1951a. "Caloric Requirements of the Obese." *Industrial Medicine and Surgery*. June; 20(6):267–71.

————. 1951b. "The Use of Fat in a Weight Reducing Diet." *Delaware State Medical Journal*. April; 23(4):79–86.

————. 1949. "Obesity in Industry—The Problem and Its Solution." *Industrial Medicine*. June: 259–60.

Peppa, M., J. Uribarri, and H. Vlassara. 2003. "Glucose, Advanced Glycation End Products, and Diabetes Complications: What Is New and What Works—Council's Voice." *Clinical Diabetes*. Oct.; 21(4):186–87.

Perkins, K. A. 1993. "Weight Gain Following Smoking Cessation." *Journal of Consulting and Clinical Psychology*. Oct.; 61(5):768–77.

Pfeiffer, R., L. McShane, M. Wargovich, et al. 2003. "The Effect of a Low-Fat, High-Fiber, Fruit and Vegetable Intervention on Rectal Mucosal Proliferation." *Cancer*. Sept. 15; 98(6):1161–68.

Phillips, R. L. 1975. "Role of Life-Style and Dietary Habits in Risk of Cancer Among Seventh-Day Adventists." *Cancer Research*. Nov.; 35(11, pt. 2):3513–22.

Phinney, S. D., A. B. Tang, C. R. Waggoner, R. G. Tezanos-Pinto, and P. A. Davis. 1991. "The Transient Hypercholesterolemia of Major Weight Loss." *American Journal of Clinical Nutrition*. June; 53(6):1404–10.

Pierson, J. 1970. "How to Stay 10 Lbs. Thinner." *Vogue*. June; 158–59, 184–85.

Pirozzo, S., C. Summerbell, C. Cameron, and P. Glasziou. 2002. "Advice on Low-Fat Diets for Obesity." *Cochrane Database of Systematic Reviews*. No. 2:CD003640.

Piscatelli, R. L., G. M. Cerchio, and S. A. Kleit. 1969. "The Ketogenic Diet in the Management of Obesity." In Wilson, ed., 1969, 185–90.

Pi-Sunyer, F. X. 2003. "A Clinical View of the Obesity Problem." *Science*. Feb. 7; 299(5608):859–60.

Pi-Sunyer, F. X., and T. B. Van Itallie. 1975. "Obesity and Diabetes." In *Diabetes Mellitus*, 4th edition, ed. K. E. Sussman and R. J. Metz. New York: American Diabetes Association, 265–70.

Plath, S. 1996. *The Bell Jar*. New York: Harper. [Originally published 1971.]

Plumb, R. K. 1962. "Diet Linked to Cut in Heart Attacks." *New York Times*. May 17; 39.

Pollak, M. N., E. S. Schernhammer, and S. E. Hankinson. 2004. "Insulin-Like Growth Factors and Neoplasia." *Nature Reviews. Cancer*. July; 4(7):505–18.

Pollan, M. 2006. *Omnivore's Dilemma: A Natural History of Four Meals*. New York: Penguin Books.

Popper, K. R. 1979. *Objective Knowledge: An Evolutionary Approach*. Revised edition. Oxford: Clarendon Press.

Potts, M. 1987. "Development of 4 Products Reflects Game Plan." *Washington Post*. July 26; H1.

Powles, J. 2001. "Commentary: Mediterranean Paradoxes Continue to Provoke." *International Journal of Epidemiology.* Oct.; 30(5):1076–77.

Powley, T. L. 1977. "The Ventromedial Hypothalamic Syndrome, Satiety, and a Cephalic Phase Hypothesis." *Psychological Review.* Jan.; 84(1):89–126.

Preble, W. E. 1915. "Obesity and Malnutrition." *Boston Medical and Surgical Journal.* Jan.–June; 172:740–44.

Prentice, G. 1923. "Cancer Among Negroes." *British Medical Journal.* Dec. 15; 1181.

Prentice, R. L., B. Caan, R. T. Chlebowski, et al. 2006. "Low-Fat Dietary Pattern and Risk of Invasive Breast Cancer: The Women's Health Initiative Randomized Controlled Dietary Modification Trial." *JAMA.* Feb. 8; 295(6):629–42.

Preston, S. H., N. Keyfitz, and R. Schoen. 1972. *Causes of Death: Life Tables for National Populations.* New York: Seminar Press.

Price, R. A., M. A. Charles, D. J. Pettitt, and W. C. Knowler. 1993. "Obesity in Pima Indians: Large Increases Among Post–World War II Birth Cohorts." *American Journal of Physical Anthropology.* Dec.; 92(4):473–79.

Prior, I. A. 1971. "The Price of Civilization." *Nutrition Today.* July/Aug.; 2–11.

Prior, I. A., R. Beaglehole, F. Davidson, and C. E. Salmond. 1978. "The Relationships of Diabetes, Blood Lipids, and Uric Acid Levels in Polynesians." *Advances in Metabolic Disorders.* 9:241–61.

Prior, I. A., B. S. Rose, and F. Davidson. 1964. "Metabolic Maladies in New Zealand Maoris." *British Medical Journal.* April 25; 1(5390):1065–69.

Putnam, J., J. Allshouse, and L. S. Kantor. 2002. "U.S. Per Capita Food Supply Trends: More Calories, Refined Carbohydrates, and Fats." *FoodReview.* Winter; 25(3):2–15.

Pyorala, K. 1979. "Relationship of Glucose Tolerance and Plasma Insulin to the Incidence of Coronary Heart Disease: Results from Two Population Studies in Finland." *Diabetes Care.* March–April; 2(2):131–41.

Qiu, W. Q., D. M. Walsh, Z. Ye, et al. 1998. "Insulin-Degrading Enzyme Regulates Extracellular Levels of Amyloid Beta-Protein by Degradation." *Journal of Biological Chemistry.* Dec. 4; 273(49):32730–38.

Quintao, E., S. M. Grundy, and E. H. Ahrens. 1971. "Effects of Dietary Cholesterol on the Regulation of Total Body Cholesterol in Man." *Journal of Lipid Research.* March; 12(2):233–47.

Rabagliati, A. 1897. *Air, Food and Exercises.* London: Baillière, Tindall & Cox.

Rabast, U., H. Kasper, and J. Schonborn. 1978. "Comparative Studies in Obese Subjects Fed Carbohydrate-Restricted and High Carbohydrate 1,000-Calorie Formula Diets." *Nutrition and Metabolism.* 22(5):269–77.

Rabast, U., J. Schonborn, and H. Kasper. 1979. "Dietetic Treatment of Obesity with Low and High-Carbohydrate Diets: Comparative Studies and Clinical Results." *International Journal of Obesity.* 1979; 3(3):201–11.

Rabinowitz, D., and K. L. Zierler. 1962. "Forearm Metabolism in Obesity and Its Response to Intra-Arterial Insulin: Characterization of Insulin Resistance and Evidence for Adaptive Hyperinsulinism." *Journal of Clinical Investigation.* Dec.; 41:2173–81.

———. 1961. "Forearm Metabolism in Obesity and Its Response to Intra-Arterial Insulin: Evidence for Adaptive Hyperinsulinism." *Lancet.* Sept. 23; 278(7204):690–92.

Ramirez, I., M. G. Tordoff, and M. I. Friedman. 1989. "Dietary Hyperphagia and Obesity: What Causes Them?" *Physiology & Behavior.* Jan.; 45(1):163–68.

Randall, H. T. 1973. "Water, Electrolytes, and Acid-Base Balance." In Goodhart and Shils, eds., 1973, 324–61.

Randle, P. J., P. B. Garland, C. N. Hales, and E. A. Newsholme. 1963. "The Glucose Fatty-

Acid Cycle: Its Role in Insulin Sensitivity and the Metabolic Disturbances of Diabetes Mellitus." *Lancet.* April 13; 281(7285):787–89.

Ranson, S. W., and S. L. Clark. 1964. *The Anatomy of the Nervous System: Its Development and Function.* 10th edition. Philadelphia: W. B. Saunders.

Ravona-Springer, R., M. Davidson, and S. Noy. 2003. "The Role of Cardiovascular Risk Factors in Alzheimer's Disease." *CNS Spectrums.* Nov.; 8(11):824–33.

Ravussin, E. 2005. "Physiology: A NEAT Way to Control Weight?" *Science.* Jan. 28; 307(5709):530–31.

———. 1993. "Energy Metabolism in Obesity." *Diabetes Care.* Jan.; 16(1 suppl.):232–38.

Ravussin, E., S. Lillioja, W. C. Knowler, et al. 1988. "Reduced Rate of Energy Expenditure as a Risk Factor for Body-Weight Gain." *New England Journal of Medicine.* Feb. 25; 318(8):467–72.

Ravussin, E., and B. A. Swinburn. 1992. "Effect of Caloric Restriction and Weight Loss on Energy Expenditure." In Wadden and Van Itallie, eds., 1992, 163–89.

Rea, A. M. 1983. *Once a River: Bird Life and Habitat Changes on the Middle Gila.* Tucson: University of Arizona Press.

Reader, G., R. Melchionna, L. E. Hinkle, et al. 1952. "Treatment of Obesity." *American Journal of Medicine.* 13(4):478–86.

Reaven, G. M. 2005. "The Insulin Resistance Syndrome: Definition and Dietary Approaches to Treatment." *Annual Review of Nutrition.* 25:391–406.

———. 1988. "Banting Lecture 1988: Role of Insulin Resistance in Human Disease." *Diabetes.* Dec.; 37(12):1595–607.

Reaven, G. M., A. Calciano, R. Cody, C. Lucas, and R. Miller. 1963. "Carbohydrate Intolerance and Hyperlipemia in Patients with Myocardial Infarction Without Known Diabetes Mellitus." *Journal of Clinical Endocrinology & Metabolism.* Oct.; 23:1013–23.

Reaven, G. M., and Y. D. Chen. 1996. "Insulin Resistance, Its Consequences, and Coronary Heart Disease: Must We Choose One Culprit?" *Circulation.* May; 93:1780–83.

Reaven, G. M., Y. D. Chen, J. Jeppesen, P. Maheux, and R. M. Krauss. 1993. "Insulin Resistance and Hyperinsulinemia in Individuals with Small, Dense Low Density Lipoprotein Particles." *Journal of Clinical Investigation.* July; 92(1):141–46.

Reaven, G. M., R. L. Lerner, M. P. Stern, and J. W. Farquhar. 1967. "Role of Insulin in Endogenous Hypertriglyceridemia." *Journal of Clinical Investigation.* Nov.; 46(11): 1756–67.

Reaven, G. M., and J. M. Olefsky. 1978. "The Role of Insulin Resistance in the Pathogenesis of Diabetes Mellitus." In *Advances in Metabolic Disorders,* vol. 9, ed. M. Miller and P. H. Bennet. New York: Academic Press, 313–33.

Rebuffé-Scrive, M. 1987. "Regional Adipose Tissue Metabolism in Women During and After Reproductive Life and in Men." In Berry et al., eds., 1987, 82–91.

Rebuffé-Scrive, M., J. Eldh, L. O. Hafstrom, and P. Björntorp. 1986. "Metabolism of Mammary, Abdominal, and Femoral Adipocytes in Women Before and After Menopause." *Metabolism.* Sept.; 35(9):792–97.

Reichley, K. B., W. H. Mueller, C. L. Hanis, et al. 1987. "Centralized Obesity and Cardiovascular Disease Risk in Mexican Americans." *American Journal of Epidemiology.* March; 125(3):373–86.

Reid, J. C. 1858. *Reid's Tramp, or, A Journal of the Incidents of Ten Months Travel Through Texas, New Mexico, Arizona, Sonora, and California.* Selma, Ala.: John Hard.

Reiser, S., E. Bohn, J. Hallfrisch, O. E. Michaels IV, M. Keeney, and E. S. Prather. 1981. "Serum Insulin and Glucose in Hyperinsulinemic Subjects Fed Three Different Levels of Sucrose." *American Journal of Clinical Nutrition.* Nov.; 34(11):2348–58.

Rendle Short, A. 1920. "The Causation of Appendicitis." *British Journal of Surgery.* 8:171–88.

Renold, A. E., and G. F. Cahill, Jr., eds. 1965. *Handbook of Physiology. Section 5. Adipose Tissue.* Washington, D.C.: American Physiological Society.

————. 1965a. "Preface." In Renold and Cahill, eds. 1965, 1–3.

————. 1965b. "Metabolism of Isolated Adipose Tissue—A Summary." In Renold and Cahill, eds., 1965, 483–90.

Renold, A. E., O. B. Crofford, W. Stauffacher, and B. Jeanreaud. 1965. "Hormonal Control of Adipose Tissue Metabolism, with Special Reference to the Effects of Insulin." *Diabetologia.* Aug.; 1(1):4–12.

Rensberger, B. 1974. "Curb on Waste Urged to Help World's Hungry." *New York Times.* Oct. 25:81.

Research Committee. 1965. "Low-Fat Diet in Myocardial Infarction: A Controlled Trial." *Lancet.* Sept. 11; 286(7411):501–4.

Reshef, L., Y. Olswang, H. Cassuto, et al. 2003. "Glyceroneogenesis and the Triglyceride/Fatty Acid Cycle." *Journal of Biological Chemistry.* Aug. 15; 278(33):30413–16.

Review Panel of the National Heart Institute. 1969. *Mass Field Trials of the Diet-Heart Question: Their Significance, Feasibility, and Applicability—Report of the Diet-Heart Review Panel of the National Heart Institute.* American Heart Association Monograph No. 28. American Heart Association.

Richards, R., and M. de Casseres. 1974. "The Problem of Obesity in Developing Countries: Its Prevalence and Morbidity." In Burland, Samuel, and Yudkin, eds., 1974, 74–84.

Richter, C. P. 1976. "Total Self-Regulatory Functions in Animal and Human Beings." In *The Psychobiology of Curt Richter,* ed. E. M. Blass. Baltimore: York Press, 194–226.

Rilliet, B. 1954. "Treatment of Obesity by a Low-Calorie Diet: Hanssen-Boller-Pennington Diet." *Praxis.* Sept. 9; 43(36):761–63.

Rinkel, M., and H. E. Himwich, eds. 1959. *Insulin Treatment in Psychiatry.* New York: Philosophical Library.

Risser, J. 1980. "Food Firms Helped Fund Diet Report." *Des Moines Register.* May 20; 1.

Ritenbaugh, C., R. E. Patterson, R. T. Chlebowski, et al. 2003. "The Women's Health Initiative Dietary Modification Trial: Overview and Baseline Characteristics of Participants." *Annals of Epidemiology.* Oct.; 13(9 suppl.):S87–97.

Rittenberg, D., and R. Schoenheimer. 1937. "Deuterium as an Indicator in the Study of Intermediary Metabolism. XI. Further Studies on the Biological Uptake of Deuterium into Organic Substances, with Special Reference to Fat and Cholesterol Formation." *Journal of Biological Chemistry.* 121:235–53.

Roberts, S. B., J. Savage, W. A. Coward, B. Chew, and A. Lucas. 1988. "Energy Expenditure and Intake in Infants Born to Lean and Overweight Mothers." *New England Journal of Medicine.* Feb. 25; 318(8):461–66.

Rocchini, A. P. 1998. "Obesity and Blood Pressure Regulation." In Bray et al., eds., 1998, 677–96.

Rodale, R. 1974. "Fiber Crowds Out Fat." *Washington Post.* May 9; E2.

————. 1973. "Eating Crude Fiber: Preventing the Onset of Heart Disease." *Washington Post.* April 12; F6.

Rodin, J. 1987. "Weight Change Following Smoking Cessation: The Role of Food Intake and Exercise." *Addictive Behaviors.* 12(4):303–17.

————. 1985. "Insulin Levels, Hunger, and Food Intake: An Example of Feedback Loops in Body Weight Regulation." *Health Psychology.* 4(1):1–24.

———. 1980. "The Externality Theory Today." In Stunkard, ed., 1980, 226–39.

———. 1979. "Pathogenesis of Obesity: Energy Intake and Expenditure." In Bray, ed., 1979, 37–68.

Rogers, A. E., and M. P. Longnecker. 1988. "Biology of Disease: Dietary and Nutritional Influences on Cancer—A Review of Epidemiologic and Experimental Data." *Laboratory Investigation*. Dec.; 59(6):729–59.

Rolls, B., and R. A. Barnett. 2000. *The Volumetrics Weight-Control Plan: Feel Full on Fewer Calories*. New York: HarperCollins.

Rony, H. R. 1940. *Obesity and Leanness*. Philadelphia: Lea & Febiger.

Root, W., and R. de Rochemont. 1995. *Eating in America: A History*. Hopewell, N.J.: Ecco Press.

Rosamond, W. D., L. E. Chambless, A. R. Folsom, et al. 1998. "Trends in the Incidence of Myocardial Infarction and in Mortality Due to Coronary Heart Disease, 1987 to 1994." *New England Journal of Medicine*. Sept. 24; 339:861–67.

Rosamond, W. D., A. R. Folsom, L. E. Chambless, et al. [ARIC Investigators.] 2001. "Atherosclerosis Risk in Communities—Coronary Heart Disease Trends in Four United States Communities: The Atherosclerosis Risk in Communities (ARIC) Study, 1987–1996." *International Journal of Epidemiology*. Oct.; 30(1 suppl.):S17–22.

Rose, G. 1985. "Sick Individuals and Sick Populations." *International Journal of Epidemiology*. March; 14(1):32–38.

———. 1981. "Strategy of Prevention: Lessons from Cardiovascular Disease." *British Medical Journal (Clinical Research and Education)*. June 6; 282(6279):1847–51.

Rose, G., H. Blackburn, A. Keys, et al. 1974. "Colon Cancer and Blood-Cholesterol." *Lancet*. Feb. 9; 303(7850):181–83.

Rose, H. E., and J. Mayer. 1968. "Activity, Calorie Intake, Fat Storage, and the Energy Balance of Infants." *Pediatrics*. Jan.; 41(1):18–29.

Rosenberg, C. 1981. Wilbur Olin Atwater. In *Dictionary of Scientific Biography*, ed. C. Gillespie. New York: Scribners, 1:325–26.

Rosenthal, B., M. Jacobson, and M. Bohm. 1976. "Professors on the Take." *Progressive*. Nov.; 42–47.

Rosenzweig, J. L. 1994. "Principles of Insulin Therapy." In Kahn and Weir, eds., 1994, 460–88.

Ross, E. B. 1987. "An Overview of Trends in Dietary Variation from Hunter-Gatherer to Modern Capitalist Societies." In Harris and Ross, eds., 1987, 7–56.

Rothwell, N. J., and M. J. Stock. 1981. "Thermogenesis: Comparative and Evolutionary Considerations." In Cioffi, James, and Van Itallie, eds., 1981, 335–44.

Rous, P. 1914. "The Influence of Diet on Transplanted and Spontaneous Mouse Tumors." *Journal of Experimental Medicine*. 20:433–51.

Rovner, S. 1988. "Off of the Heart Disease Treadmill." *Washington Post*. Jan. 19; Z8.

———. 1986. "Diets Don't Work, Obesity Experts Are Told." *Washington Post*. June 25; H10.

Rubins, H. B., S. J. Robins, D. Collins, et al. 1999. "Gemfibrozil for the Secondary Prevention of Coronary Heart Disease in Men with Low Levels of High-Density Lipoprotein Cholesterol: Veterans Affairs High-Density Lipoprotein Cholesterol Intervention Trial Study Group." *New England Journal of Medicine*. Aug. 5; 341(6):410–18.

Rubner, M. 1982. *The Laws of Energy Conservation in Nutrition*. Ed. R. J. Joy. Trans. A. Markoff and A. Sandri-White. New York: Academic Press. [Originally published 1902.]

Rucknagel, D. L., and Neel, J. V. 1961. "The Hemoglobinopathies." In *Progress in Medical Genetics*, ed. A. Steinberg. New York: Grune & Stratton, 158–260.

Russell, F. 1975. *The Pima Indians*. Tucson: University of Arizona Press. [Originally published 1905.]

Rynearson, E. H., and C. F. Gastineau. 1949. *Obesity* . . . Springfield, Ill.: Charles C. Thomas.

Sabbagh, K. 2002. *The Riemann Hypothesis: The Greatest Unsolved Problem in Mathematics*. New York: Farrar, Straus and Giroux.

Sackett, D. L. 2002. "The Arrogance of Preventive Medicine." *Canadian Medical Association Journal*. Aug. 20; 167(4):363–64.

Sai, F. T. 1967. "Drastic Change in Food Habits in Relation to Socio-Cultural Change." In *Regulation of Hunger and Satiety*, vol. 2 of *Proceedings of the Seventh International Congress of Nutrition, Hamburg 1966*. Oxford: Pergamon Press, 144–46.

Salans, L. B., S. W. Cushman, E. S. Horton, E. Danforth, Jr., and E. A. Sims. 1974. "Hormones and the Adipocyte: Factors Influencing the Metabolic Effects of Insulin and Adrenaline." In Burland, Samuel, and Yudkin, eds., 1974, 204–16.

Salcedo, J., and D. Stetten, Jr. 1943. "The Turnover of Fatty Acids in the Congenitally Obese Mouse." *Journal of Biological Chemistry*. Dec.; 151(2):413–16.

Samaha, F. F., N. Iqubal, P. Seshadri, et al. 2003. "A Low-Carbohydrate As Compared with a Low-Fat Diet in Severe Obesity." *New England Journal of Medicine*. May 22; 348(21):2074–81.

Sandler, M., and G. H. Bourne, eds. 1963. *Atherosclerosis and Its Origin*. New York: Academic Press.

Sapolsky, H. M., ed. 1986. *Consuming Fears: The Politics of Product Risks*. New York: Basic Books.

Saris, W. H., S. N. Blair, M. A. van Baak, et al. 2003. "How Much Physical Activity Is Enough to Prevent Unhealthy Weight Gain? Outcome of the IASO 1st Stock Conference and Consensus Statement." *Obesity Reviews*. May; 4(2):101–14.

Sasaki, N., R. Fukatsu, K. Tsuzuki, et al. 1998. "Advanced Glycation End Products in Alzheimer's Disease and Other Neurodegenerative Diseases." *American Journal of Pathology*. Oct.; 153(4):1149–55.

Schachter, S., and J. Rodin. 1974. *Obese Humans and Rats*. Potomac, Md.: Erlbaum Associates.

Schack-Nielsen, L., C. Molgaard, T. L. Sorensen, G. Greisen, and K. F. Michaelsen. 2006. "Secular Change in Size at Birth from 1973 to 2003: National Data from Denmark." *Obesity*. July; 14(7):1257–63.

Schaefer, O., J. A. Hildes, L. M. Medd, and D. G. Cameron. 1975. "The Changing Pattern of Neoplastic Disease in Canadian Eskimos." *Canadian Medical Association Journal*. June 21; 112(12):1399–1400.

Schatzkin, A., P. Greenwald, D. P. Byar, and C. K. Clifford. 1989. "The Dietary Fat–Breast Cancer Hypothesis Is Alive." *JAMA*. June 9; 261(22):3284–87.

Schatzkin, A., E. Lanza, D. Corle, et al. 2000. "Lack of Effect of a Low-Fat, High-Fiber Diet on the Recurrence of Colorectal Adenomas: Polyp Prevention Trial Study Group." *New England Journal of Medicine*. April 20; 342(16):1149–55.

Schettler, G., and G. Schlierf. 1974. "Obesity." In Sipple and McNutt, eds., 1974, 390–401.

Schmeck, H. M. 1964. "Special Diet Cuts Heart Cases Here." *New York Times*. Oct. 8; 45.

Schmidt-Nielsen, K., H. B. Haines, and D. B. Hackel. 1964. "Diabetes Mellitus in the Sand Rat Induced by Standard Laboratory Diets." *Science*. Feb. 14; 143:689–90.

Schneider, J. E., and G. N. Wade. 1989. "Availability of Metabolic Fuels Controls Estrous Cyclicity of Syrian Hamsters." *Science*. June 16; 244(4910):1326–28.

Schoenheimer, R. 1961. *The Dynamic State of Body Constituents*. Cambridge, Mass.: Harvard University Press.

Schornagel, H. E. 1953. "The Connection Between Nutrition and Mortality from Coronary Sclerosis During and After World War II." *Documenta de Medicina Geographica et Tropica.* June; 5(2):173–83.

Schutz, Y., and E. Jéquier. 1998. "Resting Energy Expenditure, Thermic Effect of Food, and Total Energy Expenditure." In Bray, Bouchard, and James, eds., 1998, 443–55.

Schwartz, H. 1986. *Never Satisfied: A Cultural History of Diets, Fantasies, and Fat.* New York: Doubleday.

Schwartz, R. P., and J. B. Sidbury, Jr. 1974. "Childhood Obesity." *Connecticut Medicine.* Dec.; 38(12):660–63.

Schweitzer, A. 1998. Trans. A. B. Lemke. *Out of My Life and Thought: An Autobiography.* Baltimore: Johns Hopkins University Press. [Originally published 1933.]

———. 1957. "Preface." In A. Berglas, *Cancer: Nature, Cause, and Cure.* Paris: Institut Pasteur, ix.

Sclafani, A. 1987. "Carbohydrate, Taste, Appetite, and Obesity: An Overview." *Neuroscience & Biobehavioral Reviews.* Summer; 11(2):131–53.

———. 1981a. "Extremes in Body Weight in Experimental Animal Preparations." In Cioffi, James, and Van Itallie, eds., 1981, 153–60.

———. 1981b. "The Role of Hyperinsulinemia and the Vagus Nerve in Hypothalamic Hyperphagia Reexamined." *Diabetologia.* March; 20(suppl):402–10.

———. 1980. "Dietary Obesity." In Stunkard, ed., 1980, 166–81.

Sclafani, A., and J. W. Nissenbaum. 1988. "Robust Conditioned Flavor Preference Produced by Intragastric Starch Infusions in Rats." *American Journal of Physiology.* Oct.; 255(4, pt. 2):R672–75.

Scott, E. M., and I. V. Griffith. 1957. "Diabetes Mellitus in Eskimos." *Metabolism.* July; 6(4):320–25.

Scrimshaw, N. S., and W. Dietz. 1995. "Potential Advantages and Disadvantages of Human Obesity." In de Garine and Pollock, eds., 1995, 147–62.

Sears, B., and B. Lawren. 1995. *The Zone: A Dietary Road Map.* New York: HarperCollins.

Seftel, H. C., K. J. Keeley, A. R. Walker, J. J. Theron, and D. Delange. 1965. "Coronary Heart Disease in Aged South African Bantu." *Geriatrics.* March; 20:194–205.

Segal, K. R., and F. X. Pi-Sunyer. 1989. "Exercise and Obesity." *Medical Clinics of North America.* Jan.; 73(1):217–36.

Select Committee on Nutrition and Human Needs of the United States Senate. 1977a. *Dietary Goals for the United States.* Washington, D.C.: U.S. Government Printing Office.

———. 1977b. *Dietary Goals for the United States.* 2nd edition. Washington, D.C.: U.S. Government Printing Office.

———. 1977c. *Dietary Goals for the United States—Supplemental Views.* Washington, D.C.: U.S. Government Printing Office.

———. 1977d. *Cardiovascular Disease.* Vol. 2, pt. 1, of *Diet Related to Killer Diseases;* hearings before the Select Committee on Nutrition and Human Needs of the United States Senate, Ninety-Fifth Congress, Feb. 1 and 2, 1977. Washington, D.C.: U.S. Government Printing Office.

———. 1977e. *Obesity.* Vol. 2, pt. 2, of *Diet Related to Killer Diseases;* hearings before the Select Committee on Nutrition and Human Needs of the United States Senate, Ninety-Fifth Congress, Feb. 1 and 2, 1977. Washington, D.C.: U.S. Government Printing Office.

———. 1977f. *Response to Dietary Goals of the United States: Re Meat.* Vol. 3 of *Diet Related to Killer Diseases;* hearings before the Select Committee on Nutrition and Human

Needs of the United States Senate, Ninety-Fifth Congress, March 24, 1977. Washington, D.C.: U.S. Government Printing Office.

————. 1976. *Diet Related to Killer Diseases;* hearings before the Select Committee on Nutrition and Human Needs of the United States Senate, Ninety-Fourth Congress, July 27 and 28, 1976. Washington, D.C.: U.S. Government Printing Office.

————. 1973a. *Sugar in Diet, Diabetes, and Heart Disease.* Hearing Before the Select Committee on Nutrition and Human Needs of the United States Senate, Ninety-Third Congress, pt. 2, April 30, May 1 and 2, 1973. Washington, D.C.: U.S. Government Printing Office.

————. 1973b. *Obesity and Fad Diets;* hearing Before the Select Committee on Nutrition and Human Needs of the United States Senate, Ninety-Third Congress, pt. 1, April 12, 1973. Washington, D.C.: U.S. Government Printing Office.

Shafrir, E. 1991. "Metabolism of Disaccharides and Monosaccharides with Emphasis on Sucrose and Fructose and Their Lipogenic Potential." In Gracey, Kretchmer, and Rossi, eds., 1991, 131–52.

————. 1985. "Effect of Sucrose and Fructose on Carbohydrate and Lipid Metabolism and the Resulting Consequences." In *Regulation of Carbohydrate Metabolism,* vol. 2, ed. R. Beitner. Boca Raton, Fla.: CRC Press, 95–140.

Shaper, A. G. 1967. "Blood Pressure Studies in East Africa." In *The Epidemiology of Hypertension,* ed. J. Stamler, R. Stamler, and T. N. Pullman. New York: Grune & Stratton, 139–49.

————. 1962. "Cardiovascular Studies in the Samburu Tribe of Northern Kenya." *American Heart Journal.* April; 63(4):437–42.

Shaper, A. G., P. J. Leonard, K. W. Jones, and M. Jones. 1969. "Environmental Effects on the Body Build, Blood Pressure, and Blood Chemistry of Nomadic Warriors Serving in the Army in Kenya." *East African Medical Journal.* May; 46(5):282–89.

Shapiro, F. R. 2006. *The Yale Book of Quotations.* New Haven: Yale University Press.

Shaten, B. J., L. H. Kuller, M. O. Kjelsberg, et al. 1997. "Lung Cancer Mortality After 16 Years in MRFIT Participants in Intervention and Usual-Care Groups: Multiple Risk Factor Intervention Trial." *Annals of Epidemiology.* Feb.; 7(2):125–36.

Shekelle, R. B., A. M. Shryock, O. Paul, et al. 1981. "Diet, Serum Cholesterol, and Death from Coronary Heart Disease: The Western Electric Study." *New England Journal of Medicine.* Jan. 8; 304(2):65–70.

Sheldon, W. H., and S. S. Stevens. 1942. *The Varieties of Temperament: A Psychology of Constitutional Differences.* New York: Harper & Brothers.

Shen, M. M., R. M. Krauss, F. T. Lindgren, and T. M. Forte. 1981. "Heterogeneity of Serum Low Density Lipoproteins in Normal Human Subjects." *Journal of Lipid Research.* Feb.; 22(2):236–44.

Shen, S. W., G. M. Reaven, and J. W. Farquhar. 1970. "Comparison of Impedance to Insulin-Mediated Glucose Uptake in Normal Subjects and in Subjects with Latent Diabetes." *Journal of Clinical Investigation.* Dec.; 49(12):2151–60.

Shetty, P. S. 1999. "Adaptation to Low Energy Intakes: The Responses and Limits to Low Intakes in Infants, Children and Adults." *European Journal of Clinical Nutrition.* 53(1 suppl.):s14–s33.

Shlossman, M., W. C. Knowler, D. J. Pettitt, and R. J. Genco. 1990. "Type 2 Diabetes Mellitus and Periodontal Disease." *Journal of the American Dental Association.* Oct.; 121(4):532–36.

Sidbury, J. B., Jr., and R. P. Schwartz. 1975. "A Program for Weight Reduction in Children." In Collip, ed., 1975, 65–74.

Silver, S., and J. Bauer. 1931. "Obesity, Constitutional or Endocrine?" *American Journal of the Medical Sciences.* 181:769–77.

Silverstone, J. T., and F. Lockhead. 1963. "The Value of a 'Low Carbohydrate' Diet in Obese Diabetics." *Metabolism.* Aug.; 12(8):710–13.

Simpson, R. G., A. Benedetti, G. M. Grodsky, J. H. Karam, and P. H. Forsham. 1968. "Early Phase of Insulin Release." *Diabetes.* Nov.; 17(11):684–92.

Sims, E. A. 1976. "Experimental Obesity, Dietary-Induced Thermogenesis, and Their Clinical Implications." *Clinics in Endocrinology & Metabolism.* July; 5(2):377–95.

Sims, E. A., G. A. Bray, E. Danforth, Jr., et al. 1974. "Experimental Obesity in Man. VI. The Effect of Variations in Intake of Carbohydrate on Carbohydrate, Lipid, and Cortisol Metabolism." *Hormone and Metabolic Research Supplement.* 4:70–77.

Sims, E. A., and E. Danforth, Jr. 1987. "Expenditure and Storage of Energy in Man." *Journal of Clinical Investigation.* April; 79(4):1019–25.

———. 1974. "Role of Insulin in Obesity." *Israeli Journal of Medical Sciences.* Oct.; 10(10):1222–29.

Sims, E. A., E. Danforth, Jr., E. S. Horton, G. A. Bray, J. A. Glennon, and L. B. Salans. 1973. "Endocrine and Metabolic Effects of Experimental Obesity in Man." *Recent Progress in Hormone Research.* 29:457–96.

Sims, E. A., R. F. Goldman, C. M. Gluck, E. S. Horton, P. C. Kelleher, and D. W. Rowe. 1968. "Experimental Obesity in Man." *Transactions of the Association of American Physicians.* 81:153–70.

Sims, E. A., and E. S. Horton. 1968. "Endocrine and Metabolic Adaptation to Obesity and Starvation." *American Journal of Clinical Nutrition.* Dec.; 21(12):1455–70.

Sinclair, U. 2003. *The Jungle.* Tucson: Sharp Press. [Originally published 1906.]

Singh, R., A. Barden, T. Mori, and L. Beilin. 2001. "Advanced Glycation End-Products: A Review." *Diabetologia.* Feb.; 44(2):129–46.

Singman, H. S., S. N. Berman, C. Cowell, E. Maslansky, and M. Archer. 1980. "The Anti-Coronary Club: 1957 to 1972." *American Journal of Clinical Nutrition.* June; 33(6): 1183–91.

Sipple, H. L., and K. W. McNutt, eds. 1974. *Sugars in Nutrition.* New York: Academic Press.

Slome, C., B. Gampel, J. H. Abramson, and N. Scotch. 1960. "Weight, Height, and Skinfold Thickness of Zulu Adults in Durban." *South African Medical Journal.* June 11; 34:505–9.

Smith, C. J., E. M. Manahan, and S. G. Pablo. 1994. "Food Habit and Cultural Changes Among the Pima Indians." In Joe and Young, eds., 1994, 407–33.

Smith, M. A., P. L. Richey, S. Taneda, et al. 1994. "Advanced Maillard Reaction End Products, Free Radicals, and Protein Oxidation in Alzheimer's Disease." *Annals of the New York Academy of Sciences.* June 7; 91(12):5710–14.

Smith, U. 1985. "Regional Differences in Adipocyte Metabolism and Possible Consequences in Vivo." In Hirsch and Van Itallie, eds., 1985, 77–81.

Sniderman, A., S. Shapiro, D. Marpole, B. Skinner, B. Teng, and P. O. Kwiterovich, Jr. 1980. "Association of Coronary Atherosclerosis with HyperApoBetalipoproteinemia." *Proceedings of the National Academy of Sciences.* Jan.; 77(1):604–8.

Snowdon, D. A. 2003. "Healthy Aging and Dementia—Findings from the Nun Study." *Annals of Internal Medicine.* Sept. 2; 139(5, pt. 2):450–54.

Sondike, S. B., N. Copperman, and M. S. Jacobson. 2003. "Effects of a Low-Carbohydrate Diet on Weight Loss and Cardiovascular Risk Factor in Overweight Adolescents." *Journal of Pediatrics.* March; 142(3):253–58.

Sontag, S. 1990. *Illness as Metaphor and AIDS and Its Metaphors.* New York: Picador.

Spark, R. F. 1973. "Fat Americans: They Don't Know When They're Hungry, They Don't Know When They're Full." *New York Times Magazine.* Jan. 6:10, 42–3, 50–1.

Speke, J. H. 1969. *Journal of the Discovery of the Source of the Nile.* London: J. M. Dent & Sons. [Originally published 1863.]

Spencer, B., and F. J. Gillen. 1912. *Across Australia.* Vol. 1. London: Macmillan.

Spicer, E. H. 1962. *Cycles of Conquest: The Impact of Spain, Mexico, and the United States on the Indians of the Southwest, 1533–1960.* Tucson: University of Arizona Press.

Spielman, R. S., S. S. Fajans, J. V. Neel, S. Pek, J. C. Floyd, and W. J. Oliver. 1982. "Glucose Tolerance in Two Unacculturated Indian Tribes of Brazil." *Diabetologia.* Aug.; 23(2):90–93.

Spier, L. 1978. *Yuman Tribes of the Gila River.* New York: Dover Publications.

Spock, B. 1985. *Baby and Child Care.* 5th edition. New York: Pocket Books.

———. 1976. *Baby and Child Care.* 4th edition. New York: Hawthorne Books.

———. 1968. *Baby and Child Care.* 3rd edition. New York: Meredith Press.

———. 1957. *The Common Sense Book of Baby and Child Care.* 2nd edition. New York: Duell, Sloan, and Pearce.

———. 1946. *The Common Sense Book of Baby and Child Care.* New York: Duell, Sloan, and Pearce.

Spock, B., and M. B. Rothenberg. 1992. *Dr. Spock's Baby and Child Care.* 6th edition. New York: Dutton.

Squires, S. 1985. "The Nutrition Game: Eating Less Fat, More Starchy Foods Makes Playing and Winning a Snap." *Washington Post.* March 20; 12.

Stamler, J. 1967. *Lectures on Preventive Cardiology.* New York: Grune & Stratton.

———. 1962. "The Early Detection of Heart Disease." In *Heart Disease Control,* ed. F. W. Reynolds. Ann Arbor: University of Michigan School of Public Health Continued Education Series no. 97, 48–71.

Stamler, J., D. M. Berkson, and H. A. Lindberg. 1972. "Risk Factors: Their Role in the Etiology and Pathogenesis of the Atherosclerotic Diseases." In *The Pathogenesis of Atherosclerosis,* eds. R. W. Wissler and J. C. Geer. Baltimore: Williams & Wilkins, 41–119.

Stamler, J., D. Wentworth, and J. D. Neaton. 1986. "Is Relationship Between Serum Cholesterol and Risk of Premature Death from Coronary Heart Disease Continuous and Graded? Findings in 356,222 Primary Screenees of the Multiple Risk Factor Intervention Trial (MRFIT)." *JAMA.* Nov. 28; 256(20):2823–28.

Stanford, C. 2001. *Significant Others.* New York: Basic Books.

Starbucks Coffee Company. 2006. *Nutrition by the Cup.* Seattle: Starbucks Coffee Company. Online at http://www.starbucks.com/retail/nutrition_beverages.asp

Stare, F. J. 1987. *Harvard's Department of Nutrition, 1942–1986.* Norwell, Mass.: Christopher Publishing House.

Stefansson, V. 1960a. *Cancer: Disease of Civilization? An Anthropological and Historical Study.* New York: Hill and Wang.

———. 1960b. "Food and Food Habits in Alaska and Northern Canada." In *Human Nutrition: Historic and Scientific,* ed. I. Goldstein. New York: International Universities Press, 23–60.

———. 1946. *Not by Bread Alone.* New York: Macmillan.

———. 1936. "Adventures in Diet." Reprint from *Harper's Monthly Magazine.* Chicago: Institute of American Meat Packers

Stein, J. H., K. M. West, J. M. Robey, D. F. Tirador, and G. W. McDonald. 1965. "The High Prevalence of Abnormal Glucose Tolerance in the Cherokee Indians of North Carolina." *Archives of Internal Medicine.* Dec.; 116(6):842–45.

Steinberg, D. 2005. "An Interpretive History of the Cholesterol Controversy. Part II. The Early Evidence Linking Hypercholesterolemia to Coronary Disease in Humans." *Journal of Lipid Research*. Feb.; 46(2):179–90.

———. 1997. "Low Density Lipoprotein Oxidation and Its Pathobiological Significance." *Journal of Biological Chemistry*. Aug. 22; 272(34):20963–66.

Steinberg, D., and M. Vaughan. 1965. "Release of Free Fatty Acids from Adipose Tissue in Vitro in Relation to Rates of Triglyceride Synthesis and Degradation." In Renold and Cahill, eds., 1965, 335–47.

Steiner, M. M. 1950. "The Management of Obesity in Childhood." *Medical Clinics of North America*. Jan.; 34(1):223–34.

Stemmermann, G. N., A. M. Nomura, L. K. Heilbrun, E. S. Pollack, and A. Kagan. 1981. "Serum Cholesterol and Colon Cancer Incidence in Hawaiian Japanese Men." *Journal of the National Cancer Institute*. Dec.; 67(6):1179–82.

Stene, J. A., and I. L. Roberts. 1928. "A Nutrition Study on an Indian Reservation." *Journal of the American Dietetic Association*. March; 3(4):215–22.

Stern, J. S., and P. Lowney. 1986. "Obesity: The Role of Physical Activity." In Brownell and Foreyt, eds., 1986, 145–58.

Steward, H. L., M. C. Bethea, S. S. Andrews, and L. A. Balart. 1998. *Sugar Busters! Cut Sugar to Trim Fat*. New York: Ballantine Books.

Stiebeling, H. K. 1939. "Food Habits, Old and New." In *Food and Life, Yearbook of Agriculture*. U.S. Department of Agriculture. Washington, D.C.: U.S. Government Printing Office, 124–30.

Stillman, I. M. and S. S. Baker. 1967. *The Doctor's Quick Weight Loss Diet*. Englewood Cliffs, N.J.: Prentice-Hall.

Stitt, A. W. 2001. "Advanced Glycation: An Important Pathological Event in Diabetic and Age Related Ocular Disease." *British Journal of Ophthalmology*. June; 85(6):746–53.

Stitt, A. W., R. Bucala, and H. Vlassara. 1997. "Atherogenesis and Advanced Glycation: Promotion, Progression, and Prevention. *Annals of the New York Academy of Sciences*. April 15; 811:115–27, 127–29.

Stock, A. L., and J. Yudkin. 1970. "Nutrient Intake of Subjects on Low Carbohydrate Diet Used in Treatment of Obesity." *American Journal of Clinical Nutrition*. July; 23(7): 948–52.

Stock, M., and N. Rothwell. 1982. *Obesity and Leanness: Basic Aspects*. New York: John Wiley.

Stockton, W. 1989. "When Exercise Isn't Enough." *New York Times*. Feb. 20; C11.

Stolberg, S. G. 1999. "Fiber Does Not Help Prevent Colon Cancer, Study Finds." *New York Times*. Jan. 21; A14.

Stone, N. J., R. I. Levy, D. S. Fredrickson, and J. Verter. 1974. "Coronary Artery Disease in 116 Kindred with Familial Type II Hyperlipoproteinemia." *Circulation*. March; 49(3):476–88.

Stout C., J. Marrow, E. N. Brandt, Jr., and S. Wolf. 1964. "Unusually Low Incidence of Death from Myocardial Infarction: Study of an Italian American Community in Pennsylvania." *JAMA*. June 8; 188:845–49.

Stout, R. W. 1970. "Development of Vascular Lesions in Insulin-Treated Animals Fed a Normal Diet." *British Medical Journal*. Sept. 19; 3(5724):685–87.

———. 1969. "Insulin Stimulation of Cholesterol Synthesis by Arterial Tissue." *Lancet*. Aug. 30; 294(7618):467–68.

———. 1968. "Insulin-Stimulated Lipogenesis in Arterial Tissue in Relation to Diabetes and Atheroma." *Lancet*. Sept. 28; 292(7570):702–3.

Stout, R. W., E. L. Bierman, and R. Ross. 1975. "Effect of Insulin on the Proliferation of

Cultured Primate Arterial Smooth Muscle Cells." *Circulation Research*. Feb.; 36(2): 319–27.

Stout, R. W., and J. Vallance-Owen. 1969. "Insulin and Atheroma." *Lancet*. May 31; 293(7605):1078–80.

Strang, J. M., and F. A. Evans. 1929. "The Energy Exchange in Obesity." *Journal of Clinical Investigation*. 6:277–89.

Strang, J. M., H. B. McClugage, and F. A. Evans. 1930. "Further Studies in the Dietary Correction of Obesity." *American Journal of the Medical Sciences*. 179:687–94.

Strasser, H. 1968. "A Breeding Program for Spontaneously Diabetic Experimental Animals: Psammomys Obesus (Sand Rat) and Acomys Cahirinus (Spiny Mouse)." *Laboratory Animal Care*. June; 18(3):328–38.

Stricker, E. M. 1978. "Hyperphagia." *New England Journal of Medicine*. May 4; 298(18):1010–13.

Strittmatter, W. J., A. M. Saunders, D. Schmechel, et al. 1993. "Apolipoprotein E: High-Avidity Binding to Beta-Amyloid and Increased Frequency of Type 4 Allele in Late-Onset Familial Alzheimer Disease." *Proceedings of the National Academy of Sciences*. March 1; 90(5):1977–81.

Strong, J. P., J. Rosal, R. J. Deupree, and H. C. McGill, Jr. 1966. "Diet and Serum Cholesterol Levels in Baboons." *Experimental and Molecular Pathology*. Feb.; 5(1):82–91.

Stuart, R. B. 1976. "Behavioral Control of Overeating: A Status Report." In Bray, ed., 1976b, 367–86.

Stulb, S. C., J. R. McDonough, B. G. Greenberg, and C. G. Hames. 1965. "The Relationship of Nutrient Intake and Exercise to Serum Cholesterol Levels in White Males in Evans County, Georgia." *American Journal of Clinical Nutrition*. Feb.; 16:238–42.

Stunkard, A. J., ed. 1980. *Obesity*. Philadelphia: W. B. Saunders.

———. 1980. "Introduction and Overview." In Stunkard, ed., 1980, 1–24.

———. 1976a. "Obesity and Social Environment." In Howard, ed., 1976, 178–90.

———. 1976b. *The Pain of Obesity*. Palo Alto, Calif.: Bull Publishing.

———. 1976c. "Studies on TOPS: A Self-Help Group for Obesity." In Bray, ed., 1976b, 387–92.

———. 1973. "The Obese: Background and Programs." In Mayer, ed., 1973, 29–36.

Stunkard, A., and M. McClaren-Hume. 1959. "The Results of Treatment for Obesity: A Review of the Literature and a Report of a Series." *Archives of Internal Medicine*. Jan.; 103(1):79–85.

Stunkard, A. J., and T. A. Wadden, eds. 1993. *Obesity—Theory and Therapy*. 2nd edition. New York: Raven Press.

Suárez, G., J. D. Etlinger, J. Maturana, and D. Weitman. 1995. "Fructated Protein Is More Resistant to ATP-Dependent Proteolysis Than Glucated Protein Possibly as a Result of Higher Content of Maillard Fluorophores." *Archives of Biochemistry and Biophysics*. Aug. 1; 321(1):209–13.

Suárez, G., R. Rajaram, O. L. Oronsky, and M. A. Gawinowicz. 1989. "Nonenzymatic Glycation of Bovine Serum Albumin by Fructose (Fructation): Comparison with the Maillard Reaction Initiated by Glucose." *Journal of Biological Chemistry*. March 5; 264(7):3674–79.

Sugarman, C. 1999. "Eat Fat, Get Thin? Dieters on Protein-Rich Regimens Report Great Success, but Some Doctors Question the Safety of These Low-Carb Plans." *Washington Post*. Nov. 23; Z10.

———. 1989. "Experts Agree: Eat More Fruit, Vegetables." *Washington Post*. March 2; A1.

Sullivan, P. 2004. "Ancel Keys, K Ration Creator, Dies." *Washington Post*. Nov. 24; A1.

Surkan, P. J., C. C. Hsieh, A. L. Johansson, P. W. Dickman, and S. Cnattingius. 2004.

"Reasons for Increasing Trends in Large for Gestational Age Births." *Obstetrics & Gynecology.* Oct.; 104(4):720–26.

Susic, D., J. Varagic, J. Ahn, and E. D. Frohlich. 2004. "Crosslink Breakers: A New Approach to Cardiovascular Therapy." *Current Opinion in Cardiology.* July; 19(4):336–40.

Sutin, J. 1976. "Neural Factors in the Control of Food Intake." In Bray, ed., 1976b, 1–11.

Suzuki, M., and T. Tamura. 1988. "Sweeteners: Low-Energetic and Low-Insulinogenic." In *Diet and Obesity*, ed. G. A. Bray, J. Le Blanc, S. Inoue, and S. Masahige. Tokyo: Japan Scientific Society Press, 163–73.

Swanson, J. E., D. C. Laine, W. Thomas, and J. P. Bantle. 1992. "Metabolic Effects of Dietary Fructose in Healthy Subjects." *American Journal of Clinical Nutrition.* April; 55(4):851–56.

Sytkowski, P. A., W. B. Kannel, and R. B. D'Agostino. 1990. "Changes in Risk Factors and the Decline in Mortality from Cardiovascular Disease: The Framingham Heart Study." *New England Journal of Medicine.* June 7; 322(23):1635–41.

Szanto, S., and J. Yudkin. 1969. "The Effect of Dietary Sucrose on Blood Lipids, Serum Insulin, Platelet Adhesiveness, and Body Weight in Human Volunteers." *Postgraduate Medical Journal.* Sept.; 45(527):602–7.

Sztalryd, C., J. Hamilton, B. A. Horwitz, P. Johnston, and F. B. Kraemer. 1996. "Alterations of Lipolysis and Lipoprotein Lipase in Chronically Nicotine-Treated Rats." *American Journal of Physiology.* Feb.; 270(2, pt. 1):E215–23.

Taller, H. 1961. *Calories Don't Count.* New York: Simon & Schuster.

Tan, M. H., E. G. Wilmshurst, R. E. Gleason, and J. S. Soeldner. 1973. "Effect of Posture on Serum Lipids." *New England Journal of Medicine.* Aug. 23; 289(8):416–19.

Tannebaum, A. 1959. "Nutrition and Cancer." In *The Physiopathology of Cancer*, 2nd edition, ed. F. Homburger. New York: Hoeber-Harper, 517–65.

———. 1942. "The Genesis and Growth of Tumors. III. Effects of a High-Fat Diet." *Cancer Research.* 2:468–75.

Tanner, T. H. 1869a. *A Manual of Clinical Medicine and Physical Diagnosis.* 2nd edition. Philadelphia: Henry C. Lee.

———. 1869b. *The Practice of Medicine.* 6th edition. London: Henry Renshaw.

Tanzi, R. E., and A. B. Parson. 2000. *Decoding Darkness: The Search for the Genetic Causes of Alzheimer's Disease.* Cambridge, Mass.: Perseus Publishing.

Tappy, L., and E. Jéquier. 1993. "Fructose and Dietary Thermogenesis." *American Journal of Clinical Nutrition.* Nov.; 58(5 suppl.):766s–70s.

Tarnower, H., and S. S. Baker. 1978. *The Complete Scarsdale Medical Diet.* New York: Bantam Books.

Tarr, Y. Y. 1972. *The New York Times Natural Foods Dieting Book.* New York: Weathervane Books.

Task Force Sponsored by the American Society for Clinical Nutrition. 1979. "The Evidence Relating Six Dietary Factors to the Nation's Health." *American Journal of Clinical Nutrition.* 32(suppl.):2621–748.

Taubes, G. 1998. "The Political Science of Salt." *Science.* Aug. 14; 281(5379):898–907.

———. 1996. "Looking for the Evidence in Medicine." *Science.* April 5; 272(5258):22–24.

Taylor, W. C., T. M. Pass, D. S. Shepard, and A. L. Komaroff. 1987. "Cholesterol Reduction and Life Expectancy: A Model Incorporating Multiple Risk Factors." *Annals of Internal Medicine.* April; 106(4):605–14.

Teitelbaum, P. 1955. "Sensory Control of Hypothalamic Hyperphagia." *Journal of Comparative Physiology and Psychology.* June; 48(3):156–63.

Temin, H. M. 1968. "Carcinogenesis by Avian Sarcoma Viruses. X. The Decreased

Requirement for Insulin-Replaceable Activity in Serum for Cell Multiplication." *International Journal of Cancer.* Nov. 15; 3(6):771–87.

———. 1967. "Studies on Carcinogenesis by Avian Sarcoma Viruses. VI. Differential Multiplication of Uninfected and of Converted Cells in Response to Insulin." *Journal of Cell Physiology.* June; 69(3):377–84.

Teng, B., G. R. Thompson, A. D. Sniderman, T. M. Forte, R. M. Krauss, and P. O. Kwiterovich, Jr. 1983. "Composition and Distribution of Low Density Lipoprotein Fractions in HyperApoBetalipoproteinemia, Normolipidemia, and Familial Hypercholesterolemia." *Proceedings of the National Academy of Sciences.* Nov.; 80(21): 6662–66.

Thom, T., N. Haase, W. Rosamond, et al. [American Heart Association Statistics Committee and Stroke Statistics Subcommittee.] 2006. "Heart Disease and Stroke Statistics— 2006 Update: A Report from the American Heart Association Statistics Committee and Stroke Statistics Subcommittee." *Circulation.* Feb. 14; 113(6):e85–151.

Thomas, B. M., and A. T. Miller. 1958. "Adaptation to Forced Exercise in the Rat." *American Journal of Physiology.* May; 193(2):350–54.

Thomas, W. I., and D. Thomas. 1929. *The Child in America.* 2nd edition. New York: Alfred A. Knopf.

Thompson, L., and S. Squires. 1987. "What You Should Know, How You Should Eat, When You Should Worry: Cholesterol Survival Guide." *Washington Post.* Oct. 20; Z14.

Thorpe, G. L. 1957. "Treating Overweight Patients." *JAMA.* Nov. 16; 165(11):1361–65.

Tolchin, M. 1959. "Helping the Overweight Child." *New York Times Magazine.* Oct. 11; 62, 64.

Tolstoi, E. 1929a. "The Effect of an Exclusive Meat Diet Lasting One Year on the Carbohydrate Tolerance of Two Normal Men." *Journal of Biological Chemistry.* Sept.; 83(3):747–52.

———. 1929b. "The Effect of an Exclusive Meat Diet on the Chemical Constituents of the Blood." *Journal of Biological Chemistry.* Sept.; 83(3):753–58.

Tolstoy, L. N. 2000. *Anna Karenina.* Trans. C. Garnett. New York: Modern Library Classics. [Originally published serially 1875–77.]

Torrey, J. C. 1930. "Influence of an Exclusive Meat Diet on the Human Intestinal Flora." *Proceedings of the Society for Experimental Biology and Medicine.* Dec.; 28(3):295–96.

Torrey, J. C., and E. Montu. 1931. "The Influence of an Inclusive Meat Diet on the Flora of the Human Colon." *Journal of Infectious Diseases.* 49:141–76.

Toufexis, A. 1988. "The Food You Eat May Kill You." *Time.* Aug. 8. Online at http://www.time.com/time/magazine/article/0,9171,968077,00.html.

Trollope, F. M. 1832. *Domestic Manners of the Americans.* Republished online by Project Gutenberg, http://www.gutenberg.org/dirs/1/0/3/4/10345/10345-8.txt.

Trowell, H. 1981. Hypertension, Obesity, Diabetes Mellitus, and Coronary Heart Disease." In Trowell and Burkitt, eds., 1981, 3–32.

———. 1975a. "Refined Carbohydrate Foods and Fibre." In Burkitt and Trowell, eds., 1975, 23–41.

———. 1975b. "Ischaemic Heart Disease, Atheroma, and Fibrinolysis." In Burkitt and Trowell, eds., 1975, 195–226.

———. 1975c. "Obesity in the Western World." *Plant Foods Man.* 1:157–68.

———. 1972a. "Ischemic Heart Disease and Dietary Fiber." *American Journal of Clinical Nutrition.* Sept.; 25(9):926–32.

———. 1972b. "Fiber: A Natural Hypocholesteremic Agent." *American Journal of Clinical Nutrition.* May; 25(5):464–65.

———. 1960. *Non-Infective Disease in Africa*. London: Edward Arnold.

———. 1956. "A Case of Coronary Heart Trouble in an African." *East African Medical Journal*. 33:393.

Trowell, H. C., and D. P. Burkitt, eds. 1981. *Western Diseases: Their Emergence and Prevention*. London: Edward Arnold.

———. 1981a. "Preface." In Trowell and Burkitt, eds., 1981, xiii–xvi.

———. 1981b. "Contributors' Reports." In Trowell and Burkitt, eds., 1981, 427–35.

———. 1975. "Concluding Considerations." In Burkitt and Trowell, 1975, 333–45.

Trulson, M. F., R. E. Clancy, W. J. Jessop, R. W. Childers, and F. J. Stare. 1964. "Comparisons of Siblings in Boston and Ireland." *Journal of the American Dietetic Association*. Sept.; 45:225–29.

Truswell, A. S. 1977. "Dietary Fat and Heart Disease." *Lancet*. Dec. 3; 310(8049):1173.

Truswell, A. S., J. I. Mann, and G. D. Campbell. 1971. "Serum-Lipids in Sugar-Cane Cutters." *Lancet*. March 20; 297(7699):602.

Tso, C. S. 1997. "An Overview of Ascorbic Acid Chemistry and Biochemistry." In *Vitamin C in Health and Disease*, ed. L. Packer and J. Fuchs. New York: Marcel Dekker, 25–58.

Tuia, I. 2001. "The Tokelau Connection." In *The Health of Pacific Societies: Ian Prior's Life and Work*. Aoteroa, N.Z.: Steele Roberts, 32–39.

Tulloch, J. A. 1962. *Diabetes Mellitus in the Tropics*. London: E. & S. Livingstone.

Tuma, R. 2001. "The Two Faces of Oxygen." *Science of Aging Knowledge Environment*. Oct. 3; 2001(1):oa5. Online at http://sageke.sciencemag.org./cgi/content/full/2001/1/oa5.

Tunstall Pedoe, H. 1984. "Epidemiology of Coronary Heart Disease." In *The Encylopaedia of Medical Ignorance*, ed. R. Duncan and M. Weston-Smith. Oxford: Pergamon Press, 95–106.

U.K. Department of Health. 1998. *Nutritional Aspects of the Development of Cancer. Report of the Working Group on Diet and Cancer of the Committee on Medical Aspects of Food and Nutritional Policy. Report on Health and Social Subjects 48*. London: The Stationery Office.

———. 1994. *Nutritional Aspects of Cardiovascular Disease. Report of the Cardiovascular Review Group of the Committee on Medical Aspects of Food Policy. Report on Health and Social Subjects 46*. London: Her Majesty's Stationery Office.

———. 1989. *Dietary Sugars and Human Disease. Report of the Panel on Dietary Sugars. Committee on Medical Aspects of Food Policy. Report on Health and Social Subjects 37*. London: Her Majesty's Stationery Office.

U.S. Department of Agriculture (USDA). n.d. Nutrient Database for Standard Reference. Online at http://www.nal.usda.gov/fnic/cgi-bin/nut_search.pl.

———. 2000. "Major Trends in U.S. Food Supply, 1909–99." *FoodReview*. Jan.–April; 23(1):8–15.

———. 1992. "The Food Guide Pyramid." *Home and Garden Bulletin* No. 252. Washington, D.C.: U.S. Government Printing Office.

———. 1953. *Consumption of Food in the United States, 1909–1952*. Agriculture Handbook No. 62. Washington, D.C.: USDA Bureau of Agricultural Economics.

U.S. Department of Agriculture Center for Nutrition Policy and Promotion. 1998. "Is Total Fat Consumption Really Decreasing?" *Nutrition Insights*. Insight 5, April.

USDA Economic Research Service. 2005. Food Availability Spreadsheets. Last updated Dec. 21, 2005. Downloaded Oct. 14, 2006. Online at http://www.ers.usda.gov/data/foodconsumption/FoodAvailSpreadsheets.htm.

U.S. Department of Agriculture and U.S. Department of Health, Education, and Welfare

[HEW]. 1980. "Nutrition and Your Health: Dietary Guidelines for Americans." *Home and Garden Bulletin.* No. 228. Washington, D.C.: U.S. Department of Agriculture.

U.S. Department of Commerce, Bureau of the Census. 1949. *Historical Statistics of the United States, 1789–1945.* Washington, D.C.: U.S. Government Printing Office.

U.S. Department of Health and Human Services [USDHHS]. 2001. *The Surgeon General's Call to Action to Prevent and Decrease Overweight and Obesity, 2001.* Washington D.C.: U.S. Government Printing Office.

———. 1988. *The Surgeon General's Report on Nutrition and Health.* Washington, D.C.: U.S. Government Printing Office.

U.S. Department of Health and Human Services and U.S. Department of Agriculture. 2005. *Dietary Guidelines for Americans, 2005.* 6th edition. Washington, D.C.: U.S. Government Printing Office.

U.S. Department of Health, Education, and Welfare [HEW]. 1971. *Arteriosclerosis: A Report by the National Heart and Lung Institute Task Force on Arteriosclerosis.* 2 vols. U.S. Department of Health, Education, and Welfare Publication No. (NIH) 72-137 and 72-219. Washington, D.C.: National Institutes of Health.

Vague, J. 1956. "The Degree of Masculine Differentiation of Obesities: A Factor Determining Predisposition to Diabetes, Atherosclerosis, Gout, and Uric Calculous Disease." *American Journal of Clinical Nutrition.* Jan.–Feb.; 4(1):20–34.

Van Gaal, L. F. 1998. "Dietary Treatment of Obesity." In Bray, Bouchard, and James, eds., 1998, 875–90.

Vanhanen, M., K. Koivisto, L. Moilanen, et al. 2006. "Association of Metabolic Syndrome with Alzheimer Disease: A Population-Based Study." *Neurology.* Sept. 12; 67(5): 843–47.

Van Itallie, T. B. 1980a. "Dietary Approaches to the Treatment of Obesity." In Stunkard, ed., 1980, 249–61.

———. 1980b. "Diets for Weight Reduction: Mechanisms of Action and Physiologic Effects." In Bray, ed., 1980, 15–24.

———. 1979. "Conservative Approaches to Treatment." In Bray, ed., 1979, 164–78.

———. 1978. "Dietary Approaches to the Treatment of Obesity." *Psychiatric Clinics of North America.* Dec.; 1(3):609–20. [Also referenced as A. J. Stunkard, ed., *Obesity: Basic Mechanisms and Treatment* (Philadelphia: W. B. Saunders, 1978)]

Van Itallie, T. B., and T. H. Nufert. 2003. "Ketones: Metabolism's Ugly Duckling." *Nutrition Review.* Oct.; 61(10):327–41.

Van Itallie, T. B., M. Yan, and S. A. Hashim. 1976. "Dietary Approaches to Obesity: Metabolic and Appetitive Considerations." In Howard, ed., 1976, 256–69.

Vartiainen, I., and K. Kanerva. 1947. "Arteriosclerosis and Wartime." *Annales Medicinae Internae Fenniae.* 36(3):748–58.

Vaselli, J. R., M. P. Cleary, and T. B. Van Itallie. 1984. "Obesity." In Olson et al., eds., 1984, 35–56.

Vitek, M. P., K. Bhattacharya, J. M. Glendening, et al. 1994. "Advanced Glycation End Products Contribute to Amyloidosis in Alzheimer Disease." *Proceedings of the National Academy of Sciences.* May 24; 91(11):4766–70.

Wadden, T. A., A. J. Stunkard, K. D. Brownell, and S. C. Day. 1985. "A Comparison of Two Very-Low-Calorie Diets: Protein-Sparing-Modified Fast Versus Protein-Formula-Liquid Diet." *American Journal of Clinical Nutrition.* March; 41(3):533–39.

Wadden, T. A., and T. B. Van Itallie, eds. 1992. *Treatment of the Seriously Obese Patient.* New York: Guilford Press.

Wade, G. N. 1982. "Obesity Without Overeating in Golden Hamsters." *Physiology & Behavior.* Oct.; 29(4):701–7.

Wade, G. N., and J. E. Schneider. 1992. "Metabolic Fuels and Reproduction in Female Mammals." *Neuroscience & Biobehavioral Reviews*. Summer; 16(2):235–72.

Wade, N. 1980. "Food Board's Fat Report Hits Fire." *Science*. July 11; 209(4453):248–50.

Wahlberg, F., and B. Thomasson. 1968. "Glucose Tolerance in Ischaemic Cardiovascular Disease." In Dickens, Randle, and Whelan, eds., 1968, vol. 2, 185–98.

Walker, A. R. 1964. "Overweight and Hypertension in Emerging Populations." *American Heart Journal*. Nov.; 68(5):581–85.

———. 1962. "Health Hazards in the Urbanization of the African." *American Journal of Clinical Nutrition*. Dec.; 11(6):551–53.

Walker, A. R., P. E. Cleaton-Jones, and B. D. Richardson. 1978. "Is Sugar Good for You?" *South African Medical Journal*. Oct. 7; 54(15):589–90.

Walker, A. R., and B. F. Walker. 1969. "Bowel Motility and Colonic Cancer." *British Medical Journal*. July 26; 238.

Wallace, R. B., D. B. Hunninghake, S. Reiland, et al. 1980. "Alterations of Plasma High-Density Lipoprotein Cholesterol Levels Associated with Consumption of Selected Medications: The Lipid Research Clinics Program Prevalence Study." *Circulation*. Nov.; 62(4, pt. 2):IV77–82.

Walldius, G., I. Jungner, I. Holme, A. H. Aastveit, W. Kolar, and E. Steiner. 2001. "High Apolipoprotein B, Low Apolipoprotein A-I, and Improvement in the Prediction of Fatal Myocardial Infarction (AMORIS Study): A Prospective Study." *Lancet*. Dec. 15; 358(9298):2026–33.

Wallis, C. 1984. "Hold the Eggs and Butter." *Time*. March 26; 56–63.

Warburg, O. 1956. "On the Origin of Cancer Cells." *Science*. Feb. 24; 123(3191):309–14.

Ward, A. 1911. *The Grocer's Encyclopedia*. New York: Artemus Ward. Online at http://digital.lib.msu.edu/projects/cookbooks/books/grocersencyclopedia/ency.html#ency359.gif.

Wassertheil-Smoller, S., S. L. Hendrix, M. Limacher, et al. WHI Investigators. 2003. "Effect of Estrogen Plus Progestin on Stroke in Postmenopausal Women: The Women's Health Initiative, a Randomized Trial." *JAMA*. May 28; 289(20):2673–84.

Waterlow, J. C. 1986. "Metabolic Adaptation to Low Intakes of Energy and Protein." *Annual Review of Nutrition*. 6:495–526.

Watson, G. S., E. R. Peskind, S. Asthana, et al. 2003. "Insulin Increases CSF Abeta42 Levels in Normal Older Adults." *Neurology*. June 24; 60(12):1899–903.

Webb, G. 1992. *A Pima Remembers*. Tucson: University of Arizona Press. [Originally published 1959.]

Weinberg, R. A. 2007. *The Biology of Cancer*. New York: Garland Science.

———. 1996. *Racing to the Beginning of the Road: The Search for the Origin of Cancer*. New York: Harmony Books.

Weinert, B. T., and P. S. Timiras. 2003. "Theories of Aging." *Journal of Applied Physiology*. Oct.; 95(4):1706–16.

Welborn, T. A., A. Breckenridge, A. H. Rubinstein, C. T. Dollery, and T. R. Fraser. 1966. "Serum-Insulin in Essential Hypertension and in Peripheral Vascular Disease." *Lancet*. June 18; 287(7451):1336–37.

Welborn, T. A., and K. Wearne. 1979. "Coronary Heart Disease Incidence and Cardiovascular Mortality in Busselton with Reference to Glucose and Insulin Concentrations." *Diabetes Care*. March–April; 2(2):154–60.

Wellcome Library. Cleave, "Peter" (1906–1983). Online at http://www.aim25.ac.uk/cgi-bin/search2?coll_id=4602&inst_id=20.

Werner, S. C. 1955. "Comparison Between Weight Reduction on a High-Calorie, High-Fat Diet and on an Isocaloric Regimen High in Carbohydrate." *New England Journal of Medicine*. April 21; 252(16):661–65.

Wertheimer, E. 1965. "Introduction: 'A Perspective.' " In Renold and Cahill, eds., 1965, 5–11.

Wertheimer, E., and R. Shafrir. 1960. "Influence of Hormones on Adipose Tissue as a Center of Fat Metabolism." *Recent Progress in Hormone Research.* 16:467–95.

Wertheimer, E., and R. Shapiro. 1948. "The Physiology of Adipose Tissue." *Physiology Reviews.* Oct.; 28:451–64.

Wessen, A. 2001. "Ian Prior and the Tokelau Island Migrant Studies." In *The Health of Pacific Societies: Ian Prior's Life and Work* (Aoteroa, N.Z.: Steele Roberts), 16–25.

Wessen, A. F., A. Hooper, J. Huntsman, I. A. Prior, and C. E. Salmond, eds. 1992. *Migration and Health in a Small Society: The Case of Tokelau.* Oxford: Clarendon Press.

West, K. M. 1981. "North American Indians." In Trowell and Burkitt, eds., 1981, 129–37.

Westlund, K., and R. Nicolaysen. 1972. "Ten-Year Mortality and Morbidity Related to Serum Cholesterol: A Follow-Up of 3,751 Men Aged 40–49." *Scandinavian Journal of Clinical and Laboratory Investigation.* 127(suppl.): 1–24.

Whalen, R. G. 1950. "We Think Ourselves into Fatness." *New York Times Magazine.* Dec. 3; 22, 36, 38, 40.

Whelan, E. M., and F. J. Stare. 1983. *The One-Hundred-Percent Natural, Purely Organic, Cholesterol-Free, Megavitamin, Low-Carbohydrate Nutrition Hoax.* New York: Atheneum.

White, L., H. Petrovitch, G. W. Ross, et al. 1996. "Prevalence of Dementia in Older Japanese-American Men in Hawaii: The Honolulu-Asia Aging Study." *JAMA.* Sept. 25; 276(12):955–60.

White, P., and E. P. Joslin. 1959. "The Etiology and Prevention of Diabetes." In Joslin et al., eds., 1959, 47–98.

White, P., and N. Selvey. 1974. *Let's Talk About Food.* Acton, Mass.: Publishing Sciences Group.

White, P. D. 1971. *My Life and Medicine: An Autobiographical Memoir.* Boston: Gambit Inc.
———. 1945. *Heart Disease.* 3rd edition. New York: Macmillan.

White, P. L. 1962. "Calories Don't Count." *JAMA.* March 10; 179(10):184.

Whitehead, A. N. 1980. *Nature and Life.* Chicago: University of Chicago Press.

Widdowson, E. M. 1962. "Nutritional Individuality." *Proceedings of the Nutrition Society.* March 17; 21(2):121–28.

Wilder, R. M. 1933. "The Treatment of Obesity." *International Clinics.* 4:1–21.

Wilder, R. M., and W. L. Wilbur. 1938. "Diseases of Metabolism and Nutrition." *Archives of Internal Medicine.* Feb.; 61:297–365.

Will, J. C., and T. Byers. 1996. "Does Diabetes Mellitus Increase the Requirement for Vitamin C?" *Nutrition Reviews.* July; 54(7):193–202.

Willett, W. C. 2001. *Eat, Drink, and Be Healthy: The Harvard Medical School Guide to Healthy Eating.* New York: Simon & Schuster.

Willett, W. C., D. J. Hunter, M. J. Stampfer, et al. 1992. "Dietary Fat and Fiber in Relation to Risk of Breast Cancer: An 8-Year Follow-Up." *JAMA.* Oct. 21; 268(15):2037–44.

Willett, W. C., F. Sacks, A. Trichopoulos, et al. 1995. "Mediterranean Diet Pyramid: A Cultural Model for Healthy Eating." *American Journal of Clinical Nutrition.* June; 61(6 suppl.):1402S–6S.

Willett, W. C., and M. Stampfer. 1998. "Implications of Total Energy Intake for Epidemiologic Analyses." In *Nutritional Epidemiology,* 2nd edition, ed. W. C. Willett. New York: Oxford University Press, 273–301.

Willett, W. C., M. J. Stampfer, G. A. Colditz, B. A. Rosner, C. H. Hennekens, and F. E. Speizer. 1987. "Dietary Fat and the Risk of Breast Cancer." *New England Journal of Medicine.* Jan. 1; 316(1):22–28.

Williams, R. H., W. H. Daughaday, W. F. Rogers, S. P. Asper, and B. T. Towery. 1948. "Obesity and Its Treatment, with Particular Reference to the Use of Anorexigenic Compounds." *Annals of Internal Medicine.* 29(3):510–32.

Williams, R. R., P. D. Sorlie, M. Feinleib, P. M. McNamara, W. B. Kannel, and T. R. Dawber. 1981. "Cancer Incidence by Levels of Cholesterol." *JAMA.* Jan. 16; 245(3):247–52.

Williams, W. R. 1908. *The Natural History of Cancer with Special Reference to Its Causation and Prevention.* London: William Heinemann.

Wilson, J. D., D. W. Foster, H. M. Kronenberg, and P. R. Larsen. 1998. "Principles of Endocrinology." In *Williams Textbook of Endocrinology*, 9th edition, ed. J. D. Wilson, D. W. Foster, H. M. Kronenberg, and P. R. Larsen. Philadelphia: W. B. Saunders, 1–10.

Wilson, N. L., ed. 1969. *Obesity*. Philadelphia: F. A. Davis.

Wilson, P. W., R. B. D'Agostino, D. Levy, A. M. Belanger, H. Silbershatz, and W. B. Kannel. 1998. "Prediction of Coronary Heart Disease Using Risk Factor Categories." *Circulation.* May 12; 97(18):1837–47.

Winnick, M., ed. 1975. *Childhood Obesity*. New York: John Wiley.

Witztum, J. L., and D. Steinberg. 1981. "Role of Oxidized Low Density Lipoprotein in Atherogenesis." *Journal of Clinical Investigation.* Dec.; 88(6):1785–92.

Wolever, T. M. 1997. "The Glycemic Index: Flogging a Dead Horse?" *Diabetes Care.* March; 20(3):452–56.

Woods, S. C., and D. Porte. 1976. "Insulin and the Set-Point Regulation of Body Weight." In Novin, Wyrwicka, and Bray, eds., 1976, 273–80.

Woods, S. C., J. R. Vasselli, E. Kaestner, G. A. Skarmary, P. Milburn, and M. Vitiello. 1977. "Conditioned Insulin Secretion and Meal Feeding in Rats." *Journal of Comparative Physiology and Psychology.* 91(1):128–33.

Woody, E. 1950. "Eat Well and Lose Weight." *Holiday.* June; 65–70, 157–62.

Woolsey, T. D., and I. M. Moriyama. 1948. "Statistical Studies of Heart Diseases. II. Important Factors in Heart Disease Mortality Trends." *Public Health Reports.* Sept. 24; 63(39):1247–71.

World Cancer Research Fund and American Institute for Cancer Research. 1997. *Food, Nutrition and the Prevention of Cancer: A Global Perspective.* Washington, D.C.: American Institute for Cancer Research.

World Health Organization. 2006. "The World Health Organization Notes the Women's Health Initiative Diet Modification Trial, but Reaffirms That the Fat Content of Your Diet Does Matter." Downloaded Feb. 16, 2006. Online at http://www.who.int/nmh/media/Response_statement_16_feb_06F.pdf.

———. 2004. "Global Strategy on Diet, Physical Activity and Health, Obesity and Overweight." Downloaded March 25, 2005. Online at http://www.who.int/dietphysicalactivity/publications/facts/obesity/en/.

———. 2003. "Nutrition in Transition: Globalization and Its Impact on Nutritional Patterns and Diet-Related Diseases. Updated Wed. Sept. 3, 2003. Online at http://www.who.int/nut/trans.htm.

Wright, J. D., J. Kennedy-Stephenson, C. Y. Wang, M. A. McDowell, and C. L. Johnson. 2004. "Trends in Intake of Energy and Macronutrients—United States, 1971–2000." *Morbidity and Mortality Weekly Report.* Feb. 6; 53(4):80–82.

Wu, Y., K. Cui, K. Miyoshi, et al. 2003. "Reduced Circulating Insulin-Like Growth Factor I Levels Delay the Onset of Chemically and Genetically Induced Mammary Tumors." *Cancer Research.* Aug. 1; 63(15):4384–88.

Wu, Y., S. Yakar, L. Zhao, L. Hennighausen, and D. LeRoith. 2002. "Circulating Insulin-

Like Growth Factor-I Levels Regulate Colon Cancer Growth and Metastasis." *Cancer Research.* Feb. 15; 62(4):1030–35.

Wynder, E. L., G. A. Leveille, J. H. Weisburger, and G. E. Livingston, eds. 1983. *Environmental Aspects of Cancer: The Role of Macro and Micro Components of Foods.* Westport, Conn.: Food and Nutrition Press.

Yalow, R. S., and S. A. Berson. 1960. "Immunoassay of Endogenous Plasma Insulin in Man." *Journal of Clinical Investigation.* July; 39:1157–75.

Yalow, R. S., S. M. Glick, J. Roth, and S. A. Berson. 1965. "Plasma Insulin and Growth Hormone Levels in Obesity and Diabetes." *Annals of the New York Academy of Sciences.* Oct. 8; 131(1):357–73.

Yan, S. D., X. Chen, A. M. Schmidt, et al. 1994. "Glycated Tau Protein in Alzheimer Disease: A Mechanism for Induction of Oxidant Stress." *Proceedings of the National Academy of Sciences.* Aug. 2; 91(16):7787–91.

Yancy, W. S., Jr., M. K. Olsen, J. R. Guyton, R. P. Bakst, and E. C. Westman. 2004. "A Low-Carbohydrate, Ketogenic Diet Versus a Low-Fat Diet to Treat Obesity and Hyperlipidemia: A Randomized, Controlled Trial." *Annals of Internal Medicine.* May 18; 140(10):769–77.

Yano, K., G. G. Rhoads, A. Kagan, and J. Tillotson. 1978. "Dietary Intake and the Risk of Coronary Heart Disease in Japanese Men Living in Hawaii." *American Journal of Clinical Nutrition.* July; 31(7):1270–79.

Yerushalmy, J., and H. E. Hilleboe. 1957. "Fat in the Diet and Mortality from Heart Disease: A Methodologic Note." *New York State Journal of Medicine.* July 15; 57(14): 2343–54.

Yeung, D. L. 1976. "Relationships Between Cigarette-Smoking, Oral-Contraceptives, and Plasma Vitamins A, E, and C, and Plasma Triglycerides and Cholesterol." *American Journal of Clinical Nutrition.* Nov.; 29(11):1216–21.

Yost, T. J., D. R. Jensen, B. R. Haugen, and R. H. Eckel. 1998. "Effect of Dietary Macronutrient Composition on Tissue-Specific Lipoprotein Lipase Activity and Insulin Action in Normal-Weight Subjects." *American Journal of Clinical Nutrition.* Aug.; 68(2): 296–302.

You, W., F. Jin, G. Gridley, et al. 2000. "Trends in Colorectal Cancer Rates in Urban Shanghai, 1972–1996, in Relation to Dietary Changes." *Annals of Epidemiology.* Oct. 1; 10(7):46.

Young, C. M. 1976. "Dietary Treatment of Obesity." In Bray, ed., 1976b, 361–66.

———. 1952. "Weight Reduction Using a Moderate Fat Diet. 1. Clinical Responses and Energy Metabolism." *Journal of the American Dietetic Association.* May; 28(5):410–6.

Young, C. M., E. L. Empey, V. U. Serraon, and Z. H. Pierce. 1957. "Weight Reduction in Obese Young Men: Metabolic Studies." *Journal of Nutrition.* March 10; 61(3):437–56.

Young, C. M., I. Ringler, and B. J. Greer. 1953. "Reducing and Post-Reducing Maintenance on the Moderate Fat Diet: Metabolic Studies." *Journal of the American Dietetic Association.* Sept.; 29(9):890–96.

Young, J. H. 1981. "The Long Struggle for the 1906 Law: Food and Drug Administration." *FDA Consumer.* June. Online at http://vm.cfsan.fda.gov/~lrd/history2.html.

Young, R. A. 1976. "Fat, Energy and Mammalian Survival." *American Zoologist.* 16:699–710.

Yudkin, J. 1986. *Pure, White, and Deadly.* Revised edition. New York: Viking.

———. 1974. "The Low-Carbohydrate Diet." In Burland, Samuel, and Yudkin, eds., 1974, 271–80.

———. 1972a. *Pure, White, and Deadly.* London: Davis-Poynter.

———. 1972b. *Sweet and Dangerous*. New York: P. H. Wyden.

———. 1972c. "The Low-Carbohydrate Diet in the Treatment of Obesity." *Postgraduate Medical Journal*. May; 51(5):151–54.

———. 1959. "The Causes and Cure of Obesity." *Lancet*. Dec. 19; 274(7112):1135–38.

———. 1958. *This Slimming Business*. London: MacGibbon and Kee.

———. 1957. "Diet and Coronary Thrombosis: Hypothesis and Fact." *Lancet*. July 27; 270(6987):155–62.

Yudkin, J., and M. Carey. 1960. "The Treatment of Obesity by the 'High-Fat' Diet: The Inevitability of Calories." *Lancet*. Oct. 29; 276(7157):939–41.

Yudkin, J., V. V. Kakkar, and S. Szanto. 1969. "Sugar Intake, Serum Insulin and Platelet Adhesiveness in Men with and Without Peripheral Vascular Disease." *Postgraduate Medical Journal*. Sept.; 45(527):608–11.

Yuncker, B. 1962. "The Fat Americans or Calories DO Count." *New York Post*. April 23; 25.

Zahorska-Markiewicz, B. 1980. "Weight Reduction and Seasonal Variation." *International Journal of Obesity*. 4(2):139–43.

Zekry, D., J. J. Hauw, and G. Gold. 2002. "Mixed Dementia: Epidemiology, Diagnosis, and Treatment." *Journal of the American Geriatric Society*. Aug.; 50(8):1431–38.

Zierler, K. L., and D. Rabinowitz. 1964. "Effect of Very Small Concentrations of Insulin on Forearm Metabolism: Persistence of Its Action on Potassium and Free Fatty Acids Without Its Effect on Glucose." *Journal of Clinical Investigation*. May; 43:950–62.

Zimmet, P., K. G. Alberti, and J. Shaw. 2001. "Global and Societal Implications of the Diabetes Epidemic." *Nature*. Dec. 13; 414(6865):782–87.

Zukel, W. J., O. Paul, and H. W. Schnaper. 1981. "The Multiple Risk Factor Intervention Trial (MRFIT). I. Historical Perspective." *Preventive Medicine*. July; 10(4):387–401.

Acknowledgments

It's always dangerous, when challenging beliefs that are so passionately embraced, to acknowledge that you have paid attention to the skeptics who preceded you. This can be used as evidence that you are exceedingly gullible and will believe anything you read. Nonetheless, I concede that I indeed took seriously and am grateful for the efforts of those who trod portions of this path before me: in particular, Russell Smith, Uffe Ravnskov (and his International Network of Cholesterol Skeptics), Wolfgang Lutz, James Le Fanu, and Thomas Moore on the relationship between cholesterol and heart disease; Alfred Pennington, Herman Taller, and Robert Atkins on the subject of diet and weight; and Peter Cleave and John Yudkin, who came closest to putting it all together. I read the works of these authors with skepticism, but no more or less than that of other contributors to the literature. The book that may have been most influential in altering my perspective and yet never made it into this text, for reasons of narrative flow and length rather than relevance, was Weston Price's 1939 classic *Nutrition and Physical Degeneration: A Comparison of Primitive and Modern Diets and Their Effects*.

Drafts of this book were read in part or in whole and corrections suggested by Robert Bauchwitz, John Benditt, Kenneth Carpenter, Michael Eades, Richard Feinman, Mark Friedman, Richard Hanson, David Jacobs, Cynthia Kenyon, Ron Krauss, Mitch Lazar, Jamie Robins, Bruce Schechter, Jeremy Stone, Clifford Taubes, Nina Teicholz, and Eric Westman. I am deeply grateful to all these individuals for their time, their efforts, and their acumen. Any errors in either fact or form, however, remain mine alone. I would also like to thank the literally hundreds of researchers, clinicians, and public-health authorities who took the time to speak with me at length, many of whom did so repeatedly, even though they fundamentally disagreed with articles I had already written on this subject.

I am grateful to Colin Norman and Tim Appenzeller for their invaluable help and encouragement at *Science* on the series of investigations that took me ever more deeply into the questionable practices of preventive medicine and public health. I'm grateful to Hugo Lindgren and Adam Moss, both formerly of *The New York Times Magazine*, for taking the chance on the very controversial article—"What If It's All Been a Big Fat Lie?"— that led directly to the work on this book.

I am deeply indebted to Jon Segal at Knopf for an extraordinary job of editing and for being, quite simply, everything I could ever hope for in an editor. I'd also like to thank Knopf editorial assistant Kyle McCarthy and copyeditor Terry Zaroff. I am grateful, as ever, to my agent at ICM, Kris Dahl, for two decades of unwavering support.

I would like to thank Alexis Bramos-Hantman, Jeanna Bryner, Jasmin Chua, Susan England, Emily Hager, Jeanne Lenzer, David Mahfouda, Tariq Malik, Chung Pak, Gaia Remerowski, Sandra Neufeldt, Rochelle Thomas, and Dori Zook for helping with the research and providing the legwork for this book. I can't thank Richard Ahrens enough for his translation of Bahner's 1955 discussion of lipophilia. I'm grateful to Stefan Hagen for his German connections. I'd like to thank Barry Glassner for his camaraderie, Charles Mann for his friendship and his guidance, and Marion Roach Smith, as ever, for her sisterly wisdom. I'm grateful to Ned Tanen, Kitty Hawks, and Lawrence Lederman for their unconditional support and encouragement. Finally, I'd like to thank the late, great Louie Vassilakis (1949–2004) for making one otherwise cold and cacophonous corner of Manhattan feel like home.

Index

ILLUSTRATION CREDITS

63, 64   Charts showing data from MRFIT trial. Reprinted from *The Lancet*, 328, Browner, Hulley, Kuller, Martin, and Wentworth. "Serum Cholesterol, Blood Pressure, and Mortality: Implications from a Cohort of 361,662 Men," pages 933–936. Copyright October 1986, with permission from Elsevier.

236   "Fat Louisa" photograph. Reprinted from *The Pima Indians*, Russell, page 67. Copyright 1908.

242   Photographs from Nigeria. Reprinted from Obesity Symposium, Adadevoh. "Obesity in the African." 60–73. 1974, with permission from Elsevier.

361   Photographs of lipodystrophy with lower-body obesity. Die Krankheiten des Stoffwechsels und ihre Behandlung. Copyright 1931, page 186, Die Magersucht, Grafe, Figure 20 (Photograph of O. B. Meyer). With kind permission of Springer Science and Business Media.

A NOTE ON THE TYPE

This book was set in Scala, a typeface designed by the Dutch designer Martin Majoor (b. 1960) in 1988 and released by the FontFont foundry in 1990. While designed as a fully modern family of fonts containing both a serif and a sans serif alphabet, Scala retains many refinements normally associated with traditional fonts.

Composed by North Market Street Graphics, Lancaster, Pennsylvania

Printed and bound by Berryville Graphics, Berryville, Virginia

Designed by M. Kristen Bearse